AF356383

Fractional Order Systems and Their Applications

Fractional Order Systems and Their Applications

Guest Editors

António Lopes
Liping Chen

Basel • Beijing • Wuhan • Barcelona • Belgrade • Novi Sad • Cluj • Manchester

Guest Editors

António Lopes	Liping Chen
Mechanical Engineering	School of Electrical
University of Porto	Engineering and Automation
Porto	Hefei University of Technology
Portugal	Hefei
	China

Editorial Office

MDPI AG

Grosspeteranlage 5

4052 Basel, Switzerland

This is a reprint of the Special Issue, published open access by the journal *Fractal and Fractional* (ISSN 2504-3110), freely accessible at: www.mdpi.com/journal/fractalfract/special_issues/FOSTA.

For citation purposes, cite each article independently as indicated on the article page online and using the guide below:

Lastname, A.A.; Lastname, B.B. Article Title. *Journal Name* **Year**, *Volume Number*, Page Range.

ISBN 978-3-7258-2680-3 (Hbk)

ISBN 978-3-7258-2679-7 (PDF)

https://doi.org/10.3390/books978-3-7258-2679-7

Contents

About the Editors

António Lopes

António M. Lopes holds a Ph.D. and Habilitation in mechanical engineering from the Faculty of Engineering of the University of Porto (FEUP). He works with the Associated Laboratory for Energy, Transport, and Aeronautics (LAETA), Institute of Science and Innovation, in Mechanical Engineering and Industrial Engineering (INEGI). He is the coordinator of the research group Intelligent Systems & Control at LAETA-INEGI. His research areas include automation, robotics, complex systems, and fractional order systems. He has coauthored more than 230 SCI journal papers. He is the Coeditor-in-Chief of the Journal of Environmental Accounting and Management and Journal of Machine Design and Automation Intelligence and the Section Editor-in-Chief of Fractal and Fractional. He has been included on the Top 2% Stanford List of scientists in 2021, 2022, 2023, and 2024. He is currently a Professor of Mechanical Engineering at FEUP.

Liping Chen

Liping Chen received his B.S. degree in applied mathematics from Anhui Normal University, Wuhu, China; his M.S. degree in basic mathematics from Anhui University, Hefei, China; and his Ph.D. degree from the School of Automation, Chongqing University, Chongqing, China, in 2007, 2010, and 2013, respectively. He is now a Professor at Hefei University of Technology, China. His current research interests include the state estimation and health diagnosis of lithium-ion batteries, fractional order systems, and nonlinear dynamic systems. He has published over 120 SCI-indexed papers and received four provincial-level rewards. He was selected for the Top 2% of Scientists on the Stanford List in 2021, 2022, 2023, and 2024. He serves as a commentator for the AMS, a commentator for the German zbMATH, a member of the Fractional Order Systems and Control Committee of the Chinese Society of Automation, a member of the Vehicle Control and Intelligence Committee of the Chinese Society of Automation, and a member of the Industrial Control Computer Professional Committee of the Computer Society. In addition, he is an Associate Editor of the Journal of Nonlinear, Complex and Data Science, Associate Editor of the International Journal of Dynamics and Control, and Editorial Board Member of the Journal of Machine Design and Automation Intelligence.

Preface

Fractional calculus (FC) generalizes the concepts of derivative and integral orders to non-integer orders. It was introduced by Leibniz (1646–1716) but remained a purely mathematical exercise for a long time, despite the original contributions to the field of important mathematicians, physicists, and engineers. FC has experienced rapid development in recent decades, both in mathematics and applied sciences, being recognized as an excellent tool to describe complex dynamics. Based on this, several models governing physical phenomena in the areas of science and engineering have been reformulated in light of FC for them to better reflect their non-local and frequency- and history-dependent properties. Applications of FC include modeling of diffusion, viscoelasticity, and relaxation processes in fluid mechanics; the dynamics of mechanical, electronic, and biological systems; and signal processing and control.

This reprint compiles articles from the Special Issue "Fractional Order Systems and Their Applications", which focused on original and new research results on modeling and control of fractional order systems with applications in science and engineering. It includes 13 manuscripts addressing novel issues and specific topics that illustrate the richness and applicability of fractional calculus.

António Lopes and Liping Chen
Guest Editors

fractal and fractional

Editorial

Fractional Order Systems and Their Applications

António M. Lopes [1,*] and **Liping Chen** [2]

1   LAETA/INEGI, Faculty of Engineering, University of Porto, Rua Dr. Roberto Frias, 4200-465 Porto, Portugal
2   School of Electrical Engineering and Automation, Hefei University of Technology, Hefei 230009, China; lip_chen@hfut.edu.cn
*   Correspondence: aml@fe.up.pt

Citation: Lopes, A.M.; Chen, L. Fractional Order Systems and Their Applications. *Fractal Fract.* **2022**, 6, 389. https://doi.org/10.3390/fractalfract6070389

Received: 7 July 2022
Accepted: 11 July 2022
Published: 13 July 2022

Publisher's Note: MDPI stays neutral with regard to jurisdictional claims in published maps and institutional affiliations.

Fractional calculus (FC) generalizes the concepts of derivative and integral to non-integer orders. It was introduced by Leibniz (1646–1716), but remained a purely mathematical exercise for a long time, despite the original contributions of important mathematicians, physicists, and engineers. FC experienced rapid development during the last few decades both in mathematics and applied sciences, being recognized as an excellent tool to describe complex dynamics. From this perspective, several models governing physical phenomena in the area of science and engineering have been reformulated in light of FC for better reflecting their non-local, frequency- and history-dependent properties. Applications of FC include modeling of diffusion, viscoelasticity, and relaxation processes in fluid mechanics, dynamics of mechanical, electronic and biological systems, signal processing, control, and others.

The Special Issue "Fractional Order Systems and Their Applications" focuses on original and new research results on modeling and control of fractional order systems with applications in science and engineering. It includes 13 manuscripts addressing novel issues and specific topics that illustrate the richness and applicability of fractional calculus. In the follow-up the selected manuscripts are presented in alphabetic order of their titles.

In the paper "A modified Leslie-Gower model incorporating Beddington-DeAngelis functional response, double allee effect and memory effect" [1] the authors propose a modified Leslie–Gower predator-prey model with Beddington–DeAngelis functional response and double Allee effect in the growth rate of a predator population. To consider memory effects the Caputo fractional-order derivative is used. The dynamic behavior of the model for both strong and weak Allee effect is investigated.

The manuscript "A study of coupled systems of ψ-Hilfer type sequential fractional differential equations with integro-multipoint boundary conditions" [2] investigates the existence and uniqueness of solutions for a coupled system of ψ-Hilfer type sequential fractional differential equations supplemented with nonlocal integro-multi-point boundary conditions. The results are obtained via the classical Banach and Krasnosel'skii's fixed point theorems and the Leray–Schauder alternative.

In "Asymptotic stabilization of delayed linear fractional-order systems subject to state and control constraints" [3] the asymptotic stabilization of delayed linear fractional-order systems (DLFS) subject to state and control constraints is investigated. The existence conditions for feedback controllers of DLFS subject to both state and control constraints are given. A sufficient condition for invariance of polyhedron set is established by using invariant set theory. A new Lyapunov function is constructed on the basis of the constraints, and some sufficient conditions for the asymptotic stability of DLFS are obtained. Feedback controller and the corresponding solution algorithms are also given.

In their work "Chaos control for a fractional-order jerk system via time delay feedback controller and mixed controller" [4] the authors propose a novel fractional-order jerk system. They show that, under some suitable parameters, the fractional-order jerk system displays a chaotic phenomenon. To suppress the chaotic behavior two control strategies are proposed: a time delay feedback controller; and a mixed controller, which includes a time delay feedback controller and a fractional-order PD^σ controller. A sufficient condition ensuring the stability and the creation of Hopf bifurcation for the fractional-order controlled jerk system is derived.

The paper "Control and robust stabilization at unstable equilibrium by fractional controller for magnetic levitation systems" [5] study a method to control and stabilize a levitation system in the presence of disturbance and parameter variations. The stabilization and disturbance rejection are achieved by fractional order PID, fractional order sliding mode, and fractional order Fuzzy control approaches. To design the controllers a tuning hybrid method based on GWO–PSO algorithms is applied with different performance criteria.

In the work "Controllability for fuzzy fractional evolution equations in credibility space" [6] the authors address the exact controllability for Caputo fuzzy fractional evolution equations in the credibility space from the perspective of the Liu process. As a result, the study's theoretical result can be used to create stochastic extensions in credibility space.

In "Dynamics of fractional-order digital manufacturing supply chain system and its control and synchronization" [7] a fractional-order digital manufacturing supply chain system is proposed and solved by the Adomian decomposition method. Dynamical characteristics of the system are studied by using phase portrait, bifurcation diagram, and maximum Lyapunov exponent diagram. The complexity of the system is also investigated by means of complexity measures. The importance of the fractional-order derivative in the modeling of the system is shown. Feedback controllers to control the chaotic supply chain system and synchronize two supply chain systems are proposed.

The manuscript "Existence, uniqueness, and E_q-Ulam-type stability of fuzzy fractional differential equation" [8] concerns with the existence and uniqueness of the Cauchy problem for a system of fuzzy fractional differential equation with Caputo derivative of order $q \in (1, 2]$. By using direct analytic methods, the E_q–Ulam-type results are also presented.

In "Fractals Parrondo's paradox in alternated superior complex system" [9] a kind of fractals Parrondo's paradoxial phenomenon "deiconnected + diconnected = connected" in an alternated superior complex system $z_{n+1} = \beta(z_2^n + c_i) + (1 - \beta)z_n, i = 1, 2$ is addressed. The connectivity variation in superior Julia sets is explored by analyzing the connectivity loci. The position relation between the superior Mandelbrot set and the connectivity loci is graphically investigated. Moreover, graphical examples obtained by the use of the escape-time algorithm and the derived criteria are presented.

In their paper "Fractional integral inequalities for exponentially nonconvex functions and their applications" [10] the authors define a new generic class of functions involving a certain modified Fox–Wright function. A useful identity using fractional integrals and the modified Fox–Wright function with two parameters is found. Some Hermite–Hadamard-type integral inequalities are established.

In the paper "Guaranteed cost leaderless consensus protocol design for fractional-order uncertain multi-agent systems with state and input delays" [11] addresses the guaranteed cost leaderless consensus of delayed fractional-order multi-agent systems (FOMASs) with nonlinearities and uncertainties. A guaranteed cost function for FOMAS is proposed to simultaneously consider consensus performance and energy consumption. By employing the linear matrix inequality approach and the fractional-order Razumikhin theorem, a delay-dependent and order-dependent consensus protocol is formulated for FOMASs with input delay.

In "Jacobi spectral collocation technique for time-fractional inverse heat equations" [12] a numerical solution for time-fractional inverse heat equations is proposed. The authors focus on obtaining the unknown source term along with the unknown temperature function based on an additional condition given in an integral form. The proposed scheme is based on a spectral collocation approach to obtain the two independent variables.

The manuscript "State of charge estimation of lithium-ion batteries based on fuzzy fractional-order unscented Kalman filter" [13] proposes a method to estimate the state of charge of lithium-ion batteries. The algorithm combines fuzzy inference with fractional-order unscented Kalman filter to infer the measurement noise in real time and take advantage of fractional calculus in describing the dynamic behavior of the lithium batteries.

To sum up, the guest editors hope that the selected papers will help scholars and researchers to push forward the progress in fractional calculus and its applications, namely in modeling and control of nonlinear and complex systems.

Funding: This research received no external funding.

Conflicts of Interest: The authors declare no conflict of interest.

References

1. Rahmi, E.; Darti, I.; Suryanto, A. A modified Leslie-Gower model incorporating Beddington-DeAngelis functional response, double allee effect and memory effect. *Fractal Fract.* **2021**, *5*, 84. [CrossRef]
2. Samadi, A.; Nuchpong, C.; Ntouyas, S.K.; Tariboon, J. A study of coupled systems of ψ-Hilfer type sequential fractional differential equations with integro-multipoint boundary conditions. *Fractal Fract.* **2021**, *5*, 162. [CrossRef]
3. Si, X.; Wang, Z.; Song, Z.; Zhang, Z. Asymptotic stabilization of delayed linear fractional-order systems subject to state and control constraints. *Fractal Fract.* **2022**, *6*, 67. [CrossRef]
4. Xu, C.; Liao, M.; Li, P.; Yao, L.; Qin, Q.; Shang, Y. Chaos control for a fractional-order jerk system via time delay feedback controller and mixed controller. *Fractal Fract.* **2021**, *5*, 257. [CrossRef]
5. Ataşlar-Ayyıldız, B.; Karahan, O.; Yılmaz, S. Control and robust stabilization at unstable equilibrium by fractional controller for magnetic levitation systems. *Fractal Fract.* **2021**, *5*, 101. [CrossRef]
6. Niazi, A.U.K.; Iqbal, N.; Shah, R.; Wannalookkhee, F.; Nonlaopon, K. Controllability for fuzzy fractional evolution equations in credibility space. *Fractal Fract.* **2021**, *5*, 112. [CrossRef]
7. He, Y.; Zheng, S.; Yuan, L. Dynamics of fractional-order digital manufacturing supply chain system and its control and synchronization. *Fractal Fract.* **2021**, *5*, 128. [CrossRef]
8. Niazi, A.U.K.; He, J.; Shafqat, R.; Ahmed, B. Existence, uniqueness, and E_q-Ulam-type stability of fuzzy fractional differential equation. *Fractal Fract.* **2021**, *5*, 66. [CrossRef]
9. Zhang, Y.; Wang, D. Fractals Parrondo's paradox in alternated superior complex system. *Fractal Fract.* **2021**, *5*, 39. [CrossRef]
10. Srivastava, H.M.; Kashuri, A.; Mohammed, P.O.; Baleanu, D.; Hamed, Y. Fractional integral inequalities for exponentially nonconvex functions and their applications. *Fractal Fract.* **2021**, *5*, 80. [CrossRef]
11. Tian, Y.; Xia, Q.; Chai, Y.; Chen, L.; Lopes, A.M.; Chen, Y. Guaranteed cost leaderless consensus protocol design for fractional-order uncertain multi-agent systems with state and input delays. *Fractal Fract.* **2021**, *5*, 141. [CrossRef]
12. Abdelkawy, M.A.; Amin, A.Z.; Babatin, M.M.; Alnahdi, A.S.; Zaky, M.A.; Hafez, R.M. Jacobi spectral collocation technique for time-fractional inverse heat equations. *Fractal Fract.* **2021**, *5*, 115. [CrossRef]
13. Chen, L.; Chen, Y.; Lopes, A.M.; Kong, H.; Wu, R. State of charge estimation of lithium-ion batteries based on fuzzy fractional-order unscented Kalman filter. *Fractal Fract.* **2021**, *5*, 91. [CrossRef]

fractal and fractional

MDPI

Article

A Modified Leslie–Gower Model Incorporating Beddington–DeAngelis Functional Response, Double Allee Effect and Memory Effect

Emli Rahmi [1,2], Isnani Darti [1,*], Agus Suryanto [1] and Trisilowati [1]

[1] Department of Mathematics, Faculty of Mathematics and Natural Sciences, University of Brawijaya, Malang 65145, Indonesia; emlirahmi@ung.ac.id (E.R.); suryanto@ub.ac.id (A.S.); trisilowati@ub.ac.id (T.)
[2] Department of Mathematics, Faculty of Mathematics and Natural Sciences, State University of Gorontalo, Bone Bolango 96119, Indonesia
* Correspondence: isnanidarti@ub.ac.id

Abstract: In this paper, a modified Leslie–Gower predator-prey model with Beddington–DeAngelis functional response and double Allee effect in the growth rate of a predator population is proposed. In order to consider memory effect on the proposed model, we employ the Caputo fractional-order derivative. We investigate the dynamic behaviors of the proposed model for both strong and weak Allee effect cases. The existence, uniqueness, non-negativity, and boundedness of the solution are discussed. Then, we determine the existing condition and local stability analysis of all possible equilibrium points. Necessary conditions for the existence of the Hopf bifurcation driven by the order of the fractional derivative are also determined analytically. Furthermore, by choosing a suitable Lyapunov function, we derive the sufficient conditions to ensure the global asymptotic stability for the predator extinction point for the strong Allee effect case as well as for the prey extinction point and the interior point for the weak Allee effect case. Finally, numerical simulations are shown to confirm the theoretical results and can explore more dynamical behaviors of the system, such as the bi-stability and forward bifurcation.

Keywords: Leslie–Gower; double Allee effect; Hopf bifurcation; global stability; nonlocal operator

check for **updates**

Citation: Rahmi, E.; Darti, I.; Suryanto, A.; Trisilowati. A Modified Leslie–Gower Model Incorporating Beddington–DeAngelis Functional Response, Double Allee Effect and Memory Effect. *Fractal Fract.* **2021**, *5*, 84. https://doi.org/10.3390/fractalfract5030084

Academic Editors: António M. Lopes and Liping Chen

Received: 29 June 2021
Accepted: 28 July 2021
Published: 1 August 2021

Publisher's Note: MDPI stays neutral with regard to jurisdictional claims in published maps and institutional affiliations.

1. Introduction

Modeling interaction between prey and its predator has become a dominant topic in mathematical biology due to its ubiquitous existence and fundamentality in many biological systems. Study of the dynamics of the predator-prey model could provide qualitative explanations of numerous phenomena that can occur in predator and prey interaction. One of the crucial phenomenon in ecology that influences the per capita growth rate either in the predator or prey population is the Allee effect, which describes a condition where, at low population densities, the per capita growth rate of the population has a positive dependence with its density. There are two kinds of Allee effects, namely the strong Allee effect and the weak Allee effect. In the strong Allee effect, there is a population threshold value named the Allee threshold, below which the per capita growth rate of the population is negative [1,2]. In terms of conservation biology, if the Allee threshold is larger, then it places a population at higher risk of extinction in a low-density population. Meanwhile, in the weak Allee effect, the per capita growth rate of the population always remains positive but is still reduced at low population densities [3–5]. If two or more mechanisms that generate the Allee effect works simultaneously on a single population, then it is known as the double (or multiple) Allee effect [6,7]. Biological evidence of such phenomenon from both terrestrial and aquatic habitat is given in Table 2 of [7,8] and the references cited therein.

From the mathematical point of view, some scholars have been investigating the dynamics of predator-prey models with the Allee effect in the prey population [9–13] or predator population [14–16]. The main focus of all the mentioned research studies is to investigate whether the Allee effect has a tremendous impact on the occurrence of various dynamics in the predator-prey model. For the double Allee effect phenomenon, there are some papers that study the double Allee effect in the prey, see for example [17,18]. However, most of the studies just focused on the double Allee effect in the growth of prey population, although observations are showing that the double Allee effect could be discovered in the growth of the predator population. A typical example comes from the endangered species African wild dog (*Lycaon pictus*). Their social system requires a high population density to survive and reproduce. Being a predator, the African wild dog is a generalist species with the Thomson's gazelle (*Eudorcas thomsonii*) as their common prey but it also hunts other animals such as the impala (*Aepyceros melampus*), warthog (*Phacochoerus aethiopicus*), hares, etc. They live in permanent packs of about 27 adults and pups and have to share food after killing their prey. There is also interference among predators in their hunting behavior. We refer the readers to [19] for details.

In the natural world, the presence of memory must exist in prey and predator interaction since the growth rates of prey and predator at any point depend on the history of the variables at all previous times and not only on the current state which is local to that point [20–24]. Recently, fractional calculus through the fractional derivatives has been known to provide an excellent instrument for describing the memory and hereditary properties of various materials and processes [25], such as in biology, finance, engineering, and physics (see, for example, [26–30] and the references therein). Some interesting papers regarding the Allee effect in the fractional-order predator-prey models are provided in [31,32]. In [31], Suryanto et al. have studied the local stability of the modified Leslie–Gower model with the Beddington–DeAngelis functional response and additive Allee effect in the prey population. They construct the numerical scheme that preserves the dynamics of its first-order system provided by [33]. Later, in [32], Baisad and Moonchai considered the Gause predator-prey model that includes the Allee effect in the prey population and Holling type-III functional response. They also studied the local stability and sufficient conditions of a Hopf bifurcation at the positive equilibrium point. In both papers, the dynamical behaviors are influenced by the order of the derivative.

Motivated by the above mentioned points, this paper aims to study the fractional-order Leslie–Gower predator-prey model incorporating the Beddington–DeAngelis functional response and double Allee effect. The proposed model includes the Caputo fractional-order derivative to capture the effect of memory in the growth rates of both prey and predator. From what we know, the dynamic of our proposed model that incorporates the double Allee effect in the growth of predator and memory effect under the Caputo fractional-order derivative has not been proposed and investigated by other scholars. This work may reveal an interesting ecological point of view to the importance of the double Allee effect than the single Allee effect towards the management of exploited or threatened predator population.

The remaining part of this paper is organized as follows. In Section 2, the model formulation is given. The existence, uniqueness, non-negativity, and boundedness of solutions of our model are discussed in Section 3. Then, we investigate the dynamic behaviors of the model for both weak and strong Allee effects. In Section 4, the existence of non-negative equilibrium points and the local stability of non-negative equilibrium points along with Hopf bifurcation analysis are presented. Next, the sufficient conditions for the global stability of the equilibrium points are carried out in Section 5. Numerical simulations are shown in Section 6 to verify our analytical findings as well as to numerically explore the impact of capturing rate, the Allee threshold, and the order of the fractional-order system on the dynamics of our model. Finally, we draw conclusions in Section 7.

2. Model Formulation

One of the primary directions in modelling the interaction of prey-predator populations is based on the mass conservation principle, which says that the predators can grow only as a function of what they have consumed. Under this principle, the general model of the predator-prey dynamics takes the following model [34]:

$$\frac{dx}{dt} = f(x)x - g(x,y)y,$$
$$\frac{dy}{dt} = kg(x,y)y - \mu y,$$

(1)

where $x(t)$ and $y(t)$ are, respectively, the prey and predator population densities at time t, $f(x)$ is the per capita growth rate of prey, $g(x,y)$ is the functional response, k is the predation efficiency, and μ is the predator per capita death rate. An alternative to the conservative predator-prey model (1) is to abandon the mass conservation principle. This type of model does not explicitly describe the relationship between predation rate and the reproduction rate of predator. A foremost predator-prey model in this direction is the Leslie–Gower model [35]. The Leslie–Gower model maintains the prey equation as in system (1) but applies a logistic type of model for the predator equation. By considering that the per capita growth rate of prey obeys the logistic growth and that predation follows the Beddington–DeAngelis functional response, we have the following Leslie–Gower model.

$$\frac{dx}{dt} = \hat{r}x - \hat{\beta}_1 x^2 - \frac{\hat{b}xy}{1 + \hat{c}x + \hat{q}y},$$
$$\frac{dy}{dt} = \hat{s}y - \hat{\sigma}y^2.$$

(2)

Notice that the logistic type forms in both prey and predator equations are written in the form as suggested by [36]. This typical logistic form is for avoiding paradoxes in the logistic equation [37,38]. All parameters $\hat{r}, \hat{\beta}_1, \hat{b}, \hat{c}, \hat{q}, \hat{s}, \hat{\sigma}$ in the system (2) are real and positive. The ecological meaning of the parameters are as follows: $\hat{r}$ and $\hat{\beta}_1$ are the intrinsic growth rate of the prey and the prey intraspecific competition coefficient in the absence of predation, respectively; $\hat{s}$ and $\hat{\sigma}$ are the intrinsic growth rate of the predator and the predator intraspecific competition coefficient, respectively; $\hat{b}$ and $\hat{c}$ measure the effect of capturing rate and handling time by the predator to the predation rate, respectively; $\hat{q}$ is the strength of interference among predators. The coefficient of predator intraspecific competition is assumed to depend on the prey density, i.e., $\hat{\sigma} = \frac{\hat{\beta}_2}{x}$, where $\hat{\beta}_2$ is the constant of predator intraspecific competition. Such assumption makes sense because when the prey is available in abundant $(x \to \infty)$, then there will no intraspecific competition $(\hat{\sigma} \to 0)$ and the predator can attain its maximum per capita growth rate $\hat{s}$. On the contrary, if the prey is rare $(x \to 0)$, then the intraspecific competition becomes maximum $(\hat{\sigma} \to \infty)$ and, hence, the predator will become extinct as the per capita growth rate of predator becomes $-\infty$. When a severe scarcity of prey occurs, the predator can switch to alternative populations, which causes the reduction in intraspecific competition for hunting the favourite food x. To account for such phenomenon, Aziz-Alaoui and Okiye [39] proposed a modified Leslie–Gower model by introducing an inhibition coefficient $(\hat{l})$ in the intraspecific competition due to the availability of alternative food for the predator. The intraspecific competition coefficient now becomes $\hat{\sigma} = \frac{\hat{\beta}_2}{\hat{l}+x}$. The modified Leslie–Gower model with the Beddington–DeAngelis functional response has been studied in [40,41]. Today, the Leslie–Gower type model still attracts many scholars (see [42–44] and the references therein).

In this paper, we reconsider the modified Leslie–Gower predator-prey model (2) but we assume that the intrinsic growth rate of the predator population is affected by the double Allee effect. Then the predator-prey model (2) takes the following form:

$$\frac{dx}{dt} = \hat{r}x - \hat{\beta}_1 x^2 - \frac{\hat{b}xy}{1 + \hat{c}x + \hat{q}y},$$
$$\frac{dy}{dt} = \hat{s}\left(\frac{y - \hat{m}}{y + \hat{n}}\right)y - \frac{\hat{\beta}_2}{\hat{l}+x}y^2,$$

(3)

where $\hat{m}$ is the Allee threshold, $\hat{n}$ is the auxiliary Allee effect constant with $\hat{n} > 0$ and $\hat{n} > -\hat{m}$. In the second equation of model (3), the intrinsic growth rate of the predator $\hat{s}$ is affected by double Allee effects. Without the intraspecific competition for the predator, the per capita growth rate of the predator is reduced from $\hat{s}$ to $\hat{s}\left(\frac{y-\hat{m}}{y+\hat{n}}\right)$ due to the Allee effect [45]. Therefore, the Allee effect is strong if $\hat{m} > 0$ and weak if $-\hat{n} < \hat{m} \leq 0$ [9,18].

In order to seize the entire time population growth condition, we consider a fractional-order derivative to the left-hand side of the classical derivative system (3) as follows:

$$
\begin{aligned}
D_*^\alpha x &= \hat{r}x - \hat{\beta}_1 x^2 - \frac{\hat{b}xy}{1+\hat{c}x+\hat{q}y}, \\
D_*^\alpha y &= \hat{s}\left(\frac{y-\hat{m}}{y+\hat{n}}\right)y - \frac{\hat{\beta}_2}{\hat{l}+x}y^2,
\end{aligned}
\tag{4}
$$

with initial conditions $x(0) > 0$ and $y(0) > 0$. D_*^α represents the Caputo fractional-order derivative of a real valued function f, which is defined by the following:

$$
D_*^\alpha f(t) = \frac{1}{\Gamma(n-\alpha)} \int_0^t \frac{f^n(\tau)}{(t-\tau)^{n-\alpha-1}}\, d\tau,
$$

where $\Gamma(\cdot)$ is the Gamma function and $\alpha \in (n-1,n], n \in \mathbb{Z}^+$ [25].

In order to overcome the inconsistency of time dimension between the left-hand side of system (4) with its right-hand side, we follow [46–48] to modify all of the biological parameters in the right-hand side that have time dimension ($time^{-1}$). Thus, we have a new system.

$$
\begin{aligned}
D_*^\alpha x &= \hat{r}^\alpha x - \hat{\beta}_1^\alpha x^2 - \frac{\hat{b}^\alpha xy}{1+\hat{c}x+\hat{q}y}, \\
D_*^\alpha y &= \hat{s}^\alpha\left(\frac{y-\hat{m}}{y+\hat{n}}\right)y - \frac{\hat{\beta}_2^\alpha}{\hat{l}+x}y^2.
\end{aligned}
\tag{5}
$$

For simplification, we replace system (5) with the redefined parameters as follows:

$$
\begin{aligned}
D_*^\alpha x &= rx - \beta_1 x^2 - \frac{bxy}{1+cx+qy}, \\
D_*^\alpha y &= s\left(\frac{y-m}{y+n}\right)y - \frac{\beta_2}{l+x}y^2,
\end{aligned}
\tag{6}
$$

where

$$
\hat{r}^\alpha = r, \;\; \hat{\beta}_1^\alpha = \beta_1, \;\; \hat{b}^\alpha = b, \;\; \hat{c} = c, \;\; \hat{q} = q, \;\; \hat{s}^\alpha = s, \;\; \hat{m} = m, \;\; \hat{n} = n, \;\; \hat{\beta}_2^\alpha = \beta_2, \;\; \hat{l} = l.
$$

Notice that, the authors [49] have studied the local stability and have shown numerically a Hopf bifurcation circumstance around the interior point for the case of $m = 0$. In this paper, we focus on the local and global dynamics of system (6) for both $m > 0$ and $m < 0$ cases.

3. Preliminaries Results

In this section, we bring out the fact that system (6) is biologically well-behaved by showing that the solution of system (6) exists and is unique as well being non-negative and bounded.

3.1. Existence and Uniqueness

The existence and uniqueness of the solution of system (6) are examined in the region $\Omega \times [0,T]$, where $\Omega = \{(x,y) \in \mathbb{R}^2 : \max\{|x|,|y|\} \leq M\}$ and $T < +\infty$. Let a mapping $F(Z) = (F_1(Z), F_2(Z))$ with the following.

$$F_1(Z) = rx - \beta_1 x^2 - \frac{bxy}{1 + cx + qy},$$

$$F_2(Z) = \frac{sy^2}{y + n} - \frac{msy}{y + n} - \frac{\beta_2 y^2}{l + x}. \tag{7}$$

For any $Z = (x, y)$, $\bar{Z} = (\bar{x}, \bar{y})$, $Z, \bar{Z} \in \Omega$, it follows from (7) that the following is the case:

$$\|F(Z) - F(\bar{Z})\| = |F_1(Z) - F_1(\bar{Z})| + |F_2(Z) - F_2(\bar{Z})|$$

$$= \left| r(x - \bar{x}) - \beta_1(x^2 - \bar{x}^2) - b\left(\frac{xy}{1 + cx + qy} - \frac{\bar{x}\bar{y}}{1 + c\bar{x} + q\bar{y}}\right)\right| +$$

$$\left| s\left(\frac{y^2}{y + n} - \frac{\bar{y}^2}{\bar{y} + n}\right) - ms\left(\frac{y}{y + n} - \frac{\bar{y}}{\bar{y} + n}\right) - \beta_2\left(\frac{y^2}{l + x} - \frac{\bar{y}^2}{l + \bar{x}}\right)\right|$$

$$\leq r|x - \bar{x}| + \beta_1|x^2 - \bar{x}^2| + b\left|\frac{xy(1 + c\bar{x} + q\bar{y}) - \bar{x}\bar{y}(1 + cx + qy)}{(1 + cx + qy)(1 + c\bar{x} + q\bar{y})}\right|$$

$$+ s\left|\frac{y^2(\bar{y} + n) - \bar{y}^2(y + n)}{(y + n)(\bar{y} + n)}\right| + |m|s\left|\frac{y(\bar{y} + n) - \bar{y}(y + n)}{(y + n)(\bar{y} + n)}\right|$$

$$+ \beta_2\left|\frac{y^2(l + \bar{x}) - \bar{y}^2(l + x)}{(l + x)(l + \bar{x})}\right|$$

$$= r|x - \bar{x}| + \beta_1|x + \bar{x}||x - \bar{x}|$$

$$+ b\left|\frac{(x + cx\bar{x})(y - \bar{y}) + (\bar{y} + qy\bar{y})(x - \bar{x})}{(1 + cx + qy)(1 + c\bar{x} + q\bar{y})}\right|$$

$$+ s\left|\frac{(y\bar{y} + n(y + \bar{y}))(y - \bar{y})}{(y + n)(\bar{y} + n)}\right| + |m|s\left|\frac{n(y - \bar{y})}{(y + n)(\bar{y} + n)}\right|$$

$$+ \beta_2\left|\frac{(l + x)(y + \bar{y})(y - \bar{y}) - y^2(x - \bar{x})}{(l + x)(l + \bar{x})}\right|$$

$$\leq r|x - \bar{x}| + \beta_1|x + \bar{x}||x - \bar{x}| + b|(x + cx\bar{x})(y - \bar{y})|$$

$$+ b|(\bar{y} + qy\bar{y})(x - \bar{x})| + \frac{s}{n^2}|(y\bar{y} + n(y + \bar{y}))(y - \bar{y})|$$

$$+ \frac{|m|s}{n}|y - \bar{y}| + \frac{\beta_2}{l^2}|(l + x)(y + \bar{y})(y - \bar{y}) - y^2(x - \bar{x})|$$

$$\leq r|x - \bar{x}| + 2\beta_1 M|x - \bar{x}| + (bM + bcM^2)|y - \bar{y}|$$

$$+ (bM + bqM^2)|x - \bar{x}| + \left(\frac{sM^2}{n^2} + \frac{2sM}{n}\right)|y - \bar{y}| + \frac{|m|s}{n}|y - \bar{y}|$$

$$+ \left(\frac{2\beta_2 M}{l} + \frac{2\beta_2 M^2}{l^2}\right)|y - \bar{y}| + \frac{\beta_2 M^2}{l^2}|x - \bar{x}|$$

$$= L_1|x - \bar{x}| + L_2|y - \bar{y}|$$

$$\leq L\|Z - \bar{Z}\|$$

where

$$L_1 = r + (2\beta_1 + b)M + \left(bq + \frac{\beta_2}{l^2}\right)M^2,$$

$$L_2 = \frac{|m|s}{n} + \left(b + \frac{2s}{n} + \frac{2\beta_2}{l}\right)M + \left(bc + \frac{s}{n^2} + \frac{2\beta_2}{l^2}\right)M^2,$$

$$L = \max\{L_1, L_2\}.$$

Since $F(Z)$ satisfies the Lipschitz condition with respect to Z, it follows from Theorem 3.7 in [50] that there exists a unique solution $Z(t)$ of system (6) with initial condition $Z(0) = (x(0), y(0))$. Consequently, we have the following theorem.

Theorem 1. *For each $Z(0) = (x(0), y(0)) \in \Omega$, then initial value problem of system (6) has a unique solution $Z(t) \in \Omega$ which is defined for all $t \geq 0$.*

3.2. Non-Negativity and Boundedness

In order to prove that all solutions of system (6) are non-negative and bounded, let $\mathbb{R}^2_+ = \{W = (x, y)^T \in \mathbb{R}^2 | x(t) \geq 0, y(t) \geq 0\}$ be the non-negative quadrant on the $xy-$ plane. In the case of the biological significance, we must ensure that when the initial condition starts in $\mathbb{R}^2_+$, then the solution of system (6) remains in $\mathbb{R}^2_+$ for all $t \geq t_0$.

Theorem 2. *If $x(t_0) \geq 0$ and $y(t_0) \geq 0$, then all solutions of the system (6) are non-negative and uniformly bounded.*

Proof. Let $W(t_0) = \begin{pmatrix} x(t_0) \\ y(t_0) \end{pmatrix} \in \mathbb{R}^2_+$ and assume that $W(t) = \begin{pmatrix} x(t) \\ y(t) \end{pmatrix}$ for $t \geq t_0$ be the solutions of system (6).

Suppose that assumption is false, then there exists $t^* > t_0$ such that $W(t) > 0$ for $t_0 \leq t < t^*$, $W(t^*) = 0$, and $W(t^*_+) < 0$ for $t^*_+ > t^*$. From system (6), one has the following.

$$D^\alpha_* W(t)|_{t=t^*} = 0. \tag{8}$$

Based on Lemma 1 in [51], we have $W(t^*_+) = 0$, which contradicts with $W(t^*_+) < 0$ for $t^*_+ > t^*$. Therefore, we conclude $W(t) \geq 0$ for all $t \geq 0$.

Next, we prove the boundedness of all solutions of system (6). From the first equation of system (6), we have the following.

$$
\begin{aligned}
D^\alpha_* x(t) + x(t) &= rx - \beta_1 x^2 - \frac{bxy}{1 + cx + qy} + x \\
&= -\beta_1 x^2 + (1+r)x - \frac{bxy}{1 + cx + qy} \\
&= -\beta_1 \left(x - \frac{(1+r)}{2\beta_1} \right)^2 + \frac{(1+r)^2}{4\beta_1} - \frac{bxy}{1 + cx + qy} \\
&\leq \frac{(1+r)^2}{4\beta_1}.
\end{aligned}
$$

By Lemma 3 in [51], we have the following:

$$x(t) \leq \left(x(t_0) - \frac{(1+r)^2}{4\beta_1} \right) E_\alpha[-(t - t_0)^\alpha] + \frac{(1+r)^2}{4\beta_1} \rightarrow \frac{(1+r)^2}{4\beta_1}, t \rightarrow \infty, \tag{9}$$

where E_α is the Mittag–Leffler function. Therefore, $x(t)$ with initial condition $x(t_0)$ are confined to the region Ω_1 where the following is the case.

$$\Omega_1 = \left\{ x(t) \leq \frac{(1+r)^2}{4\beta_1} + \epsilon_1 = \gamma_1, \epsilon_1 > 0 \right\}. \tag{10}$$

From the second equation of system (6), we have the following.

$$D^\alpha_* y(t) + sy(t) = \frac{sy^2}{y+n} - \frac{msy}{y+n} - \frac{\beta_2 y^2}{1+x} + sy.$$

We also have $x(t) \leq \gamma_1$ from (10), then the following obtains.

$$D_*^\alpha y(t) + sy(t) \leq \frac{sy^2}{y+n} - \frac{msy}{y+n} - \frac{\beta_2 y^2}{1+\gamma_1} + sy$$

$$\leq \frac{sy^2}{y} - \frac{msy}{y+n} - \frac{\beta_2 y^2}{1+\gamma_1} + sy$$

$$= -\frac{\beta_2 y^2}{1+\gamma_1} + 2sy - \frac{msy}{y+n}$$

$$= -\frac{\beta_2}{1+\gamma_1}\left(y - \frac{(1+\gamma_1)s}{\beta_2}\right)^2 + \frac{(1+\gamma_1)s^2}{\beta_2} - \frac{msy}{y+n}$$

$$\leq \frac{(1+\gamma_1)s^2}{\beta_2}$$

Again by using Lemma 3 in [51], for the strong Allee effect ($m > 0$), we have the following.

$$y(t) \leq \left(y(t_0) - \frac{(1+\gamma_1)s}{\beta_2}\right) E_\alpha[-s(t-t_0)^\alpha] + \frac{(1+\gamma_1)s}{\beta_2} \to \frac{(1+\gamma_1)s}{\beta_2}, t \to \infty. \tag{11}$$

However, for the weak Allee effect ($m < 0$), we have the following.

$$y(t) \leq \left(y(t_0) - \left(\frac{(1+\gamma_1)s}{\beta_2} - m\right)\right) E_\alpha[-s(t-t_0)^\alpha]$$

$$+ \left(\frac{(1+\gamma_1)s}{\beta_2} - m\right) \to \left(\frac{(1+\gamma_1)s}{\beta_2} - m\right), t \to \infty. \tag{12}$$

Therefore, the solution of $y(t)$ with initial condition $y(t_0)$ are confined to region Ω_2 where

$$\Omega_2 = \{y(t) \leq \gamma_2\}, \tag{13}$$

and where the following is the case.

$$\gamma_2 = \begin{cases} \frac{(1+\gamma_1)s}{\beta_2} + \epsilon_2, & \epsilon_2 > 0, \quad m > 0 \\ \left(\frac{(1+\gamma_1)s}{\beta_2} - m\right) + \epsilon_2, & \epsilon_2 > 0, \quad m < 0 \end{cases}$$

$\square$

4. Equilibrium Points and Their Local Stability

In this section, the equilibrium points and existence conditions are obtained and their local stability is analyzed by using the Matignon condition [25] for the weak ($m < 0$) and the strong ($m > 0$) Allee effect, respectively.

(1) The equilibrium points of system (6) for the weak Allee effect ($m < 0$) are as follows:

    (a)   Both prey and predator extinction point $W_0 = (0,0)$, which always exists;

    (b)   The predator extinction point $W_1 = (\frac{r}{\beta_1}, 0)$, which always exists;

    (c)   The prey extinction point $W_2 = (0, \bar{y}_w)$ where we have the following.

$$\bar{y}_w = \frac{\frac{ls}{\beta_2} - n}{2} + \frac{\sqrt{(\frac{ls}{\beta_2} - n)^2 - 4\frac{mls}{\beta_2}}}{2}$$

Denote W_2 as always existing.

(d)   The interior point $\hat{W} = (\hat{x}_w, \hat{y}_w)$ where $\hat{x}_w = \dfrac{\beta_2 \hat{y}_w (\hat{y}_w + n)}{s(\hat{y}_w - m)} - l$ and $\hat{y}_w$ are all positive roots of the quartic equation (14):

$$a_1 y^4 + a_2 y^3 + a_3 y^2 + a_4 y + a_5 = 0, \tag{14}$$

where

$$
\begin{aligned}
a_1 &= \beta_1 \beta_2 qs + c\beta_1 \beta_2^2, \\
a_2 &= (b - qr - lq\beta_1)s^2 + (\beta_1\beta_2(1 + nq - mq - 2cl) - cr\beta_2)s + 2c\beta_1\beta_2^2 n, \\
a_3 &= (c\beta_1 l^2 + 2lmq\beta_1 + lcr + 2mqr - 2bm - l\beta_1 - r)s^2 \\
&\quad + (\beta_1\beta_2(2clm - 2cln - nmq - m + n) + cr\beta_2(m - n))s + c\beta_1\beta_2^2 n^2, \\
a_4 &= (-2cl^2 m\beta_1 - lm^2 q\beta_1 - 2clmr - m^2 qr + bm^2 + 2lm\beta_1 + 2mr)s^2 \\
&\quad + (2clmn\beta_1\beta_2 + cmnr\beta_2 - mn\beta_1\beta_2)s, \\
a_5 &= m^2(cl - 1)(l\beta_1 + r)s^2.
\end{aligned}
$$

(2)   The equilibrium points of system (6) for the strong Allee effect $(m > 0)$ are as follows:

(a)   Both prey and predator extinction point $S_0 = (0,0)$, which always exists;

(b)   The predator extinction point $S_1 = (\frac{r}{\beta_1}, 0)$, which always exists;

(c)   The prey extinction point $S_{2,3} = (0, \bar{y}_s)$ where $\bar{y}_s$ is the positive solution of the quadratic equation $y^2 - \left(\frac{ls}{\beta_2} - n\right)y + \frac{mls}{\beta_2} = 0$. The existence of $S_{2,3}$ is described as follows:

(i)   If $\left(\dfrac{ls}{\beta_2} - n\right)^2 < 4\dfrac{mls}{\beta_2}$, then the prey extinction point does not exist.

(ii)   If $\left(\dfrac{ls}{\beta_2} - n\right)^2 = 4\dfrac{mls}{\beta_2}$ and $n < \dfrac{ls}{\beta_2}$, then there exists a unique prey extinction point, $S_2 = S_3 = \left(0, \dfrac{1}{2}\left(\dfrac{ls}{\beta_2} - n\right)\right)$.

(iii)   If $\left(\dfrac{ls}{\beta_2} - n\right)^2 > 4\dfrac{mls}{\beta_2}$, then there exist two prey extinction points, i.e., the following is the case.

$$S_{2,3} = \left(0, \frac{\frac{ls}{\beta_2} - n}{2} \pm \frac{\sqrt{(\frac{ls}{\beta_2} - n)^2 - 4\frac{mls}{\beta_2}}}{2}\right)$$

(d)   The interior point $\hat{S} = (\hat{x}_s, \hat{y}_s)$ exists if $\hat{y}_s > m$ where $\hat{x}_s = \dfrac{\beta_2 \hat{y}_s (\hat{y}_s + n)}{s(\hat{y}_s - m)} - l$ and $\hat{y}_s$ are also all positive roots of the quartic Equation (14).

In order to study the local stability of system (6) around an equilibrium point (x^*, y^*), we consider the following Jacobian matrix J of system (6), which is given by the following.

$$J(x^*, y^*) = \begin{bmatrix} J_{1,1} & J_{1,2} \\ J_{2,1} & J_{2,2} \end{bmatrix} \tag{15}$$

$$J_{1,1} = r - 2\beta_1 x^* - \frac{by^*}{1 + cx^* + qy^*} + \frac{bcx^* y^*}{(1 + cx^* + qy^*)^2}$$

$$J_{1,2} = -\frac{bx^*}{1 + cx^* + qy^*} + \frac{bqx^* y^*}{(1 + cx^* + qy^*)^2}$$

$$J_{2,1} = \frac{\beta_2 y^{*2}}{(l + x^*)^2}$$

$$J_{2,2} = \frac{s(y^* - m)}{y^* + n} + sy^*\left(\frac{1}{y^* + n} - \frac{y^* - m}{(y^* + n)^2}\right) - \frac{2\beta_2 y^*}{l + x^*}.$$

Theorem 3. *The stability properties of trivial and axial equilibrium points of system (6) for the weak Allee effect $(m < 0)$ are as follows:*

(a) $W_0 = (0,0)$ *is always unstable.*

(b) $W_1 = (\frac{r}{\beta_1}, 0)$ *is always a saddle point.*

(c) $W_2 = (0, \bar{y}_w)$ *is locally asymptotically stable if $r < \dfrac{b\bar{y}_w}{1 + q\bar{y}_w}$ and $m + n < \dfrac{\beta_2}{ls}(\bar{y}_w + n)^2$.*

Proof.

(a)   By substituting W_0 to (15), we obtain the Jacobian matrix.

$$J(W_0) = \begin{bmatrix} r & 0 \\ 0 & -\dfrac{ms}{n} \end{bmatrix}. \tag{16}$$

Therefore, the eigenvalues of (16) are $\lambda_1 = r > 0$ and $\lambda_2 = -\dfrac{ms}{n} > 0$, since $|\arg(\lambda_1)| = |\arg(\lambda_2)| = 0 < \frac{\alpha\pi}{2}$, E_0 is always unstable by the Matignon condition [25].

(b)   By substituting W_1 to (15), we obtain the Jacobian matrix.

$$J(W_1) = \begin{bmatrix} -r & -\dfrac{br}{cr + \beta_1} \\ 0 & -\dfrac{ms}{n} \end{bmatrix}. \tag{17}$$

The Jacobian matrix (17) has eigenvalues $\lambda_1 = -r < 0$ and $\lambda_2 = -\dfrac{ms}{n} > 0$, showing that $|\arg(\lambda_1)| = \pi > \frac{\alpha\pi}{2}$ and $|\arg(\lambda_2)| = 0 < \frac{\alpha\pi}{2}$. Hence, W_1 is always a saddle point.

(c)   By evaluating (15) at W_2, we obtain the following.

$$J(W_2) = \begin{bmatrix} r - \dfrac{b\bar{y}_w}{1 + q\bar{y}_w} & 0 \\ \dfrac{\beta_2 \bar{y}_w^2}{l^2} & \left(\dfrac{s(m + n)}{(\bar{y}_w + n)^2} - \dfrac{\beta_2}{l}\right)\bar{y}_w \end{bmatrix}. \tag{18}$$

The eigenvalues of (18) are as follows.

$$\lambda_1 = r - \frac{b\bar{y}_w}{1 + q\bar{y}_w} \quad \text{and} \quad \lambda_2 = \left(\frac{s(m + n)}{(\bar{y}_w + n)^2} - \frac{\beta_2}{l}\right)\bar{y}_w.$$

Thus, $|\arg(\lambda_1)| = \pi > \dfrac{\alpha\pi}{2}$ and $|\arg(\lambda_2)| = \pi > \dfrac{\alpha\pi}{2}$, whenever $r < \dfrac{b\bar{y}_w}{1 + q\bar{y}_w}$ and $m + n < \dfrac{\beta_2}{ls}(\bar{y}_w + n)^2$.

$\square$

Theorem 4. *Suppose that $m < 0$ (weak Allee effect) and the following is the case.*

$$\chi_{1w} = -\left(\beta_1 \hat{x}_w + \frac{\beta_2 \hat{y}_w}{l + \hat{x}_w}\right) + \left(\frac{bc\hat{x}_w \hat{y}_w}{(1 + c\hat{x}_w + q\hat{y}_w)^2} + \frac{s(m + n)\hat{y}_w}{(\hat{y}_w + n)^2}\right)$$

$$\chi_{2w} = \beta_1 \hat{x}_w \hat{y}_w\left(\frac{\beta_2}{l + \hat{x}_w} - \frac{s(m + n)}{(\hat{y}_w + n)^2}\right)$$

$$+ \frac{bs\hat{x}_w}{(1 + c\hat{x}_w + q\hat{y}_w)^2} \left(\frac{\beta_2 c(m+n)\hat{y}_w^2 + s(1 + c\hat{x}_w)(\hat{y}_w - m)^2}{\beta_2(\hat{y}_w + n)^2} - \frac{\beta_2 c\hat{y}_w^2}{s(l + \hat{x}_w)} \right)$$

$$\alpha^* = \frac{2}{\pi} \left| \tan^{-1} \frac{\sqrt{4\chi_{2w} - (\chi_{1w})^2}}{\chi_{1w}} \right|.$$

The interior point $\hat{W} = (\hat{x}_w, \hat{y}_w)$ is locally asymptotically stable if the following is the case:

(i) $\chi_{1w}^2 \geq 4\chi_{2w}$, $\chi_{1w} < 0$, *and* $\chi_{2w} > 0$.
(ii) $\chi_{1w}^2 < 4\chi_{2w}$, *and if* $\chi_{1w} < 0$, *or* $\chi_{1w} > 0$ *and* $\alpha < \alpha^*$.

Proof. By evaluating (15) at the interior equilibrium point $\hat{W} = (\hat{x}_w, \hat{y}_w)$, we obtain the following.

$$J(\hat{W}) = \begin{bmatrix} -\beta_1 \hat{x}_w + \dfrac{bc\hat{x}_w \hat{y}_w}{(1 + c\hat{x}_w + q\hat{y}_w)^2} & -\dfrac{b\hat{x}_w(1 + c\hat{x}_w)}{(1 + c\hat{x}_w + q\hat{y}_w)^2} \\ \dfrac{s^2(\hat{y}_w - m)^2}{\beta_2(\hat{y}_w + n)^2} & \dfrac{s(m+n)\hat{y}_w}{(\hat{y}_w + n)^2} - \dfrac{\beta_2 \hat{y}_w}{l + \hat{x}_w} \end{bmatrix} \tag{19}$$

The Jacobian matrix (19) has polynomial characteristic $\lambda^2 - \chi_{1w}\lambda + \chi_{2w} = 0$. By utilizing the Routh–Hurwitz criterion for Caputo fractional-order [52], it follows that $\hat{W}$ is locally asymptotically stable if condition (*i*) or (*ii*) is satisfied. $\square$

Theorem 5. *The stability properties of trivial and axial equilibrium points of system (6) for strong Allee effect ($m > 0$) are as follows:*

(a) $S_0 = (0,0)$ *is a saddle point.*
(b) $S_1 = (\frac{r}{\beta_1}, 0)$ *is always locally asymptotically stable.*
(c) $S_{2,3} = (0, \bar{y}_s)$ *is locally asymptotically stable if* $r < \dfrac{b\bar{y}_s}{1 + q\bar{y}_s}$ *and* $m + n < \dfrac{\beta_2}{ls}(\bar{y}_s + n)^2.$

Proof.

(a) By substituting S_0 to (15), we obtain the following.

$$J(S_0) = \begin{bmatrix} r & 0 \\ 0 & -\dfrac{ms}{n} \end{bmatrix}. \tag{20}$$

It is clear that the eigenvalues of (20) are $\lambda_1 = r > 0$ and $\lambda_2 = -\dfrac{ms}{n} < 0$, and $|\arg(\lambda_1)| = 0 < \frac{\alpha\pi}{2}$ and $|\arg(\lambda_2)| = \pi > \frac{\alpha\pi}{2}$. Thus, S_0 is a saddle point.

(b) The Jacobian matrix (15) evaluated at S_1 is the following:

$$J(S_1) = \begin{bmatrix} -r & -\dfrac{br}{cr + \beta_1} \\ 0 & -\dfrac{ms}{n} \end{bmatrix}, \tag{21}$$

where its eigenvalues are $\lambda_1 = -r < 0$ and $\lambda_2 = -\dfrac{ms}{n} < 0$, since $|\arg(\lambda_{1,2})| = \pi > \frac{\alpha\pi}{2}$, S_1 is always locally asymptotically stable.

(c) By evaluating (15) at $S_{2,3}$, we acquire the following.

$$J(S_{2,3}) = \begin{bmatrix} r - \dfrac{b\bar{y}_s}{1 + q\bar{y}_c} & 0 \\ \dfrac{\beta_2 \bar{y}_s^2}{l^2} & \left(\dfrac{s(m+n)}{(\bar{y}_s + n)^2} - \dfrac{\beta_2}{l} \right)\bar{y}_s \end{bmatrix}. \tag{22}$$

The eigenvalues of (22) are as follows.

$$\lambda_1 = r - \frac{b\bar{y}_s}{1 + q\bar{y}_s} \quad \text{and} \quad \lambda_2 = \left(\frac{s(m+n)}{(\bar{y}_s + n)^2} - \frac{\beta_2}{l} \right) \bar{y}_s. \tag{23}$$

Therefore $S_{2,3}$ is locally asymptotically stable if $r < \dfrac{b\bar{y}_s}{1 + q\bar{y}_s}$ and $m + n < \dfrac{\beta_2}{ls}(\bar{y}_w + n)^2$
because in this case $|\arg(\lambda_1)| = |\arg(\lambda_2)| = \pi > \dfrac{\alpha\pi}{2}$.
$\square$

Theorem 6. *For the case of strong Allee effect ($m > 0$), the interior point $\hat{S} = (\hat{x}_s, \hat{y}_s)$ is locally asymptotically stable if the following is the case:*

(i) $\quad \chi_{1s}^2 \geq 4\chi_{2s}, \chi_{1s} < 0, \text{ and } \chi_{2s} > 0.$

(ii) $\quad \chi_{1s}^2 < 4\chi_{2s}, \text{ and if } \chi_{1s} < 0, \text{ or } \chi_{1s} > 0 \text{ and } \alpha < \alpha^*;$

where the following is the case.

$$\chi_{1s} = -\left(\beta_1 \hat{x}_s + \frac{\beta_2 \hat{y}_s}{l + \hat{x}_s} \right) + \left(\frac{bc\hat{x}_s\hat{y}_s}{(1 + c\hat{x}_s + q\hat{y}_s)^2} + \frac{s(m+n)\hat{y}_s}{(\hat{y}_s + n)^2} \right)$$

$$\chi_{2s} = \beta_1 \hat{x}_s \hat{y}_s \left(\frac{\beta_2}{l + \hat{x}_s} - \frac{s(m+n)}{(\hat{y}_s + n)^2} \right)$$

$$+ \frac{bs\hat{x}_s}{(1 + c\hat{x}_s + q\hat{y}_s)^2} \left(\frac{\beta_2 c(m+n)\hat{y}_s^2 + s(1 + c\hat{x}_s)(\hat{y}_s - m)^2}{\beta_2(\hat{y}_s + n)^2} - \frac{\beta_2 c\hat{y}_s^2}{s(l + \hat{x}_s)} \right)$$

$$\alpha^* = \frac{2}{\pi} \left| \tan^{-1} \frac{\sqrt{4\chi_{2s} - (\chi_{1s})^2}}{\chi_{1s}} \right|.$$

Proof. The Jacobian matrix (15) at interior equilibrium point $\hat{S} = (\hat{x}_s, \hat{y}_s)$ is given by the following.

$$J(\hat{S}) = \begin{bmatrix} -\beta_1 \hat{x}_s + \dfrac{bc\hat{x}_s\hat{y}_s}{(1 + c\hat{x}_s + q\hat{y}_s)^2} & -\dfrac{b\hat{x}_s(1 + c\hat{x}_s)}{(1 + c\hat{x}_s + q\hat{y}_s)^2} \\ \dfrac{s^2(\hat{y}_s - m)^2}{\beta_2(\hat{y}_s + n)^2} & \dfrac{s(m+n)\hat{y}_s}{(\hat{y}_s + n)^2} - \dfrac{\beta_2 \hat{y}_s}{l + \hat{x}_s} \end{bmatrix}. \tag{24}$$

The Jacobian matrix (24) has polynomial characteristic $\lambda^2 - \chi_{1s}\lambda + \chi_{2s} = 0$. By using the Routh–Hurwitz criterion for Caputo fractional-order [52], the stability condition is completely proven. $\square$

Hopf bifurcation on a fractional-order system occurs when the Jacobian matrix evaluated at an equilibrium point has two complex conjugate eigenvalues and there is a limit-cycle when the stability of that system changes. Here, we use the conditions for the existence of a Hopf bifurcation which was introduced by [53]. According to Theorems 4 and 6, the stability of the interior equilibrium point for both weak and strong Allee effects is influenced by the order of the fractional derivative (α). Thus, we can establish the condition for the existence of a Hopf bifurcation at the interior point as α passes through the critical value α^* in the following theorem.

Theorem 7 (Existence of Hopf bifurcation [53]). *Let $\chi_{1w}^2 < 4\chi_{2w}$ (or $\chi_{1s}^2 < 4\chi_{2s}$) and $\chi_{1w} > 0$ (or $\chi_{1s} > 0$). System (6) undergoes a Hopf bifurcation around the interior point $\hat{W}$ (or $\hat{S}$) when α crosses α^*.*

Proof. Based on Theorem 6, when $\chi_{1s}^2 < 4\chi_{2s}$ and $\chi_{1s} > 0$, the eigenvalues of system (6) at $\hat{S}$ are a pair of complex conjugate numbers with the real parts are positive. We also confirm that $\phi_{1,2}(\alpha^*) = 0$ and $\frac{d\phi(\alpha)}{d\alpha}|_{\alpha=\alpha^*} \neq 0$ where $\phi_i(\alpha) = \alpha\frac{\pi}{2} - \min_{1 \leq i \leq 2}|\arg(\lambda_i(\alpha))|$. Based on

Theorem 3 in [53], the equilibrium point $\hat{S}$ undergoes a Hopf bifurcation when α crosses α^*. The similar proof works for the weak Allee effect case. $\square$

5. Global Stability

5.1. System with Weak Allee Effect

We know from the previous analysis that in the case of the weak Allee effect, the prey extinction point $W_2 = (0, \bar{y}_w)$ and the interior point $\hat{W} = (\hat{x}_w, \hat{y}_w)$ are conditionally locally asymptotically stable. In the following, we study the global asymptotic stability of those equilibrium points.

Theorem 8. *If* $-n\left(\dfrac{\bar{y}_w}{\gamma_2 + n} - 1\right) \leq m \leq -\dfrac{\gamma_2 + n}{s\bar{y}_w}\left(\dfrac{r^2}{4\beta_1} + \dfrac{\beta_2 \bar{y}_w^2}{l}\right)$, *then* $W_2 = (0, \bar{y}_w)$ *is globally asymptotically stable.*

Proof. We consider the following positive definite Lyapunov function.

$$V_1(x,y) = x + y - \bar{y}_w - \bar{y}_w \ln \frac{y}{\bar{y}_w}$$

Calculating the α-order derivative of $V_1(x,y)$ along the solution of system (6) and applying Lemma 3.1 in [54], we obtain the following.

$$D_*^\alpha V_1(x,y) \leq D_*^\alpha x + \frac{y - \bar{y}_w}{y} D_*^\alpha y$$

$$= rx - \beta_1 x^2 - \frac{bxy}{1 + cx + qy} + (y - \bar{y}_w)\left(\frac{s(y - m)}{y + n} - \frac{\beta_2 y}{l + x}\right)$$

$$= -\beta_1\left(x - \frac{r}{2\beta_1}\right)^2 + \frac{r^2}{4\beta_1} - \frac{bxy}{1 + cx + qy} + \frac{sy^2}{y + n} - \frac{msy}{y + n} - \frac{s\bar{y}_w y}{y + n} + \frac{ms\bar{y}_w}{y + n}$$

$$- \frac{\beta_2 \bar{y}_w y}{l + x} + \frac{\beta_2 \bar{y}_w^2}{l + x} - \frac{\beta_2}{l + x}(y - \bar{y}_w)^2$$

$$\leq \frac{r^2}{4\beta_1} + \frac{sy^2}{y + n} - \frac{msy}{y + n} - \frac{s\bar{y}_w y}{y + n} + \frac{ms\bar{y}_w}{y + n} + \frac{\beta_2 \bar{y}_w^2}{l + x} - \frac{\beta_2}{l + x}(y - \bar{y}_w)^2$$

$$\leq \frac{r^2}{4\beta_1} + sy - \frac{msy}{n} - \frac{s\bar{y}_w y}{\gamma_2 + n} + \frac{ms\bar{y}_w}{\gamma_2 + n} + \frac{\beta_2 \bar{y}_w^2}{l} - \frac{\beta_2}{l + x}(y - \bar{y}_w)^2$$

$$= \left(\frac{r^2}{4\beta_1} + \frac{ms\bar{y}_w}{\gamma_2 + n} + \frac{\beta_2 \bar{y}_w^2}{l}\right) + s\left(1 - \frac{m}{n} - \frac{\bar{y}_w}{\gamma_2 + n}\right)y - \frac{\beta_2}{l + x}(y - \bar{y}_w)^2$$

Since $-n\left(\dfrac{\bar{y}_w}{\gamma_2 + n} - 1\right) \leq m \leq -\dfrac{\gamma_2 + n}{s\bar{y}_w}\left(\dfrac{r^2}{4\beta_1} + \dfrac{\beta_2 \bar{y}_w^2}{l}\right)$, we obtain the following.

$$D_*^\alpha V_1 \leq -\frac{\beta_2}{l + x}(y - \bar{y}_w)^2.$$

In this case, $D_*^\alpha V_1(x,y) \leq 0$ for all $(x,y) \in \mathbb{R}_+^2$ and $D_*^\alpha V_1(x,y) = 0$ implies that $y = \bar{y}_w$. Substituting $y = \bar{y}_w$ to the second equation of system (6) we obtain the following.

$$0 = D_*^\alpha \bar{y}_w = s\bar{y}_w\left(\frac{\bar{y}_w - m}{\bar{y}_w + n}\right) - \frac{\beta_2 \bar{y}_w^2}{l + x}. \tag{25}$$

The unique solution of (25) is $x = 0$, which shows that singleton $\{W_2\}$ is the only invariant set on which $D_*^\alpha V_1(x,y) = 0$. By Lemma 4.6 in [55], it is proven that W_2 is globally asymptotically stable. $\square$

Theorem 9. *The interior point $\hat{W}$ of model (6) is globally asymptotically stable if the following is the case:*

(i) $\quad -n\left(\dfrac{\hat{y}_w}{\gamma_2+n}+\dfrac{\beta_2\hat{y}_w}{s(l+\gamma_1)}-1\right)\le m\le-\dfrac{\beta_2(\gamma_2+n)\hat{y}_w}{ls}$ *and;*

(ii) $\quad b<\min\left\{2\beta_1\dfrac{(1+c\hat{x}_w+q\hat{y}_w)(1+c\gamma_1+q\gamma_2)}{1+c\hat{x}_w+2c\hat{y}_w(1+c\gamma_1+q\gamma_2)},\ \dfrac{2\beta_1\beta_2}{r}\dfrac{(1+c\gamma_1+q\gamma_2)}{(1+c\hat{x}_w)(l+\gamma_1)}\right\}.$

Proof. Consider the following positive definite Lyapunov function.

$$V_2(x,y)=\frac{r}{\beta_1}(1+c\hat{x}_w+q\hat{y}_w)\left(x-\hat{x}_w-\hat{x}_w\ln\frac{x}{\hat{x}_w}\right)+\left(y-\hat{y}_w-\hat{y}_w\ln\frac{y}{\hat{y}_w}\right).$$

By taking the α-order derivative of $V_2(x,y)$ along the solution of system (6) and applying Lemma 3.1 in [54], one has the following.

$$\begin{aligned}
D_*^\alpha V_2 &\le \frac{r}{\beta_1}(1+c\hat{x}_w+q\hat{y}_w)\frac{x-\hat{x}_w}{x}D_*^\alpha x+\frac{y-\hat{y}_w}{y}D_*^\alpha y\\[4pt]
&=\frac{r}{\beta_1}(1+c\hat{x}_w+q\hat{y}_w)(x-\hat{x}_w)\left(r-\beta_1 x-\frac{by}{1+cx+qy}\right)\\[4pt]
&\quad+(y-\hat{y}_w)\left(\frac{s(y-m)}{y+n}-\frac{\beta_2 y}{l+x}\right)\\[4pt]
&=-r(1+c\hat{x}_w+q\hat{y}_w)(x-\hat{x}_w)^2-\frac{br(1+c\hat{x}_w)(x-\hat{x}_w)(y-\hat{y}_w)}{\beta_1(1+cx+qy)}\\[4pt]
&\quad+\frac{bcr\hat{y}_w(x-\hat{x}_w)^2}{\beta_1(1+cx+qy)}+\frac{sy^2}{y+n}-\frac{msy}{y+n}-\frac{s\hat{y}_wy}{y+n}+\frac{ms\hat{y}_w}{y+n}-\frac{\beta_2\hat{y}_wy}{l+x}\\[4pt]
&\quad+\frac{\beta_2\hat{y}_w^2}{l+x}-\beta_2\frac{(y-\hat{y}_w)^2}{l+x}\\[4pt]
&\le\left(\frac{bcr}{\beta_1}\hat{y}_w-r(1+c\hat{x}_w+q\hat{y}_w)\right)(x-\hat{x}_w)^2\\[4pt]
&\quad+\frac{br}{\beta_1}\frac{(1+c\hat{x}_w)}{(1+c\gamma_1+q\gamma_2)}\frac{(x-\hat{x}_w)^2+(y-\hat{y}_w)^2}{2}+sy-\frac{msy}{n}\\[4pt]
&\quad-\frac{s\hat{y}_wy}{\gamma_2+n}+\frac{ms\hat{y}_w}{\gamma_2+n}-\frac{\beta_2\hat{y}_wy}{l+\gamma_1}+\frac{\beta_2\hat{y}_w^2}{l}-\beta_2\frac{(y-\hat{y}_w)^2}{l+\gamma_1}\\[4pt]
&=\left(\frac{bcr}{\beta_1}\hat{y}_w-r(1+c\hat{x}_w+q\hat{y}_w)+\frac{br(1+c\hat{x}_w)}{2\beta_1(1+c\gamma_1+q\gamma_2)}\right)(x-\hat{x}_w)^2\\[4pt]
&\quad-\left(\frac{\beta_2}{l+\gamma_1}-\frac{br(1+c\hat{x}_w)}{2\beta_1(1+c\gamma_1+q\gamma_2)}\right)(y-\hat{y}_w)^2\\[4pt]
&\quad+\left(s-\frac{ms}{n}-\frac{s\hat{y}_w}{\gamma_2+n}-\frac{\beta_2\hat{y}_w}{l+\gamma_1}\right)y+\left(\frac{ms\hat{y}_w}{\gamma_2+n}+\frac{\beta_2\hat{y}_w^2}{l}\right).
\end{aligned}$$

It is clear that if

$$-n\left(\frac{\hat{y}_w}{\gamma_2+n}+\frac{\beta_2\hat{y}_w}{s(l+\gamma_1)}-1\right)\le m\le-\frac{\beta_2(\gamma_2+n)\hat{y}_w}{ls}$$

and

$$b<\min\left\{2\beta_1\frac{(1+c\hat{x}_w+q\hat{y}_w)(1+c\gamma_1+q\gamma_2)}{(1+c\hat{x}_w+2c\hat{y}_w(1+c\gamma_1+q\gamma_2))},\ \frac{2\beta_1\beta_2}{r}\frac{(1+c\gamma_1+q\gamma_2)}{(1+c\hat{x}_w)(l+\gamma_1)}\right\},$$

then $D_*^\alpha V_2(x,y)\le 0$ for all $(x,y)\in\mathbb{R}_+^2$. Moreover, $D_*^\alpha V_2(x,y)=0$ implies that $(x,y)=(\hat{x},\hat{y})$. Based on that, the only invariant set on which $D_*^\alpha V_2(x,y)=0$ is the singleton $\{\hat{W}\}$. By Lemma 4.6 in [55], it follows that $\hat{W}$ is globally asymptotically stable. $\quad\square$

5.2. System with Strong Allee Effect

Previous analysis shows that for the system with a strong Allee effect, the predator extinction point $S_1 = (\frac{r}{\beta_1}, 0)$ is always locally asymptotically stable. Hence, in the following, we study the global asymptotic stability only for the equilibrium point $S_1 = (\frac{r}{\beta_1}, 0)$.

Theorem 10. *If* $m \geq \left(\dfrac{br}{\beta_1 s} + 1 \right)(\gamma_2 + n)$, *then the predator extinction point* $S_1 = \left(\dfrac{r}{\beta_1}, 0 \right)$ *of system (6) is globally asymptotically stable.*

Proof. Define a positive definite Lyapunov function at S_1.

$$V_3(x, y) = \left(x - \frac{r}{\beta_1} - \frac{r}{\beta_1} \ln \frac{\beta_1 x}{r} \right) + y.$$

According to Lemma 3.1. in [54], the α-order derivative of $V_3(x, y)$ along the solution of system (6) satisfies the following.

$$
\begin{aligned}
D_*^\alpha V_3(x, y) &\leq \frac{x - \frac{r}{\beta_1}}{x} D_*^\alpha x + D_*^\alpha y \\
&\quad - \frac{x - \frac{r}{\beta_1}}{x}\left(r - \beta_1 x - \frac{by}{1 + cx + qy} \right)x + y\left(\frac{s(y - m)}{y + n} - \frac{\beta_2 y}{1 + x} \right) \\
&= -\beta_1\left(x - \frac{r}{\beta_1} \right)^2 - \frac{bxy}{1 + cx + qy} + \frac{bry}{\beta_1(1 + cx + qy)} + s\frac{y^2}{y + n} - ms\frac{y}{y + n} \\
&\quad - \beta_2\frac{y^2}{1 + x} \\
&\leq -\beta_1\left(x - \frac{r}{\beta_1} \right)^2 + \frac{bry}{\beta_1} + sy - ms\frac{y}{y + n}.
\end{aligned}
$$

Based on Theorem 2, we have $y(t) \leq \gamma_2$ and, thus, we have the following.

$$
\begin{aligned}
D_*^\alpha V_3(x, y) &\leq -\beta_1\left(x - \frac{r}{\beta_1} \right)^2 + \frac{bry}{\beta_1} + sy - ms\frac{y}{\gamma_2 + n} \\
&= -\beta_1\left(x - \frac{r}{\beta_1} \right)^2 + \left(\frac{br}{\beta_1} + s - \frac{ms}{\gamma_2 + n} \right)y.
\end{aligned}
$$

We note that if $m \geq \left(\dfrac{br}{\beta_1 s} + 1 \right)(\gamma_2 + n)$, then $D_*^\alpha V_3(x, y) \leq 0$ for all $(x, y) \in \mathbb{R}_+^2$. Furthermore, $D_*^\alpha V_3(x, y) = 0$ implies that $(x, y) = (\frac{r}{\beta_1}, 0)$.

Hence, the only invariant set on which $D_*^\alpha V_3(x, y) = 0$ is the singleton $\{S_1\}$. By Lemma 4.6 in [55], the predator extinction point S_1 is globally asymptotically stable. $\square$

6. Numerical Simulations

In this section, some numerical simulations of system (6) are presented to verify the analytical results such as the stability of equilibrium points and a Hopf bifurcation and to explore the dynamical behavior of the system (6) for both weak and strong Allee effects, respectively. The recent development of the numerical schemes for the fractional-order system such as the Grünwald–Letnikov method [31,56], the predictor-corrector method [57–59], the homotopy perturbation method [60,61], and the Laplace adomian decomposition method [62] have been used to solve the fractional-order differential equations. Here, we apply the fractional-order predictor-corrector method provided by Diethelm et al. [63] to obtain numerical solutions of the system (6). Based on that, we divide this section into three subsections that demonstrate the effects of the capturing rate to the predation rate (b), the Allee threshold (m), and the order of fractional system (α) on the stability of equilibrium points. Since parameter values based on real-life observations are not available, we use hypothetical parameters that correspond to the analytical results.

6.1. The Influence of the Capturing Rate (b)

To understand how the capturing rate (b) could influence the dynamics of system (6) in the weak Allee effect case ($m < 0$), we use the following hypothetical parameter values.

$$r = 0.5, \ \beta_1 = 0.05, \ c = 1, \ q = 0.1, \ s = 0.1, \ l = 1, \ \beta_2 = 0.05, \ m = -1, \ \text{and} \ n = 3. \quad (26)$$

The bifurcation diagrams for both predator and prey populations controlled by $b \in [0.2, 0.6]$ and $\alpha = 0.98$ are depicted in Figure 1a. When $b < b_{w1}^* \approx 0.24103$, the interior point $\hat{W}$ is asymptotically stable. For example, we plot a phase portrait of the system (6) in Figure 1b for $b = 0.20$. In this case, one can easily compute that $\chi_{1w}^2 - 4\chi_{2w} = -0.06299 < 0$ and $\chi_{1w} = -0.08727 < 0$. According to Theorem 4, the interior point $\hat{W} = (4.30171, 8.80734)$ is asymptotically stable. The stability of the interior point $\hat{W}$ is also achieved in the interval $b \in (b_{w2}^* = 0.53846, b_{w3}^* = 0.55487)$. To show such typical behavior, we plot a phase portrait of the system (6) in Figure 1d for $b = 0.54$. Furthermore, our numerical simulations also show the existence of limit-cycles solution enclosing the interior point $\hat{W}$ for $b \in (b_{w1}^*, b_{w2}^*)$ as provided in the green area in Figure 1a. As an example, we plot some numerical solutions of the system (6) with $b = 0.27$ in Figure 1c. It is observed that all solutions of system (6) converge to a limit-cycle and the interior point $\hat{W} = (2.76631, 5.82563)$ losses its stability. This situation shows that the system (6) undergoes a Hopf bifurcation with respect to b. When $b > b_{w3}^*$, the interior point $\hat{W}$ does not exist anymore and the prey extinction point $W_2 = (0, 1)$ becomes stable via forward bifurcation. The numerical solutions of system (6), which show the stability of $W_2 = (0, 1)$, are shown in Figure 1e.

Next, we numerically investigate the effect of the capturing rate (b) under the strong Allee effect case ($m > 0$) using the following hypothetical parameters.

$$r = 0.5, \ \beta_1 = 0.1, \ c = 1, \ q = 0.1, \ s = 0.1, \ l = 1, \ \beta_2 = 0.05, \ m = 0.4, \ \text{and} \ n = 0.2. \quad (27)$$

Here, we also take $\alpha = 0.98$. We notice that the predator extinction point $S_1 = (5, 0)$ is always asymptotically stable, see Theorem 5b. Then, by varying b from 0.2 to 0.6, we plot the bifurcation diagrams for both predator and prey population in Figure 2a. It is found that the system (6) in this case has similar qualitative behavior as in the previous simulation (see Figures 1 and 2), with the exception of the stability properties of the predator extinction point $S_1 = (5, 0)$. In the first simulation, the predator extinction point is always unstable while, in the latter case, it is always locally asymptotically stable. The local stability properties of the predator extinction point can be clearly observed from the phase portraits depicted in Figure 2b–e.

Remark 1. *Based on Figures 1 and 2, the capturing rate (b) has great influence to both prey and predator population densities. For parameter values as in (26) and (27), for both weak and strong Allee effects, the densities of both prey and predator populations decrease with increasing capturing rates. In particular, increasing the capturing rate such that $b > b_{w3}^*$ (for the weak Allee effect case) or $b > b_{s3}^*$ (for the strong Allee effect case) may stabilize the prey extinction point $(0, 1)$ as stated in Theorems 3c and 5c. This means that the prey population will become extinct while the predator still survives. We also notice that for the case of strong Allee effect ($m > 0$), the system (6) may exhibit a bistability phenomenon, see Figure 2b,d,e. Hence, the system (6) is highly sensitive to the initial condition when the Allee effect is strong. From the ecological point of view, this result is quite interesting. We can still maintain the existence of both prey and predator even under conditions of a strong Allee effect, that is, as long as we have a sufficiently large initial predator density. Furthermore, Figure 1b,d,e confirm the global asymptotic stability behavior of system (6) for the weak Allee effect case.*

6.2. The Impacts of the Allee Threshold (m)

We next study numerically the impacts of the Allee threshold (m) to the dynamics of system (6) using the following hypothetic values of parameters.

$$r = 0.5, \ \beta_1 = 0.1, \ b = 0.4, \ c = 1, \ q = 0.1, \ s = 0.1, \ l = 1, \ \beta_2 = 0.05, \ \text{and} \ n = 3. \tag{28}$$

In Figure 3a we show the bifurcation diagrams of both predator and prey populations for parameter $m \in [-2, 1]$ and $\alpha = 0.98$. It is shown that when $m < m_1^* \approx -1.73333$, the prey extinction point W_2 is asymptotically stable. If we increase m such that $m > m_1^* - 1.73333$, then the prey extinction point W_2 loses its stability when an asymptotically stable interior point $\hat{W}$ appears. This shows that the system (6) exhibits a forward bifurcation. The interior point $\hat{W}$ remains asymptotically stable for $m \in (m_1^*, m_2^*)$. Moreover, Figure 3a shows the occurence of a Hopf bifurcation when m passes through $m_2^* \approx -1.50769$. Indeed, in the interval $m \in (m_2^*, m_3^* = 0)$, there is no stable equilibrium point and there exists a limit cycle around the interior point $\hat{W}$; see the green area in Figure 3a. Our further observation shows that the system (6) with $m \in (m_3^*, m_4^* \approx 0.82356)$ has a bistability phenomenon where both interior point $\hat{S}$ and the predator extinction point $S_1 = (5, 0)$ are locally asymptotically stable. If we increase m such that $m > m_4^*$, the interior point $\hat{S}$ loses its existence and the predator extinction point S_1 becomes a unique stable equilibrium point of the system (6).

In order to provide a better view of the dynamics of the system (6) with parameter values (28) and $\alpha = 0.98$, we ploted several phase-portraits for different values of m in Figure 3b–e. Figure 3b shows the phase-portrait of the system (6) with $m = -1.79 < m_1^*$. Here, we have $0.5 = r < \frac{b\bar{y}_w}{1 + q\bar{y}_w} = 0.50960$ and $1.21 = m + n < \frac{\beta_2}{ls}(\bar{y}_w + n)^2 = 9.94580$. According to Theorem 3c, the prey extintion point $W_2 = (0, 1.46)$ is asymptotically stable. Figure 3b confirms this behavior where all numerical solutions converge to the prey extinction point $W_2 = (0, 1.46)$. The phase portrait of the system (6) with $m = -0.46 > m_2^*$ is depicted in Figure 3c. In this respect, we have $\chi_{1w} = 0.03488 > 0$ and $0.87258 = \alpha^* < \alpha = 0.98$. Thus, the stability condition of the interior point $\hat{W} = (1.21751, 2.31593)$ in Figure 3c, where all numerical solutions of the system (6) are convergent to a limit-cycle. The appearence of a stable limit-cycle indicates the presence of a Hopf bifurcation in the system. Next, we consider a case of $m \in (m_3^*, m_4^*)$ by choosing $m = 0.4$. The phase-portrait in Figure 3d shows that numerical solutions are convergent to the predator extinction point $S_1 = (5, 0)$ or to the interior point $\hat{S}_1 = (2.24039, 2.40121)$, depending on the initial conditions. Hence, the system (6) exhibits a bistability phenomenon. Finally, if we take $m = 0.90 > m_4^*$, then we have situation where the interior point $\hat{S}$ disappears and the predator extinction point $S_1 = (5, 0)$ becomes the only equilibrium point which is asymptotically stable. This behavior is plotted in Figure 3e.

Note that, when the Allee effect is weak, i.e., when $-n < m < 0$, the predator always exists as depicted in the interval $[-2, m_3^*]$ in Figure 3a. In this case, the predator growth rate is always positive and the prey population could suffer from extinction as the m value decreases. On the other hand, when the Allee effect is strong, i.e., when $m > 0$, there is a condition where the predator is always extinct as depicted by the blue solid line in the interval $[m_3^*, 1]$ in Figure 3a. However, generally speaking, the system (6) could have a positive or negative growth rate on the predator population since there is a bistability phenomenon in that interval as explained before. These situations show that our system (6) may provide the condition for the existence of the predator population which is affected by the double Allee effect.

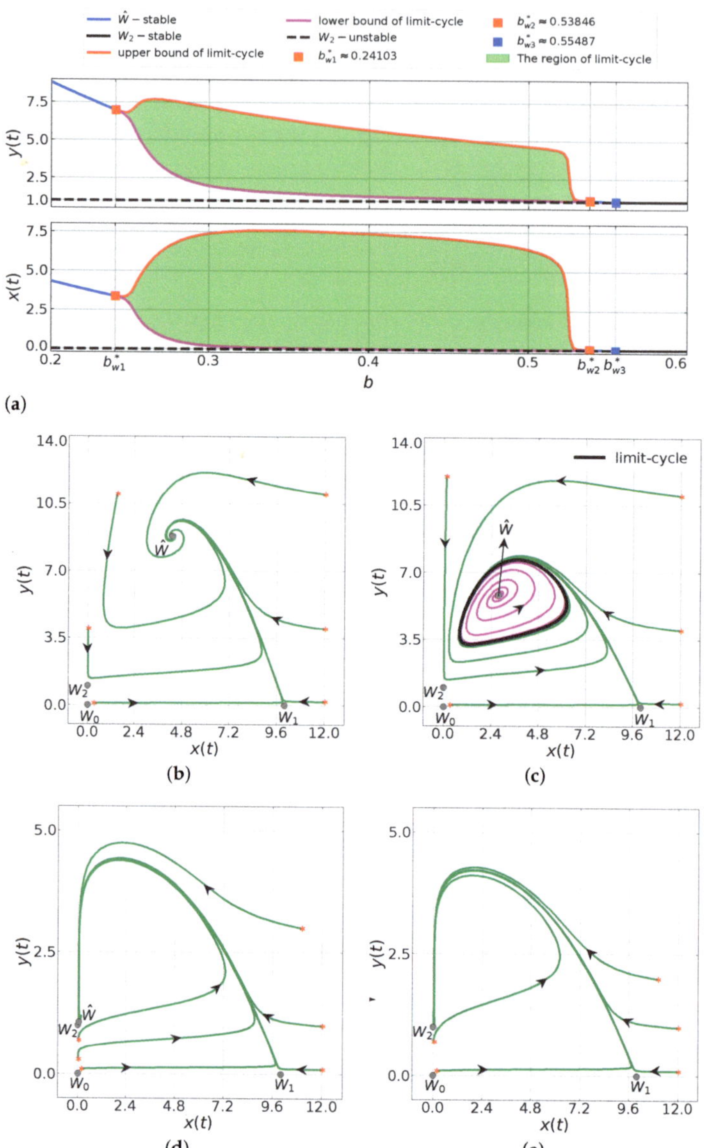

Figure 1. Dynamics of system (6) with parameter values (26) and $\alpha = 0.98$. (**a**) Bifurcation diagram of system (6) driven by b for the case of the weak Allee effect ($m = -1 < 0$). (**b–e**) Phase-portrait of system (6) with (**b**) $b = 0.20$, (**c**) $b = 0.27$, (**d**) $b = 0.54$, and (**e**) $b = 0.57$. The predator extinction point is always unstable.

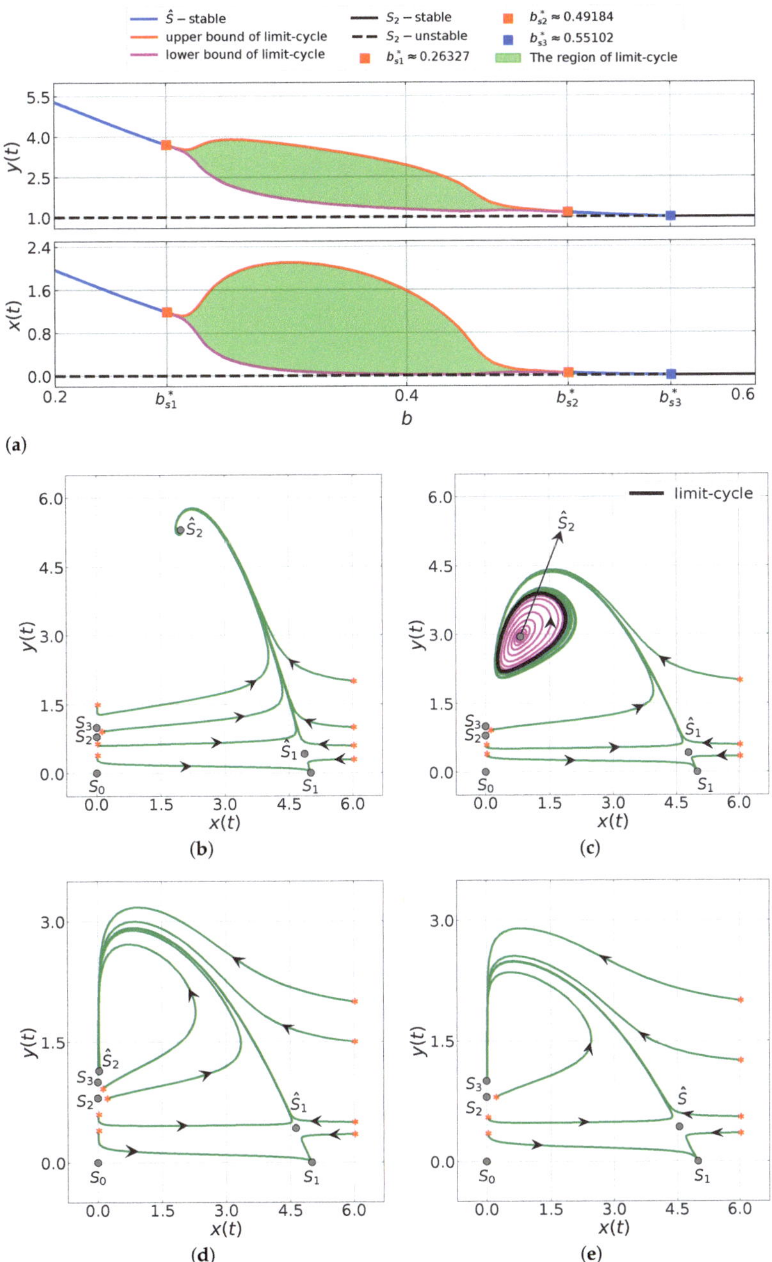

Figure 2. Dynamics of system (6) with parameter values (27) and $\alpha = 0.98$. (**a**) Bifurcation diagram of system (6) driven by b for the case of strong Allee effect ($m = 0.4 > 0$). (**b–e**) Phase-portrait of system (6) with (**b**) $b = 0.20$, (**c**) $b = 0.40$, (**d**) $b = 0.50$, and (**e**) $b = 0.60$. The predator extinction point is always locally stable.

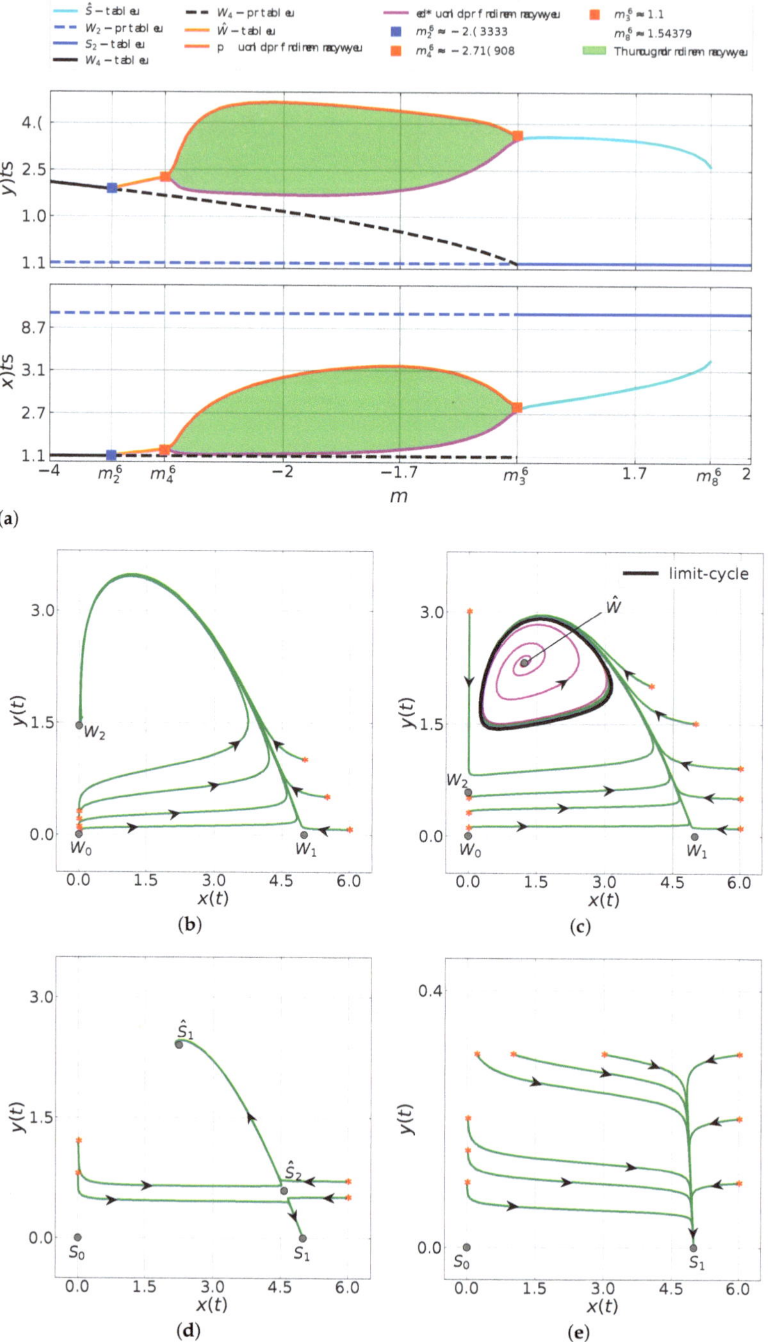

Figure 3. Dynamics of system (6) with parameter values (28) and $\alpha = 0.98$. (**a**) Bifurcation diagram of system (6) driven by m. (**b**–**e**) Phase-portrait of system (6) with (**b**) $m = -1.79$, (**c**) $m = -0.46$, (**d**) $m = 0.40$, and (**e**) $m = 0.90$. The predator extinction point is unstable for the weak Allee effect case ($m < 0$) and is locally asymptotically stable for the strong Allee effect case ($m < 0$).

6.3. The Effects of the Order of Fractional System (α)

In the following simulations, we study the influence of the order of the fractional derivative (α) by considering the system (6) with the following hypothetical parameter values.

$$
\begin{aligned}
r &= 0.5, & \beta_1 &= 0.1, & b &= 0.4, & c &= 1, & q &= 0.1, \\
s &= 0.1, & l &= 1, & \beta_2 &= 0.05, & m &= 0.5, & n &= 0.2.
\end{aligned}
\tag{29}
$$

Based on parameter values (29) and Theorem 7, we can find a critical value $\alpha^* \approx 0.80631$ such that the interior point is asymptotically stable if $\alpha < \alpha^*$ and it is unstable if $\alpha > \alpha^*$. In order to observe such behavior, we plot the bifurcation diagram for $\alpha \in [0.7, 1.0]$, see Figure 4a. It is observed that the interior point $\hat{S} = (0.38825, 1.80915)$ is asymptotically stable when $\alpha < \alpha^*$. If $\alpha > \alpha^*$, then the interior point $\hat{S}$ loses its stability and all numerical solutions converge to a limit-cycle via a Hopf bifurcation. The Hopf bifurcation can also be observed from the phase-portraits shown in Figure 4b for $\alpha = 0.73 < \alpha^*$ and Figure 4c for $\alpha = 0.89 > \alpha^*$.

In Figure 4a, we observe that the order of the fractional derivative does not affect the stability of the interior point as long as $0 < \alpha < \alpha^*$. To verify this property, we plot in Figure 5 the time series of both prey and predator populations for $\alpha = 0.5, 0.6, 0.7, 0.75$. It is shown that all solutions are indeed convergent to the interior point $\hat{S} = (0.38825, 1.80915)$, but the solution of system (6) with a higher–order fractional derivative has faster convergence.

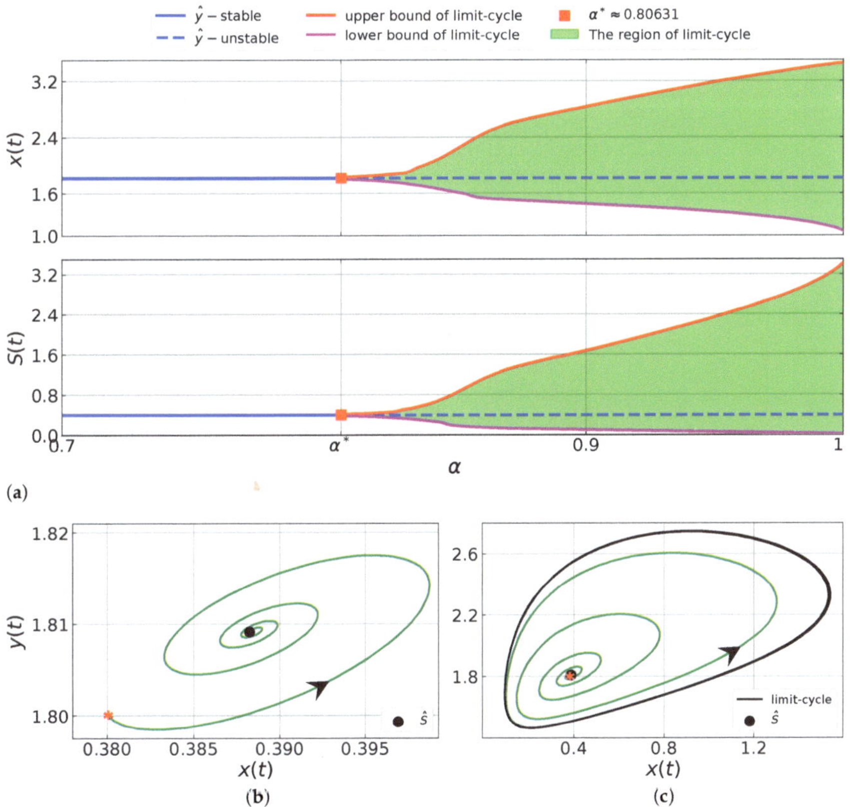

Figure 4. Dynamics of system (6) with parameter values (29). (**a**) Bifurcation diagram of system (6) driven by α. (**b,c**) Phase-portrait of system (6) with (**b**) $\alpha = 0.73$ and (**c**) $\alpha = 0.89$.

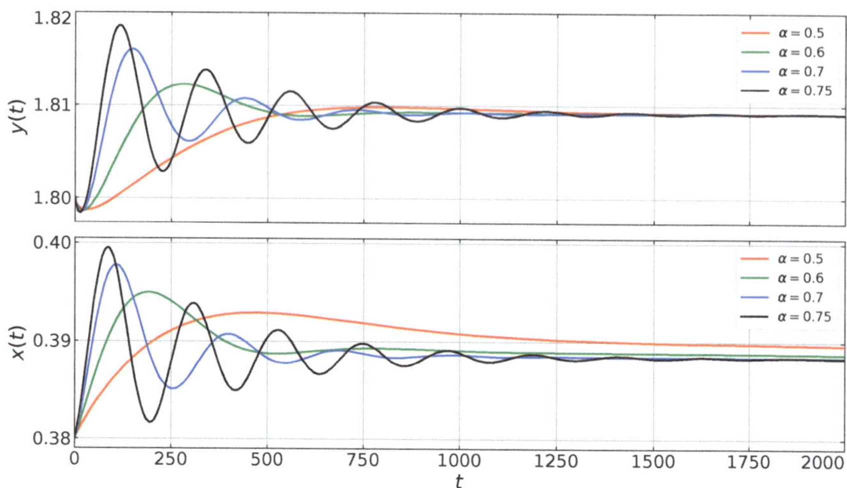

Figure 5. Time series of system (6) with parameter values (29) and the initial conditions taken around the interior point $\hat{S} = (0.38825, 1.80915)$.

7. Conclusions

In this paper, a fractional-order Leslie–Gower predator-prey model with Beddington–DeAngelis functional response and double Allee effect in the predator population is proposed and the dynamics of the model has been analyzed. First, the existence, uniqueness, non-negativity, and boundedness of the solution have been proven. We then determined all possible non-negative equilibrium points and their local and global stability properties. We found that the model has four types of biologically feasible equilibrium. The extinction of both prey and predator point is always unstable for both weak and strong Allee effect cases. The predator extinction point is always stable for the strong Allee effect case, but it is always unstable for the weak Allee effect case. The prey extinction point and the interior point are conditionally stable. Our numerical simulations showed that for the case of the weak Allee effect, there is a capturing rate threshold b^* such that for $b > b^*$ the prey population is extinct while the predator population still survives. However, for the case of the strong Allee effect, the situation is also dependent on the initial value. Here, if the initial predator population is relatively low then the predator will become extinct but the prey will survive. Additionally, we also proved the existence of a Hopf bifurcation about the interior point driven by the order of the fractional derivative (α) and the critical α^* of this bifurcation has been determined analytically. The occurrence of the Hopf bifurcation has been confirmed by our numerical simulations. The existence of Hopf bifurcation controlled by α has also been observed in [31,32]. This shows that the unstable interior point of the first order system (i.e., the system is convergent to a limit cycle) may become stable in the fractional order system (i.e., the system converges to the interior point).

Author Contributions: Conceptualization, I.D. and E.R.; methodology, I.D. and A.S.; software, E.R.; validation, I.D. and T.; formal analysis, E.R., I.D. and A.S.; investigation, E.R. and A.S.; resources, I.D.; data curation, E.R. and T.; writing—original draft preparation, E.R.; writing—review and editing, I.D., A.S. and T.; visualization, E.R. and A.S.; supervision, I.D., A.S. and T.; project administration, A.S.; funding acquisition, I.D. All authors have read and agreed to the published version of the manuscript.

Funding: This research is funded by FMIPA-UB via PNBP-University of Brawijaya according to DIPA-UB number DIPA-023.17.2.677512/2021 and the Professor and Doctoral Research Grant 2021 under contract number 1604/UN10.F09/PN/2021.

Institutional Review Board Statement: Not applicable.

Informed Consent Statement: Not applicable.

Data Availability Statement: Not applicable.

Acknowledgments: The authors thank the editors and the reviewers for their critical suggestions, contributions, and comments that greatly improved this manuscript.

Conflicts of Interest: The authors declare no conflicts of interest in this paper.

References

1. Cushing, J.M. Backward bifurcations and strong Allee effects in matrix models for the dynamics of structured populations. *J. Biol. Dyn.* **2014**, *8*, 57–73. [CrossRef]
2. Buffoni, G.; Groppi, M.; Soresina, C. Dynamics of predator-prey models with a strong Allee effect on the prey and predator-dependent trophic functions. *Nonlinear Anal. Real World Appl.* **2016**, *30*, 143–169. [CrossRef]
3. Sasmal, S.K.; Chattopadhyay, J. An eco-epidemiological system with infected prey and predator subject to the weak Allee effect. *Math. Biosci.* **2013**, *246*, 260–271. [CrossRef]
4. Saifuddin, M.; Biswas, S.; Samanta, S.; Sarkar, S.; Chattopadhyay, J. Complex dynamics of an eco-epidemiological model with different competition coefficients and weak Allee in the predator. *Chaos Solitons Fractals* **2016**, *91*, 270–285. [CrossRef]
5. Pal, S.; Sasmal, S.K.; Pal, N. Chaos control in a discrete-time predator-prey model with weak Allee effect. *Int. J. Biomath.* **2018**, *11*, 1–26. [CrossRef]
6. Arancibia-Ibarra, C.; Flores, J.; Bode, M.; Pettet, G.; Van Heijster, P. A modified May–Holling–Tanner predator-prey model with multiple Allee effects on the prey and an alternative food source for the predator. *Discret. Contin. Dyn. Syst.* **2021**, *26*, 943–962. [CrossRef]
7. Berec, L.; Angulo, E.; Courchamp, F. Multiple Allee effects and population management. *Trends Ecol. Evol.* **2007**, *22*, 185–191. [CrossRef] [PubMed]
8. Pavlová, V.; Berec, L.; Boukal, D.S. Caught between two Allee effects: Trade-off between reproduction and predation risk. *J. Theor. Biol.* **2010**, *264*, 787–798. [CrossRef] [PubMed]
9. Feng, P.; Kang, Y. Dynamics of a modified Leslie–Gower model with double Allee effects. *Nonlinear Dyn.* **2015**, *80*, 1051–1062. [CrossRef]
10. Lai, L.; Zhu, Z.; Chen, F. Stability and bifurcation in a predator–prey model with the additive Allee effect and the fear effect. *Mathematics* **2020**, *8*, 1280. [CrossRef]
11. Martínez-Jeraldo, N.; Aguirre, P. Allee effect acting on the prey species in a Leslie–Gower predation model. *Nonlinear Anal. Real World Appl.* **2019**, *45*, 895–917. [CrossRef]
12. Zhang, C.; Yang, W. Dynamic behaviors of a predator–prey model with weak additive Allee effect on prey. *Nonlinear Anal. Real World Appl.* **2020**, *55*, 103137. [CrossRef]
13. Xiao, Z.; Xie, X.; Xue, Y. Stability and bifurcation in a Holling type II predator–prey model with Allee effect and time delay. *Adv. Differ. Equ.* **2018**, *2018*, 288. [CrossRef]
14. Teixeira Alves, M.; Hilker, F.M. Hunting cooperation and Allee effects in predators. *J. Theor. Biol.* **2017**, *419*, 13–22. [CrossRef] [PubMed]
15. Guan, X.; Liu, Y.U.; Xie, X. Stability analysis of a Lotka-Volterra type predator-prey system with Allee effect on the predator species. *Commun. Math. Biol. Neurosci.* **2018**, *2018*, 9. [CrossRef]
16. Bodine, E.N.; Yust, A.E. Predator–prey dynamics with intraspecific competition and an Allee effect in the predator population. *Lett. Biomath.* **2017**, *4*, 23–38. [CrossRef]
17. Pal, P.J.; Saha, T. Qualitative analysis of a predator-prey system with double Allee effect in prey. *Chaos Solitons Fractals* **2015**, *73*, 36–63. [CrossRef]
18. Singh, M.K.; Bhadauria, B.S.; Singh, B.K. Bifurcation analysis of modified Leslie-Gower predator-prey model with double Allee effect. *Ain Shams Eng. J.* **2018**, *9*, 1263–1277. [CrossRef]
19. Courchamp, F.; Clutton-Brock, T.; Grenfell, B. Multipack dynamics and the Allee effect in the African wild dog, Lycaon pictus. *Anim. Conserv.* **2000**, *3*, 277–285. [CrossRef]
20. El-Shahed, M.; Ahmed, A.M.; Abdelstar, I.M.E. Dynamics of a plant-herbivore model with fractional order. *Progr. Fract. Differ. Appl.* **2017**, *3*, 59–67. [CrossRef]
21. Suryanto, A.; Darti, I.; Panigoro, H.S.; Kilicman, A. A fractional-order predator–prey model with ratio-dependent functional response and linear harvesting. *Mathematics* **2019**, *7*, 1100. [CrossRef]
22. Khoshsiar Ghaziani, R.; Alidousti, J.; Eshkaftaki, A.B. Stability and dynamics of a fractional order Leslie–Gower prey–predator model. *Appl. Math. Model.* **2016**, *40*, 2075–2086. [CrossRef]
23. Panigoro, H.S.; Suryanto, A.; Kusumawinahyu, W.M.; Darti, I. Continuous threshold harvesting in a Gause-type predator-prey model with fractional-order. *AIP Conf. Proc.* **2020**, *2264*, 040001. [CrossRef]
24. Panigoro, H.S.; Suryanto, A.; Kusumawinahyu, W.M.; Darti, I. Dynamics of an eco-epidemic predator–prey model involving fractional derivatives with power-law and Mittag–Leffler kernel. *Symmetry* **2021**, *13*, 785. [CrossRef]
25. Petras, I. *Fractional-Order Nonlinear Systems*; Springer: London, UK, 2011.
26. Kumar, S.; Kumar, R.; Cattani, C.; Samet, B. Chaotic behaviour of fractional predator-prey dynamical system. *Chaos Solitons Fractals* **2020**, *135*, 109811. [CrossRef]

27. Singh, J.; Ganbari, B.; Kumar, D.; Baleanu, D. Analysis of fractional model of guava for biological pest control with memory effect. *J. Adv. Res.* **2021**, in press. [CrossRef]

28. El-Ajou, A.; Al-Smadi, M.; Oqielat, M.N.; Momani, S.; Hadid, S. Smooth expansion to solve high-order linear conformable fractional PDEs via residual power series method: Applications to physical and engineering equations. *Ain Shams Eng. J.* **2020**, *11*, 1243–1254. [CrossRef]

29. Ali, H.M.; Habib, M.A.; Miah, M.M.; Akbar, M.A. Solitary wave solutions to some nonlinear fractional evolution equations in mathematical physics. *Heliyon* **2020**, *6*, e03727. [CrossRef]

30. Shi, G.; Gao, J. European option pricing problems with fractional uncertain processes. *Chaos Solitons Fractals* **2021**, *143*, 110606. [CrossRef]

31. Suryanto, A.; Darti, I.; Anam, S. Stability analysis of a fractional order modified Leslie-Gower model with additive Allee effect. *Int. J. Math. Math. Sci.* **2017**, *2017*, 8273430 . [CrossRef]

32. Baisad, K.; Moonchai, S. Analysis of stability and Hopf bifurcation in a fractional Gauss-type predator–prey model with Allee effect and Holling type-III functional response. *Adv. Differ. Equ.* **2018**, *2018*, 82. [CrossRef]

33. Indrajaya, D.; Suryanto, A.; Alghofari, A.R. Dynamics of modified Leslie-Gower predator-prey model with Beddington-DeAngelis functional response and additive Allee effect. *Int. J. Ecol. Dev.* **2016**, *31*, 60–71.

34. Ginzburg, L.R. Assuming reproduction to be a function of consumption raises doubts about some popular predator-prey models. *J. Anim. Ecol.* **1998**, *67*, 325–327. [CrossRef]

35. Leslie, P.H.; Gower, J.C. The properties of a stochastic model for the predator-prey type of interaction between two species. *Biometrika* **1960**, *47*, 219–234. [CrossRef]

36. Mallet, J. The struggle for existence: How the notion of carrying capacity, K, obscures the links between demography, Darwinian evolution, and speciation. *Evol. Ecol. Res* **2012**, *14*, 627–665.

37. Gabriel, J.P.; Saucy, F.; Bersier, L.-F. Paradoxes in the logistic equation? *Ecol. Model.* **2005**, *185*, 147–151. [CrossRef]

38. Arditi, R.; Bersier, L.-F.; Rohr, R.P. The perfect mixing paradox and the logistic equation: Verhulst vs. Lotka. *Ecosphere* **2016**, *7*, e01599. [CrossRef]

39. Aziz-Alaoui, M.A.; Okiye, M.D. Boundedness and Global Stability for a Predator-Prey Model with Modified and Holling-Type II Schemes. *Appl. Math. Lett.* **2002**, *16*, 1069–1075. [CrossRef]

40. Yu, S. Global stability of a modified Leslie-Gower model with Beddington-DeAngelis functional response. *Adv. Differ. Equ.* **2014**, *2014*, 84. [CrossRef]

41. Vera-Damián, Y.; Vidal, C.; González-Olivares, E. Dynamics and bifurcations of a modified Leslie-Gower–type model considering a Beddington-DeAngelis functional response. *Math. Methods Appl. Sci.* **2019**, *42*, 3179–3210. [CrossRef]

42. Melese, D.; Feyissa, S. Stability and bifurcation analysis of a diffusive modified Leslie-Gower prey-predator model with prey infection and Beddington DeAngelis functional response. *Heliyon* **2021**, *7*, e06193. [CrossRef]

43. Arancibia-Ibarra, C.; Flores, J. Dynamics of a Leslie–Gower predator–prey model with Holling type II functional response, Allee effect and a generalist predator. *Math. Comput. Simul.* **2021**, *188*, 1–22. [CrossRef]

44. Tiwari, B.; Raw, S.N. Dynamics of Leslie–Gower model with double Allee effect on prey and mutual interference among predators. *Nonlinear Dyn.* **2021**, *1229–1257*, 1229–1257. [CrossRef]

45. Zhou, S.R.; Liu, Y.F.; Wang, G. The stability of predator–prey systems subject to the Allee effects. *Theor. Popul. Biol.* **2005**, *67*, 23–31. [CrossRef] [PubMed]

46. Diethelm, K. A fractional calculus based model for the simulation of an outbreak of dengue fever. *Nonlinear Dyn.* **2013**, *71*, 613–619. [CrossRef]

47. Moustafa, M.; Mohd, M.H.; Ismail, A.I.; Abdullah, F.A. Dynamical analysis of a fractional-order eco-epidemiological model with disease in prey population. *Adv. Differ. Equ.* **2020**, *2020*, 48. [CrossRef]

48. Panigoro, H.S.; Suryanto, A.; Kusumawinahyu, W.M.; Darti, I. A Rosenzweig–MacArthur model with continuous threshold harvesting in predator involving fractional derivatives with power law and mittag–leffler kernel. *Axioms* **2020**, *9*, 122. [CrossRef]

49. Rahmi, E.; Darti, I.; Suryanto, A.; Trisilowati; Panigoro, H.S. Stability analysis of a fractional-order Leslie-Gower model with Allee effect in predator. *J. Phys. Conf. Ser.* **2021**, *1821*, 012051. [CrossRef]

50. Li, Y.; Chen, Y.Q.; Podlubny, I. Stability of fractional-order nonlinear dynamic systems: Lyapunov direct method and generalized Mittag-Leffler stability. *Comput. Math. Appl.* **2010**, *59*, 1810–1821. [CrossRef]

51. Li, H.L.; Zhang, L.; Hu, C.; Jiang, Y.L.; Teng, Z. Dynamical analysis of a fractional-order predator-prey model incorporating a prey refuge. *J. Appl. Math. Comput.* **2017**, *54*, 435–449. [CrossRef]

52. Ahmed, E.; El-Sayed, A.; El-Saka, H.A. On some Routh–Hurwitz conditions for fractional order differential equations and their applications in Lorenz, Rössler, Chua and Chen systems. *Phys. Lett. A* **2006**, *358*, 1–4. [CrossRef]

53. Li, X.; Wu, R. Hopf bifurcation analysis of a new commensurate fractional-order hyperchaotic system. *Nonlinear Dyn.* **2014**, *78*, 279–288. [CrossRef]

54. Vargas-De-León, C. Volterra-type Lyapunov functions for fractional-order epidemic systems. *Commun. Nonlinear. Sci. Numer. Simul.* **2015**, *24*, 75–85. [CrossRef]

55. Huo, J.; Zhao, H.; Zhu, L. The effect of vaccines on backward bifurcation in a fractional order HIV model. *Nonlinear Anal. Real World Appl.* **2015**, *26*, 289–305. [CrossRef]

56. Arenas, A.J.; González-Parra, G.; Chen-Charpentier, B.M. Construction of nonstandard finite difference schemes for the SI and SIR epidemic models of fractional order. *Math. Comput. Simul* **2016**, *121*, 48–63. [CrossRef]
57. Barman, D.; Roy, J.; Alrabaiah, H.; Panja, P.; Mondal, S.P.; Alam, S. Impact of predator incited fear and prey refuge in a fractional order prey predator model. *Chaos Solitons Fractals* **2020**, *142*, 110420. [CrossRef]
58. Das, M.; Samanta, G.P. A prey-predator fractional order model with fear effect and group defense. *Int. J. Dyn. Control* **2020**, *9*, 334–349. [CrossRef]
59. Alidousti, J.; Ghafari, E. Dynamic behavior of a fractional order prey-predator model with group defense. *Chaos Solitons Fractals* **2020**, *134*, 109688. [CrossRef]
60. Jena, R.M.; Chakraverty, S.; Baleanu, D.; Jena, S.K. Analysis of time-fractional dynamical model of romantic and interpersonal relationships with non-singular kernels: A comparative study. *Math. Meth. Appl. Sci.* **2020**, *44*, 2183–2199. [CrossRef]
61. Ullah, I.; Ahmad, S.; Rahman, M.; Arfan, M. Investigation of fractional order tuberculosis (TB) model via Caputo derivative. *Chaos Solitons Fractals* **2020**, *142*, 110479. [CrossRef]
62. Mahmood, S.; Shah, R.; Khan, H.; Arif, M. Laplace adomian decomposition method for multi dimensional time fractional model of Navier-Stokes equation. *Symmetry* **2019**, *11*, 149. [CrossRef]
63. Diethelm, K.; Ford, N.J.; Freed, A.D. A predictor-corrector approach for the numerical solution of fractional differential equations. *Nonlinear Dyn.* **2002**, *29*, 3–22. [CrossRef]

fractal and fractional

MDPI

Article

A Study of Coupled Systems of ψ-Hilfer Type Sequential Fractional Differential Equations with Integro-Multipoint Boundary Conditions

Ayub Samadi [1,†] , **Cholticha Nuchpong** [2,†] , **Sotiris K. Ntouyas** [3,4,†] **and Jessada Tariboon** [5,*,†]

1   Department of Mathematics, Miyaneh Branch, Islamic Azad University, Miyaneh 1477893855, Iran; ayubtoraj1366@gmail.com
2   Thai-German Pre-Engineering School, College of Industrial Technology, King Mongkut's University of Technology North Bangkok, Bangkok 10800, Thailand; cholticha.nuch@gmail.com
3   Department of Mathematics, University of Ioannina, 451 10 Ioannina, Greece; sntouyas@uoi.gr
4   Nonlinear Analysis and Applied Mathematics (NAAM)-Research Group, Department of Mathematices, Faculty of Science, King Abdulaziz University, P.O. Box 80203, Jeddah 21589, Saudi Arabia
5   Intelligent and Nonlinear Dynamic Innovations, Department of Mathematics, Faculty of Applied Science, King Mongkut's University of Technology North Bangkok, Bangkok 10800, Thailand
*   Correspondence: jessada.t@sci.kmutnb.ac.th
†   These authors contributed equally to this work.

Abstract: In this paper, the existence and uniqueness of solutions for a coupled system of ψ-Hilfer type sequential fractional differential equations supplemented with nonlocal integro-multi-point boundary conditions is investigated. The presented results are obtained via the classical Banach and Krasnosel'skiǐ's fixed point theorems and the Leray–Schauder alternative. Examples are included to illustrate the effectiveness of the obtained results.

Keywords: ψ-Hilfer fractional derivative; Riemann–Liouville fractional derivative; Caputo fractional derivative; system of fractional differential equations

check for **updates**

Citation: Samadi, A.; Nuchpong, C.; Ntouyas, S.K.; Tariboon, J. A Study of Coupled Systems of ψ-Hilfer Type Sequential Fractional Differential Equations with Integro-Multipoint Boundary Conditions. *Fractal Fract.* **2021**, 5, 162. https://doi.org/10.3390/fractalfract5040162

Academic Editors: António M. Lopes and Liping Chen

Received: 18 September 2021
Accepted: 1 October 2021
Published: 10 October 2021

Publisher's Note: MDPI stays neutral with regard to jurisdictional claims in published maps and institutional affiliations.

1. Introduction

Fractional calculus, as an extension of usual integer calculus, is a forceful tool to express real-world problems rather than integer-order differentiations, so that this idea has wide applications in various fields such as, mathematics, physics, engineering, biology, finance, economy and other sciences (see [1–3] and related references therein). Accordingly, many researchers have studied initial and boundary value problems for fractional differential equations (see [4–11]). Additionally, fractional differential equations involving coupled systems have nonlocal natures and applications in many real = world process. The investigation of types of integral and differential operators and the relationship between these operators plays a key role in studying fractional differential equations. Fractional operators of a function concerning another function were introduced by Kilbas et al. [5]. Later, Almeida [12] introduced the notion of the ψ-Caputo fractional operator. For some applications of ψ operator, we refer to the papers [13–15]. Hilfer [16] extended both Riemann–Liouville and Caputo fractional derivatives by presenting a family of derivative operators. Different models based on Hilfer fractional derivative have been considered in [17–21], and references cited therein. Many applications of Hilfer fractional differential equations can be found in many fields of mathematics, physics, etc (see [22–24]). The study of boundary value problems for Hilfer-fractional differential equations of order in $(1, 2]$, and nonlocal boundary conditions was initiated in [25] by studying the boundary value problem of the form:

$$\begin{cases} {}^{H}D^{\alpha_1,\beta_1}u(z) = h(z,u(z)), \quad z \in [c,d], \ 1 < \alpha_1 \leq 2, \ 0 \leq \beta_1 \leq 1, \\ u(c) = 0, \quad u(d) = \sum_{i=1}^{m} \varepsilon_i I^{\phi_i} u(\xi_i), \quad \phi_i > 0, \varepsilon_i \in \mathbb{R}, \xi_i \in [c,d], \end{cases} \tag{1}$$

where ${}^{H}D^{\alpha_1,\beta_1}$ is the Hilfer fractional derivative of order α_1, $1 < \alpha_1 \leq 2$, and parameter β_1, $0 \leq \beta_1 \leq 1$, $f : [c,d] \times \mathbb{R} \longrightarrow \mathbb{R}$ is a continuous function, $c \geq 0$ and I^{ϕ_i} is the Riemann–Liouville fractional integral of order $\phi_i, i = 1,2,\ldots,m$. Several existence and uniqueness results were proved by using a variety of fixed point theorems.

Wongcharoen et al. [26] studied a system of Hilfer-type fractional differential equations of the form

$$\begin{cases} {}^{H}D^{\alpha_1,\beta_1}u(z) = f_1(z,u(z),v(z)), \quad z \in [c,d], \\ {}^{H}D^{\bar{\alpha}_1,\beta_1}v(z) = g_1(z,u(z),v(z)), \quad z \in [c,d], \\ u(c) = 0, \quad u(d) = \sum_{i=1}^{m} \bar{\theta}_i I^{\phi_i} v(\xi_i), \\ v(c) = 0, \quad v(d) = \sum_{j=1}^{n} \bar{\zeta}_j I^{\bar{\phi}_j} u(z_i), \end{cases} \tag{2}$$

where ${}^{H}D^{\alpha_1,\beta_1;\psi}$ and ${}^{H}D^{\bar{\alpha}_1,\beta_1;\psi}$ are the Hilfer fractional derivatives of orders α_1 and $\bar{\alpha}_1$, $1 < \alpha_1, \bar{\alpha}_1 < 2$ and parameter β_1, $0 \leq \beta_1 \leq 1$, $f_1, g_1 : [c,d] \times \mathbb{R} \times \mathbb{R} \longrightarrow \mathbb{R}$ are continuous functions, $c \geq 0$, $\bar{\theta}_i, \bar{\zeta}_i \in \mathbb{R}$, and $I^{\phi_i}, I^{\bar{\phi}_j}$ are the Riemann–Liouville fractional integrals of order $\phi_i > 0, \bar{\phi}_j > 0, i = 1,2,\ldots,m, j = 1,2,\ldots,n$.

Sitho et al. [27] proved the existence and uniqueness of solutions for the following class of boundary value problems consisting of fractional-order ψ-Hilfer-type differential equations supplemented with nonlocal integro-multipoint boundary conditions of the form:

$$\begin{cases} {}^{H}D^{\alpha_1,\beta_1;\psi}u(z) = h(z,u(z)), \quad z \in [c,d], \\ u(c) = 0, \quad u(d) = \sum_{i=1}^{n} \mu_i \int_{c}^{\eta_i} \psi'(s)ds + \sum_{j=1}^{m} \lambda_j u(\xi_j), \end{cases} \tag{3}$$

where ${}^{H}D^{\alpha_1,\beta_1;\psi}$ is the ψ-Hilfer fractional derivative operator of order α_1, $1 < \alpha_1 < 2$ and parameter β_1, $0 \leq \beta_1 \leq 1$, $f : [c,d] \times \mathbb{R} \to \mathbb{R}$ is a continuous function, $c \geq 0$, $\mu_i, \lambda_j \in \mathbb{R}$, $\eta_i, \xi_j \in (c,d), i = 1,2,\ldots,n, j = 1,2,\ldots m$ and ψ is a positive increasing function on $(c,d]$, which has a continuous derivative $\psi'(t)$ on (c,d).

Recently, in [28], the boundary value problem (3) was extended to sequential ψ-Hilfer-type fractional differential equations involving integral multi-point boundary conditions of the form

$$\begin{cases} \left({}^{H}D^{\alpha_1,\beta_1;\psi} + k\,{}^{H}D^{\alpha_1-1,\beta_1;\psi}\right)u(z) = f(z,u(z)), \quad k \in \mathbb{R}, \ z \in [c,d], \\ u(c) = 0, \quad u(d) = \sum_{i=1}^{n} \mu_i \int_{a}^{\eta_i} \psi'(s)u(s)ds + \sum_{j=1}^{m} \theta_j u(\xi_j), \end{cases} \tag{4}$$

where the notations are the same as those of problem (3).

In the present research, inspired by the published articles in this direction, we study the existence and uniqueness of solutions for the following coupled system of sequential ψ-Hilfer-type fractional differential equations with integro-multi-point boundary conditions of the form

$$\begin{cases} ({}^{H}D^{\alpha_1,\beta_1;\psi} + k\,{}^{H}D^{\alpha_1-1,\beta_1;\psi})u(z) = f(z,u(z),v(z)), \quad z \in [c,d], \\ ({}^{H}D^{\bar{\alpha}_1,\beta_1;\psi} + k\,{}^{H}D^{\bar{\alpha}_1-1,\beta_1;\psi})v(z) = g(z,u(z),v(z)), \quad z \in [c,d], \\ u(c) = 0, \quad u(d) = \sum_{i=1}^{n} \mu_i \int_{c}^{\eta_i} \psi'(s)v(s)ds + \sum_{j=1}^{m} \theta_j v(\xi_j), \\ v(c) = 0, \quad v(d) = \sum_{r=1}^{p} \nu_r \int_{c}^{\varsigma_r} \psi'(s)u(s)ds + \sum_{s=1}^{q} \tau_s u(\sigma_s), \end{cases} \tag{5}$$

where $^{H}D^{\alpha_1,\beta_1;\psi}$ and $^{H}D^{\overline{\alpha}_1,\beta_1;\psi}$ are the ψ-Hilfer fractional derivatives of orders α_1 and $\overline{\alpha}_1$, $1 < \alpha_1, \overline{\alpha}_1 < 2$ and parameter β_1, $0 \leq \beta_1 \leq 1$, $f, g : [c,d] \times \mathbb{R} \times \mathbb{R} \longrightarrow \mathbb{R}$ are continuous functions, $c \geq 0$, $\mu_i, \theta_j, v_r, \tau_s \in \mathbb{R}^+$, $\eta_i, \xi_j, \varsigma_r, \sigma_s \in (c,d)$, $i = 1,2,\ldots,n$, $j = 1,2,\ldots,m$, $r = 1,2,\ldots,p$, $s = 1,2,\ldots,q$ and ψ is an increasing and positive monotone function on $(c,d]$ having a continuous derivative ψ' on (c,d).

The classical fixed point theorems are applied in order to obtain our main existence and uniqueness results. Thus, the Banach fixed point theorem is applied to obtain the uniqueness result, while Leray–Schuader alternative and Krasnosel'skiĭ's fixed point theorems are the basic tools used to present the existence results.

The rest of the paper is organized as follows: we recall some primitive concepts in Section 2. In Section 3, an auxiliary lemma is proved, which is a basic tool in proving the main results of the paper, which are presented in Section 4. The main results are supported by numerical examples.

2. Preliminaries

In this section, some basic concepts in connection to fractional calculus and fixed point theory are assigned. Throughout the paper, by $\mathcal{X} = C([c,d],\mathbb{R})$, we denote the Banach space of all continuous mappings from $[c,d]$ to $\mathbb{R}$ endowed with the norm $\|x\| = \sup\{|x(t)|; t \in [c,d]\}$. It is clear that the space $\mathcal{X} \times \mathcal{X}$, endowed with the norm $\|(x,y)\| = \|x\| + \|y\|$, is a Banach space.

Definition 1 ([2]). *Let (c,d), $(-\infty \leq c < d \leq \infty)$ $\alpha > 0$ and $\psi(z)$ be a positive increasing function on $(c,d]$, with continuous derivative $\psi'(z)$ on (c,d). The ψ-Riemann–Liouville fractional integral of a function h with respect to a another function ψ on $[c,d]$ is defined by*

$$I^{\alpha,\psi}h(z) = \frac{1}{\Gamma(\alpha)} \int_c^z \psi'(s)(\psi(z) - \psi(s))^{\alpha-1}h(s)ds, \; z > c > 0,$$

where $\Gamma(.)$ is the Euler Gamma function.

Definition 2 ([29]). *Let $\psi \in C^n([c,d],\mathbb{R})$ with $\psi'(z) \neq 0$ and $\eta > 0$, $n \in \mathbb{N}$. The Riemann–Liouville derivatives of a function h with connection to another function ψ of order η is represented as*

$$\begin{aligned}
D^{\eta;\psi}h(z) &= \left(\frac{1}{\psi'(z)}\frac{d}{dz}\right)^n I_{c+}^{n-\eta;\psi}, \\
&= \frac{1}{\Gamma(n-\eta)}\left(\frac{1}{\psi'(z)}\frac{d}{dz}\right)^n \int_c^z \psi'(s)(\psi(z) - \psi(s))^{n-\eta-1}h(s)ds,
\end{aligned}$$

where $n = [\eta] + 1$, $[\eta]$ represent the integer part of real number η.

Definition 3 ([29]). *Assume that $n - 1 < \eta < n$ with $n \in \mathbb{N}$ and $[c,d]$ is the interval so that $-\infty \leq c < d \leq \infty$ and $h, \psi \in C^n([c,d],\mathbb{R})$ are two functions, such that ψ is increasing and $\psi'(z) \neq 0$ for all $z \in [c,d]$. The ψ-Hilfer fractional derivative of a function h of order η and type $0 \leq \overline{\eta} \leq 1$ is defined by*

$$^{H}D_{c+}^{\eta,\overline{\eta};\psi}h(z) = I_{c+}^{\overline{\eta}(n-\eta);\psi}\left(\frac{1}{\psi'(z)}\frac{d}{dz}\right)^n I_{c+}^{(1-\overline{\eta})(n-\eta);\psi}h(z) = I_{c+}^{\gamma-\eta;\psi}D_{c+}^{\gamma;\psi}h(z),$$

where $n = [\eta] + 1$, $[\eta]$ represents the integer part of the real number η with $\gamma = \eta + \overline{\eta}(n - \eta)$.

Lemma 1 ([29]). *If $h \in C^n(J,\mathbb{R})$, $n - 1 < \eta < n$, $0 \leq \overline{\eta} \leq 1$ and $\gamma = \eta + \overline{\eta}(n - \eta)$, then*

$$I_{c+}^{\eta;\psi}\left(^{H}D_{c+}^{\eta,\overline{\eta};\psi}h\right)(z) = h(z) - \sum_{k=1}^{n} \frac{(\psi(z) - \psi(a))^{\gamma-k}}{\Gamma(\gamma-k+1)} \nabla_{\psi}^{[n-k]} I_{c+}^{(1-\overline{\eta})(n-\eta);\psi}h(c),$$

for all $z \in J$, where $\nabla_\psi^{[n]} h(z) = \left(\frac{1}{\psi'(z)} \frac{d}{dz} \right)^n h(z)$.

Finally, we summarize the fixed point theorems used to prove the main results in this paper. X is a Banach space in each theorem.

Lemma 2. *(Banach fixed point theorem [30]). Let D be a closed set in X and $T : D \to D$ satisfies*

$$|Tu - Tv| \le \lambda |u - v|, \text{ for some } \lambda \in (0,1), \text{ and for all } u, v \in D.$$

Then T admits one fixed point in D.

Lemma 3. *(Leray–Schauder alternative [31]). Let the set Ω be closed bounded convex in X and O an open set contained in Ω with $0 \in O$. Then, for the continuous and compact $T : \bar{U} \to \Omega$, either:*
(a) T admits a fixed point in $\bar{U}$, or
(aa)   There exists $u \in \partial U$ and $\mu \in (0,1)$ with $u = \mu T(u)$.

Lemma 4. *(Krasnosel'skiĭ fixed point theorem [32]). Let M be a closed, bounded, convex and nonempty subset of a Banach space X. Let A, B be operators such that (i) $Ax + By \in M$ where $x, y \in M$, (ii) A is compact and continuous and (iii) B is a contraction mapping. Then there exists $z \in M$ such that $z = Az + Bz$.*

3. An Auxiliary Result

We prove the following auxiliary lemma, concerning a linear variant of the coupled system (5), which is useful to present the coupled system (5) as a fixed point problem.

Lemma 5. *Let $c \ge 0, 1 < \alpha_1, \bar{\alpha}_1 < 2, 0 \le \beta_1 \le 1, \gamma = \alpha_1 + 2\beta_1 - \alpha_1\beta_1, \gamma_1 = \bar{\alpha}_1 + 2\beta_1 - \bar{\alpha}_1\beta_1$ and $\Lambda \ne 0$.*
Then, for $h_1, h_2 \in C([c, d], \mathbb{R})$, the unique solution of the coupled system

$$\begin{cases} (^H D^{\alpha_1,\beta_1;\psi} + k\,^H D^{\alpha_1-1,\beta_1;\psi})u(z) = h_1(z), \; z \in [c,d], \\ (^H D^{\bar{\alpha}_1,\beta_1;\psi} + k\,^H D^{\bar{\alpha}_1-1,\beta_1;\psi})v(z) = h_2(z), \; z \in [c,d], \\ u(c) = 0, \; u(d) = \sum_{i=1}^{n} \mu_i \int_c^{\eta_i} \psi'(s)v(s)ds + \sum_{j=1}^{m} \theta_j v(\xi_j), \\ v(c) = 0, \; v(d) = \sum_{r=1}^{p} v_r \int_c^{\varsigma_r} \psi'(s)u(s)ds + \sum_{s=1}^{q} \tau_s u(\sigma_s), \end{cases} \tag{6}$$

is given as

$$\begin{aligned} u(z) = {} & -k \int_c^z \psi'(s)u(s)ds + I_{c+}^{\alpha_1;\psi} h_1(z) \\ & + \frac{(\psi(z) - \psi(c))^{\gamma-1}}{\Lambda\Gamma(\gamma)} \Big[\Delta \Big(\sum_{i=1}^{n} \mu_i \int_c^{\eta_i} \psi'(s) I^{\bar{\alpha}_1;\psi} h_2(s)ds + \sum_{j=1}^{m} \theta_j I^{\bar{\alpha}_1;\psi} h_2(\xi_j) \\ & - k \sum_{j=1}^{m} \theta_j \int_c^{\xi_j} \psi'(s)v(s)ds - k \sum_{i=1}^{n} \mu_i \int_c^{\eta_i} \psi'(s) \int_c^s \psi'(t)v(t)dtds \\ & + k \int_c^d \psi'(s)u(s)ds - I_{c+}^{\alpha_1;\psi} h_1(d) \Big) + B \Big(\sum_{r=1}^{p} v_r \int_c^{\varsigma_r} \psi'(s) I^{\alpha_1;\psi} h_1(s)ds \\ & + \sum_{s=1}^{q} \tau_s I^{\alpha_1;\psi} h_1(\sigma_s) - k \sum_{s=1}^{q} \tau_s \int_c^{\sigma_s} \psi'(s)u(s)ds \\ & - k \sum_{r=1}^{p} v_r \int_a^{\varsigma_r} \psi'(s) \int_c^s \psi'(t)u(t)dtds \end{aligned}$$

$$+k\int_c^d \psi'(s)v(s)ds - I_{c+}^{\bar{\alpha}_1;\psi}h_2(d)\Big)\Big], \tag{7}$$

and

$$
\begin{aligned}
v(z) = & -k\int_c^z \psi'(s)v(s)ds + I_{c+}^{\bar{\alpha}_1;\psi}h_2(z) \\
& +\frac{(\psi(z)-\psi(c))^{\gamma_1-1}}{\Lambda\Gamma(\gamma_1)}\Big[A\Big(\sum_{r=1}^p v_r\int_c^{\varsigma_r}\psi'(s)I^{\alpha_1;\psi}h_1(s)ds + \sum_{s=1}^q \tau_s I^{\alpha_1;\psi}h_1(\sigma_s) \\
& -k\sum_{s=1}^q \tau_s\int_c^{\sigma_s}\psi'(s)u(s)ds - k\sum_{r=1}^p v_r\int_c^{\varsigma_r}\psi'(s)\int_c^s\psi'(t)u(t)dtds \\
& +k\int_c^c \psi'(s)v(s)ds - I_{c+}^{\bar{\alpha}_1;\psi}h_2(d)\Big) + \Gamma\Big(\sum_{i=1}^n \mu_i\int_c^{\eta_i}\psi'(s)I^{\bar{\alpha}_1;\psi}h_2(s)ds \\
& +\sum_{j=1}^m \theta_j I^{\bar{\alpha}_1;\psi}h_2(\xi_j) - k\sum_{j=1}^m \theta_j\int_c^{\xi_j}\psi'(s)v(s)ds \\
& -k\sum_{i=1}^n \mu_i\int_c^{\eta_i}\psi'(s)\int_c^s\psi'(t)v(t)dtds \\
& +k\int_c^d \psi'(s)u(s)ds - I_{c+}^{\alpha_1;\psi}h_1(d)\Big)\Big], \tag{8}
\end{aligned}
$$

where

$$
\begin{aligned}
A &= \frac{(\psi(d)-\psi(c))^{\gamma-1}}{\Gamma(\gamma)}, \\
B &= \sum_{i=1}^n \mu_i\int_c^{\eta_i}\frac{\psi'(s)(\psi(s)-\psi(a))^{\gamma_1-1}}{\Gamma(\gamma_1)}ds + \sum_{j=1}^m \theta_j\frac{(\psi(\xi_j)-\psi(c))^{\gamma_1-1}}{\Gamma(\gamma_1)}, \\
\Gamma &= \sum_{r=1}^p v_r\int_c^{\varsigma_r}\frac{\psi'(s)(\psi(s)-\psi(a))^{\gamma-1}}{\Gamma(\gamma)}ds + \sum_{s=1}^q \tau_s\frac{(\psi(\sigma_s)-\psi(c))^{\gamma-1}}{\Gamma(\gamma)}, \\
\Delta &= \frac{(\psi(d)-\psi(c))^{\gamma_1-1}}{\Gamma(\gamma_1)}, \tag{9}
\end{aligned}
$$

and

$$\Lambda = A\Delta - B\Gamma.$$

Proof. Taking the operator I^α on both sides of equations in (6) and using Lemma 1, we conclude that

$$
\begin{aligned}
u(z) &= c_0\frac{(\psi(z)-\psi(c))^{-(2-\alpha_1)(1-\beta_1)}}{\Gamma(1-(2-\alpha_1)(1-\beta_1))} + c_1\frac{(\psi(z)-\psi(c))^{1-(2-\alpha_1)(1-\beta_1)}}{\Gamma(2-(2-\alpha_1)(1-\beta_1))} \\
&\quad -k\int_c^z \psi'(s)u(s)ds + I_{c+}^{\alpha_1;\psi}h_1(z) \\
&= c_0\frac{(\psi(z)-\psi(c))^{\gamma-2}}{\Gamma(\gamma-1)} + c_1\frac{(\psi(z)-\psi(c))^{\gamma-1}}{\Gamma(\gamma)} \\
&\quad -k\int_c^z \psi'(s)u(s)ds + I_{c+}^{\alpha_1;\psi}h_1(z), \\
v(z) &= d_0\frac{(\psi(z)-\psi(c))^{\gamma_1-2}}{\Gamma(\gamma_1-1)} + d_1\frac{(\psi(z)-\psi(c))^{\gamma_1-1}}{\Gamma(\gamma_1)} \\
&\quad -k\int_c^z \psi'(s)v(s)ds + I_{c+}^{\bar{\alpha}_1;\psi}h_2(z).
\end{aligned}
$$

Hence, due to $u(c), v(c) = 0$, we obtain $c_0, d_0 = 0$. Consequently,

$$u(z) = c_1 \frac{(\psi(z) - \psi(c))^{\gamma-1}}{\Gamma(\gamma)} - k \int_c^z \psi'(s)u(s)ds + I_{c+}^{\alpha_1;\psi}h_1(z), \tag{10}$$

$$v(z) = d_1 \frac{(\psi(z) - \psi(c))^{\gamma_1-1}}{\Gamma(\gamma_1)} - k \int_c^z \psi'(s)v(s)ds + I_{c+}^{\bar{\alpha}_1;\psi}h_2(z). \tag{11}$$

From $u(d) = \sum_{i=1}^n \mu_i \int_c^{\eta_i} \psi'(s)v(s)ds + \sum_{j=1}^m \theta_j v(\xi_j)$ and $v(d) = \sum_{r=1}^p v_r \int_c^{\varsigma_r} \psi'(s)u(s)ds + \sum_{s=1}^q \tau_s u(\sigma_s)$, we have

$$c_1 \frac{(\psi(d) - \psi(c))^{\gamma-1}}{\Gamma(\gamma)} - k \int_c^d \psi'(s)u(s)ds + I_{c+}^{\alpha_1;\psi}h_1(d)$$

$$= d_1 \sum_{i=1}^n \mu_i \int_c^{\eta_i} \frac{\psi'(s)(\psi(s) - \psi(c))^{\gamma_1-1}}{\Gamma(\gamma_1)}ds + \sum_{i=1}^n \mu_i \int_c^{\eta_i} \psi'(s) I^{\bar{\alpha}_1;\psi}h_2(s)ds$$

$$+ d_1 \sum_{j=1}^m \theta_j \frac{(\psi(\xi_j) - \psi(c))^{\gamma_1-1}}{\Gamma(\gamma_1)} - k \sum_{j=1}^m \theta_j \int_c^{\xi_j} \psi'(s)v(s)ds + \sum_{j=1}^m \theta_j I^{\bar{\alpha}_1;\psi}h_2(\xi_j)$$

$$- k \sum_{i=1}^n \mu_i \int_c^{\eta_i} \psi'(s) \int_c^s \psi'(t)v(t)dtds,$$

and

$$d_1 \frac{(\psi(d) - \psi(c))^{\gamma_1-1}}{\Gamma(\gamma_1)} - k \int_c^d \psi'(s)v(s)ds + I_{c+}^{\bar{\alpha}_1;\psi}h_2(d)$$

$$= c_1 \sum_{r=1}^p v_r \int_a^{\varsigma_r} \frac{\psi'(s)(\psi(s) - \psi(a))^{\gamma-1}}{\Gamma(\gamma)}ds + \sum_{r=1}^p v_r \int_c^{\varsigma_r} \psi'(s) I^{\alpha_1;\psi}h_1(s)ds$$

$$+ c_1 \sum_{s=1}^q \tau_s \frac{(\psi(\sigma_s) - \psi(c))^{\gamma-1}}{\Gamma(\gamma)} - k \sum_{s=1}^q \tau_s \int_a^{\sigma_s} \psi'(s)u(s)ds + \sum_{s=1}^q \tau_s I^{\alpha_1;\psi}h_1(\sigma_s)$$

$$- k \sum_{r=1}^p v_r \int_a^{\varsigma_r} \psi'(s) \int_c^s \psi'(t)u(t)dtds,$$

or

$$Ac_1 - Bd_1 = P,$$
$$-\Gamma c_1 + \Delta d_1 = Q,$$

where A, B, Γ, Δ are defined by (9) and

$$P = \sum_{i=1}^n \mu_i \int_c^{\eta_i} \psi'(s) I^{\bar{\alpha}_1;\psi}h_2(s)ds + \sum_{j=1}^m \theta_j I^{\bar{\alpha}_1;\psi}h_2(\xi_j) - k \sum_{j=1}^m \theta_j \int_c^{\xi_j} \psi'(s)v(s)ds$$

$$- k \sum_{i=1}^n \mu_i \int_c^{\eta_i} \psi'(s) \int_c^s \psi'(t)v(t)dtds + k \int_c^d \psi'(s)u(s)ds - I_{c+}^{\alpha_1;\psi}h_1(d),$$

and

$$Q = \sum_{r=1}^p v_r \int_c^{\varsigma_r} \psi'(s) I^{\alpha_1;\psi}h_1(s)ds + \sum_{s=1}^q \tau_s I^{\alpha_1;\psi}h_1(\sigma_s) - k \sum_{s=1}^q \tau_s \int_c^{\sigma_s} \psi'(s)u(s)ds$$

$$- k \sum_{r=1}^p v_r \int_c^{\varsigma_r} \psi'(s) \int_c^s \psi'(t)u(t)dtds + k \int_c^d \psi'(s)v(s)ds - I_{c+}^{\bar{\alpha}_1;\psi}h_2(d).$$

By solving the above system, we find

$$c_1 = \frac{\Delta P + BQ}{\Lambda}, \quad d_1 = \frac{AQ + \Gamma P}{\Lambda}.$$

Substituting the values of c_1 and d_1 into Equations (10) and (11), respectively, we obtain the solutions (7) and (8). The converse is obtained by direct computation. The proof is finished. $\square$

4. Existence and Uniqueness Results

Keeping in mind Lemma 5, we define an operator $\mathcal{P} : \mathcal{X} \times \mathcal{X} \to \mathcal{X} \times \mathcal{X}$ by

$$\mathcal{P}(u,v)(z) := \left(\mathcal{P}_1(u,v)(z), \mathcal{P}_2(u,v)(z)\right), \tag{12}$$

where

$$
\begin{aligned}
\mathcal{P}_1(u,v)(z) =\ & -k \int_c^z \psi'(s)u(s)ds + I_{c+}^{\alpha_1;\psi} f_{uv}(z) \\
& + \frac{(\psi(z)-\psi(c))^{\gamma-1}}{\Lambda\Gamma(\gamma)}\Big[\Delta\Big(\sum_{i=1}^n \mu_i \int_c^{\eta_i} \psi'(s) I^{\bar{\alpha}_1;\psi} g_{uv}(s)ds + \sum_{j=1}^m \theta_j I^{\bar{\alpha}_1;\psi} g_{uv}(\xi_j) \\
& -k\sum_{j=1}^m \theta_j \int_c^{\xi_j} \psi'(s)v(s)ds - k\sum_{i=1}^n \mu_i \int_c^{\eta_i} \psi'(s) \int_c^s \psi'(t)v(t)dtds \\
& +k\int_c^d \psi'(s)u(s)ds - I_{c+}^{\alpha_1;\psi} f_{uv}(d)\Big) + B\Big(\sum_{r=1}^p v_r \int_c^{\varsigma_r} \psi'(s) I^{\alpha_1;\psi} f_{uv}(s)ds \\
& + \sum_{s=1}^q \tau_s I^{\alpha_1;\psi} f_{uv}(\sigma_s) - k\sum_{s=1}^q \tau_s \int_c^{\sigma_s} \psi'(s)u(s)ds \\
& -k\sum_{r=1}^p v_r \int_c^{\varsigma_r} \psi'(s) \int_c^s \psi'(t)u(t)dtds \\
& +k\int_c^d \psi'(s)v(s)ds - I_{c+}^{\bar{\alpha}_1;\psi} g_{uv}(d)\Big)\Big],
\end{aligned}
\tag{13}
$$

and

$$
\begin{aligned}
\mathcal{P}_2(u,v)(z) =\ & -k \int_c^z \psi'(s)v(s)ds + I_{c+}^{\bar{\alpha}_1;\psi} g_{uv}(z) \\
& + \frac{(\psi(z)-\psi(c))^{\gamma_1-1}}{\Lambda\Gamma(\gamma_1)}\Big[A\Big(\sum_{r=1}^p v_r \int_c^{\varsigma_r} \psi'(s) I^{\alpha_1;\psi} f_{uv}(s)ds + \sum_{s=1}^q \tau_s I^{\alpha_1;\psi} h_1(\sigma_s) \\
& -k\sum_{s=1}^q \tau_s \int_c^{\sigma_s} \psi'(s)u(s)ds - k\sum_{r=1}^p v_r \int_c^{\varsigma_r} \psi'(s) \int_c^s \psi'(t)u(t)dtds \\
& +k\int_c^d \psi'(s)v(s)ds - I_{c+}^{\bar{\alpha}_1;\psi} g_{uv}(d)\Big) + \Gamma\Big(\sum_{i=1}^n \mu_i \int_c^{\eta_i} \psi'(s) I^{\bar{\alpha}_1;\psi} g_{uv}(s)ds \\
& + \sum_{j=1}^m \theta_j I^{\bar{\alpha}_1;\psi} g_{uv}(\xi_j) - k\sum_{j=1}^m \theta_j \int_c^{\xi_j} \psi'(s)v(s)ds \\
& -k\sum_{i=1}^n \mu_i \int_c^{\eta_i} \psi'(s) \int_c^s \psi'(t)v(t)dtds \\
& +k\int_c^d \psi'(s)u(s)ds - I_{c+}^{\alpha_1;\psi} f_{uv}(d)\Big)\Big],
\end{aligned}
\tag{14}
$$

where

$$f_{uv}(z) = f(z, u(z), v(z)), \quad g_{uv}(z) = g(z, u(z), v(z)).$$

For the sake of convenience, we use the following notations:

$$
\begin{aligned}
\mathcal{A}_1 \; = \; & \frac{(\psi(d)-\psi(c))^{\alpha_1}}{\Gamma(\alpha_1+1)} + \frac{(\psi(d)-\psi(c))^{\gamma-1}}{|\Lambda|\Gamma(\gamma)}\Big[|\Delta|\frac{(\psi(d)-\psi(c))^{\alpha_1}}{\Gamma(\alpha_1+1)} \\
& + |B|\Big(\sum_{r=1}^{p} v_r \frac{(\psi(\varsigma_r)-\psi(c))^{\alpha_1+1}}{\Gamma(\alpha_1+2)} + \sum_{s=1}^{q} \tau_s \frac{(\psi(\sigma_s)-\psi(c))^{\alpha_1}}{\Gamma(\alpha_1+1)}\Big)\Big],
\end{aligned}
\tag{15}
$$

$$
\begin{aligned}
\mathcal{A}_2 \; = \; & \frac{(\psi(d)-\psi(c))^{\gamma-1}}{|\Lambda|\Gamma(\gamma)}\Big[|\Delta|\Big(\sum_{i=1}^{n} \mu_i \frac{(\psi(\eta_i)-\psi(c))^{\bar{\alpha}_1+1}}{\Gamma(\bar{\alpha}_1+2)} \\
& + \sum_{j=1}^{m} \theta_j \frac{(\psi(\xi_j)-\psi(c))^{\bar{\alpha}_1}}{\Gamma(\bar{\alpha}_1+1)}\Big) + |B|\frac{(\psi(d)-\psi(c))^{\bar{\alpha}_1}}{\Gamma(\bar{\alpha}_1+1)}\Big],
\end{aligned}
\tag{16}
$$

$$
\begin{aligned}
\mathcal{A}_3 \; = \; & |k|(\psi(d)-\psi(c)) + \frac{(\psi(d)-\psi(c))^{\gamma-1}}{|\Lambda|\Gamma(\gamma)}\Big[|\Delta||k|(\psi(d)-\psi(c)) \\
& + |B|\Big(|k|\sum_{s=1}^{q} \tau_s(\psi(\sigma_s)-\psi(c)) + \frac{1}{2}|k|\sum_{r=1}^{p} v_r(\psi(\varsigma_r)-\psi(c))^2\Big)\Big],
\end{aligned}
\tag{17}
$$

$$
\begin{aligned}
\mathcal{A}_4 \; = \; & \frac{(\psi(d)-\psi(c))^{\gamma-1}}{|\Lambda|\Gamma(\gamma)}\Big[|\Delta|\Big(|k|\sum_{j=1}^{m} \theta_j(\psi(\xi_j)-\psi(c)) \\
& + \frac{1}{2}|k|\sum_{i=1}^{n} \mu_i(\psi(\eta_i)-\psi(c))^2\Big) + |B||k|(\psi(d)-\psi(c))\Big],
\end{aligned}
\tag{18}
$$

$$
\begin{aligned}
\mathcal{B}_1 \; = \; & \frac{(\psi(d)-\psi(c))^{\gamma_1-1}}{|\Lambda|\Gamma(\gamma_1)}\Big[|A|\Big(v_r \sum_{i=1}^{n} \frac{(\psi(\varsigma_i)-\psi(c))^{\alpha_1}}{\Gamma(\alpha_1+1)} + \sum_{s=1}^{q} \tau_s \frac{(\psi(\sigma_s)-\psi(c))^{\alpha_1}}{\Gamma(\alpha_1+1)}\Big) \\
& + |\Gamma|\frac{(\psi(d)-\psi(c))^{\alpha_1}}{\Gamma(\alpha_1+1)}\Big],
\end{aligned}
\tag{19}
$$

$$
\begin{aligned}
\mathcal{B}_2 \; = \; & \frac{(\psi(d)-\psi(c))^{\bar{\alpha}_1}}{\Gamma(\bar{\alpha}_1+1)} + \frac{(\psi(d)-\psi(c))^{\gamma_1-1}}{|\Lambda|\Gamma(\gamma_1)}\Big[|A|\frac{(\psi(d)-\psi(c))^{\bar{\alpha}_1}}{\Gamma(\bar{\alpha}_1+1)} \\
& + |\Gamma|\Big(\sum_{i=1}^{n} \mu_i \frac{(\psi(\eta_i)-\psi(c))^{\bar{\alpha}_1+1}}{\Gamma(\bar{\alpha}_1+2)} + \sum_{j=1}^{m} \theta_j \frac{(\psi(\xi_j)-\psi(c))^{\bar{\alpha}_1}}{\Gamma(\bar{\alpha}_1+1)}\Big)\Big],
\end{aligned}
\tag{20}
$$

$$
\begin{aligned}
\mathcal{B}_3 \; = \; & \frac{(\psi(d)-\psi(c))^{\gamma_1-1}}{|\Lambda|\Gamma(\gamma_1)}\Big[|A|\Big(|k|\sum_{j=1}^{m} \tau_s(\psi(\sigma_s)-\psi(c)) \\
& + \frac{1}{2}|k|\sum_{r=1}^{p} v_r(\psi(\varsigma_r)-\psi(c))^2\Big) + |\Gamma||k|(\psi(d)-\psi(c))\Big],
\end{aligned}
\tag{21}
$$

$$
\begin{aligned}
\mathcal{B}_4 \; = \; & |k|(\psi(d)-\psi(c)) + \frac{(\psi(d)-\psi(c))^{\gamma_1-1}}{|\Lambda|\Gamma(\gamma_1)}\Big[|A||k|(\psi(d)-\psi(c)) \\
& + |\Gamma|\Big(|k|\sum_{j=1}^{m} \theta_j)(\psi(\xi_j)-\psi(c)) + \frac{1}{2}|k|\sum_{i=1}^{n} \mu_i(\psi(\eta_i)-\psi(c))^2\Big)\Big].
\end{aligned}
\tag{22}
$$

4.1. Existence and Uniqueness Result via Banach Fixed Point Theorem

Here, by using the Banach contraction mapping principle, we prove an existence and uniqueness result.

Theorem 1. *Assume that $\Lambda \neq 0$ and $f, g : [c,d] \times \mathbb{R}^2 \longrightarrow \mathbb{R}$ two functions satisfying the condition:*
(H_1) *there exist positive real constants ℓ_1, ℓ_2 such that, for all $z \in [c,d]$ and $u_i, v_i \in \mathbb{R}, i = 1,2$ we have*

$$
\begin{aligned}
|f(z, u_1, v_1) - f(z, u_2, v_2)| &\leq \ell_1\big(|u_1 - u_2| + |v_1 - v_2|\big), \\
|g(z, u_1, v_1) - g(z, u_2, v_2)| &\leq \ell_2\big(|u_1 - u_2| + |v_1 - v_2|\big).
\end{aligned}
$$

Then, system (5) admits a unique solution on $[c,d]$ provided that

$$
\ell_1(\mathcal{A}_1 + \mathcal{B}_1) + \ell_2(\mathcal{A}_2 + \mathcal{B}_2) + \mathcal{A}_3 + \mathcal{A}_4 + \mathcal{B}_3 + \mathcal{B}_4 < 1,
\tag{23}
$$

where $\mathcal{A}_i, \mathcal{B}_i, i = 1,2,3,4$ are given by (15)–(18) and (19)–(22) respectively.

Proof. We transform system (5) into a fixed point problem, $(u,v)(z) = \mathcal{P}(u,v)(z)$, where the operator $\mathcal{P}$ is defined as in (12). Applying the Banach contraction mapping principle, we show that the operator $\mathcal{P}$ has a unique fixed point, which is the unique solution of system (5).

Let $\sup_{z \in [c,d]} |f(z,0,0)| := \overline{M} < \infty$ and $\sup_{z \in [c,d]} |g(z,0,0)| := \overline{N} < \infty$. Next, we set $B_r := \{(u,v) \in \mathcal{X} \times \mathcal{X} : \|(u,v)\| \leq r\}$ with

$$r \geq \frac{\overline{M}(\mathcal{A}_1 + \mathcal{B}_1) + \overline{N}(\mathcal{A}_2 + \mathcal{B}_2)}{1 - [\ell_1(\mathcal{A}_1 + \mathcal{B}_1) + \ell_2(\mathcal{A}_2 + \mathcal{B}_2) + \mathcal{A}_3 + \mathcal{A}_4 + \mathcal{B}_3 + \mathcal{B}_4]}. \tag{24}$$

Observe that B_r is a bounded, closed, and convex subset of $\mathcal{X}$.
First, we show that $\mathcal{P}B_r \subset B_r$.

For any $(u,v) \in B_r$, $z \in [c,d]$, using the condition (H_1), we have

$$
\begin{aligned}
|f_{uv}(z)| = |f(z,u(z),v(z))| &\leq |f(z,u(z),v(z)) - f(z,0,0)| + |f(z,0,0)| \\
&\leq \ell_1(|u(z)| + |v(z)|) + \overline{M} \\
&\leq \ell_1(\|u\| + \|v\|) + \overline{M} \leq \ell_1 r + \overline{M},
\end{aligned}
$$

and

$$|g_{uv}(z)| = |g(z,u(z),v(z))| \leq \ell_2(\|u\| + \|v\|) + \overline{N} \leq \ell_2 r + \overline{N}.$$

Then, we obtain

$$
\begin{aligned}
&|\mathcal{P}_1(u,v)(z)| \\
\leq\ & |k| \int_c^d \psi'(s)|u(s)|ds + I_{c+}^{\alpha_1;\psi}|f_{uv}|(z) \\
&+ \frac{(\psi(d) - \psi(c))^{\gamma-1}}{|\Lambda|\Gamma(\gamma)} \Bigg[|\Delta| \Bigg(\sum_{i=1}^n \mu_i \int_c^{\eta_i} \psi'(s) I^{\overline{\alpha}_1;\psi}|g_{uv}|(s)ds + \sum_{j=1}^m \theta_j I^{\overline{\alpha}_1;\psi}|g_{uv}|(\xi_j) \\
&+ |k| \sum_{j=1}^m \theta_j \int_c^{\xi_j} \psi'(s)|v(s)|ds + |k| \sum_{i=1}^n \mu_i \int_c^{\eta_i} \psi'(s) \int_c^s \psi'(t)|v(t)|dtds \\
&+ |k| \int_c^d \psi'(s)|u(s)|ds + I_{c+}^{\alpha_1;\psi}|f_{uv}|(d) \Bigg) + |B| \Bigg(\sum_{r=1}^p v_r \int_c^{\varsigma_r} \psi'(s) I^{\alpha_1;\psi}|f_{uv}|(s)ds \\
&+ \sum_{s=1}^q \tau_s I^{\alpha_1;\psi}|f_{uv}|(\sigma_s) + |k| \sum_{s=1}^q \tau_s \int_c^{\sigma_s} \psi'(s)|u(s)|ds \\
&+ |k| \sum_{r=1}^p v_r \int_a^{\varsigma_r} \psi'(s) \int_c^s \psi'(t)|u(t)|dtds + |k| \int_c^d \psi'(s)|v(s)|ds + I_{c+}^{\overline{\alpha}_1\psi}|g_{uv}|(d) \Bigg) \Bigg] \\
\leq\ & |k|(\psi(d) - \psi(c))\|u\| + \frac{(\psi(d) - \psi(c))^{\alpha_1}}{\Gamma(\alpha_1 + 1)}(\ell_1 r + \overline{M}) \\
&+ \frac{(\psi(d) - \psi(c))^{\gamma-1}}{|\Lambda|\Gamma(\gamma)} \Bigg[|\Delta| \Bigg(\sum_{i=1}^n \mu_i \frac{(\psi(\eta_i) - \psi(c))^{\overline{\alpha}_1+1}}{\Gamma(\overline{\alpha}_1 + 2)}(\ell_2 r + \overline{N}) \\
&+ \sum_{j=1}^m \theta_j \frac{(\psi(\xi_j) - \psi(c))^{\overline{\alpha}_1}}{\Gamma(\overline{\alpha}_1 + 1)}(\ell_2 r + \overline{N}) + |k| \sum_{j=1}^m \theta_j(\psi(\xi_j) - \psi(c))\|v\| \\
&+ \frac{1}{2}|k| \sum_{i=1}^n \mu_i(\psi(\eta_i) - \psi(c))^2\|v\| + |k|(\psi(d) - \psi(c))\|u\| \\
&+ \frac{(\psi(d) - \psi(c))^{\alpha_1}}{\Gamma(\alpha_1 + 1)}(\ell_1 r + \overline{M}) \Bigg) + |B| \Bigg(\sum_{r=1}^p v_r \frac{(\psi(\varsigma_r) - \psi(c))^{\alpha_1+1}}{\Gamma(\alpha_1 + 2)}(\ell_1 r + \overline{M}) \\
&+ \sum_{s=1}^q \tau_s \frac{(\psi(\sigma_s) - \psi(c))^{\alpha_1}}{\Gamma(\alpha_1 + 1)}(\ell_1 r + \overline{M}) + |k| \sum_{s=1}^q \tau_s(\psi(\sigma_s) - \psi(c))\|u\|
\end{aligned}
$$

$$
\begin{aligned}
&+\frac{1}{2}|k|\sum_{r=1}^{p} v_r(\psi(\varsigma_r)-\psi(c))^2\|u\|+|k|(\psi(d)-\psi(c)\|v\| \\
&\quad +\frac{(\psi(d)-\psi(c))^{\overline{\alpha}_1}}{\Gamma(\overline{\alpha}_1+1)}(\ell_2 r+\overline{N}))\Big] \\
\leq\;& (\ell_1 r+\overline{M})\Big\{\frac{(\psi(d)-\psi(c))^{\alpha_1}}{\Gamma(\alpha_1+1)}+\frac{(\psi(d)-\psi(c))^{\gamma-1}}{|\Lambda|\Gamma(\gamma)}\Big[|\Delta|\frac{(\psi(d)-\psi(c))^{\alpha_1}}{\Gamma(\alpha_1+1)} \\
&\quad +|B|\Big(\sum_{r=1}^{p} v_r\frac{(\psi(\varsigma_r)-\psi(c))^{\alpha_1+1}}{\Gamma(\alpha_2+2)}+\sum_{s=1}^{q}\tau_s\frac{(\psi(\sigma_s)-\psi(c))^{\alpha_1}}{\Gamma(\alpha_1+1)}\Big)\Big]\Big\} \\
&+(\ell_2 r+\overline{N})\Big\{\frac{(\psi(d)-\psi(c))^{\gamma-1}}{|\Lambda|\Gamma(\gamma)}\Big[|\Delta|\Big(\sum_{i=1}^{n}\mu_i\frac{(\psi(\eta_i)-\psi(c))^{\overline{\alpha}_1+1}}{\Gamma(\overline{\alpha}_1+2)} \\
&\quad +\sum_{j=1}^{m}\theta_j\frac{(\psi(\xi_j)-\psi(c))^{\overline{\alpha}_1}}{\Gamma(\overline{\alpha}_1+1)}\Big)+|B|\frac{(\psi(d)-\psi(c))^{\overline{\alpha}_1}}{\Gamma(\overline{\alpha}_1+1)}\Big]\Big\}+r\Big\{|k|(\psi(d)-\psi(c)) \\
&\quad +\frac{(\psi(d)-\psi(c))^{\gamma-1}}{|\Lambda|\Gamma(\gamma)}\Big[|\Delta||k|(\psi(d)-\psi(c))+|B|\Big(|k|\sum_{s=1}^{q}\tau_s(\psi(\sigma_s)-\psi(c)) \\
&\quad +\frac{1}{2}|k|\sum_{r=1}^{p} v_r(\psi(\varsigma_r)-\psi(c))^2\Big)\Big]\Big\}+r\frac{(\psi(d)-\psi(c))^{\gamma-1}}{|\Lambda|\Gamma(\gamma)}\Big[|\Delta|\Big(|k|\sum_{j=1}^{m}\theta_j(\psi(\xi_j)-\psi(c)) \\
&\quad +\frac{1}{2}|k|\sum_{i=1}^{n}\mu_i(\psi(\eta_i)-\psi(c))^2\Big)+|B||k|(\psi(d)-\psi(c))\Big] \\
=\;& \mathcal{A}_1(\ell_1 r+\overline{M})+\mathcal{A}_2(\ell_2 r+\overline{N})+r(\mathcal{A}_3+\mathcal{A}_4).
\end{aligned}
$$

Hence
$$
\|\mathcal{P}_1(u,v)\|\leq \mathcal{A}_1(\ell_1 r+\overline{M})+\mathcal{A}_2(\ell_2 r+\overline{N})+r(\mathcal{A}_3+\mathcal{A}_4).
$$

Similarly, we find that
$$
\|\mathcal{P}_2(u,v)\|\leq \mathcal{B}_1(\ell_1 r+\overline{M})+\mathcal{B}_2(\ell_2 r+\overline{N})+r(\mathcal{B}_3+\mathcal{B}_4).
$$

Consequently, we have
$$
\begin{aligned}
\|\mathcal{P}(x,y)\|\;\leq\;& \Big[\ell_1(\mathcal{A}_1+\mathcal{B}_1)+\ell_2(\mathcal{A}_2+\mathcal{B}_2)+\mathcal{A}_3+\mathcal{A}_4+\mathcal{B}_3+\mathcal{B}_4\Big]r \\
&+(\mathcal{A}_1+\mathcal{B}_1)\overline{M}+(\mathcal{A}_2+\mathcal{B}_2)\overline{N}\leq r,
\end{aligned}
$$

which implies that $\mathcal{P}B_r\subset B_r$.

Next we show that $\mathcal{P}:\mathcal{X}\times\mathcal{X}\to\mathcal{X}\times\mathcal{X}$ is a contraction.

Using condition (H_1), for any (u_1,v_1), $(u_2,v_2)\in\mathcal{X}\times\mathcal{X}$ and for each $z\in[c,d]$, we have

$$
\begin{aligned}
&|\mathcal{P}_1(u_1,v_1)(z)-\mathcal{P}_1(u_2,v_2)(z)| \\
\leq\;& |k|\int_c^d \psi'(s)|u_1(s)-u_2(s)|ds+I_{c+}^{\alpha_1;\psi}|f_{u_1 v_1}-f_{u_2 v_2}|(z) \\
&+\frac{(\psi(d)-\psi(c))^{\gamma-1}}{|\Lambda|\Gamma(\gamma)}\Big[|\Delta|\Big(\sum_{i=1}^{n}\mu_i\int_c^{\eta_i}\psi'(s)I^{\overline{\alpha}_1;\psi}|g_{u_1 v_1}-g_{u_2 v_2}|(s)ds \\
&+\sum_{j=1}^{m}\theta_j I^{\overline{\alpha}_1;\psi}|g_{u_1 v_1}-g_{u_2 v_2}|(\xi_j)+|k|\sum_{j=1}^{m}\theta_j\int_c^{\xi_j}\psi'(s)|v_1(s)-v_2(s)|ds \\
&+|k|\sum_{i=1}^{n}\mu_i\int_c^{\eta_i}\psi'(s)\int_c^{s}\psi'(s)|v(t)|dtds+|k|\int_c^d\psi'(t)|u_1(s)-u_2(s)|ds
\end{aligned}
$$

$$+ I_{c^+}^{\alpha_1;\psi}|f_{u_1v_1} - f_{u_2v_2}|(b)\Big) + |B|\Big(\sum_{r=1}^{p} v_r \int_{c}^{\varsigma_r} \psi'(s) I^{\alpha_1;\psi}|f_{u_1v_1} - f_{u_2v_2}|(s)ds$$

$$+ \sum_{s=1}^{q} \tau_s I^{\alpha_1;\psi}|f_{u_1v_1} - f_{u_2v_2}|(\sigma_s) + |k| \sum_{s=1}^{q} \tau_s \int_{c}^{\sigma_s} \psi'(s)|u_1(s) - u_2(s)|ds$$

$$+ |k| \sum_{r=1}^{p} v_r \int_{c}^{\varsigma_r} \psi'(s) \int_{c}^{s} \psi'(t)|u(t)|dtds + |k| \int_{c}^{d} \psi'(s)|v_1(s) - v_2(s)|ds$$

$$+ I_{c^+}^{\overline{\alpha}_1\psi}|g_{u_1v_1} - g_{u_2v_2}|(d)\Big)\Big]$$

$$\leq\quad |k|(\psi(d) - \psi(c))\|u_1 - u_2\| + \frac{(\psi(d) - \psi(c))^{\alpha_1}}{\Gamma(\alpha_1 + 1)}\ell_1(\|u_1 - u_2\| + \|v_1 - v_2\|)$$

$$+ \frac{(\psi(d) - \psi(c))^{\gamma-1}}{|\Lambda|\Gamma(\gamma)}\Big[|\Delta|\Big(\sum_{i=1}^{n} \mu_i \frac{(\psi(\eta_i) - \psi(c))^{\overline{\alpha}_1+1}}{\Gamma(\overline{\alpha}_1 + 2)}\ell_2(\|u_1 - u_2\| + \|v_1 - v_2\|)$$

$$+ \sum_{j=1}^{m} \theta_j \frac{(\psi(\xi_j) - \psi(c))^{\overline{\alpha}_1}}{\Gamma(\overline{\alpha}_1 + 1)}(\ell_2 r + N) + |k| \sum_{j=1}^{m} \theta_j(\psi(\xi_j) - \psi(c))\|v_1 - v_2\|$$

$$+ \frac{1}{2}|k| \sum_{i=1}^{n} \mu_i(\psi(\eta_i) - \psi(c))^2\|v_1 - v_2\| + |k|(\psi(d) - \psi(c))\|u_1 - u_2\|$$

$$+ \frac{(\psi(d) - \psi(c))^{\alpha_1}}{\Gamma(\alpha_1 + 1)}\ell_1(\|u_1 - u_2\| + \|v_1 - v_2\|)\Big)$$

$$+ |B|\Big(\sum_{r=1}^{p} v_r \frac{(\psi(\varsigma_r) - \psi(c))^{\alpha_1+1}}{\Gamma(\alpha_1 + 2)}\ell_1(\|u_1 - u_2\| + \|v_1 - v_2\|)$$

$$+ \sum_{s=1}^{q} \tau_s \frac{(\psi(\sigma_s) - \psi(c))^{\alpha_1}}{\Gamma(\alpha_1 + 1)}\ell_1(\|u_1 - u_2\| + \|v_1 - v_2\|)$$

$$+ |k| \sum_{s=1}^{q} \tau_s(\psi(\sigma_s) - \psi(c))\|u_1 - u_2\| + \frac{1}{2}|k| \sum_{r=1}^{p} v_r(\psi(\varsigma_r) - \psi(c))^2\|u_1 - u_2\|$$

$$+ |k|(\psi(d) - \psi(c)\|v_1 - v_2\| + \frac{(\psi(d) - \psi(c))^{\overline{\alpha}_1}}{\Gamma(\overline{\alpha}_1 + 1)}\ell_2(\|u_1 - u_2\| + \|v_1 - v_2\|)\Big)\Big]$$

$$\leq\quad \ell_1(\|u_1 - u_2\| + \|v_1 - v_2\|)\Big\{\frac{(\psi(d) - \psi(c))^{\alpha_1}}{\Gamma(\alpha_1 + 1)}$$

$$+ \frac{(\psi(d) - \psi(c))^{\gamma-1}}{|\Lambda|\Gamma(\gamma)}\Big[|\Delta|\frac{(\psi(d) - \psi(c))^{\alpha_1}}{\Gamma(\alpha_1 + 1)} + |B|\Big(\sum_{r=1}^{p} v_r \frac{(\psi(\varsigma_r) - \psi(c))^{\alpha_1+1}}{\Gamma(\alpha_1 + 2)}$$

$$+ \sum_{s=1}^{q} \tau_s \frac{(\psi(\sigma_s) - \psi(c))^{\alpha_1}}{\Gamma(\alpha_1 + 1)}\Big)\Big]\Big\}$$

$$+ \ell_2(\|u_1 - u_2\| + \|v_1 - v_2\|)\Big\{\frac{(\psi(d) - \psi(c))^{\gamma-1}}{|\Lambda|\Gamma(\gamma)}\Big[|\Delta|\Big(\sum_{i=1}^{n} \mu_i \frac{(\psi(\eta_i) - \psi(c))^{\overline{\alpha}_1+1}}{\Gamma(\overline{\alpha}_1 + 2)}$$

$$+ \sum_{j=1}^{m} \theta_j \frac{(\psi(\xi_j) - \psi(c))^{\overline{\alpha}_1}}{\Gamma(\overline{\alpha}_1 + 1)}\Big) + |B|\frac{(\psi(d) - \psi(c))^{\overline{\alpha}_1}}{\Gamma(\overline{\alpha}_1 + 1)}\Big]\Big\}$$

$$+ \|u_1 - u_2\|\Big\{|k|(\psi(d) - \psi(c)) + \frac{(\psi(d) - \psi(c))^{\gamma-1}}{|\Lambda|\Gamma(\gamma)}\Big[|\Delta||k|(\psi(d) - \psi(c))$$

$$+ |B|\Big(|k| \sum_{s=1}^{q} \tau_s(\psi(\sigma_s) - \psi(c)) + \frac{1}{2}|k| \sum_{r=1}^{p} v_r(\psi(\varsigma_r) - \psi(c))^2\Big)\Big]\Big\}$$

$$+ \frac{(\psi(b) - \psi(a))^{\gamma-1}}{|\Lambda|\Gamma(\gamma)}\Big[|\Delta|\Big(|k| \sum_{j=1}^{m} \theta_j(\psi(\xi_j) - \psi(a))$$

$$+ \frac{1}{2}|k| \sum_{i=1}^{n} \mu_i(\psi(\eta_i) - \psi(c))^2\Big) + |B||k|(\psi(d) - \psi(c))\Big]\|v_1 - v_2\|$$

$$\begin{aligned}
&= (\ell_1\mathcal{A}_1 + \ell_2\mathcal{A}_2)(\|u_1 - u_2\| + \|v_1 - v_2\|) + \mathcal{A}_3\|u_1 - u_2\| + \mathcal{A}_4\|v_1 - v_2\|. \\
&\leq (\ell_1\mathcal{A}_1 + \ell_2\mathcal{A}_2) + \mathcal{A}_3 + \mathcal{A}_4)(\|v_1 - v_2\| + \|v_1 - v_2\|),
\end{aligned}$$

and therefore

$$\|\mathcal{P}_1(u_1,v_1) - \mathcal{P}_1(u_2,v_2)\| \leq (\ell_1\mathcal{A}_1 + \ell_2\mathcal{A}_2) + \mathcal{A}_3 + \mathcal{A}_4)(\|u_1 - u_2\| + \|v_1 - v_2\|). \tag{25}$$

Similarly, we find that

$$\|\mathcal{P}_2(u_1,v_1) - \mathcal{P}_2(u_2,v_2)\| \leq (\ell_1\mathcal{B}_1 + \ell_2\mathcal{B}_2) + \mathcal{B}_3 + \mathcal{B}_4)(\|u_1 - u_2\| + \|v_1 - v_2\|). \tag{26}$$

From (25) and (26), it yields

$$\begin{aligned}
\|\mathcal{P}(u_1,u_1) - \mathcal{P}(u_2,u_2)\| &\leq \left[\ell_1(\mathcal{A}_1 + \mathcal{B}_1) + \ell_2(\mathcal{A}_2 + \mathcal{B}_2) + \mathcal{A}_3 + \mathcal{A}_4 + \mathcal{B}_3 + \mathcal{B}_4\right] \\
&\quad \times \left(\|u_1 - u_2\| + \|v_1 - v_2\|\right).
\end{aligned}$$

Since $\ell_1(\mathcal{A}_1 + \mathcal{B}_1) + \ell_2(\mathcal{A}_2 + \mathcal{B}_2) + \mathcal{A}_3 + \mathcal{A}_4 + \mathcal{B}_3 + \mathcal{B}_4 < 1$, by (23), the operator $\mathcal{P}$ is a contraction. Therefore, using the Banach contraction mapping principle (Lemma 1), the operator $\mathcal{P}$ has a unique fixed point. Hence, system (5) has a unique solution on $[c,d]$. The proof is completed. $\square$

4.2. Existence Result via Leray-Schauder Alternative

The Leray–Schauder alternative (Lemma 3) is used in the proof of our first existence result.

Theorem 2. *Let* $\Lambda \neq 0$, *and* $f, g : [c,d] \times \mathbb{R}^2 \to \mathbb{R}$ *be continuous functions. Assume that:*

(H_2) *There exist real constants* $u_i, v_i \geq 0$ *for* $i = 1, 2$ *and* $u_0, v_0 > 0$ *such that for all* $u, v \in \mathbb{R}$, *we have*

$$\begin{aligned}
|f(z, u(z), v(z))| &\leq u_0 + u_1|u| + u_2|v|, \\
|g(z, u(z), v(z))| &\leq v_0 + v_1|u| + v_2|v|.
\end{aligned}$$

If $(\mathcal{A}_1 + \mathcal{B}_1)u_1 + (\mathcal{A}_2 + \mathcal{B}_2)v_1 + \mathcal{A}_3 + \mathcal{B}_3 < 1$ *and* $(\mathcal{A}_1 + \mathcal{B}_1)u_2 + (\mathcal{A}_2 + \mathcal{B}_2)v_2 + \mathcal{A}_4 + \mathcal{B}_4 < 1$, *where* $\mathcal{A}_i, \mathcal{B}_i$ *for* $i = 1, 2$ *are given by* (15)–(18) *and* (19)–(22), *respectively, then the system* (5) *admits at least one solution on* $[c,d]$.

Proof. Obviously, the operator $\mathcal{P}$ is continuous, due to the continuity of the functions f, g on $[c,d] \times \mathbb{R}^2$. Now, show that the operator $\mathcal{P} : \mathcal{X} \times \mathcal{X} \to \mathcal{X} \times \mathcal{X}$ is completely continuous. Let $B_r \subset \mathcal{X} \times \mathcal{X}$ be a bounded set, where $B_r = \{(u,v) \in \mathcal{X} \times \mathcal{X} : \|(u,v)\| \leq r\}$. Then, for any $(u,v) \in B_r$, there exist positive real numbers W_1 and W_2 such that $|f_{uv}(z)| = |f(z, u(t), v(z))| \leq W_1$ and $|g_{uv}(z)| = |g(z, u(z), v(z))| \leq W_2$.

Thus, for each $(u,v) \in B_r$ we have

$$\begin{aligned}
&|\mathcal{P}_1(u,v)(z)| \\
&\leq |k|\int_c^d \psi'(s)|u(s)|ds + I_{c+}^{\alpha_1;\psi}|f_{uv}(z)| \\
&\quad + \frac{(\psi(d) - \psi(c))^{\gamma-1}}{|\Lambda|\Gamma(\gamma)}\left[|\Delta|\left(\sum_{i=1}^n \mu_i \int_c^{\eta_i} \psi'(s)I^{\bar{\alpha}_1;\psi}|g_{uv}(s)|ds + \sum_{j=1}^m \theta_j I^{\bar{\alpha}_1;\psi}|g_{uv}(\xi_j)|\right.\right. \\
&\quad + |k|\sum_{j=1}^m \theta_j \int_c^{\xi_j} \psi'(s)|v(s)|ds + |k|\sum_{i=1}^n \mu_i \int_c^{\eta_i} \psi'(s)\int_c^s \psi'(t)|v(t)|dtds \\
&\quad + |k|\int_c^d \psi'(s)|u(s)|ds + I_{c+}^{\alpha_1;\psi}|f_{uv}|(d)\bigg) + |B|\left(\sum_{r=1}^p v_r \int_c^{\varsigma_r} \psi'(s)I^{\alpha_1;\psi}|f_{uv}|(s)ds\right.
\end{aligned}$$

$$
\begin{aligned}
&+ \sum_{s=1}^{q} \tau_s I^{\alpha_1;\psi} |f_{uv}|(\sigma_s) + |k| \sum_{s=1}^{q} \tau_s \int_{c}^{\sigma_s} \psi'(s) |u(s)| ds \\
&+ |k| \sum_{r=1}^{p} v_r \int_{c}^{\varsigma_r} \psi'(s) \int_{c}^{s} \psi'(t) |u(t)| dt ds \\
&+ |k| \int_{c}^{d} \psi'(s) |v(s)| ds + I_{c+}^{\bar{\alpha}_1 \psi} |g_{uv}|(d) \Big) \Big] \\
\leq\ & |k|(\psi(d) - \psi(c)) \|u\| + \frac{(\psi(d) - \psi(c))^{\alpha_1}}{\Gamma(\alpha_1 + 1)} W_1 \\
&+ \frac{(\psi(d) - \psi(c))^{\gamma-1}}{|\Lambda| \Gamma(\gamma)} \Big[|\Delta| \Big(\sum_{i=1}^{n} \mu_i \frac{(\psi(\eta_i) - \psi(c))^{\bar{\alpha}_1 + 1}}{\Gamma(\bar{\alpha}_1 + 2)} W_2 + \sum_{j=1}^{m} \theta_j \frac{(\psi(\xi_j) - \psi(c))^{\bar{\alpha}_1}}{\Gamma(\bar{\alpha}_1 + 1)} W_2 \\
&+ |k| \sum_{j=1}^{m} \theta_j (\psi(\xi_j) - \psi(c)) \|v\| + \frac{1}{2} |k| \sum_{i=1}^{n} \mu_i (\psi(\eta_i) - \psi(c))^2 \|v\| \\
&+ |k|(\psi(d) - \psi(c)) \|u\| + \frac{(\psi(d) - \psi(c))^{\alpha_1}}{\Gamma(\alpha_1 + 1)} W_1 \Big) + |B| \Big(\sum_{r=1}^{p} v_r \frac{(\psi(\varsigma_r) - \psi(c))^{\alpha_1 + 1}}{\Gamma(\alpha_1 + 2)} W_1 \\
&+ \sum_{s=1}^{q} \tau_s \frac{(\psi(\sigma_s) - \psi(c))^{\alpha_1}}{\Gamma(\alpha_1 + 1)} W_1 + |k| \sum_{s=1}^{q} \tau_s (\psi(\sigma_s) - \psi(c)) \|u\| \\
&+ \frac{1}{2} |k| \sum_{r=1}^{n} v_r (\psi(\varsigma_r) - \psi(c))^2 \|u\| + |k|(\psi(d) - \psi(c) \|v\| + \frac{(\psi(d) - \psi(c))^{\bar{\alpha}_1}}{\Gamma(\bar{\alpha}_1 + 1)} W_2 \Big) \Big] \\
\leq\ & W_1 \Big\{ \frac{(\psi(d) - \psi(c))^{\alpha_1}}{\Gamma(\alpha_1 + 1)} + \frac{(\psi(d) - \psi(c))^{\gamma-1}}{|\Lambda| \Gamma(\gamma)} \Big[|\Delta| \frac{(\psi(d) - \psi(c))^{\alpha_1}}{\Gamma(\alpha_1 + 1)} \\
&+ |B| \Big(\sum_{r=1}^{n} v_r \frac{(\psi(\varsigma_r) - \psi(a))^{\alpha_1 + 1}}{\Gamma(\alpha_1 + 2)} + \sum_{s=1}^{q} \tau_s \frac{(\psi(\sigma_s) - \psi(c))^{\alpha_1}}{\Gamma(\alpha_1 + 1)} \Big) \Big] \Big\} \\
&+ W_2 \Big\{ \frac{(\psi(d) - \psi(c))^{\gamma-1}}{|\Lambda| \Gamma(\gamma)} \Big[|\Delta| \Big(\sum_{i=1}^{n} \mu_i \frac{(\psi(\eta_i) - \psi(c))^{\bar{\alpha}_1 + 1}}{\Gamma(\bar{\alpha}_1 + 2)} \\
&+ \sum_{j=1}^{m} \theta_j \frac{(\psi(\xi_j) - \psi(c))^{\bar{\alpha}_1}}{\Gamma(\bar{\alpha}_1 + 1)} \Big) + |B| \frac{(\psi(d) - \psi(c))^{\bar{\alpha}_1}}{\Gamma(\bar{\alpha}_1 + 1)} \Big] \Big\} + r \Big\{ |k|(\psi(d) - \psi(c)) \\
&+ \frac{(\psi(d) - \psi(c))^{\gamma-1}}{\Gamma(\gamma)} \Big[|\Delta| |k| (\psi(d) - \psi(c)) + |B| \Big(|k| \sum_{s=1}^{q} \tau_s (\psi(\sigma_s) - \psi(c)) \\
&+ \frac{1}{2} |k| \sum_{r=1}^{p} v_r (\psi(\varsigma_r) - \psi(c))^2 \Big) \Big] \Big\} + r \frac{(\psi(d) - \psi(c))^{\gamma-1}}{|\Lambda| \Gamma(\gamma)} \Big[|\Delta| \Big(|k| \sum_{j=1}^{m} \theta_j (\psi(\xi_j) - \psi(c)) \\
&+ \frac{1}{2} |k| \sum_{i=1}^{n} \mu_i (\psi(\eta_i) - \psi(c))^2 \Big) + |B| |k| (\psi(d) - \psi(c)) \Big] \\
=\ & \mathcal{A}_1 W_1 + \mathcal{A}_2 W_2 + r(\mathcal{A}_3 + \mathcal{A}_4),
\end{aligned}
$$

which yields

$$
\|\mathcal{P}_1(u,v)\| \leq \mathcal{A}_1 W_1 + \mathcal{A}_2 W_2 + r(\mathcal{A}_3 + \mathcal{A}_4).
$$

Similarly, we obtain that

$$
\|\mathcal{P}_2(u,v)\| \leq \mathcal{B}_1 W_1 + \mathcal{B}_2 W_2 + r(\mathcal{B}_3 + \mathcal{B}_4).
$$

Hence, from the above inequalities, we find that the operator $\mathcal{P}$ is uniformly bounded, since

$$
\|\mathcal{P}(u,v)\| \leq (\mathcal{A}_1 + \mathcal{B}_1) W_1 + (\mathcal{B}_1 + \mathcal{B}_2) W_2 + r(\mathcal{A}_3 + \mathcal{A}_4 + \mathcal{B}_3 + \mathcal{B}_4).
$$

Next, we prove that the operator $\mathcal{P}$ is equicontinuous. Let $\tau_1, \tau_2 \in [c,d]$ with $\tau_1 < \tau_2$. Then, we have

$$
|\mathcal{P}_1(u,v)(\tau_2) - \mathcal{P}_1(u,v)(\tau_1)|
$$

$$
\leq \left| I_{c+}^{\alpha_1;\psi} f_{uv}(\tau_2) - I_{c+}^{\alpha_1;\psi} f_{uv}z(\tau_1) \right|
$$

$$
+ \frac{(\psi(\tau_2)-\psi(c))^{\gamma-1} - (\psi(\tau_1)-\psi(c))^{\gamma-1}}{|\Lambda|\Gamma(\gamma)} \left[|\Delta| \left(\sum_{i=1}^{n} \mu_i \int_c^{\eta_i} \psi'(s) I^{\bar\alpha_1;\psi}|g_{uv}(s)|ds \right.\right.
$$

$$
+ \sum_{j=1}^{m} \theta_j I^{\bar\alpha_1;\psi}|g_{uv}(\xi_j)| + |k| \sum_{j=1}^{m} \theta_j \int_c^{\xi_j} \psi'(s)|v(s)|ds
$$

$$
+ |k| \sum_{i=1}^{n} \mu_i \int_c^{\eta_i} \psi'(s) \int_c^{s} \psi'(t)|v(t)|dtds + |k| \int_c^d \psi'(s)|u(s)|ds + I_{c+}^{\alpha_1;\psi}|f_{uv}|(d) \right)
$$

$$
+ |B| \left(\sum_{r=1}^{p} v_r \int_c^{\varsigma_r} \psi'(s) I^{\alpha_1;\psi}|f_{uv}|(s)ds + \sum_{s=1}^{q} \tau_s I^{\alpha_1;\psi}|f_{uv}|(\sigma_s) \right.
$$

$$
+ |k| \sum_{s=1}^{q} \tau_s \int_c^{\sigma_s} \psi'(s)|u(s)|ds + |k| \sum_{r=1}^{p} v_r \int_c^{\varsigma_r} \psi'(s) \int_c^{s} \psi'(t)|u(t)|dtds
$$

$$
\left.\left. + |k| \int_c^d \psi'(s)|v(s)|ds + I_{c+}^{\bar\alpha_1\psi}|g_{uv}|(d) \right) \right]
$$

$$
\leq W_1 \left| \int_c^{\tau_1} \psi'(s) \frac{(\psi(\tau_2)-\psi(s))^{\alpha_1-1} - (\psi(\tau_1)-\psi(s))^{\alpha_1-1}}{\Gamma(\alpha_1)} ds \right.
$$

$$
\left. + \int_{\tau_1}^{\tau_2} \psi'(s) \frac{(\psi(\tau_2)-\psi(s))^{\alpha_1-1}}{\Gamma(\alpha_1)} ds \right|
$$

$$
+ \frac{(\psi(\tau_2)-\psi(c))^{\gamma-1} - (\psi(\tau_1)-\psi(c))^{\gamma-1}}{|\Lambda|\Gamma(\gamma)} \left[|\Delta| \left(\sum_{i=1}^{n} \mu_i \frac{(\psi(\eta_i)-\psi(c))^{\bar\alpha_1+1}}{\Gamma(\bar\alpha_1+2)} W_2 \right.\right.
$$

$$
+ \sum_{j=1}^{m} \theta_j \frac{(\psi(\xi_j)-\psi(c))^{\bar\alpha_1}}{\Gamma(\bar\alpha_1+1)} W_2 + |k| \sum_{j=1}^{m} \theta_j(\psi(\xi_j)-\psi(c))\|v\|
$$

$$
+ \frac{1}{2}|k| \sum_{i=1}^{n} \mu_i(\psi(\eta_i)-\psi(c))^2\|v\| + |k|(\psi(d)-\psi(c))\|u\|
$$

$$
\left. + \frac{(\psi(d)-\psi(c))^{\alpha_1}}{\Gamma(\alpha_1+1)} W_1 \right) + |B| \left(\sum_{r=1}^{p} v_r \frac{(\psi(\varsigma_r)-\psi(c))^{\alpha_1+1}}{\Gamma(\alpha_1+2)} W_1 \right.
$$

$$
+ \sum_{s=1}^{q} \tau_s \frac{(\psi(\sigma_s)-\psi(c))^{\alpha_1}}{\Gamma(\alpha_1+1)} W_1 + |k| \sum_{s=1}^{q} \tau_s(\psi(\sigma_s)-\psi(c))\|u\|
$$

$$
\left.\left. + \frac{1}{2}|k| \sum_{r=1}^{p} v_r(\psi(\varsigma_r)-\psi(c))^2\|u\| + |k|(\psi(d)-\psi(c)\|v\| + \frac{(\psi(d)-\psi(c))^{\bar\alpha_1}}{\Gamma(\bar\alpha_1+1)} W_2 \right) \right]
$$

$$
\leq \frac{W_1}{\Gamma(\alpha_1+1)} \left[2(\psi(\tau_2)-\psi(\tau_1))^{\alpha_1} + (\psi(\tau_2)-\psi(c))^{\alpha_1} - (\psi(\tau_1)-\psi(c))^{\alpha_1} \right]
$$

$$
+ \frac{[(\psi(\tau_2)-\psi(c))^{\gamma-1} - (\psi(\tau_1)-\psi(c))^{\gamma-1}]}{|\Lambda|\Gamma(\gamma)} \left\{ W_1|\Delta| \frac{(\psi(d)-\psi(c))^{\alpha_1}}{\Gamma(\alpha_1+1)} \right.
$$

$$
+ |B|W_1 \left(\sum_{r=1}^{p} v_r \frac{(\psi(\varsigma_r)-\psi(c))^{\alpha_1+1}}{\Gamma(\alpha_1+2)} + \sum_{s=1}^{q} \tau_s \frac{(\psi(\sigma_s)-\psi(c))^{\alpha_1}}{\Gamma(\alpha_1+1)} \right)
$$

$$
+ W_2 \left[|\Delta| \left(\frac{(\psi(d)-\psi(c))^{\gamma-1}}{|\Lambda|\Gamma(\gamma)} + \sum_{j=1}^{m} \theta_j \frac{(\psi(\xi_j)-\psi(c))^{\bar\alpha_1}}{\Gamma(\bar\alpha_1+1)} \right) + |B| \frac{(\psi(d)-\psi(c))^{\bar\alpha_1}}{\Gamma(\bar\alpha_1+1)} \right]
$$

$$
+ r \left[|\Delta||k|(\psi(d)-\psi(c)) + |B| \left(|k| \sum_{s=1}^{q} \tau_s(\psi(\sigma_s)-\psi(c)) \right.\right.
$$

$$+\frac{1}{2}|k|\sum_{r=1}^{p}v_r(\psi(\varsigma_r)-\psi(c))^2\Big)\Big]+r\Big[|\Delta|\Big(|k|\sum_{j=1}^{m}\theta_j(\psi(\xi_j)-\psi(c))$$

$$+\frac{1}{2}|k|\sum_{i=1}^{n}\mu_i(\psi(\eta_i)-\psi(c))^2\Big)+|B||k|(\psi(d)-\psi(c))\Big]\Big\}.$$

Therefore, we obtain

$$|\mathcal{P}_1(u,v)(\tau_2)-\mathcal{P}_1(u,v)(\tau_1)|\to 0,\quad\text{as }\tau_1\to\tau_2.$$

Analogously, we can obtain the following inequality:

$$|\mathcal{P}_2(u,v)(\tau_2)-\mathcal{P}_2(u,v)(\tau_1)|\to 0,\quad\text{as }\tau_1\to\tau_2.$$

Hence, the set $\mathcal{P}\Phi$ is equicontinuous. Accordingly, the Arzelá–Ascoli theorem implies that the operator $\mathcal{P}$ is completely continuous.

Finally, we show the boundedness of the set $\Xi=\{(u,v)\in\mathcal{X}\times\mathcal{X}:(u,v)=\mu\mathcal{P}(u,v),\ 0\le\mu\le 1\}$. Let any $(u,v)\in\Xi$, then $(u,v)=\mu\mathcal{P}(u,v)$. We have, for all $z\in[c,d]$,

$$u(z)=\mu\mathcal{P}_1(u,v)(z),\quad v(z)=\mu\mathcal{P}_2(u,v)(z).$$

Then, we obtain

$$\|u\|\ \le\ (u_0+u_1\|u\|+u_2\|v\|)\mathcal{A}_1+(v_0+v_1\|u\|+v_2\|v\|)\mathcal{A}_2+\|u\|(\mathcal{A}_3+\mathcal{B}_3),$$
$$\|v\|\ \le\ (u_0+u_1\|u\|+u_2\|v\|)\mathcal{B}_1+(v_0+v_1\|u\|+v_2\|v\|)\mathcal{B}_2+\|v\|(\mathcal{A}_4+\mathcal{B}_4),$$

which imply that

$$\|u\|+\|v\|\ \le\ (\mathcal{A}_1+\mathcal{B}_1)u_0+(\mathcal{A}_2+\mathcal{B}_2)v_0+\Big[(\mathcal{A}_1+\mathcal{B}_1)u_1+(\mathcal{A}_2+\mathcal{B}_2)v_1$$
$$+\mathcal{A}_3+\mathcal{B}_3\Big]\|u\|+\Big[(\mathcal{A}_1+\mathcal{B}_1)u_2+(\mathcal{A}_2+\mathcal{B}_2)v_2+\mathcal{A}_4+\mathcal{B}_4\Big]\|v\|.$$

Thus, we obtain

$$\|(u,v)\|\le\frac{(\mathcal{A}_1+\mathcal{B}_1)u_0+(\mathcal{A}_2+\mathcal{B}_2)v_0}{M^*},\tag{27}$$

where $M^*=\min\{1-(\mathcal{A}_1+\mathcal{B}_1)u_1-(\mathcal{A}_2+\mathcal{B}_2)v_1-(\mathcal{A}_3+\mathcal{B}_3),1-(\mathcal{A}_1+\mathcal{B}_1)u_2-(\mathcal{A}_2+\mathcal{B}_2)v_2-(\mathcal{A}_4+\mathcal{B}_4)\}$, which shows that the set Ξ is bounded. Therefore, via the Leray–Schauder alternative (Lemma 3), the operator $\mathcal{P}$ has at least one fixed point. Hence, we deduce that problem (5) admits a solution on $[c,d]$, which completes the proof. $\square$

4.3. Existence Result via Krasnosel'skiǐ's Fixed Point Theorem

Now we apply Krasnosel'skiǐ's fixed point theorem (Lemma 4) to prove our second existence result.

Theorem 3. *Let $\Lambda\neq 0$ and $f,g:[c,d]\times\mathbb{R}^2\longrightarrow\mathbb{R}$ be continuous functions which satisfy the condition (H_1) in Theorem 1. In addition, we assume that there exist two positive constants Z_1,Z_2 such that, for all $z\in[c,d]$ and $u_i,v_i\in\mathbb{R},i=1,2$, we have*

$$|f(z,u_1,v_1)|\le Z_1$$
$$|f(z,u_1,v_1)|\le Z_2.\tag{28}$$

Moreover, assume that $\mathcal{A}_3+\mathcal{A}_4<1,\mathcal{B}_3+\mathcal{B}_4<1$ and $\left[\dfrac{(d-c)^{\alpha_1}}{\Gamma(\alpha_1+1)}\ell_1+\dfrac{(d-c)^{\bar{\alpha}_1}}{\Gamma(\bar{\alpha}_1+1)}\ell_2\right]<1$.
Then, problem (5) admits at least one solution on $[c,d]$.

Proof. Let the operator $\mathcal{P}$, defined by (12), be decomposed into four operators as

$$\begin{aligned}
\mathcal{M}(u,v)(z) \;=\;& k\int_c^t \psi'(s)u(s)ds \\
&+\frac{(\psi(z)-\psi(c))^{\gamma-1}}{\Lambda\Gamma(\gamma)}\Big[\Delta\Big(\sum_{i=1}^n \mu_i \int_c^{\eta_i}\psi'(s)I^{\bar\alpha_1;\psi}g_{uv}(s)ds + \sum_{j=1}^m \theta_j I^{\bar\alpha_1;\psi}g_{uv}(\xi_j) \\
&-k\sum_{j=1}^m \theta_j \int_c^{\xi_j}\psi'(s)v(s)ds - k\sum_{i=1}^n \mu_i \int_c^{\eta_i}\psi'(s)\int_c^s \psi'(t)v(t)dtds \\
&+k\int_c^d \psi'(s)u(s)ds - I_{c+}^{\alpha_1;\psi}f_{uv}(d)\Big) + B\Big(\sum_{r=1}^p v_r \int_c^{\zeta_r}\psi'(s)I^{\alpha_1;\psi}f_{uv}(s)ds \\
&+\sum_{s=1}^q \tau_s I^{\alpha_1;\psi}f_{uv}(\sigma_s) - k\sum_{s=1}^q \tau_s \int_c^{\sigma_s}\psi'(s)u(s)ds \\
&-k\sum_{r=1}^p v_r \int_c^{\zeta_r}\psi'(s)\int_c^s \psi'(t)u(t)dtds \\
&+k\int_c^d \psi'(s)v(s)ds - I_{c+}^{\bar\alpha_1;\psi}g_{uv}(d)\Big)\Big], \\
\mathcal{N}(u,v)(z) \;=\;& I_{c+}^{\alpha_1;\psi}f_{uv}(z), \\
\mathcal{T}(u,v)(z) \;=\;& -k\int_c^z \psi'(s)v(s)ds \\
&+\frac{(\psi(z)-\psi(c))^{\gamma_1-1}}{\Lambda\Gamma(\gamma_1)}\Big[A\Big(\sum_{r=1}^p v_r \int_c^{\zeta_r}\psi'(s)I^{\alpha_1;\psi}f_{uv}(s)ds + \sum_{s=1}^q \tau_s I^{\alpha_1;\psi}h_1(\sigma_s) \\
&-k\sum_{s=1}^q \tau_s \int_c^{\sigma_s}\psi'(s)u(s)ds - k\sum_{r=1}^p v_r \int_c^{\zeta_r}\psi'(s)\int_c^s \psi'(t)u(t)dtds \\
&+k\int_c^d \psi'(s)v(s)ds - I_{c+}^{\bar\alpha_1;\psi}g_{uv}(d)\Big) + \Gamma\Big(\sum_{i=1}^n \mu_i \int_c^{\eta_i}\psi'(s)I^{\bar\alpha_1;\psi}g_{uv}(s)ds \\
&+\sum_{j=1}^m \theta_j I^{\bar\alpha_1;\psi}g_{uv}(\xi_j) - k\sum_{j=1}^m \theta_j \int_c^{\xi_j}\psi'(s)v(s)ds \\
&-k\sum_{i=1}^n \mu_i \int_c^{\eta_i}\psi'(s)\int_c^s \psi'(t)v(t)dtds \\
&+k\int_c^d \psi'(s)u(s)ds - I_{c+}^{\alpha_1;\psi}f_{uv}(d)\Big)\Big], \\
\mathcal{R}(u,v)(z) \;=\;& I_{c+}^{\bar\alpha_1;\psi}g_{uv}(z).
\end{aligned} \tag{29}$$

Hence, $\mathcal{P}_1(u,v)(z) = \mathcal{M}(u,v)(z) + \mathcal{N}(u,v)(z)$ and $\mathcal{P}_2(u,v)(z) = \mathcal{T}(u,v)(z) + \mathcal{R}(u,v)(z)$. Let $B_\delta = \{(u,v) \in \mathcal{X} \times \mathcal{X}; \|(u,v)\| \le \delta\}$, in which

$$\delta \ge \max\left\{\frac{\mathcal{A}_1 Z_1 + \mathcal{A}_2 Z_2}{1-(\mathcal{A}_3+\mathcal{A}_4)}, \frac{\mathcal{B}_1 Z_1 + \mathcal{B}_2 Z_2}{1-(\mathcal{B}_3+\mathcal{B}_4)}\right\}.$$

First, we show that $\mathcal{P}_1(x,y) + \mathcal{P}_2(u,v) \in B_\delta$ for all $(x,y),(u,v) \in B_\delta$. As in the proof of Theorem 1, we have

$$\begin{aligned}
\mid \mathcal{M}(x,y)(z) + \mathcal{N}(u,v)(z) \mid &\le \mathcal{A}_1 Z_1 + \mathcal{A}_2 Z_2 + (\mathcal{A}_3+\mathcal{A}_4)\delta \le \delta, \\
\mid \mathcal{R}(x,y)(z) + \mathcal{S}(u,v)(z) \mid &\le \mathcal{B}_1 Z_1 + \mathcal{B}_2 Z_2 + (\mathcal{B}_3+\mathcal{B}_4)\delta \le \delta.
\end{aligned} \tag{30}$$

Accordingly, $\mathcal{P}_1(x,y) + \mathcal{P}_2(u,v) \in B_\delta$ and the condition (i) of Lemma 4 is satisfied. In the next step, we show that the operator $(\mathcal{N},\mathcal{R})$ is a contraction mapping. For $(x,y),(u,v) \in B_\delta$, we obtain

$$\begin{aligned}
\mid \mathcal{N}(x,y)(z) - \mathcal{N}(u,v)(z) \mid \; &\le I^{\alpha_1}\mid f_{x,y} - f_{u,v}\mid(z) \\
&\le \ell_1(\|x-u\| + \|y-v\|)I^{\alpha_1}(1)(d) \\
&\le \ell_1 \frac{(d-c)^{\alpha_1}}{\Gamma(\alpha_1+1)}(\|x-u\| + \|y-v\|),
\end{aligned} \tag{31}$$

and

$$| \mathcal{R}(x,y)(z) - \mathcal{R}(u,v)(z) | \quad \leq I^{\bar{\alpha}_1} | g_{x,y} - g_{u,v} | (z)$$
$$\leq \ell_2(\|x - u\| + \|y - v\|)I^{\alpha_1}(1)(d)$$
$$\leq \ell_2 \frac{(d - c)^{\bar{\alpha}_1}}{\Gamma(\bar{\alpha}_1 + 1)}(\|x - u\| + \|y - v\|). \tag{32}$$

In view of (31) and (32), we obtain

$$\|(\mathcal{N},\mathcal{R})(x,y) - (\mathcal{N},\mathcal{R})(u,v)\|$$
$$\leq \left[\frac{(d - c)^{\alpha_1}}{\Gamma(\alpha_1 + 1)}\ell_1 + \frac{(d - c)^{\bar{\alpha}_1}}{\Gamma(\bar{\alpha}_1 + 1)}\ell_2 \right](\|x - u\| + \|y - v\|). \tag{33}$$

Since $\dfrac{(d - c)^{\alpha_1}}{\Gamma(\alpha_1 + 1)}\ell_1 + \dfrac{(d - c)^{\bar{\alpha}_1}}{\Gamma(\bar{\alpha}_1 + 1)}\ell_2 < 1$, the operator $(\mathcal{N},\mathcal{R})$ is a contraction and we conclude that the condition (iii) of Lemma 4 is satisfied. In the next step, we verify the condition (ii) of Lemma 4 for the operator $(\mathcal{M},\mathcal{T})$. By using the continuity of the functions f, g, one can see that the operator $(\mathcal{M},\mathcal{T})$ is continuous. On the other hand, for any $(u,v) \in B_\delta$, as in the proof of Theorem 1, we have

$$| \mathcal{M}(u,v)(z) | \leq \left(\mathcal{A}_1 - \frac{(\psi(d) - \psi(c))^{\alpha_1}}{\Gamma(\alpha_1 + 1)} \right)Z_1 + \mathcal{A}_2 Z_2 + (\mathcal{A}_3 + \mathcal{A}_4)\delta = P^*,$$

$$| \mathcal{T}(u,v)(z) | \leq \mathcal{B}_1 Z_1 + \left(\mathcal{B}_2 - \frac{(\psi(d) - \psi(c))^{\bar{\alpha}_1}}{\Gamma(\bar{\alpha}_1 + 1)} \right)Z_2 + (\mathcal{B}_3 + \mathcal{B}_4)\delta = Q^*. \tag{34}$$

Accordingly, we have $\|(\mathcal{M},\mathcal{T})(u,v)\| \leq P^* + Q^*$, which implies that $(\mathcal{M},\mathcal{T})B_\delta$ is uniformly bounded. Finally, it is shown that the set $(\mathcal{M},\mathcal{T})B_\delta$ is equicontinuous. For this aim, let $\tau_1, \tau_2 \in [c,d]$ with $\tau_1 < \tau_2$. For any $(u,v) \in B_\delta$, similar to the proofs of equicontinuous for the operators $\mathcal{P}_1$ and $\mathcal{P}_2$ in the Theorem 2, we can show that $|\mathcal{M}(u,v)(\tau_2) - \mathcal{M}(u,v)(\tau_1)|, |\mathcal{T}(u,v)(\tau_2) - \mathcal{S}(u,v)(\tau_1)| \longrightarrow 0$ as $\tau_1 \longrightarrow \tau_2$. Consequently, the set $(\mathcal{M},\mathcal{T})B_\delta$ is equicontinuous, and by applying the Arzelá–Ascoli theorem, the operator $(\mathcal{M},\mathcal{T})$ will be compact on B_δ. Therefore, by applying Lemma 4, problem (5) has at least one solution on $[c,d]$. This completes the proof. $\square$

Example 1. *Consider the coupled system of ψ-Hilfer-type sequential fractional differential equations with integro-multipoint boundary conditions:*

$$\begin{cases} \left({}^H D^{\frac{5}{4},\frac{1}{2};(1-e^{-2t})} + k^H D^{\frac{1}{4},\frac{1}{2};(1-e^{-2t})} \right)x(t) = f(t,x(t),y(t)), \quad t \in \left[\frac{1}{11}, \frac{12}{11} \right], \\[2mm] \left({}^H D^{\frac{7}{4},\frac{1}{2};(1-e^{-2t})} + k^H D^{\frac{3}{4},\frac{1}{2};(1-e^{-2t})} \right)y(t) = g(t,x(t),y(t)), \quad t \in \left[\frac{1}{11}, \frac{12}{11} \right], \\[2mm] x\left(\frac{1}{11} \right) = 0, \quad x\left(\frac{12}{11} \right) = \frac{1}{7}\int_{\frac{1}{11}}^{\frac{5}{11}} e^{-2s}y(s)ds + \frac{2}{13}y\left(\frac{4}{11} \right) + \frac{3}{17}y\left(\frac{8}{11} \right) \\[2mm] + \frac{4}{19}y\left(\frac{10}{11} \right) + \frac{5}{23}y(1), \quad y\left(\frac{1}{11} \right) = 0, \quad y\left(\frac{12}{11} \right) = \frac{6}{29}\int_{\frac{1}{11}}^{\frac{3}{11}} e^{-2s}x(s)ds \\[2mm] + \frac{7}{31}\int_{\frac{1}{11}}^{\frac{7}{11}} e^{-2s}x(s)ds + \frac{8}{37}x\left(\frac{2}{11} \right) + \frac{9}{41}x\left(\frac{6}{11} \right) + \frac{10}{43}x\left(\frac{9}{11} \right). \end{cases} \tag{35}$$

Here $\alpha_1 = 5/4$, $\bar{\alpha}_1 = 7/4$, $\beta_1 = 1/2$, $\psi(t) = (1 - e^{-2t})$, $\psi'(t) = 2e^{-2t}$, $c = 1/11$, $d = 12/11$, $\mu_1 = 1/14$, $\eta_1 = 5/11$, $\theta_1 = 2/13$, $\theta_2 = 3/17$, $\theta_3 = 4/19$, $\theta_4 = 5/23$, $\xi_1 = 4/11$, $\xi_2 = 8/11$, $\xi_3 = 10/11$, $\xi_4 = 1$, $v_1 = 3/29$, $v_2 = 7/62$, $\varsigma_1 = 3/11$, $\varsigma_2 = 7/11$, $\tau_1 = 8/37$, $\tau_2 = 9/41$, $\tau_3 = 10/43$, $\sigma_1 = 2/11$, $\sigma_2 = 6/11$, $\sigma_3 = 9/11$, $n = 1$, $m = 4$, $p = 2$,

$q = 3$. We find that $\gamma = 13/8$, $\gamma_1 = 15/8$, $A \approx 0.9090586723$, $B \approx 0.5135134292$, $\Gamma \approx 0.4618499072$, $\Delta \approx 0.7876865883$, $\Lambda \approx 0.4788871945$, $\mathcal{A}_1 \approx 1.685246952$, $\mathcal{A}_2 \approx 0.6395331644$, $\mathcal{A}_3 \approx 0.1399736659$, $\mathcal{A}_4 \approx 0.09267043416$, $\mathcal{B}_1 \approx 0.9074990107$, $\mathcal{B}_2 \approx 1.026284484$, $\mathcal{B}_3 \approx 0.06726442842$, $\mathcal{B}_4 \approx 0.1432037148$.

(i) If the nonlinear unbounded functions f and g are given by

$$f(z, u, v) = \frac{11 e^{-\left(z - \frac{1}{11}\right)}}{2(11z + 87)} \left(\frac{u^2 + 2|u|}{1 + |u|}\right) + \frac{1}{17}\left(\cos^2 z + 1\right) \sin|v| + \frac{1}{7}, \tag{36}$$

$$g(z, u, v) = \frac{\tan^{-1}|u|}{9(1 + \sin^4 z)} + \frac{11}{10(11z + 43)}\left(\frac{3v^2 + 4|v|}{1 + |v|}\right) + \frac{3}{5}, \tag{37}$$

then we can verify the Lipchitz conditions as

$$|f(z, u_1, v_1) - f(z, u_2, v_2)| \le \frac{1}{8}(|u_1 - u_2| + |v_1 - v_2|)$$

$$|g(z, u_1, v_1) - g(z, u_2, v_2)| \le \frac{1}{9}(|u_1 - u_2| + |v_1 - v_2|),$$

in which Lipchitz constants $\ell_1 = 1/8$ and $\ell_2 = 1/9$. In addition, we can compute that

$$\ell_1(\mathcal{A}_1 + \mathcal{B}_1) + \ell_2(\mathcal{A}_2 + \mathcal{B}_2) + \mathcal{A}_3 + \mathcal{A}_4 + \mathcal{B}_3 + \mathcal{B}_4 \approx 0.9522963385 < 1.$$

Therefore, all assumptions of Theorem 1 are fulfilled and the conclusion of Theorem 1 can be applied—that the coupled system of ψ-Hilfer-type sequential fractional differential equations with integro-multipoint boundary conditions (35) with (36)-(37) has a unique solution on $[1/11, 12/11]$.

(ii) Consider the nonlinear functions f and g given by

$$f(z, u, v) = \frac{1}{z + 3} + \frac{1}{6}\left(\frac{u^{16}}{1 + |u|^{15}}\right) + \frac{1}{7}|v|e^{-u^4}, \tag{38}$$

$$g(z, u, v) = \frac{\cos^2 \pi z + 1}{3} + \frac{1}{14}\left(1 + \sin^4 v^8\right)|u| + \frac{1}{5}\left(\frac{|v|^{23}}{2 + v^{22}}\right). \tag{39}$$

Observe that the above two functions f and g are bounded by

$$|f(z, u, v)| \le \frac{11}{34} + \frac{1}{6}|u| + \frac{1}{7}|v|,$$

$$|g(z, u, v)| \le \frac{2}{3} + \frac{1}{7}|u| + \frac{1}{5}|v|.$$

Thus, we choose constants from Theorem 2 by $u_0 = 11/34$, $v_0 = 2/3$, $u_1 = 1/6$, $v_1 = 1/7$, $u_2 = 1/7$ and $v_2 = 1/5$. By direct computation, we have $(\mathcal{A}_1 + \mathcal{B}_1)u_1 + (\mathcal{A}_2 + \mathcal{B}_2)v_1 + \mathcal{A}_3 + \mathcal{B}_3 \approx 0.8773363712 < 1$ and $(\mathcal{A}_1 + \mathcal{B}_1)u_2 + (\mathcal{A}_2 + \mathcal{B}_2)v_2 + \mathcal{A}_4 + \mathcal{B}_4 \approx 0.9394299590 < 1$. Applying Theorem 2, we deduce that the boundary value problem (35) with (38) and (39) has at least one solution on $[1/11, 12/11]$.

(iii) Let the nonlinear bounded functions f and g defined by

$$f(z, u, v) = \frac{11}{24}z + \frac{1}{2} + \frac{9}{16}\left(\frac{|u|}{1 + |u|}\right) + \frac{1}{2}\sin|v|, \tag{40}$$

$$g(z, u, v) = 1 + \cos \pi z + \frac{7}{10}\tan^{-1}|u| + \frac{4}{5}\left(\frac{|v|}{1 + |v|}\right). \tag{41}$$

It is obvious that these two functions are bounded since

$$|f(z, u, v)| \le \frac{33}{16}, \quad \text{and} \quad |g(z, u, v)| \le \frac{7}{2}.$$

In addition, the condition (H_1) in Theorem 1 is satisfied with $\ell_1 = 9/16$ and $\ell_2 = 4/5$. Hence, we obtain $\mathcal{A}_3 + \mathcal{A}_4 \approx 0.2326441001 < 1$, $\mathcal{B}_3 + \mathcal{B}_4 \approx 0.2104681432 < 1$ and

$$\left[\frac{(d-c)^{\alpha_1}}{\Gamma(\alpha_1 + 1)} \ell_1 + \frac{(d-c)^{\overline{\alpha}_1}}{\Gamma(\overline{\alpha}_1 + 1)} \ell_2 \right] \approx 0.9938694509 < 1.$$

Therefore, problem (35) with (40) and (41) has at least one solution on $[1/11, 12/11]$ by using the benefit of Theorem 3. Finally, we give a remark that Theorem 1 cannot be used for this problem because

$$\ell_1(\mathcal{A}_1 + \mathcal{B}_1) + \ell_2(\mathcal{A}_2 + \mathcal{B}_2) + \mathcal{A}_3 + \mathcal{A}_4 + \mathcal{B}_3 + \mathcal{B}_4 \approx 3.234185965 > 1.$$

5. Conclusions

In this paper, we investigated a coupled system of fractional differential equations involving ψ-Hilfer fractional derivatives, supplemented with integro-multi-point boundary conditions. Firstly, we proved the equivalence between a linear variant of the system (5) and the fractional integral Equations (7) and (8). After that, the existence of a unique solution for the system (5) was proved by using Banach's fixed point theorem. The Leray–Schauder alternative and Krasnosel'skii's fixed point theorem were used to obtain the existence of solutions for the system (5). Moreover, examples were constructed to illustrate our main results. The obtained results are new and enrich the literature on coupled systems for nonlinear ψ-Hilfer fractional differential equations. The used methods are standard, but their configuration on the problem (5) is new.

Author Contributions: Conceptualization, A.S., S.K.N. and J.T.; methodology, A.S., C.N., S.K.N. and J.T.; formal analysis, A.S., C.N., S.K.N. and J.T.; funding acquisition, J.T. All authors have read and agreed to the published version of the manuscript.

Funding: This research was funded by King Mongkut's University of Technology North Bangkok. Contract No. KMUTNB-62-KNOW-30.

Institutional Review Board Statement: Not applicable.

Informed Consent Statement: Not applicable.

Conflicts of Interest: The authors declare no conflict of interest.

References

1. Mainardi, F. Boundary Value Problems for Hilfer Type Sequential Fractional Differential Equations and Inclusions with Integral Multi-Poin Boundary Conditions. In *Fractals and Fractional Calculus in Continuum Mechanics*; Carpinteri, A., Mainardi, F., Eds.; Springer: Berlin, Germany, 1997; pp. 291–348
2. Magin, R.L. *Fractional Calculus in Bioengineering*; Begell House Publishers: Chicago, IL, USA, 2006.
3. Fallahgoul, H.A.; Focardi, S.M.; Fabozzi, F.J. *Fractional Calculus and Fractional Processes with Applications to Financial Economics, Theory and Application*; Elsevier/Academic Press: London, UK, 2017.
4. Diethelm, K. *The Analysis of Fractional Differential Equations*; Lecture Notes in Mathematics; Springer: New York, NY, USA, 2010.
5. Kilbas, A.A.; Srivastava, H.M.; Trujillo, J.J. *Theory and Applications of the Fractional Differential Equations*; North-Holland Mathematics Studies; Elsevier: Amsterdam, The Netherlands, 2006; Volume 204.
6. Lakshmikantham, V.; Leela, S.; Devi, J.V. *Theory of Fractional Dynamic Systems*; Cambridge Scientific Publishers: Cambridge, UK, 2009.
7. Miller, K.S.; Ross, B. *An Introduction to the Fractional Calculus and Differential Equations*; John Wiley: New York, NY, USA, 1993.
8. Podlubny, I. *Fractional Differential Equations*; Academic Press: New York, NY, USA, 1999.
9. Samko, S.G.; Kilbas, A.A.; Marichev, O.I. *Fractional Integrals and Derivatives*; Gordon and Breach Science: Yverdon, Switzerland, 1993.
10. Ahmad, B.; Alsaedi, A.; Ntouyas, S.K.; Tariboon, J. *Hadamard Type Fractional Differential Equations, Inclusions and Inequalities*; Springer: Cham, Switzerland, 2017.
11. Zhou, Y. *Basic Theory of Fractional Differential Equations*; World Scientific: Singapore, 2014.
12. Almeida, R.A. Caputo fractional derivative of a function with respect to another function. *Commun. Nonlinear Sci. Numer. Simul.* **2017**, *44*, 460–481. [CrossRef]

13. Abdo, M.S.; Panchal, S.K.; Saeed, A.M. Fractional boundary value problem with ψ-Caputo fractional derivative. *Proc. Math. Sci.* **2019**, *129*, 1–14. [CrossRef]
14. Vivek, D.; Elsayed, E.; Kanagarajan, K. Theory and analysis of ψ-fractional differential equations with boundary conditions. *Commun. Pure Appl. Anal.* **2018**, *22*, 401–414.
15. Wahash, H.A.; Abdo, M.S.; Saeed, A.M.; Panchal, S.K. Singular fractional differential equations with ψ-Caputo operator and modified Picard's iterative method. *Appl. Math. E Notes* **2020**, *20*, 215–229.
16. Hilfer, R. *Applications of Fractional Calculus in Physics*; World Scientific: Singapore, 2000.
17. Hilfer, R. Experimental evidence for fractional time evolution in glass forming materials. *J. Chem. Phys.* **2002**, *284*, 399–408. [CrossRef]
18. Hilfer, R.; Luchko, Y.; Tomovski, Z. Operational method for the solution of fractional differential equations with generalized Riemann-Liouvill fractional derivatives. *Frac. Calc. Appl. Anal.* **2009**, *12*, 299–318.
19. Furati, K.M.; Kassim, N.D.; Tatar, N.E. Existence and uniqueness for a problem involving Hilfer fractional derivative. *Comput. Math. Appl.* **2012**, *12*, 1616–1626. [CrossRef]
20. Gu, H.; Trujillo, J.J. Existence of mild solution for evolution equation with Hilfer fractional derivative. *Appl. Math. Comput.* **2015**, *257*, 344–354. [CrossRef]
21. Wang, J.; Zhang, Y. Nonlocal initial value problems for differential equations with Hilfer fractional derivative. *Appl. Math. Comput.* **2015**, *266*, 850–859. [CrossRef]
22. Soong, T.T. *Random Differential Equations in Science and Engineering*; Academic Press: New York, NY, USA, 1973.
23. Kavitha, K.; Vijayakumar, V.; Udhayakumar, R.; Nisar, K.S. Results on the existence of Hilfer fractional neutral evolution equations with infinite delay via measures of noncompactness. *Math. Methods Appl. Sci.* **2021**, *44*, 1438–1455. [CrossRef]
24. Subashini, R.; Jothimani, K.; Nisar, K.S.; Ravichandran, C. New results on nonlocal functional integro-differential equations via Hilfer fractional derivative. *Alex. Eng. J.* **2020**, *59*, 2891–2899. [CrossRef]
25. Asawasamrit, S.; Kijjathanakorn, A.; Ntouyas, S.K.; Tariboon, J. Nonlocal boundary value problems for Hilfer fractional differential equations. *Bull. Korean Math. Soc.* **2018**, *55*, 1639–1657.
26. Wongcharoen, A.; Ntouyas, S.K.; Tariboon, J. On coupled systems for Hilfer fractional differential equations with nonlocal integral boundary conditions. *J. Math.* **2020**, *2020*, 2875152 [CrossRef]
27. Sitho, S.; Ntouyas, S.K.; Samadi, A.; Tariboon, J. Boundary value problems for ψ-Hilfer type sequential fractional differential equations and inclusions with integral multi-point boundary conditions. *Mathematics* **2021**, *9*, 1001. [CrossRef]
28. Phuangthong, N.; Ntouyas, S.K.; Tariboon, J.; Nonlaopon, K. Nonlocal sequential boundary value problems for Hilfer type fractional integro-differential equations and inclusions. *Mathematics* **2021**, *9*, 615. [CrossRef]
29. Sousa, J.V.; de Oliveira, E.C. On the ψ-Hilfer fractional derivative. *Commun. Nonlinear Sci. Numer. Simul.* **2018**, *60*, 72–91. [CrossRef]
30. Deimling, K. *Nonlinear Functional Analysis*; Springer: New York, NY, USA, 1985.
31. Granas, A.; Dugundji, J. *Fixed Point Theory*; Springer: New York, NY, USA, 2003.
32. Krasnosel'skiĭ, M.A. Two remarks on the method of successive approximations. *UspekhiMat. Nauk* **1955**, *10*, 123–127.

Article

Asymptotic Stabilization of Delayed Linear Fractional-Order Systems Subject to State and Control Constraints

Xindong Si , **Zhen Wang** * , **Zhibao Song and Ziye Zhang**

College of Mathematics and Systems Science, Shandong University of Science and Technology, Qingdao 266590, China; sixindongsk@163.com (X.S.); szb879381@163.com (Z.S.); zhangzy02@126.com (Z.Z.)
* Correspondence: wangzhen_sd@126.com

Abstract: Studies have shown that fractional calculus can describe and characterize a practical system satisfactorily. Therefore, the stabilization of fractional-order systems is of great significance. The asymptotic stabilization problem of delayed linear fractional-order systems (DLFS) subject to state and control constraints is studied in this article. Firstly, the existence conditions for feedback controllers of DLFS subject to both state and control constraints are given. Furthermore, a sufficient condition for invariance of polyhedron set is established by using invariant set theory. A new Lyapunov function is constructed on the basis of the constraints, and some sufficient conditions for the asymptotic stability of DLFS are obtained. Then, the feedback controller and the corresponding solution algorithms are given to ensure the asymptotic stability under state and control input constraints. The proposed solution algorithm transforms the asymptotic stabilization problem into a linear/nonlinear programming (LP/NP) problem which is easy to solve from the perspective of computation. Finally, three numerical examples are offered to illustrate the effectiveness of the proposed method.

Keywords: delayed linear fractional-order systems; feedback controller; positive invariant set; asymptotic stabilization

Citation: Si, X.; Wang, Z.; Song, Z.; Zhang, Z. Asymptotic Stabilization of Delayed Linear Fractional-Order Systems Subject to State and Control Constraints. *Fractal Fract.* **2022**, *6*, 67. https://doi.org/10.3390/fractalfract6020067

Academic Editors:António M. Lopes and Liping Chen

Received: 24 December 2021
Accepted: 25 January 2022
Published: 27 January 2022

Publisher's Note: MDPI stays neutral with regard to jurisdictional claims in published maps and institutional affiliations.

1. Introduction

Fractional calculus almost appeared at the same time as classic calculus, but it has not been paid more attention to due to its lack of application background and difficult calculation. Fractional calculus has experienced rapid development during the last few decades both in mathematics and applied sciences. It has been recognized as an excellent tool to describe modern complex dynamics [1,2]. From this perspective, some models governing physical phenomena have been reformulated in light of fractional calculus to better reflect their non-local, frequency- and history-dependent properties. With the rapid development of computer technology, fractional calculus is widely used in many fields, such as image processing [3], fluid mechanics [4], and environmental science [5]. Time delay often occurs in different practical systems, and the delayed fractional-order system can better describe these phenomena [6–8].

However, time delay will lead to system performance degradation, poor stability and even failure to work [9,10]. In fact, many scholars have extensively studied the stability of delayed fractional-order systems. For instance, the finite-time stability was discussed in [11–14]. Using the Laplace transform method, the globally asymptotic stability was studied in [15]. The Mittag-Leffler stability was discussed in [16,17]. The asymptotic stability was investigated using the frequency domain method [18], integral inequality method [19], linear matrix inequality (LMI) method [20], and Lyapunov function method [21,22].

On the other hand, for security reasons or physical constraints, the control input, the state, and/or output variables must be bounded in practice. That is to say, the hard constraints of these variables should be considered. At this time, how to design a feedback controller to ensure the stability of the system under constraints is an interesting topic in system theory and synthesis. For constrained integer-order systems, state feedback controllers were designed to study the asymptotic stability of linear discrete-time systems and linear continuous-time systems under constraints in [23,24]. Based on the LMI method, the optimization problem of state feedback controllers for linear discrete-time systems and continuous-time systems was further analyzed in [25,26]. A more effective method to solve the state feedback controller was developed based on the invariance of polyhedron set in [27,28]. As for unconstrained fractional-order systems, in [29], the global stabilization of fractional-order neural network was investigated by using the positive-system-based method. In [30], the stabilization of fractional-order T-S fuzzy systems was discussed by using LMI method. In [31], based on Lyapunov functions, a state feedback controller was designed to study the stabilization of fractional-order nonlinear systems. However, there are few studies concerning the stabilization problems of fractional-order system with constraints, except [32]. In [32], the stabilization problem of fractional-order linear systems with control input constraints was addressed by using the invariance of polyhedron set, but state constraints are not considered there.

Based on the above discussions, it is not only necessary but also more challenging to study the stabilization problem of DLFS under constraints. In this article, the stabilization problem of DLFS with state and control constraints is studied. Our main contributions include: (1) The sufficient conditions that ensure the state constraint set and/or the control constraint set are positive invariant sets (PIS) are established by using the invariant set theory; (2) A new Lyapunov function is constructed on the basis of the constraints, and the asymptotic stability conditions for DLFS are obtained; and (3) A feedback controller and its solution algorithm are proposed to make DLFS under the state and control input constraints asymptotically stable.

The article is organized as follows. Some preliminaries and the problem formulation are given in Section 2. In Section 3, existence conditions for the PIS and the feedback controller are developed. A feedback controller and its solution algorithm are proposed in Section 4 to ensure the DLFS under state and control constraints asymptotically stable. To illustrative the effectiveness of the proposed method, three examples are presented in Section 5. Section 6 contains some conclusions.

Notations: In this article, $\mathbb{R}_+$ denotes the set of positive real numbers, $\mathbb{R}^n$ denotes n dimensional real vector space, $\mathbb{R}^{n \times n}$ denotes $n \times n$ real matrices. z stands for a complex number and $\mathrm{Re}(z)$ stands for the real part of z. ρ_i represents the ith element of vector ρ, and Q_i represents the ith row of matrix Q, Q_{ij} represents the ith row and jth column element of matrix Q. For $\rho \in \mathbb{R}^n, \rho \geqslant 0(\rho > 0)$ means $\rho_i \geqslant 0(\rho_i > 0)$. For $a \in \mathbb{R}^n, b \in \mathbb{R}^n, a \geqslant b(a > b)$ means $a_i \geqslant b_i(a_i > b_i)$. $A > 0$ indicates that each entry of the matrix A is nonnegative.

2. Preliminaries and Problem Formulation

2.1. Preliminaries

Consider the DLFS:

$$\begin{cases} {}_0^C D_t^\alpha x(t) = Ax(t) + A_0 x(t - \tau) + Bu(t), & t > 0, \\ x(t) = x_0, & -\tau \leqslant t \leqslant 0, \end{cases} \tag{1}$$

where $0 < \alpha \leqslant 1, \tau = $ constant is the time delay, $A \in \mathbb{R}^{n \times n}$ and $A_0 \in \mathbb{R}^{n \times n}$ are the system matrices, $B \in \mathbb{R}^{n \times m}$ is the input matrix, $x(t) \in \mathbb{R}^n$ is the state, and $u(t) \in \mathbb{R}^m$ is the control input.

Definition 1 ([12]). *The Caputo fractional derivative of $x(t)$ is defined as*

$$ {}_0^C D_t^\alpha x(t) = \frac{1}{\Gamma(1 - \alpha)} \int_0^t (t - s)^{-\alpha} x'(s)ds, $$

where the order $0 < \alpha \leqslant 1$ and $\Gamma(z) = \int_0^\infty e^{-t} t^{z-1} dt, \ \mathrm{Re}(z) \in \mathbb{R}_+$.

Definition 2 ([33]). *The fractional integral of $x(t)$ is defined as*

$$_0I_t^\alpha x(t) = \frac{1}{\Gamma(\alpha)} \int_0^t (t-s)^{\alpha-1} x(s)\,ds,$$

where the order $\alpha \in \mathbb{R}_+$.

By the above Definition, we obtain the solution of system (1) as

$$\begin{aligned} x(t) &= x_0 + {}_0I_t^\alpha({}_0^C D_t^\alpha x(t)) \\ &= x_0 + \frac{1}{\Gamma(\alpha)} \int_0^t (t-s)^{\alpha-1}[Ax(s) + A_0x(s-\tau) + Bu(s)]\,ds. \end{aligned}$$

Definition 3 ([34]). *A continuous function $\beta(x) : [0, +\infty) \longrightarrow [0, +\infty)$ is said to be a class-κ function if the function $\beta(x)$ is strictly increasing and $\beta(0) = 0$.*

Definition 4 ([29]). *A nonempty set P is called the positive invariant sets (PIS) if and only if*

$$x_0 \in P \ \text{ implies } \ x(t) \in P, \quad \text{for } t > 0,$$

where $x(t)$ is the trajectory starting with the initial value x_0.

Lemma 1 ([34]). *Assume that there exist class-κ functions $\beta_i, i = 1, 2, 3$ and a continuously differentiable function $V(x(t))$ such that:*

$$\beta_1(\|x(t)\|) \leqslant V(x(t)) \leqslant \beta_2(\|x(t)\|), \tag{2}$$

and

$$_0^C D_t^\alpha V(x(t)) \leqslant -\beta_3(\|x(t)\|), \tag{3}$$

where the order $0 < \alpha < 1$, then system (1) is guaranteed to be stable. If $\beta_3(s) > 0$ for $s > 0$, then system (1) is guaranteed to be asymptotically stable.

Lemma 2 ([35]). *The polyhedron set $P(Q, \rho) = \{x(t) \in \mathbb{R}^n : Qx(t) \leqslant \rho\}, Q \in \mathbb{R}^{q \times n},$ $\rho \in \mathbb{R}^q, \rho > 0$, and $P(K, \omega) = \{x(t) \in \mathbb{R}^n : Kx(t) \leqslant \omega\}, K \in \mathbb{R}^{m \times n}, \omega \in \mathbb{R}^m, \omega > 0$ have the relation*

$$P(Q, \rho) \subseteq P(K, \omega)$$

if and only if there exists $L \in \mathbb{R}^{m \times q}, L > 0$ such that

$$\begin{cases} LQ = K, \\ L\rho \leqslant \omega. \end{cases} \tag{4}$$

2.2. Problem Formulation

In this article, the following assumptions are needed.

Assumption 1. *The state variables are constrained by the polyhedron set*

$$P(Q, \rho) = \{x(t) \in \mathbb{R}^n : Qx(t) \leqslant \rho\}, \tag{5}$$

where $Q \in \mathbb{R}^{q \times n}, q > n$, and $\rho \in \mathbb{R}^q, \rho > 0$, and the polyhedron set $P(Q, \rho)$ is closed and nonempty.

Assumption 2. *The control input $u(t)$ satisfies the following constraints*

$$-w_1 \leqslant u(t) \leqslant w_2, \tag{6}$$

where $w_1 > 0$ and $w_2 > 0$ are real vectors.

The asymptotic stabilization problem of DLFS (1) is to find a state feedback controller $u(t) = Kx(t)$ that makes all trajectories starting with the initial value x_0 asymptotically stable and also satisfy the state constraints (5) and the control constraints (6).

If there exists $u(t) = Kx(t)$ for system (1), then, by linear constraints (6), we obtain

$$P(K, -w_1, w_2) = \{x(t) \in \mathbb{R}^n : -w_1 \leqslant Kx(t) \leqslant w_2\}. \tag{7}$$

Let $\overline{A} = A + BK$; clearly, $u(t) = Kx(t)$ is the solution of the asymptotic stabilization problem of system (1) if and only if the closed-loop system

$$\begin{cases} {}_0^C D_t^\alpha x(t) = \overline{A}x(t) + A_0 x(t - \tau), & t > 0, \\ x(t) = x_0, & -\tau \leqslant t \leqslant 0 \end{cases} \tag{8}$$

is asymptotically stable, and the trajectory $x(t)$ starting with x_0 satisfy $x(t) \in P(Q, \rho)$ and $x(t) \in P(K, -w_1, w_2)$ for any $t \geqslant 0$.

3. Main Results

3.1. Existence Conditions for the Feedback Controller with the Constraints

Theorem 1. *The controller $u(t) = Kx(t), K \in \mathbb{R}^{m \times n}$ is the solution of the asymptotic stabilization problem of system (1) if*

(i)   *There exists a PIS denoted by $M \in \mathbb{R}^n$ for the system (8) satisfies $M \subseteq P(Q, \rho)$ and $M \subseteq P(K, -w_1, w_2)$;*
(ii) *There exists a Lyapunov function that makes the system (8) asymptotically stable.*

Proof. Condition (i) ensures that there exists a PIS

$$M = \{x_0 \in \mathbb{R}^n : \forall x_0 \in M, x(t) \triangleq x(t; x_0) \in M\}$$

such that $M \subseteq P(Q, \rho)$ and $M \subseteq P(K, -w_1, w_2)$, where $x(t; x_0)$ is the trajectory starting with x_0. Then we can obtain $Qx(t) \leqslant \rho, -w_1 \leq Kx(t) \leqslant w_2$. Furthermore, the condition (ii) ensures that system (8) is asymptotically stable. Hence, the controller $u(t) = Kx(t)$ is the solution of the asymptotic stabilization problem of system (1). $\square$

When considering both the state and control constraints, let $P(Q, \rho)$ be the PIS of system (8), that is, $P(Q, \rho)$ is equal to M, then we obtain:

Corollary 1. *The controller $u(t) = Kx(t), K \in \mathbb{R}^{m \times n}$ is the solution of the asymptotic stabilization problem of system (1) if*

(i)   *$P(Q, \rho)$ is a PIS of the system (8) and $P(Q, \rho) \subseteq P(K, -w_1, w_2)$;*
(ii) *There exists a Lyapunov function that makes the system (8) asymptotically stable.*

Moreover, if considering only the control constraints, let $P(K, -w_1, w_2)$ be the PIS of system (8), that is, $P(K, -w_1, w_2)$ is equal to M, we obtain:

Corollary 2. *The controller $u(t) = Kx(t), K \in \mathbb{R}^{m \times n}$ is the solution of the asymptotic stabilization problem of system (1) if*

(i)   *$P(K, -w_1, w_2)$ is a PIS of the system (8);*
(ii) *There exists a Lyapunov function that makes the system (8) asymptotically stable.*

3.2. PIS and Stability Conditions for System (8)

Theorem 2. *If there exists a matrix $K \in \mathbb{R}^{m \times n}$, real matrices $F, F_0 \in \mathbb{R}^{q \times q}$ and a scalar $\varepsilon > 0$ such that*

$$\begin{cases} Q(A + BK) = FQ, \\ QA_0 = F_0 Q, \\ (F + F_0)\rho \leqslant -\varepsilon\rho, \end{cases} \tag{9}$$

then the system (8) is asymptotically stable and the polyhedron set $P(Q, \rho)$ is a PIS.

Proof. Choose the Lyapunov function of the form

$$V(x(t)) = \max\left\{ \max\left(\frac{Q_1 x(t)}{\rho_1}, 0 \right), \cdots, \max\left(\frac{Q_q x(t)}{\rho_q}, 0 \right) \right\}.$$

If, for $1 \leqslant i \leqslant q$, it satisfies $\frac{Q_i x(t)}{\rho_i} < 0$, we obtain $V(x(t)) = 0$. However, due to the fact that $Q_i x(t)$ cannot always be negative, this case does not exist, so $V(x(t)) > 0$. Hence, there must exist an i, such that $0 < Q_i x(t) \leqslant \rho_i$, that is to say, $0 < V(x(t)) \leqslant 1$; then, according to (9), we obtain

$$
\begin{aligned}
{}^C_0 D^\alpha_t V(x(t)) &= \tfrac{Q_i}{\rho_i} {}^C_0 D^\alpha_t x(t) = \tfrac{Q_i}{\rho_i}[Ax(t) + A_0 x(t-\tau) + Bu(t)] \\
&= \tfrac{Q_i}{\rho_i}[(A + BK)x(t) + A_0 x(t-\tau)] \\
&= \tfrac{1}{\rho_i}[(FQ)_i x(t) + (F_0 Q)_i x(t-\tau)] \\
&= \tfrac{1}{\rho_i}\left(\sum_{j=1}^{n} \sum_{k=1}^{q} F_{ik} Q_{kj} x_j(t) + \sum_{j=1}^{n} \sum_{k=1}^{q} F_{0ik} Q_{kj} x_j(t) \right) \\
&= \tfrac{1}{\rho_i}\left(\sum_{k=1}^{q} F_{ik} \sum_{j=1}^{n} Q_{kj} x_j(t) + \sum_{k=1}^{q} F_{0ik} \sum_{j=1}^{n} Q_{kj} x_j(t) \right) \\
&\leqslant \tfrac{1}{\rho_i}\left(\sum_{k=1}^{q} F_{ik}\rho_k + \sum_{k=1}^{q} F_{0ik}\rho_k \right) \\
&= \tfrac{1}{\rho_i}[(F + F_0)\rho]_i \\
&\leqslant \tfrac{1}{\rho_i}(-\varepsilon\rho_i) \leqslant -\varepsilon V(x(t)) < 0.
\end{aligned}
\tag{10}
$$

By Lemma 1, we conclude that the system (8) is asymptotically stable.

Assuming that $x_0 \in P(Q, \rho)$, i.e., $Qx_0 \leq \rho$ and $\overline{A} = A + BK$, according to the Definition of $x(t)$, we obtain

$$
\begin{aligned}
Qx(t) &= Qx_0 + Q\left[\tfrac{1}{\Gamma(\alpha)} \int_0^t (t-s)^{\alpha-1}\left[\overline{A}x(s) + A_0 x(s-\tau) \right] ds \right] \\
&= Qx_0 + \tfrac{1}{\Gamma(\alpha)} \int_0^t (t-s)^{\alpha-1}[FQx(s) + F_0 Qx(s-\tau)]ds \\
&\leqslant \rho + \tfrac{1}{\Gamma(\alpha)} \int_0^t (t-s)^{\alpha-1}[FQx(s) + F_0 Qx(s-\tau)]ds \\
&\leqslant \rho + \tfrac{1}{\Gamma(\alpha)} \int_0^t (t-s)^{\alpha-1}(F\rho + F_0\rho)ds \\
&\leqslant \rho - \tfrac{\varepsilon\rho}{\Gamma(\alpha)} \int_0^t (t-s)^{\alpha-1}ds \\
&= \rho - \tfrac{\varepsilon\rho t^\alpha}{\alpha\Gamma(\alpha)} \leqslant \rho,
\end{aligned}
$$

hence the polyhedron set $P(Q, \rho)$ is a PIS of system (8). □

Theorem 3. *If there exists a matrix $K \in \mathbb{R}^{m \times n}$, real matrices $F, F_0 \in \mathbb{R}^{q \times q}$, and a scalar $\varepsilon > 0$ such that*

$$
\begin{cases}
K(A + BK) = FK, \\
KA_0 = F_0 K, \\
(\widehat{F} + \widehat{F_0})\widehat{w} \leqslant -\varepsilon\widehat{w},
\end{cases}
\tag{11}
$$

where $\widehat{w} = \begin{bmatrix} w_2 \\ w_1 \end{bmatrix}$, $\widehat{F} = \begin{bmatrix} F^+ & F^- \\ F^- & F^+ \end{bmatrix}$, *and*

$$
F_{ij}^+ = \begin{cases} F_{ij}, & \text{if } i = j, \\ \max(F_{ij}, 0), & \text{if } i \neq j, \end{cases} \qquad
F_{ij}^- = \begin{cases} 0, & \text{if } i = j, \\ \max(-F_{ij}, 0), & \text{if } i \neq j, \end{cases}
\tag{12}
$$

then the system (8) is asymptotically stable and the polyhedron set $P(K, -w_1, w_2)$ is a PIS.

Proof. From (12), we have

$$F = F^+ - F^-. \tag{13}$$

Replace F in (11) with (13), let $\overline{A} = A + BK$; then, we have

$$\begin{aligned}
K\overline{A} &= FK = (F^+ - F^-)K, \\
-K\overline{A} &= F(-K) = (F^+ - F^-)(-K) = (F^- - F^+)K.
\end{aligned}$$

Hence,

$$\begin{bmatrix} K \\ -K \end{bmatrix} \overline{A} = \begin{bmatrix} F^+ & F^- \\ F^- & F^+ \end{bmatrix} \begin{bmatrix} K \\ -K \end{bmatrix}. \tag{14}$$

Let $Q = \begin{bmatrix} K \\ -K \end{bmatrix}$, $\rho = \widehat{w} = \begin{bmatrix} w_2 \\ w_1 \end{bmatrix}$, that is, the set $P(K, -w_1, w_2)$ can be reformulated as the form of $P(Q, \rho)$. From (14), there exists a matrix. $\widehat{F} = \begin{bmatrix} F^+ & F^- \\ F^- & F^+ \end{bmatrix}$, such that

$$Q\overline{A} = \widehat{F}Q. \tag{15}$$

In the same way, we have

$$QA_0 = \widehat{F}_0 Q. \tag{16}$$

On the other hand, according to $(\widehat{F} + \widehat{F}_0)\widehat{w} \leqslant -\varepsilon\widehat{w}$, we obtain

$$(\widehat{F} + \widehat{F}_0)\rho \leqslant -\varepsilon\rho. \tag{17}$$

According to (11), we obtain (15)–(17). Then, by Theorem 2, the system (8) is asymptotically stable and the polyhedron set $P(Q, \rho)$ is a PIS. Hence, we conclude that the system (8) is asymptotically stable and the polyhedron set $P(K, -w_1, w_2)$ is a PIS. $\quad\square$

Theorem 4. *The polyhedron sets $P(Q, \rho)$ and $P(K, -w_1, w_2)$ have the relation*

$$P(Q, \rho) \subseteq P(K, -w_1, w_2)$$

if and only if there exists $L \in \mathbb{R}^{2m \times 2q}, L > 0$ such that

$$\begin{cases} L\begin{pmatrix} Q \\ Q \end{pmatrix} = \begin{pmatrix} K \\ -K \end{pmatrix}, \\ L\begin{pmatrix} \rho \\ \rho \end{pmatrix} \leqslant \begin{pmatrix} w_2 \\ w_1 \end{pmatrix}. \end{cases} \tag{18}$$

Proof. According to Lemma 2, $P(Q, \rho) \subseteq P(K, w_2)$ is equivalent to $L_1 \in \mathbb{R}^{m \times q}, L_1 > 0$ such that $L_1 Q = K$ and $L_1 \rho \leqslant w_2$. On the other hand, $P(Q, \rho) \subseteq P(-K, w_1)$ is equivalent to $L_2 \in \mathbb{R}^{m \times q}, L_2 > 0$ such that $L_2 Q = -K$ and $L_2 \rho \leqslant w_1$.

Hence, $P(Q, \rho) \subseteq P(K, -w_1, w_2)$ if and only if there exists $L = \begin{pmatrix} L_1 \\ L_2 \end{pmatrix} > 0, L \in R^{2m \times 2q}$ such that (18) holds. $\quad\square$

Remark 1. *Theorem 2 proposes a sufficient condition that ensures the state constraint set $P(Q, \rho)$ is a PIS and the system (8) is asymptotically stable. On this basis, Theorem 3 proposes a sufficient condition that ensures the control constraint set $P(K, -w_1, w_2)$ is a PIS and the system (8) is asymptotically stable. Theorem 4 gives a sufficient and necessary condition to ensure $P(Q, \rho) \subseteq P(K, -w_1, w_2)$. When considering both state and control constraints. One can use Theorem 2 and Theorem 4 to find a feedback controller for the asymptotic stabilization problem of system (1) according to Corollary 1. When considering only control constraints. One can use Theorem 3 to find a feedback controller for the asymptotic stabilization problem of system (1) according to Corollary 2.*

4. Design Algorithms

In this section, two solution algorithms are designed for the state feedback controllers.

Case 1: Considering both the state and control constraints, by Theorem 2, Theorem 4, and Corollary 1, the solution to matrix inequalities (9) and (18) is the solution to the feedback controller of the asymptotic stabilization problem for system (1). However, for system (1), the rate of convergence to equilibrium is an important index. The largest ε can ensure the fastest convergence rate to the equilibrium. This can be settled by solving the following LP problem with objective function

$$S(K, F, F_0, L, \varepsilon) = \varepsilon \tag{19}$$

and constraints

$$\begin{cases} Q(A + BK) = FQ, \\ QA_0 = F_0 Q, \\ (F + F_0)\rho \leqslant -\varepsilon\rho, \\ \varepsilon > 0, \\ L\begin{pmatrix} Q \\ Q \end{pmatrix} = \begin{pmatrix} K \\ -K \end{pmatrix}, \\ L\begin{pmatrix} \rho \\ \rho \end{pmatrix} \leqslant \begin{pmatrix} w_2 \\ w_1 \end{pmatrix}. \end{cases} \tag{20}$$

Case 2: Considering only the control constraints, by Theorem 3 and Corollary 2, the solution to matrix inequality (11) is the solution to the feedback controller of the asymptotic stabilization problem for system (1). This can be found by solving the following NP problem with objective function

$$S(K, F, F_0, \varepsilon) = \varepsilon \tag{21}$$

and constraints

$$\begin{cases} K(A + BK) = FK, \\ KA_0 = F_0 K, \\ (\widehat{F} + \widehat{F}_0)\widehat{w} \leqslant -\varepsilon\widehat{w}, \\ \varepsilon > 0. \end{cases} \tag{22}$$

Consider the maximum rate of convergence, from the positive definite function

$$V(x(t)) = \max\left\{ \max\left(\frac{Q_1 x(t)}{\rho_1}, 0 \right), \cdots, \max\left(\frac{Q_q x(t)}{\rho_q}, 0 \right) \right\},$$

we obtain

$$\begin{aligned} {}^C_0 D^\alpha_t V(x(t)) &= \frac{Q_i}{\rho_i} {}^C_0 D^\alpha_t x(t) = \frac{Q_i}{\rho_i}[Ax(t) + A_0 x(t - \tau) + Bu(t)] \\ &= \frac{Q_i}{\rho_i}[(A + BK)x(t) + A_0 x(t - \tau)] \\ &= \frac{1}{\rho_i}[(FQ)_i x(t) + (F_0 Q)_i x(t - \tau)] \\ &= \frac{1}{\rho_i}\left(\sum_{j=1}^{n} \sum_{k=1}^{q} F_{ik} Q_{kj} x_j(t) + \sum_{j=1}^{n} \sum_{k=1}^{q} F_{0ik} Q_{kj} x_j(t) \right) \\ &= \frac{1}{\rho_i}\left(\sum_{k=1}^{q} F_{ik} \sum_{j=1}^{n} Q_{kj} x_j(t) + \sum_{k=1}^{q} F_{0ik} \sum_{j=1}^{n} Q_{kj} x_j(t) \right) \\ &\leqslant \frac{1}{\rho_i}\left(\sum_{k=1}^{q} F_{ik}\rho_k + \sum_{k=1}^{q} F_{0ik}\rho_k \right) \\ &= \frac{1}{\rho_i}[(F + F_0)\rho]_i \\ &\leqslant \frac{1}{\rho_i}(-\varepsilon\rho_i) \leqslant -\varepsilon V(x(t)) < 0. \end{aligned} \tag{23}$$

Therefore, maximizing ε is to maximize the rate of convergence.

Remark 2. *When the system parameters A, A_0, and B and the constraint parameters Q, ρ, and $\hat{w}$ are fixed, from (23) we can see that the parameter ε is closely related to the rate of convergence. If $0 < \varepsilon < 1$, the largest ε can ensure the fastest convergence rate to the equilibrium.*

5. Numerical Examples

Example 1. *Consider the delayed fractional-order electrical circuit [36] shown in Figure 1 with the given resistance $R = 1\,\Omega$, inductance $L = 0.4167\,\text{mH}$, capacitance $C = 1.67\,\text{mF}$, delay element and source voltage u.*

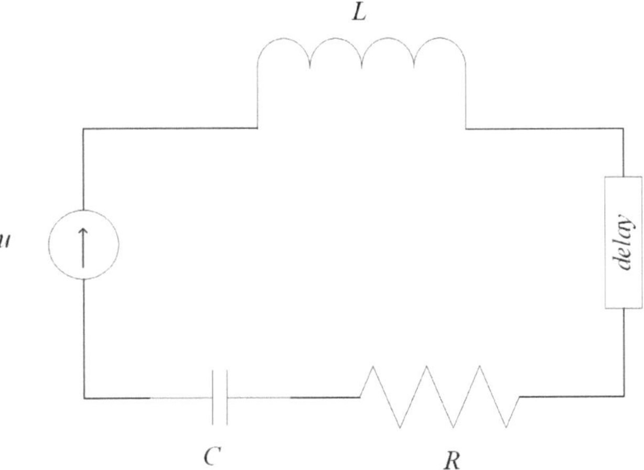

Figure 1. The delayed fractional-order electrical circuit in [36].

Let $v(t)$ represent the voltage across the capacitor and $i(t)$ represent the current passing through the inductor, using Kirchhoff's laws, we can write the circuit equations

$$\begin{bmatrix} \frac{d^{\alpha}v(t)}{dt^{\alpha}} \\ \frac{d^{\alpha}i(t)}{dt^{\alpha}} \end{bmatrix} = A \begin{bmatrix} v(t) \\ i(t) \end{bmatrix} + A_0 \begin{bmatrix} v(t-\tau) \\ i(t-\tau) \end{bmatrix} + Bu(t)$$

where $A = \begin{bmatrix} 0 & \frac{1}{C} \\ -\frac{1}{L} & -\frac{R}{L} \end{bmatrix}$, $A_0 = \begin{bmatrix} 0 & -\frac{1}{R} \\ -\frac{C}{L} & 0 \end{bmatrix}$, $B = \begin{bmatrix} 0 \\ \frac{1}{L} \end{bmatrix}$.

Let $x_1(t) = v(t), x_2(t) = i(t)$. Then, the above circuit equations can be written in the form of the DLFS (1), with $\alpha = 0.7, \tau = 0.1, A = \begin{bmatrix} 0 & 0.6 \\ -2.4 & -2.4 \end{bmatrix}, A_0 = \begin{bmatrix} 0 & -1 \\ -4 & 0 \end{bmatrix}$, and $B = \begin{bmatrix} 0 \\ 2.4 \end{bmatrix}$.

Suppose the control input $P(K, -\underline{w}, \overline{w})$ satisfies the constraint

$$-100 \leqslant Kx(t) \leqslant 50. \tag{24}$$

Without the control input, system (1) is unstable; for the initial condition $(4,1)^T$, the time response of the state is shown in Figure 2.

The asymptotic stabilization problem of DLFS (1) is to find a controller $u(t) = Kx(t)$ that makes all trajectories starting with the initial condition x_0 asymptotically stable and meanwhile satisfy the control constraints (24).

According to Case 2, solving the NP problem with the objective function

$$S(K, F, F_0, \varepsilon) = \varepsilon$$

and under the constraints

$$\begin{cases} K\left(\begin{bmatrix} 0 & 0.6 \\ -2.4 & -2.4 \end{bmatrix} + \begin{bmatrix} 0 \\ 2.4 \end{bmatrix} K \right) = FK \\ K \begin{bmatrix} 0 & -1 \\ -4 & 0 \end{bmatrix} = F_0 K \\ \left(\begin{bmatrix} F^+ & F^- \\ F^- & F^+ \end{bmatrix} + \begin{bmatrix} F_0^+ & F_0^- \\ F_0^- & F_0^+ \end{bmatrix} \right) \begin{bmatrix} 50 \\ 100 \end{bmatrix} \leqslant -\varepsilon \begin{bmatrix} 50 \\ 100 \end{bmatrix} \\ \varepsilon > 0. \end{cases}$$

By calculation, we obtain $\varepsilon = 0.8$, $K = [2 \quad 1]$, $F = 1.2$, and $F_0 = -2$.

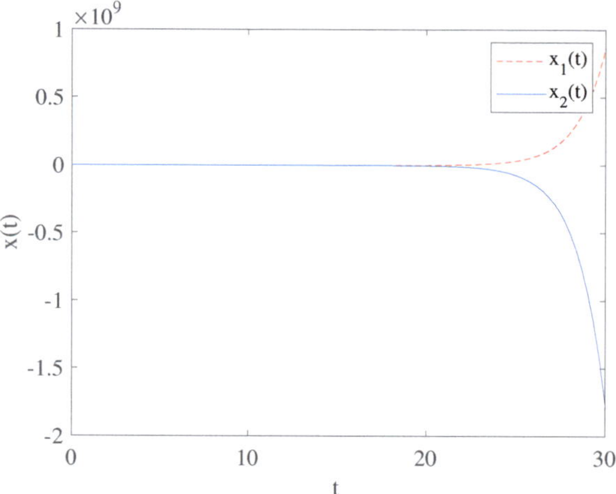

Figure 2. The time response of the system state without feedback control.

It indicates that there exists a $u(t) = [2 \quad 1]x(t)$ such that all trajectories originating from the initial condition are asymptotically stable to the origin, and meanwhile the corresponding trajectory satisfies the control constraints (24).

For the initial condition $(4, 1)^T$, the time response of the state is shown in Figure 3.

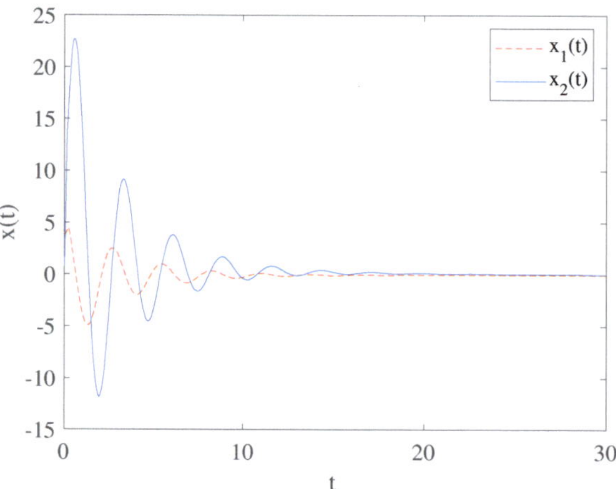

Figure 3. The time response of the system state with feedback control.

Additionally, $u(t)$ satisfies the control constraints (24), which can be seen from Figure 4.

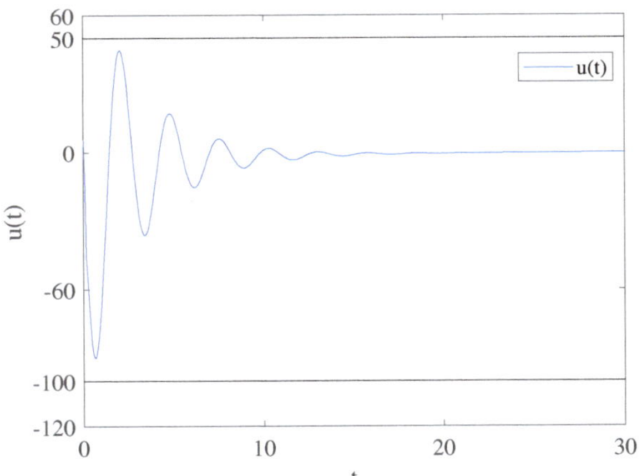

Figure 4. The control input $u(t)$.

So, $u(t) = [2 \quad 1]x(t)$ is a solution to the asymptotic stabilization problem of DLFS (1).

Example 2. *Consider another delayed fractional-order electrical circuits, which can be written in the form of the DLFS (1), with $\alpha = 0.7, \tau = 0.5, A = \begin{bmatrix} -3.25 & 10 \\ -1 & 7.1 \end{bmatrix}, A_0 = \begin{bmatrix} -1 & -2.2 \\ 0 & -2.1 \end{bmatrix}$, and $B = \begin{bmatrix} 1 \\ 0.8 \end{bmatrix}$.*

The state constraints $P(Q, \rho)$ are

$$
\begin{aligned}
2x_2(t) &\leqslant 0.8, \\
-x_1(t) + 2x_2(t) &\leqslant 2.4, \\
-0.5x_1(t) + 4x_2(t) &\geqslant -1.8,
\end{aligned}
\tag{25}
$$

where $Q = \begin{bmatrix} 0 & 2 \\ -1 & 2 \\ 0.5 & -4 \end{bmatrix} \in R^{3\times 2}$ and $\rho = \begin{bmatrix} 0.8 \\ 2.4 \\ 1.8 \end{bmatrix} \in R^3, \rho > 0$. It can be easily seen that the polyhedron set $P(Q, \rho)$ is closed and nonempty.

Suppose the control input $P(K, -\underline{w}, \overline{w})$ satisfies the constraint

$$
-6 \leqslant Kx(t) \leqslant 9.
\tag{26}
$$

Without the control input, system (1) is unstable, the time response of the state with the initial condition $(6.8, 0.4)^T$ is shown in Figure 5.

The asymptotic stabilization problem of DLFS (1) is to find a controller $u(t) = Kx(t)$ that makes all trajectories starting from the initial value x_0 asymptotically stable and meanwhile satisfy the state constraints (25) and the control constraints (26).

According to Case 1, solving the LP problem with the objective function

$$
S(K, F, F_0, L, \varepsilon) = \varepsilon
$$

and under the constraints

$$\begin{cases}
\begin{bmatrix} 0 & 2 \\ -1 & 2 \\ 0.5 & -4 \end{bmatrix} \left(\begin{bmatrix} -3.25 & 10 \\ -1 & 7.1 \end{bmatrix} + \begin{bmatrix} 1 \\ 0.8 \end{bmatrix} K \right) = F \begin{bmatrix} 0 & 2 \\ -1 & 2 \\ 0.5 & -4 \end{bmatrix}, \\[16pt]
\begin{bmatrix} 0 & 2 \\ -1 & 2 \\ 0.5 & -4 \end{bmatrix} \begin{bmatrix} -1 & -2.2 \\ 0 & -2.1 \end{bmatrix} = F_0 \begin{bmatrix} 0 & 2 \\ -1 & 2 \\ 0.5 & -4 \end{bmatrix}, \\[16pt]
(F + F_0) \begin{bmatrix} 0.8 \\ 2.4 \\ 1.8 \end{bmatrix} \leqslant -\varepsilon \begin{bmatrix} 0.8 \\ 2.4 \\ 1.8 \end{bmatrix}, \\[16pt]
L \begin{bmatrix} 0 & -1 & 0.5 & 0 & -1 & 0.5 \\ 2 & 2 & -4 & 2 & 2 & -4 \end{bmatrix}^T = \begin{bmatrix} K \\ -K \end{bmatrix}, \\[12pt]
L \begin{bmatrix} 0.8 & 2.4 & 1.8 & 0.8 & 2.4 & 1.8 \end{bmatrix}^T \leqslant \begin{bmatrix} 9 \\ 6 \end{bmatrix}, \\[12pt]
\varepsilon > 0.
\end{cases}$$

By computation, we obtain

$$\varepsilon = 0.9,$$

$$K = \begin{bmatrix} 1.25 & -10 \end{bmatrix},$$

$$F = \begin{bmatrix} -0.9 & 0 & 0 \\ 1.1 & -2 & 0 \\ 0 & 0.73 & -0.53 \end{bmatrix},$$

$$F_0 = \begin{bmatrix} -2.1 & 0 & 0 \\ 0 & -1 & 0 \\ 0 & -0.55 & -2.1 \end{bmatrix},$$

and

$$L = \begin{bmatrix} 0 & 0 & 0 & 0 & 0 & 2.5 \\ 0 & 0 & 0 & 3.75 & 1.25 & 0 \end{bmatrix}.$$

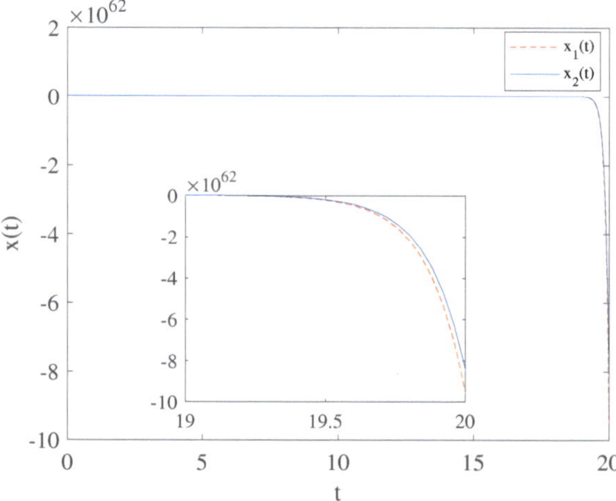

Figure 5. The time response of the system state without feedback control.

It suggests that there exists $u(t) = 1.25x_1(t) - 10x_2(t)$ such that all trajectories starting from the initial condition are asymptotically stable to the origin, and meanwhile the corresponding state trajectory satisfies the state constraints (25) and the control constraints (26).

For the initial condition $(6.8, 0.4)^T$, the time response of the state is shown in Figure 6.

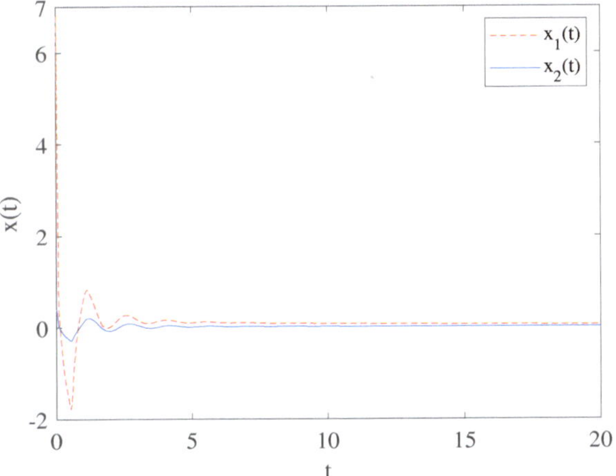

Figure 6. The time response of the system state with feedback control.

For the initial condition $(6.8, 0.4)^T$*, the corresponding phase trajectory which satisfies the state constraints (25) and the control constraints (26), is shown in Figure 7.*

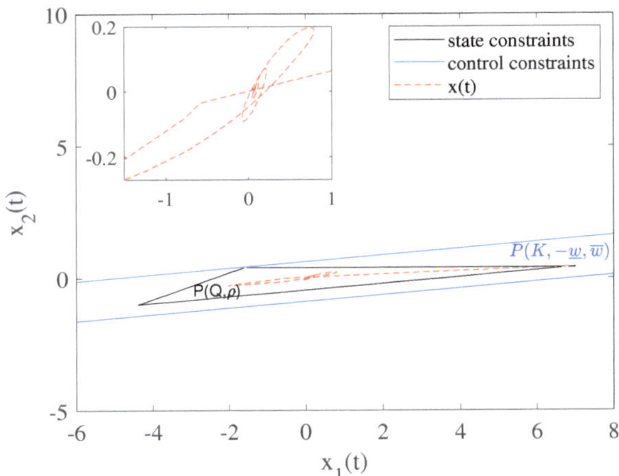

Figure 7. The phase trajectory with the initial condition $(6.8, 0.4)^T$.

Furthermore, $u(t)$ *satisfies the control constraints (26), which can be seen in Figure 8.*

Hence, $u(t) = 1.25x_1(t) - 10x_2(t)$ *is a solution to the asymptotic stabilization problem of DLFS (1).*

Clearly, from Example 2, it can be seen that the control input is not saturated because the state constraint set is a PIS. Next, we give an example without state constraints to verify the saturation of the input control.

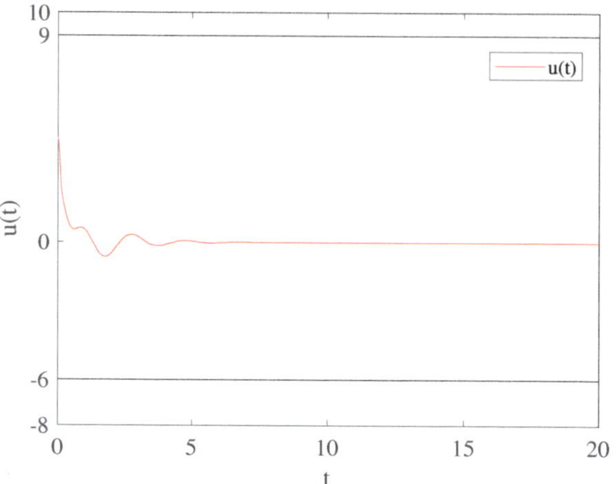

Figure 8. The control input $u(t)$.

Example 3. *Consider the DLFS (1), with* $\alpha = 0.8, \tau = 0.8, A = \begin{bmatrix} 1 & -3 \\ -1 & 5 \end{bmatrix}, A_0 = \begin{bmatrix} -1 & 0 \\ 0 & -2 \end{bmatrix},$
and $B = \begin{bmatrix} 1 & 1 \\ -1 & 0.5 \end{bmatrix}.$

Suppose the control input $P(K, -w_1, w_2)$ *satisfies the constraint*

$$\begin{bmatrix} -1 \\ -2 \end{bmatrix} \leqslant Kx(t) \leqslant \begin{bmatrix} 2 \\ 4 \end{bmatrix}. \tag{27}$$

Without the control input, system (1) is unstable; for the initial condition $(-1.84, -0.64)^T$,
the time response of the state is shown in Figure 9.

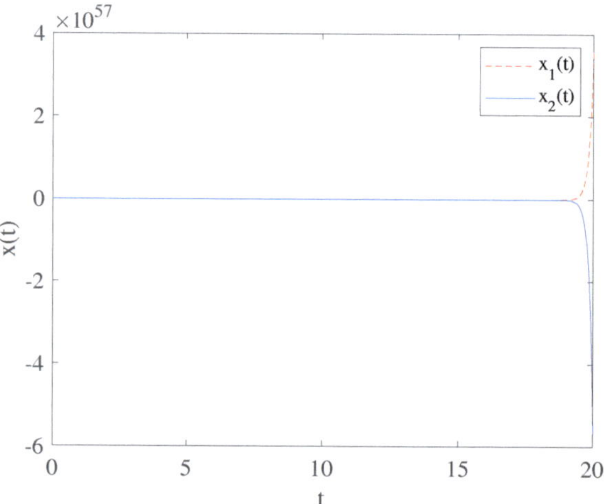

Figure 9. The time response of the system state without feedback control.

The asymptotic stabilization problem of DLFS (1) is to find a controller $u(t) = Kx(t)$ *that makes all trajectories starting with the initial condition* x_0 *asymptotically stable and meanwhile satisfy the control constraints (27).*

According to Case 2, solving the NP problem with the objective function

$$S(K, F, F_0, \varepsilon) = \varepsilon$$

and under the constraints

$$\begin{cases} K\left(\begin{bmatrix} 1 & -3 \\ -1 & 5 \end{bmatrix} + \begin{bmatrix} 1 & 1 \\ -1 & 0.5 \end{bmatrix} K\right) = FK \\[2mm] K\begin{bmatrix} -1 & 0 \\ 0 & -2 \end{bmatrix} = F_0 K \\[2mm] \left(\begin{bmatrix} F^+ & F^- \\ F^- & F^+ \end{bmatrix} + \begin{bmatrix} F_0^+ & F_0^- \\ F_0^- & F_0^+ \end{bmatrix}\right) \begin{bmatrix} 2 \\ 4 \\ 1 \\ 2 \end{bmatrix} \leqslant -\varepsilon \begin{bmatrix} 2 \\ 4 \\ 1 \\ 2 \end{bmatrix} \\[2mm] \varepsilon > 0. \end{cases}$$

By calculation, we obtain

$$\varepsilon = 0.8908,$$

$$K = \begin{bmatrix} -2.051 & 7.444 \\ -1.255 & -2.629 \end{bmatrix},$$

$$F = \begin{bmatrix} -4.11 & 0.43 \\ 0.33 & -1.96 \end{bmatrix}$$

and

$$F_0 = \begin{bmatrix} -1.634 & 1.0362 \\ 0.2239 & -1.3660 \end{bmatrix}.$$

It indicates that there exists a $u(t) = \begin{bmatrix} -2.051 & 7.444 \\ -1.255 & -2.629 \end{bmatrix} x(t)$ such that all trajectories originating from the initial condition are asymptotically stable to the origin, and meanwhile the corresponding trajectory satisfies the control constraints (27).

For the initial condition $(-1.84, -0.64)^T$, the time response of the state is shown in Figure 10.

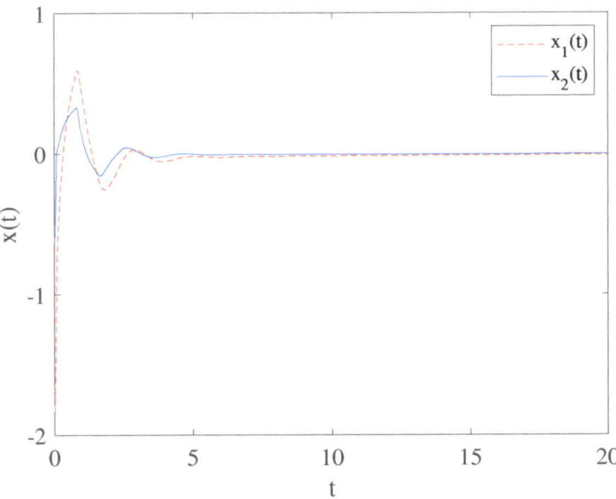

Figure 10. The time response of the system state with feedback control.

For the initial condition $(-1.84, -0.64)^T$, the corresponding phase trajectory is shown in Figure 11. It can be seen that the control constraints (27) is satisfied.

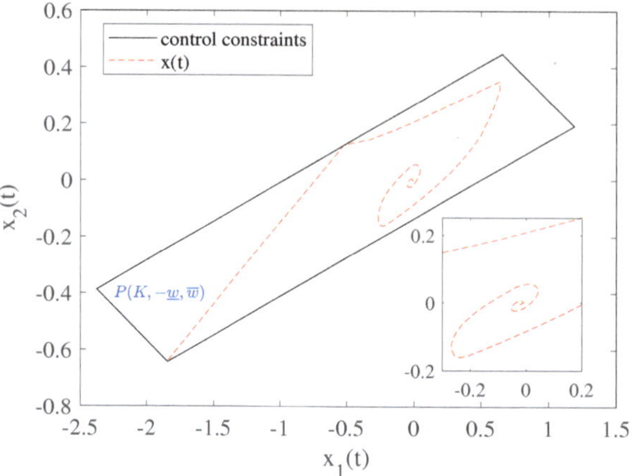

Figure 11. The phase trajectory with the initial condition $(-1.84, -0.64)^T$.

In addition, $u(t)$ satisfies the control constraints (27), which can be seen from Figure 12.

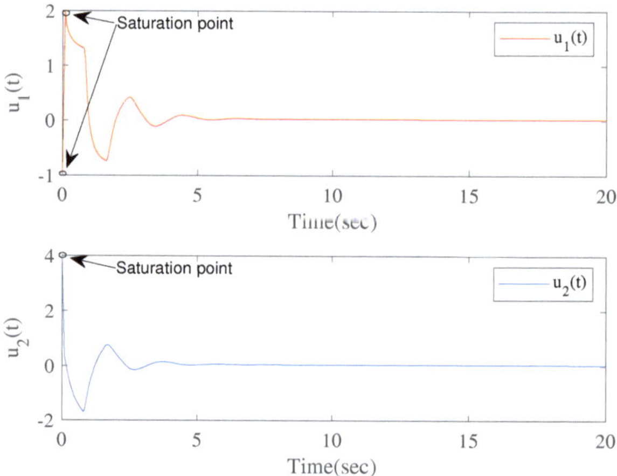

Figure 12. The control input $u(t)$.

So, $u(t) = \begin{bmatrix} -2.051 & 7.444 \\ -1.255 & -2.629 \end{bmatrix} x(t)$ *is a solution to the asymptotic stabilization problem of* DLFS (1).

6. Conclusions

The asymptotic stabilization problem of DLFS subject to state and control constraints is studied. Based on the invariant set theory and fractional-order Lyapunov stability theory, the feedback controller and the corresponding solution algorithms that ensure the asymptotic stability of the DLFS under constraints are given. Numerical examples show that the proposed method is effective. Stability of delayed fractional-order neural network systems with constraints are a very interesting topic. In the future, we will study the stabilization of delayed fractional-order neural network systems subject to constraints by using the invariant set method.

Author Contributions: Conceptualization, X.S. and Z.W.; methodology, X.S. and Z.W.; software, X.S.; validation, X.S., Z.W., Z.S. and Z.Z.; investigation, Z.W.; writing—original draft preparation, X.S.; writing—review and editing, Z.W., Z.S. and Z.Z.; supervision, Z.W.; funding acquisition, Z.W. All authors have read and agreed to the published version of the manuscript.

Funding: This work was supported by the National Natural Science Foundation of China under Grants 61973199, 62003794, 62173214, the Shandong Provincial Natural Science Foundation ZR2020QF050, ZR2021MF003 and the Taishan Scholar Project of Shandong Province of China.

Institutional Review Board Statement: Not applicable.

Informed Consent Statement: Not applicable.

Data Availability Statement: Data sharing not applicable to this article as no datasets were generated or analysed during the current study.

Acknowledgments: We would like to express our great appreciation to the editors and reviewers.

Conflicts of Interest: The authors declare no conflicts of interest.

Abbreviations

The following abbreviations are used in this manuscript:

DLFS	delayed linear fractional-order systems
PIS	positive invariant sets
LP	linear programming
NP	nonlinear programming
LMI	linear matrix inequality

References

1. Chen, L.; Chen, Y.; Lopes, A.M.; Kong, H.; Wu, R. State of charge estimation of lithium-ion batteries based on fuzzy fractional-order unscented kalman filter. *Fractal Fract.* **2021**, *5*, 91.
2. Tian, Y.; Xia, Q.; Chai, Y.; Chen, L.; Lopes, A.M.; Chen, Y. Guaranteed cost leaderless consensus protocol design for fractional-order uncertain multi-agent systems with state and input delays. *Fractal Fract.* **2021**, *5*, 141.
3. Gonzalez-Acuna, R.G.; Davila, A.; Gutierrez-Vega, J.C. Optical flow of non-integer order in particle image velocimetry techniques. *Signal Process.* **2019**, *155*, 317–322. [CrossRef]
4. Lashkarian, E.; Hejazi, S.R.; Habibi, N.; Motamednezhad, A. Symmetry properties, conservation laws, reduction and numerical approximations of time-fractional cylindrical-burgers equation. *Commun. Nonlinear Sci. Numer. Simul.* **2019**, *67*, 176–191. [CrossRef]
5. Yin, C.; Dadras, S.; Cheng, Y.H.; Huang, X.G.; Cao, J.W.; Malek, H. Multidimensional fractional-order newton-based extremum seeking for online light-energy saving technique of lighting system. *IEEE Trans. Ind. Electron.* **2019**, *67*, 8576–8586. [CrossRef]
6. Mohsenipour, R.; Jegarkandi, M.F. Robust stability analysis of fractional-order interval systems with multiple time delays. *Int. J. Robust Nonlinear Control* **2019**, *29*, 1823–1839. [CrossRef]
7. Sakthivel, R.; Mohanapriya, S.; Ahn, C.K.; Karimi, H.R. Output tracking control for fractional-order positive switched systems with input time delay. *IEEE Trans. Circuits Syst. II-Express Briefs* **2019**, *66*, 1013–1017. [CrossRef]
8. Jia, J.; Huang, X.; Li, Y.; Cao, J. Global stabilization of fractional-order memristor-based neural networks with time delay. *IEEE Trans. Neural Netw. Learn. Syst.* **2020**, *31*, 997–1009. [CrossRef]
9. Wang, Z.; Wang, X.; Xia, J.; Shen, H.; Meng, B. Adaptive sliding mode output tracking control based-FODOB for a class of uncertain fractional-order nonlinear time-delayed systems. *Sci. China-Technol. Sci.* **2020**, *63*, 1854–1862. [CrossRef]
10. Liu, H.; Pan, Y.; Cao, J.; Zhou, Y.; Wang, H. Positivity and stability analysis for fractional-order delayed systems: A T-S fuzzy model approach. *IEEE Trans. Fuzzy Syst.* **2020**, *29*, 927–939. [CrossRef]
11. Du, F.; Lu, J. Finite-time stability of neutral fractional order time delay systems with Lipschitz nonlinearities. *Appl. Math. Comput.* **2020**, *375*, 125079. [CrossRef]
12. Naifar, O.; Nagy, A.M.; Makhlouf, A.B.; Kharrat, M. Finite-time stability of linear fractional-order time-delay systems. *Int. J. Robust Nonlinear Control* **2018**, *29*, 180–187. [CrossRef]
13. Liu, P.; Zeng, Z.; Wang, J. Asymptotic and finite-time cluster synchronization of coupled fractional-order neural networks with time delay. *IEEE Trans. Neural Netw. Learn. Syst.* **2020**, *31*, 4956–4967. [CrossRef]
14. Thanh, N.T.; Phat, V.N.; Niamsup, P. New finite-time stability analysis of singular fractional differential equations with time-varying delay. *Fract. Calc. Appl. Anal.* **2020**, *23*, 504–519. [CrossRef]
15. Deng, W.; Li, C.; Lu, J. Stability analysis of linear fractional differential system with multiple time delays. *Nonlinear Dyn.* **2007**, *48*, 409–416. [CrossRef] [PubMed]

16. You, X.; Song, Q.; Zhao, Z. Global Mittag-Leffler stability and synchronization of discrete-time fractional-order complex-valued neural networks with time delays. *Neural Netw.* **2020**, *122*, 382–394. [CrossRef]

17. Hu, T.; He, Z.; Zhang, X.; Zhong, S. Finite-time stability for fractional-order complex-valued neural networks with time delay. *Appl. Math. Comput.* **2020**, *365*, 124715. [CrossRef]

18. Bonnet, C.; Partington, J.R. Stabilization of some fractional delay systems of neutral type. *Automatica* **2007**, *43*, 2047–2053. [CrossRef] [PubMed]

19. He, B.; Zhou, H.; Kou, C.; Chen, Y.Q. New integral inequalities and asymptotic stability of fractional-order systems with unbounded time delay. *Nonlinear Dyn.* **2018**, *94*, 1523–1534. [CrossRef]

20. Chen, L.; Wu, R.; Cheng, Y.; Chen, Y.Q. Delay-dependent and order-dependent stability and stabilization of fractional-order linear systems with time-varying delay. *IEEE Trans. Circuits Syst. II-Express Briefs* **2020**, *67*, 1064–1068. [CrossRef]

21. Tavazoei, M.S.; Badri, V. Stability analysis of fractional order time-delay systems: Constructing new lyapunov functions from those of integer order counterparts. *IET Control Theory Appl.* **2019**, *13*, 2476–2481. [CrossRef]

22. Nie, X.; Liu, P.; Liang, J.; Cao, J. Exact coexistence and locally asymptotic stability of multiple equilibria for fractional-order delayed Hopfield neural networks with Gaussian activation function. *Neural Netw.* **2021**, *142*, 690–700. [CrossRef]

23. Hennet, J.C.; Beziat, J.P. A class of invariant regulators for the discrete-time linear constrained regulation problem. *Automatica* **1991**, *27*, 549–554.

24. Vassilaki, M.; Bitsoris, G. Constrained regulation of linear continuous-time dynamical systems. *Syst. Control Lett.* **1989**, *13*, 247–252. [CrossRef] [PubMed]

25. Bitsoris, G.; Olaru, S. Further results on the linear constrained regulation problem. In Proceedings of the 21st Mediterranean Conference on Control and Automation, Crete, Greece, 25–28 June 2013; pp. 824–830. [CrossRef]

26. Bitsoris, G.; Olaru, S.; Vassilaki, M. On the linear constrained regulation problem for continuous-time systems. In Proceedings of the 19th World Congress International Federation of Automatic Control, Cape Town, South Africa, 24–29 August 2014; pp. 4004–4009. [CrossRef]

27. Castelan, E.B. On invariant polyhedra of continuous-time linear systems. *IEEE Trans. Autom. Control* **1991**, *38*, 1680–1685.

28. Blanchini, F. Set invariance in control. *Automatica* **1999**, *35*, 1747–1767.

29. Jia, J.; Wang, F.; Zeng, Z. Global stabilization of fractional-order memristor-based neural networks with incommensurate orders and multiple time-varying delays: A positive-system-based approach. *Nonlinear Dyn.* **2021**, *104*, 2303–2329. [CrossRef]

30. Zhang, X.; Zhao, Z. Normalization and stabilization for rectangular singular fractional order T-S fuzzy systems. *Fuzzy Sets Syst.* **2019**, *381*, 140–153. [CrossRef]

31. Zhang, Z.; Zhang, J. Asymptotic stabilization of general nonlinear fractional-order systems with multiple time delays. *Nonlinear Dyn.* **2020**, *102*, 605–619. [CrossRef]

32. Lim, Y.H.; Ahn, H.S. On the positive invariance of polyhedral sets in fractional-order linear systems. *Automatica* **2013**, *49*, 3690–3694. [CrossRef]

33. Yan, L.; Chen, Y.Q.; Podlubny, I. Stability of fractional-order nonlinear dynamic systems: Lyapunov direct method and generalized Mittag-Leffler stability. *Comput. Math. Appl.* **2010**, *59*, 1810–1821.

34. Wen, Y.; Zhou, X.F.; Zhang, Z.; Liu, S. Lyapunov method for nonlinear fractional differential systems with delay. *Nonlinear Dyn.* **2015**, *82*, 1015–1025. [CrossRef]

35. Athanasopoulos, N.; Bitsoris, G.; Vassilaki, M. Stabilization of bilinear continuous-time systems. In Proceedings of the 18th Mediterranean Conference on Control and Automation, MED'10, Marrakech, Morocco, 23–25 June 2010; pp. 442–447.

36. Kaczorek, T.; Rogowski, K. *Fractional Linear Systems and Electrical Circuits*; Springer: New York, NY, USA, 2014; p. 65.

fractal and fractal

Article

Chaos Control for a Fractional-Order Jerk System via Time Delay Feedback Controller and Mixed Controller

Changjin Xu [1,2,*], Maoxin Liao [3], Peiluan Li [4], Lingyun Yao [5], Qiwen Qin [6] and Youlin Shang [4]

[1] Guizhou Key Laboratory of Economics System Simulation, Guizhou University of Finance and Economics, Guiyang 550025, China
[2] Guizhou Key Laboratory of Big Data Statistical Analysis, Guiyang 550025, China
[3] School of Mathematics and Physics, University of South China, Hengyang 421001, China; maoxinliao@163.com
[4] School of Mathematics and Statistics, Henan University of Science and Technology, Luoyang 471023, China; lpllpl_lpl@163.com (P.L.); mathshang@sina.com (Y.S.)
[5] Library, Guizhou University of Finance and Economics, Guiyang 550025, China; lingyunyao2015@126.com
[6] School of Economics, Guizhou University of Finance and Economics, Guiyang 550025, China; hnqinqiwen@163.com
* Correspondence: xcj403@126.com

Abstract: In this study, we propose a novel fractional-order Jerk system. Experiments show that, under some suitable parameters, the fractional-order Jerk system displays a chaotic phenomenon. In order to suppress the chaotic behavior of the fractional-order Jerk system, we design two control strategies. Firstly, we design an appropriate time delay feedback controller to suppress the chaos of the fractional-order Jerk system. The delay-independent stability and bifurcation conditions are established. Secondly, we design a suitable mixed controller, which includes a time delay feedback controller and a fractional-order PD^{σ} controller, to eliminate the chaos of the fractional-order Jerk system. The sufficient condition ensuring the stability and the creation of Hopf bifurcation for the fractional-order controlled Jerk system is derived. Finally, computer simulations are executed to verify the feasibility of the designed controllers. The derived results of this study are absolutely new and possess potential application value in controlling chaos in physics. Moreover, the research approach also enriches the chaos control theory of fractional-order dynamical system.

Keywords: fractional-order Jerk system; chaos; hopf bifurcation; stability; time delay feedback controller; fractional-order PD^{σ} controller

Citation: Xu, C.; Liao, M.; Li, P.; Yao, L.; Qin, Q.; Shang, Y. Chaos Control for a Fractional-Order Jerk System via Time Delay Feedback Controller and Mixed Controller. *Fractal Fract.* **2021**, *5*, 257. https://doi.org/10.3390/fractalfract5040257

Academic Editors: António M. Lopes and Liping Chen

Received: 4 November 2021
Accepted: 29 November 2021
Published: 5 December 2021

Publisher's Note: MDPI stays neutral with regard to jurisdictional claims in published maps and institutional affiliations.

1. Introduction

In the natural world, a great deal of natural phenomena display chaotic behavior. Usually, there exist chaotic phenomena in various areas such as weather, climate, economy and finance, neural networks, biological systems, fluid mechanics and so on [1–5]. The chaotic behavior sensitively depends on the initial value of the original system. That is to say, the dynamic behavior of the system will change greatly when the initial value of the system changes slightly. This kind of complex dynamical behavior may be undesirable in numerous physical sciences, biological techniques and engineering technology, since we cannot predict the long-term development law of these systems. Based on this reason, it is important for us to seek control techniques to suppress the chaotic behavior and make the system generate our desired dynamical properties. This aspect has become a problem of focus in recent years [6]. Designing valid control mechanisms to realize the target of chaos control is very vital for both theoretical study and actual applications [7,8]. For so long, there have been many techniques to suppress the chaos of the chaotic systems. For example, Zheng [8] designed an adaptive feedback controller to control the chaotic behavior of a chaotic system. Ott, Grebogi and Yorke [9] proposed an OGY control technique to suppress the chaos of a chaotic system in 1990. Li et al. [10] controlled the chaos of a brushless DC motor via a nonlinear state feedback controller. Du et al. [11] suppressed the chaos of an economic model by virtue of phase space compression. For more related works on this theme, one can see [12–15].

Physically speaking, Jerk can be regarded as the third derivative of position with regard to the time t [16]. Usually, it can be expressed as:

$$\frac{d^3 w_1(t)}{dt} = \mathcal{J}\left(w_1, \frac{dw_1(t)}{dt}, \frac{d^2 w_1(t)}{dt}\right),$$ (1)

which is called the Jerk equation. Set

$$\begin{cases} w_2(t) = \dfrac{dw_1(t)}{dt}, \\ w_3(t) = \dfrac{d^2 w_1(t)}{dt}, \end{cases}$$ (2)

then system (1) can be rewritten as the following form:

$$\begin{cases} \dfrac{dw_1(t)}{dt} = w_2(t), \\ \dfrac{dw_2(t)}{dt} = w_3(t), \\ \dfrac{dw_3(t)}{dt} = \mathcal{J}(w_1, w_2, w_3). \end{cases}$$ (3)

In 2021, Liu et al. [16] established the following Jerk system:

$$\begin{cases} \dfrac{dw_1(t)}{dt} = w_2(t), \\ \dfrac{dw_2(t)}{dt} = w_3(t), \\ \dfrac{dw_3(t)}{dt} = -\alpha_1 w_1(t) - \alpha_2 w_2(t) - \alpha_3 w_3(t) + \alpha_4 w_3^2 + \alpha_5 w_1(t) w_2(t), \end{cases}$$ (4)

where $\alpha_i (i = 1, 2, 3, 4, 5)$ represents the real number. By virtue of Hopf bifurcation theory, Lyapunov exponents and bifurcation figures, Liu et al. [16] explored the chaotic dynamics for the model (3). For details, one can see [16].

We would like to mention that the work of Liu et al. [16] merely focused on an integer-order differential system and it does not involve the fractional-order case. Recent research has demonstrated that the fractional-order dynamical model is regarded as a more rational tool for depicting the authentic natural phenomena in the object world since it has greater superiority than integer-order ones. The advantage of the fractional-order dynamical model lies in the memory trait and hereditary peculiarity of plentiful materials and evolution processes [17–20]. The fractional-order dynamical system displays an immense application prospect in many subjects in artificial intelligence, neural networks, mechanics, economics, all sorts of physical waves, bioscience and so on [21–25]. Nowadays, a number of fractional-order differential systems have been built and rich achievements have been reported. The exploration of Hopf bifurcation of fractional-order differential systems has especially attracted great attention from many scholars. For instance, Djilali et al. [26] probed the Turing–Hopf bifurcation for a fractional-order mussel-algae model; Xiao et al. [27] handled the Hopf bifurcation control issue for fractional-order small-world networks; Xu et al. [28] revealed the effect of leakage delay on bifurcation for fractional-order complex-valued neural networks; Huang et al. [29] discussed the Hopf bifurcation for fractional-order multi-delayed neural networks. In detail, we refer readers to [30–44].

Stimulated by the discussion above and based on system (4), we establish the following fractional-order Jerk system:

$$
\begin{cases}
\dfrac{d^{\sigma} w_1(t)}{dt^{\sigma}} = w_2(t), \\[2mm]
\dfrac{d^{\sigma} w_2(t)}{dt^{\sigma}} = w_3(t), \\[2mm]
\dfrac{d^{\sigma} w_3(t)}{dt^{\sigma}} = -\alpha_1 w_1(t) - \alpha_2 w_2(t) - \alpha_3 w_3(t) + \alpha_4 w_3^2 + \alpha_5 w_1(t) w_2(t),
\end{cases}
\tag{5}
$$

where $\sigma \in (0, 1]$. The study shows that the fractional-order Jerk system (5) will display chaotic behavior when $\sigma = 0.94, \alpha_1 = 2, \alpha_2 = 1, \alpha_3 = 1.2, \alpha_4 = 0.5, \alpha_5 = 0.9$. The simulation results can be seen in Figure 1.

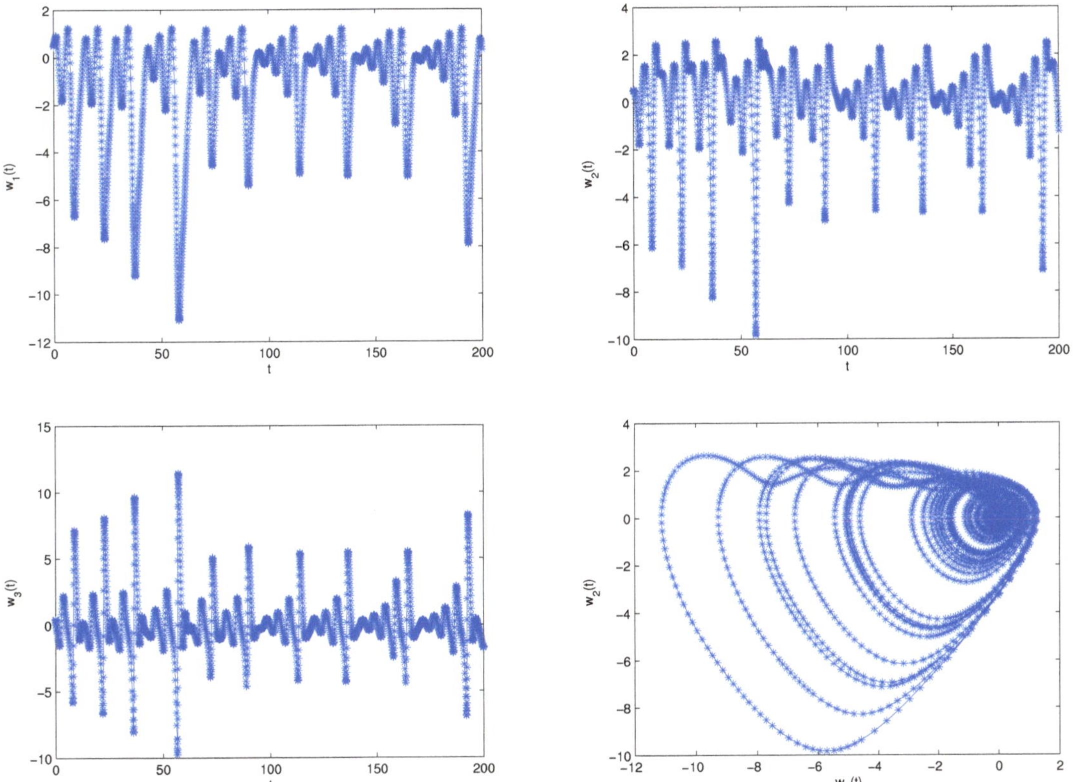

Figure 1. *Cont.*

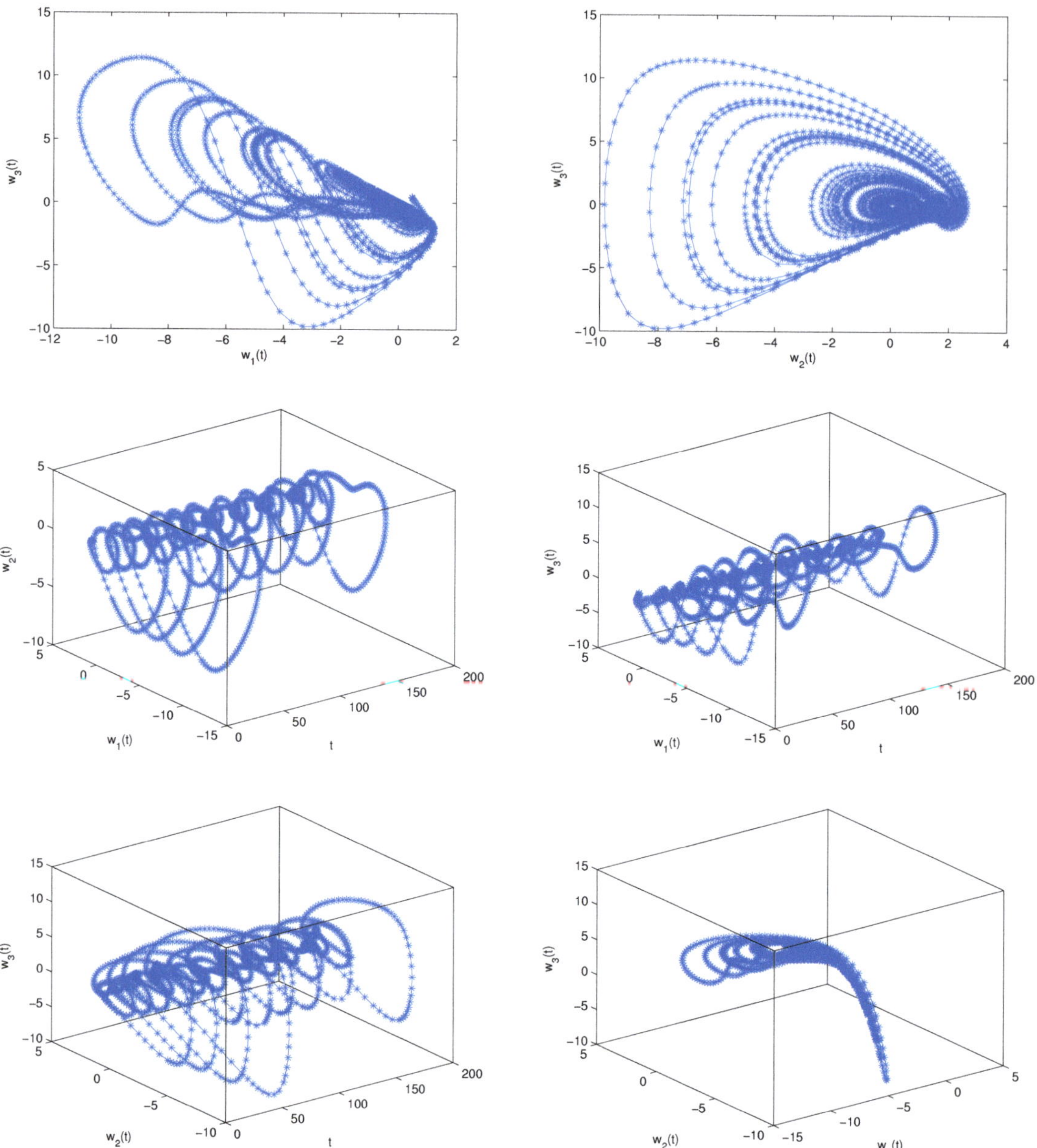

Figure 1. The numerical simulation plots of system (5) under the parameter values $\sigma = 0.94, \alpha_1 = 2, \alpha_2 = 1, \alpha_3 = 1.2,$ $\alpha_4 = 0.5, \alpha_5 = 0.9.$

In the present study, we are going to focus on the following two aspects:

1. Control the chaotic behavior of system (5) via designing a suitable time delay feedback controller;
2. Control the chaotic behavior of system (5) via designing an appropriate mixed controller which includes time delay feedback controller and fractional-order PD^{σ} con-

troller. Up to now, there have been very few papers that deal with the chaos control via this mixed controller.

The main highlights of this research can be summarized as follows:

- Based on the previous publications, we build a novel Jerk system.
- A suitable time delay feedback controller is successfully designed to suppress the chaotic behavior of Jerk system (5);
- A suitable mixed controller which includes time delay feedback controller and fractional-order PD^σ controller is successfully designed to suppress the chaotic behavior of the Jerk system (5);
- The research idea can also be applied to deal with the chaos control issue for numerous other fractional-order differential systems in many areas.

This work can be planned as follows. Basic knowledge about fractional-order dynamical systems is presented in Section 2. In Section 3, we investigate the chaos control of system (5) via a time delay feedback controller. In Section 4, we discuss the chaos control of system (5) via designing a mixed controller, which includes a time delay feedback controller and fractional-order PD^σ controller. In Section 5, Matlab simulation figures are given to support the derived results. Section 6 ends this study.

Remark 1. *The fractional-order system (5) is derived from (4) by replacing the integer-order derivatives with fractional orders. System (5) can describe the memory trait and hereditary peculiarity of the state variables more precisely.*

2. Preliminary Theory

In this segment, we state the necessary definitions and lemmas about the fractional-order dynamical system.

Definition 1 ([35]). *Define the Riemann–Liouville fractional integral of order σ for the function $g(\epsilon)$ as follows:*

$$\mathcal{I}^\sigma g(\epsilon) = \frac{1}{\Gamma(\sigma)} \int_{\epsilon_0}^{\epsilon} (\epsilon - \varsigma)^{\sigma-1} g(\chi) d\varsigma,$$

where $\epsilon > \epsilon_0, \sigma > 0$ and $\Gamma(\varsigma) = \int_0^\infty s^{\varsigma-1} e^{-s} ds.$

Definition 2 ([35]). *The Caputo-tpye fractional-order derivative of order σ for the function $g(\varsigma) \in ([\varsigma_0, \infty), R)$ is given by:*

$$\mathcal{D}^\sigma g(\varsigma) = \frac{1}{\Gamma(\iota - \sigma)} \int_{\varsigma_0}^{\varsigma} \frac{g^{(\iota)}(s)}{(r - s)^{\sigma-\iota+1}} ds,$$

where $\varsigma \geq \varsigma_0$ and ι stands for a positive integer ($\sigma \in [\iota - 1, \iota)$). Especially, when $\sigma \in (0, 1)$, then

$$\mathcal{D}^\sigma g(\varsigma) = \frac{1}{\Gamma(1 - \sigma)} \int_{\varsigma_0}^{\varsigma} \frac{g'(s)}{(\varsigma - s)^\sigma} ds.$$

Lemma 1 ([36]). *Give the fractional-order model: $\mathcal{D}^\sigma u = \mathcal{P}u, u(0) = u_0$, where $\sigma \in (0, 1)$, $u \in R^m, \mathcal{H} \in R^{m \times m}$. Let $\mu_j (j = 1, 2, \cdots, m)$ be the root of the characteristic equation of $\mathcal{D}^\sigma u = \mathcal{P}u$, then the equilibrium point of system $\mathcal{D}^\sigma u = \mathcal{P}u$ is locally asymptotically stable provided that $|arg(\mu_j)| > \frac{\sigma\pi}{2}(j = 1, 2, \cdots, m)$ and the equilibrium point of system $\mathcal{D}^\sigma u = \mathcal{P}u$ is stable, provided that $|arg(\mu_j)| > \frac{\sigma\pi}{2}(j = 1, 2, \cdots, m)$ and each critical eigenvalue satisfying $|arg(\mu_j)| = \frac{\sigma\pi}{2}(j = 1, 2, \cdots, m)$ has geometric multiplicity one.*

3. Chaos Control via Time Delay Feedback Controller

In this segment, we shall design a suitable controller to suppress the chaotic behavior of the chaotic Jerk system. Following the idea of Yu and Chen [37], we design the time delay feedback controller as follows:

$$\zeta(t) = \varrho_1[w_3(t - \vartheta) - w_3(t)], \tag{6}$$

where ϱ_1 represents feedback gain coefficient and ϑ is a delay. Adding the time delay feedback controller (6) to the second equation of system (5), one can get:

$$\begin{cases} \dfrac{d^\sigma w_1(t)}{dt^\sigma} = w_2(t), \\[2mm] \dfrac{d^\sigma w_2(t)}{dt^\sigma} = w_3(t) + \varrho_1[w_3(t - \vartheta) - w_3(t)], \\[2mm] \dfrac{d^\sigma w_3(t)}{dt^\sigma} = -\alpha_1 w_1(t) - \alpha_2 w_2(t) - \alpha_3 w_3(t) + \alpha_4 w_3^2 + \alpha_5 w_1(t) w_2(t). \end{cases} \tag{7}$$

System (7) comes from adding a perturbation term to the second equation of system (5). Clearly, system (7) owns one equilibrium point $W_1(0,0,0)$. Clearly, we can easily obtain the following linear system of (7) near the equilibrium point $W_1(0,0,0)$:

$$\begin{cases} \dfrac{d^\sigma w_1(t)}{dt^\sigma} = w_2(t), \\[2mm] \dfrac{d^\sigma w_2(t)}{dt^\sigma} = (1 - \varrho_1)w_3(t) + \varrho_1 w_3(t - \vartheta), \\[2mm] \dfrac{d^\sigma w_3(t)}{dt^\sigma} = -\alpha_1 w_1(t) - \alpha_2 w_2(t) - \alpha_3 w_3(t). \end{cases} \tag{8}$$

The characteristic equation of (8) owns the following expression:

$$\det \begin{bmatrix} s^\sigma & -1 & 0 \\ 0 & s^\sigma & (\varrho_1 - 1) + \varrho_1 e^{-s\vartheta} \\ \alpha_1 & \alpha_2 & s^\varrho + \alpha_3 \end{bmatrix} = 0. \tag{9}$$

Then,

$$\mathcal{Q}_1(s) + \mathcal{Q}_2(s)e^{-s\vartheta} = 0, \tag{10}$$

where

$$\begin{cases} \mathcal{Q}_1(s) = s^{3\sigma} + c_1 s^{2\sigma} + c_2 s^\sigma + c_3, \\ \mathcal{Q}_2(s) = c_4 s^\sigma + c_5, \end{cases} \tag{11}$$

where

$$\begin{cases} c_1 = \alpha_3, \\ c_2 = \alpha_2(1 - \varrho_1), \\ c_3 = \alpha_1(\varrho_1 - 1), \\ c_4 = -\alpha_2 \varrho_1, \\ c_5 = -\alpha_1 \varrho_1. \end{cases} \tag{12}$$

If $\vartheta = 0$, then (10) becomes:

$$\lambda^3 + c_1 \lambda^2 + (c_2 + c_4)\lambda + c_3 + c_5 = 0. \tag{13}$$

Suppose that

$$(\mathcal{S}_1) \begin{cases} c_1 > 0, \\ c_1(c_2 + c_4) > c_3 + c_5, \\ (c_3 + c_5)[c_1(c_2 + c_4) - (c_3 + c_5)] > 0 \end{cases}$$

is true, then all the roots $\lambda_1, \lambda_2, \lambda_3$ of (13) obey $|\arg(\lambda_1)| > \frac{\sigma\pi}{2}, |\arg(\lambda_2)| > \frac{\sigma\pi}{2}$ and $|\arg(\lambda_3)| > \frac{\sigma\pi}{2}$. Applying Lemma 1, we can conclude that the zero equilibrium point $W_1(0,0,0)$ of system (7) is locally asymptotically stable for $\vartheta = 0$.

Suppose that $s = iv = v\left(\cos\frac{\pi}{2} + i\sin\frac{\pi}{2}\right)$ is the root of Equation (10). Then,

$$
\begin{cases}
\mathcal{C}_1 \cos v\vartheta + \mathcal{C}_2 \sin v\vartheta = \mathcal{D}_1, \\
\mathcal{C}_2 \cos v\vartheta - \mathcal{C}_1 \sin v\vartheta = \mathcal{D}_2,
\end{cases}
\tag{14}
$$

where

$$
\begin{cases}
\mathcal{C}_1 = \rho_1 v^\sigma + \rho_2, \\
\mathcal{C}_2 = \rho_3 v^\sigma, \\
\mathcal{D}_1 = \rho_4 v^{3\sigma} + \rho_5 v^{2\sigma} + \rho_6 v^\sigma + \rho_7, \\
\mathcal{D}_2 = \rho_8 v^{3\sigma} + \rho_9 v^{2\sigma} + \rho_{10} v^\sigma,
\end{cases}
\tag{15}
$$

where

$$
\begin{cases}
\rho_1 = c_4 \cos\dfrac{\sigma\pi}{2}, \\[2mm]
\rho_2 = c_5, \\[2mm]
\rho_3 = c_4 \sin\dfrac{\sigma\pi}{2}, \\[2mm]
\rho_4 = -\cos\dfrac{3\sigma\pi}{2}, \\[2mm]
\rho_5 = -c_1 \cos\sigma\pi, \\[2mm]
\rho_6 = -c_2 \cos\dfrac{\sigma\pi}{2}, \\[2mm]
\rho_7 = -c_3, \\[2mm]
\rho_8 = -\sin\dfrac{3\sigma\pi}{2}, \\[2mm]
\rho_9 = -c_1 \sin\sigma\pi, \\[2mm]
\rho_{10} = -c_2 \sin\dfrac{\sigma\pi}{2}.
\end{cases}
\tag{16}
$$

By virtue of (14), one has

$$
\cos v\vartheta = \frac{\mathcal{D}_1 \mathcal{C}_1 + \mathcal{D}_2 \mathcal{C}_2}{\mathcal{C}_1^2 + \mathcal{C}_2^2}
\tag{17}
$$

and

$$
\mathcal{C}_1^2 + \mathcal{C}_2^2 = \mathcal{D}_1^2 + \mathcal{D}_2^2.
\tag{18}
$$

It follows from (18) that

$$
\tau_1 v^{6\sigma} + \tau_2 v^{5\sigma} + \tau_3 v^{4\sigma} + \tau_4 v^{3\sigma} + \tau_5 v^{2\sigma} + \tau_6 v^\sigma + \tau_7 = 0,
\tag{19}
$$

where

$$
\begin{cases}
\tau_1 = \rho_4^2 + \rho_8^2, \\
\tau_2 = 2(\rho_4\rho_5 + \rho_8\rho_9), \\
\tau_3 = \rho_5^2 + \rho_9^2 + 2(\rho_4\rho_6 + \rho_8\rho_{10}), \\
\tau_4 = 2(\rho_4\rho_7 + \rho_5\rho_6 + \rho_9\rho_{10}), \\
\tau_5 = \rho_6^2 + \rho_{10}^2 - \rho_1^2 - \rho_3^2 + 2\rho_5\rho_7, \\
\tau_6 = 2(\rho_6\rho_7 - \rho_1\rho_2), \\
\tau_7 = \rho_7^2 - \rho_2^2.
\end{cases}
\tag{20}
$$

Denote

$$
\Xi_1(v) = \tau_1 v^{6\sigma} + \tau_2 v^{5\sigma} + \tau_3 v^{4\sigma} + \tau_4 v^{3\sigma} + \tau_5 v^{2\sigma} + \tau_6 v^\sigma + \tau_7.
\tag{21}
$$

If

$$
(\mathcal{S}_2) \ \ |\rho_7| < |\rho_2|
$$

holds, since $\lim_{v\to+\infty} \Xi_1(v) = +\infty$, then Equation (19) has at least one positive real root. Thus, Equation (10) has at least one pair of pure roots. By means of Sun et al. [40], we can easily build the following conclusion.

Lemma 2. *(1) Assume that $\tau_k > 0 (k = 1, 2, \cdots, 7)$, then Equation (10) has no root with zero real parts for $\vartheta \geq 0$. (2) Assume that $(\mathcal{S}_2)$ is fulfilled and $\tau_k > 0 (k = 1, 2, \cdots, 6)$, then Equation (10) has a pair of purely imaginary roots $\pm i v_0$ if $\vartheta = \vartheta_0^{(i)} (i = 1, 2, \cdots,)$ where*

$$\vartheta_0^{(i)} = \frac{1}{v_0}\left[\arccos\left(\frac{\mathcal{D}_1\mathcal{C}_1 + \mathcal{D}_2\mathcal{C}_2}{\mathcal{C}_1^2 + \mathcal{C}_2^2}\right) + 2i\pi\right], \tag{22}$$

where $i = 0, 1, 2, \cdots$, and $v_0 > 0$ represents the unique zero of $\Xi_1(v)$.

Set $\vartheta_0 = \vartheta_0^{(0)}$. Now we make the following hypothesis:

$$(\mathcal{S}_3) \quad \mathcal{A}_{1R}\mathcal{A}_{2R} + \mathcal{A}_{1I}\mathcal{A}_{2I} > 0,$$

where

$$
\left\{
\begin{aligned}
\mathcal{A}_{1R} &= 3\sigma v_0^{3\sigma-1}\cos\frac{(3\sigma-1)\pi}{2} + 2\sigma c_1 v_0^{2\sigma-1}\cos\frac{(2\sigma-1)\pi}{2} \\
&\quad + \sigma c_2 v_0^{\sigma-1}\cos\frac{(\sigma-1)\pi}{2} + \sigma c_4\left[v_0^{\sigma-1}\cos\frac{(\sigma-1)\pi}{2}\cos v_0\vartheta_0\right. \\
&\quad \left. + v_0^{\sigma-1}\sin\frac{(\sigma-1)\pi}{2}\sin v_0\vartheta_0\right], \\
\mathcal{A}_{1I} &= 3\sigma v_0^{3\sigma-1}\sin\frac{(3\sigma-1)\pi}{2} + 2\sigma c_1 v_0^{2\sigma-1}\sin\frac{(2\sigma-1)\pi}{2} \\
&\quad + \sigma c_2 v_0^{\sigma-1}\sin\frac{(\sigma-1)\pi}{2} + \sigma c_4\left[v_0^{\sigma-1}\sin\frac{(\sigma-1)\pi}{2}\cos v_0\vartheta_0\right. \\
&\quad \left. - v_0^{\sigma-1}\cos\frac{(\sigma-1)\pi}{2}\sin v_0\vartheta_0\right], \\
\mathcal{A}_{2R} &= \left(c_4 v_0^\sigma\cos\frac{\sigma\pi}{2} + c_5\right)v_0\sin v_0\vartheta_0 - \left(c_4 v_0^\sigma\sin\frac{\sigma\pi}{2}\right)v_0\cos v_0\vartheta_0, \\
\mathcal{A}_{2I} &= \left(c_4 v_0^\sigma\cos\frac{\sigma\pi}{2} + c_5\right)v_0\cos v_0\vartheta_0 + \left(c_4 v_0^\sigma\sin\frac{\sigma\pi}{2}\right)v_0\sin v_0\vartheta_0.
\end{aligned}
\right. \tag{23}
$$

Lemma 3. *Suppose that $s(\vartheta) = \eta_1(\vartheta) + i\eta_2(\vartheta)$ is the root of (10) near $\vartheta = \vartheta_0$ satisfying $\eta_1(\vartheta_0) = 0, \eta_2(\vartheta_0) = v_0$, then $\mathrm{Re}\left(\frac{ds}{d\vartheta}\right)\Big|_{\vartheta=\vartheta_0, v=v_0} > 0$.*

Proof. In view of (10), one gets

$$\left(\frac{ds}{d\vartheta}\right)^{-1} = \frac{\mathcal{A}_1(s)}{\mathcal{A}_2(s)} - \frac{\vartheta}{s}, \tag{24}$$

where

$$
\left\{
\begin{aligned}
\mathcal{A}_1(s) &= 3\sigma s^{3\sigma-1} + 2\sigma c_1 s^{2\sigma-1} + \sigma c_2 s^{\sigma-1} + \sigma c_4 s^{\sigma-1}e^{-s\vartheta}, \\
\mathcal{A}_2(s) &= s e^{-s\vartheta}(c_4 s^\sigma + c_5).
\end{aligned}
\right. \tag{25}
$$

By (24), we have

$$\mathrm{Re}\left[\left(\frac{ds}{d\vartheta}\right)^{-1}\right]_{\vartheta=\vartheta_0, v=v_0} = \mathrm{Re}\left[\frac{\mathcal{A}_1(s)}{\mathcal{A}_2(s)}\right]_{\vartheta=\vartheta_0, v=v_0} = \frac{\mathcal{A}_{1R}\mathcal{A}_{2R} + \mathcal{A}_{1I}\mathcal{A}_{2I}}{\mathcal{A}_{2R}^2 + \mathcal{A}_{2I}^2}. \tag{26}$$

Applying $(\mathcal{S}_3)$, one derives

$$\operatorname{Re}\left[\left(\frac{ds}{d\vartheta}\right)^{-1}\right]_{\vartheta=\vartheta_0,\, v=v_0} > 0,$$

which finishes the proof. $\square$

By virtue of Lemma 1, the following conclusion can be easily derived.

Theorem 1. *Suppose that $(\mathcal{S}_1)$–$(\mathcal{S}_3)$ hold true, then the zero equilibrium point $W_1(0,0,0)$ of system (5) is locally asymptotically stable provided that ϑ lies in the interval $[0,\vartheta_0)$ and system (5) will generate a Hopf bifurcation around the zero equilibrium point $W_1(0,0,0)$ for $\vartheta = \vartheta_0$.*

Remark 2. *From Theorem 1, we can easily know that the delay stability region of system (5) is $[0,\vartheta_{0*})$ and the critical value of the onset of Hopf bifurcation of system (5) is ϑ_0.*

4. Chaos Control via Fractional-Order PD^σ Controller

In this segment, we shall design an appropriate controller to suppress the chaotic behavior of the chaotic Jerk system. Following the idea of Ding et al. [37–39], we design a mixed controller which includes time delay feedback controller and fractional-order PD^σ controller as follows:

The time delay feedback controller is given by

$$\zeta(t) = \varrho_2[w_3(t - \vartheta) - w_3(t)], \tag{27}$$

where ϱ_2 represents feedback gain coefficient and ϑ is a delay.

The fractional-order PD^σ controller is given by:

$$\varsigma(t) = \mu_p w_1(t - \vartheta) + \mu_d \frac{d^\sigma w_1(t)}{dt^\sigma}, \tag{28}$$

where μ_p and $\mu_d \neq 1$ stands for the proportional control coefficient and the derivative control coefficient, respectively, ϑ stands for the time delay. Adding (27) and (28) to the second equation and the first equation of system (5), we get

$$\begin{cases} \dfrac{d^\sigma w_1(t)}{dt^\sigma} = w_2(t) + \mu_p w_1(t - \vartheta) + \mu_d \dfrac{d^\sigma w_1(t)}{dt^\sigma}, \\[2mm] \dfrac{d^\sigma w_2(t)}{dt^\sigma} = w_3(t) + \varrho_2[w_3(t - \vartheta) - w_3(t)], \\[2mm] \dfrac{d^\sigma w_3(t)}{dt^\sigma} = -\alpha_1 w_1(t) - \alpha_2 w_2(t) - \alpha_3 w_3(t) + \alpha_4 w_3^2 + \alpha_5 w_1(t) w_2(t). \end{cases} \tag{29}$$

System (29) comes from adding two perturbation terms to the first equation and the second equation of system (5). It follows from (29) that

$$\begin{cases} \dfrac{d^\sigma w_1(t)}{dt^\sigma} = \dfrac{1}{1 - \mu_d} w_2(t) + \dfrac{\mu_p}{1 - \mu_d} w_1(t - \vartheta), \\[2mm] \dfrac{d^\sigma w_2(t)}{dt^\sigma} = w_3(t) + \varrho_2[w_3(t - \vartheta) - w_3(t)], \\[2mm] \dfrac{d^\sigma w_3(t)}{dt^\sigma} = -\alpha_1 w_1(t) - \alpha_2 w_2(t) - \alpha_3 w_3(t) + \alpha_4 w_3^2 + \alpha_5 w_1(t) w_2(t). \end{cases} \tag{30}$$

The linear system of (30) near the zero equilibrium point is given by:

$$
\begin{cases}
\dfrac{d^\sigma w_1(t)}{dt^\sigma} = \dfrac{1}{1-\mu_d} w_2(t) + \dfrac{\mu_p}{1-\mu_d} w_1(t-\vartheta), \\[2ex]
\dfrac{d^\sigma w_2(t)}{dt^\sigma} = (1-\varrho_2)w_3(t) + \varrho_2 w_3(t-\vartheta), \\[2ex]
\dfrac{d^\sigma w_3(t)}{dt^\sigma} = -\alpha_1 w_1(t) - \alpha_2 w_2(t) - \alpha_3 w_3(t) + \alpha_4 w_3^2 + \alpha_5 w_1(t)w_2(t).
\end{cases}
\tag{31}
$$

The characteristic equation of (31) is given by:

$$
\det\begin{bmatrix}
s^\sigma - \dfrac{\mu_p}{1-\mu_d}e^{-s\vartheta} & -\dfrac{1}{1-\mu_d} & 0 \\[2ex]
0 & s^\sigma & (\varrho_2-1)-\varrho_2 e^{-s\vartheta} \\[2ex]
\alpha_1 & \alpha_2 & s^\sigma + \alpha_3
\end{bmatrix} = 0.
\tag{32}
$$

Then,

$$
\mathcal{V}_1(s) + \mathcal{V}_2(s)e^{-s\vartheta} + \mathcal{V}_3(s)e^{-2s\vartheta} = 0,
\tag{33}
$$

where

$$
\begin{cases}
\mathcal{V}_1(s) = s^{3\sigma} + d_1 s^{2\sigma} + d_2 s^\sigma + d_3, \\
\mathcal{V}_2(s) = d_4 s^{2\sigma} + d_5 s^\sigma + d_6, \\
\mathcal{V}_3(s) = d_7,
\end{cases}
\tag{34}
$$

where

$$
\begin{cases}
d_1 = \alpha_3, \\
d_2 = \alpha_2(1-\varrho_2), \\
d_3 = \dfrac{\alpha_1(\varrho_2-1)}{1-\nu_d}, \\[2ex]
d_4 = -\dfrac{\nu_p}{1-\nu_d}, \\[2ex]
d_5 = \alpha_2\varrho_2 - \dfrac{\alpha_3}{1-\nu_d}, \\[2ex]
d_6 = \dfrac{\alpha_2(\varrho_2-1)\mu_p - \alpha_1\varrho_2}{1-\nu_d}, \\[2ex]
d_7 = \dfrac{\alpha_2\varrho_2\mu_p}{1-\nu_p}.
\end{cases}
\tag{35}
$$

Equation (33) can be rewritten as:

$$
\mathcal{V}_1(s)e^{s\vartheta} + \mathcal{V}_2(s) + \mathcal{V}_3(s)e^{-s\vartheta} = 0,
\tag{36}
$$

When $\vartheta = 0$, then (36) takes the following form:

$$
\lambda^3 + (d_1+d_4)\lambda^2 + (d_2+d_5)\lambda + d_3 + d_6 + d_7 = 0.
\tag{37}
$$

If

$$
(\mathcal{S}_4)\begin{cases}
d_1 + d_4 > 0, \\
(d_1+d_4)(d_2+d_5) > d_3 + d_6 + d_7, \\
(d_3+d_6+d_7)[(d_1+d_4)(d_2+d_5) - (d_3+d_6+d_7)] > 0
\end{cases}
$$

holds, then all the roots $\lambda_1, \lambda_2, \lambda_3$ of (37) obey $|\arg(\lambda_1)| > \frac{\sigma\pi}{2}$, $|\arg(\lambda_2)| > \frac{\sigma\pi}{2}$ and $|\arg(\lambda_3)| > \frac{\sigma\pi}{2}$. Applying Lemma 1, we get that the zero equilibrium point $W_1(0,0,0)$ of system (29) is locally asymptotically stable for $\vartheta = 0$.

Suppose that $s = i\theta = \theta\left(\cos\frac{\pi}{2} + i\sin\frac{\pi}{2}\right)$ is the root of Equation (36). Then,

$$
\begin{cases}
\mathcal{M}_1\cos\theta\vartheta + \mathcal{M}_2\sin\theta\vartheta = \mathcal{M}_3, \\
\mathcal{N}_1\cos\theta\vartheta + \mathcal{N}_2\sin\theta\vartheta = \mathcal{N}_3,
\end{cases}
\tag{38}
$$

where

$$
\begin{cases}
\mathcal{M}_1 = e_1\theta^{3\sigma} + e_2\theta^{2\sigma} + e_3\theta^{\sigma} + e_4, \\
\mathcal{M}_2 = e_5\theta^{3\sigma} + e_6\theta^{2\sigma} + e_7\theta^{\sigma} + e_8, \\
\mathcal{M}_3 = e_9\theta^{2\sigma} + e_{10}\theta^{\sigma} + e_{11}, \\
\mathcal{N}_1 = f_1\theta^{3\sigma} + f_2\theta^{2\sigma} + f_3\theta^{\sigma} + f_4, \\
\mathcal{N}_2 = f_5\theta^{3\sigma} + f_6\theta^{2\sigma} + f_7\theta^{\sigma} + f_8, \\
\mathcal{N}_3 = f_9\theta^{2\sigma} + f_{10}\theta^{\sigma},
\end{cases}
\tag{39}
$$

where

$$
\begin{cases}
e_1 = \cos\dfrac{3\theta\pi}{2}, \\
e_2 = d_1 \cos\sigma\pi, \\
e_3 = d_2 \cos\dfrac{\sigma\pi}{2}, \\
e_4 = d_3 + d_7, \\
e_5 = -\sin\dfrac{3\sigma\pi}{2}, \\
e_6 = -d_1 \sin\sigma\pi, \\
e_7 = -d_2 \sin\dfrac{\sigma\pi}{2}, \\
e_8 = d_7, \\
e_9 = -d_4 \cos\theta\pi, \\
e_{10} = -d_5 \cos\dfrac{\sigma\pi}{2}, \\
e_{11} = -d_6, \\
f_1 = \sin\dfrac{3\theta\pi}{2} \\
f_2 = d_1 \sin\sigma\pi, \\
f_3 = d_2 \sin\dfrac{\sigma\pi}{2}, \\
f_4 = d_7, \\
f_5 = \cos\dfrac{3\theta\pi}{2}, \\
f_6 = d_1 \cos\sigma\pi, \\
f_7 = d_2 \cos\dfrac{\sigma\pi}{2}, \\
f_8 = d_3 - d_7, \\
f_9 = -d_4 \sin\theta\pi,, \\
f_{10} = -d_5 \sin\dfrac{\sigma\pi}{2}.
\end{cases}
\tag{40}
$$

It follows from (38) that:

$$
\begin{cases}
\cos\theta\vartheta = \dfrac{\mathcal{M}_3\mathcal{N}_2 - \mathcal{N}_3\mathcal{M}_2}{\mathcal{M}_1\mathcal{N}_2 - \mathcal{N}_1\mathcal{M}_2}, \\
\sin\theta\vartheta = \dfrac{\mathcal{N}_3\mathcal{M}_1 - \mathcal{N}_1\mathcal{M}_3}{\mathcal{M}_1\mathcal{N}_2 - \mathcal{N}_1\mathcal{M}_2}.
\end{cases}
\tag{41}
$$

By means of (41), one derives:

$$
(\mathcal{M}_3\mathcal{N}_2 - \mathcal{N}_3\mathcal{M}_2)^2 + (\mathcal{N}_3\mathcal{M}_1 - \mathcal{N}_1\mathcal{M}_3)^2 = (\mathcal{M}_1\mathcal{N}_2 - \mathcal{N}_1\mathcal{M}_2)^2,
\tag{42}
$$

which leads to

$$
\begin{aligned}
&\xi_1\theta^{12\sigma} + \xi_2\theta^{11\sigma} + \xi_3\theta^{10\sigma} + \xi_4\theta^{9\sigma} + \xi_5\theta^{8\sigma} + \xi_6\theta^{7\sigma} + \xi_7\theta^{6\sigma} \\
&+ \xi_8\theta^{5\sigma} + \xi_9\theta^{4\sigma} + \xi_{10}\theta^{3\sigma} + \xi_{11}\theta^{2\sigma} + \xi_{12}\theta^{\sigma} + \xi_{13} = 0,
\end{aligned}
\tag{43}
$$

where

$$
\left\{
\begin{aligned}
\xi_1 =\ & (e_1 f_5 - f_1 e_5)^2, \\
\xi_2 =\ & 2(e_1 f_5 - f_1 e_5)(e_1 f_6 + e_2 f_5 - f_1 e_6 - f_2 e_5), \\
\xi_3 =\ & (e_1 f_6 + e_2 f_5 - f_1 e_6 - f_2 e_5)^2 + 2(e_1 f_5 - f_1 e_5) \\
& \times (e_1 f_7 + e_2 f_6 + e_3 f_5 - f_1 e_7 - f_2 e_6 - f_3 e_5), \\
& - (e_9 f_5 - f_9 e_5)^2 - (e_1 f_9 - f_9 e_1)^2, \\
\xi_4 =\ & 2(e_1 f_5 - f_1 e_5)(e_1 f_8 + e_2 f_7 + e_3 f_6 + e_4 f_3 - f_1 e_8 \\
& - f_2 e_7 - f_3 e_6 - f_4 e_5) + 2(e_1 f_6 + e_2 f_5 - f_1 e_6 - f_2 e_5) \\
& \times (e_1 f_7 + e_2 f_6 + e_3 f_5 - f_1 e_7 - f_2 e_6 - f_3 e_5) \\
& - 2(e_9 f_5 - f_9 e_5)(e_9 f_6 + e_{10} f_5 - f_9 e_6 - f_{10} e_5) \\
& - 2(e_1 f_9 - f_9 e_1)(e_2 f_9 + e_1 f_{10} - f_1 e_{10} - f_2 e_9), \\
\xi_5 =\ & (e_1 f_7 + e_2 f_6 + e_3 f_5 - f_1 e_7 - f_2 e_6 - f_3 e_5)^2 \\
& + 2(e_1 f_5 - f_1 e_5)(e_2 f_8 + e_3 f_7 + e_4 f_6 - f_2 e_8 \\
& - f_3 e_7 - f_4 e_6) + 2(e_1 f_6 + e_2 f_5 - f_1 e_6 - f_2 e_5) \\
& \times (e_1 f_8 + e_2 f_7 + e_3 f_6 + e_4 f_3 - f_1 e_8 - f_2 e_7 \\
& - f_3 e_6 - f_4 e_5) - (e_9 f_6 + e_{10} f_5 - f_9 e_6 - f_{10} e_5)^2 \\
& - 2(e_9 f_5 - f_9 e_5)(e_9 f_7 + e_{10} f_6 + e_{11} f_5 - f_9 e_7 - f_{10} e_6) \\
& - (e_2 f_9 + e_1 f_{10} - f_1 e_{10} - f_2 e_9)^2 - 2(e_1 f_9 - f_9 e_1) \\
& \times (e_3 f_9 + e_2 f_{10} - f_1 e_{11} - f_2 e_{10} - f_3 e_9), \\
\xi_6 =\ & 2(e_1 f_5 - f_1 e_5)(e_3 f_8 + e_4 f_7 - f_3 e_8 - f_4 e_7) \\
& + 2(e_1 f_6 + e_2 f_5 - f_1 e_6 - f_2 e_5)(e_2 f_8 + e_3 f_7 + e_4 f_6 \\
& - f_2 e_8 - f_3 e_7 - f_4 e_6) + 2(e_1 f_7 + e_2 f_6 + e_3 f_5 \\
& - f_1 e_7 - f_2 e_6 - f_3 e_5)(e_1 f_8 + e_2 f_7 + e_3 f_6 + e_4 f_3 - f_1 e_8 \\
& - f_2 e_7 - f_3 e_6 - f_4 e_5) - 2(e_9 f_5 - f_9 e_5)(e_9 f_8 + e_{10} f_7 \\
& + e_{11} f_6 - f_9 e_8 - f_{10} e_7) - 2(e_9 f_7 + e_{10} f_6 + e_{11} f_5 \\
& - f_9 e_7 - f_{10} e_6)(e_9 f_8 + e_{10} f_7 + e_{11} f_6 - f_9 e_8 - f_{10} e_7) \\
& - 2(e_1 f_9 - f_9 e_1)(e_4 f_9 + e_3 f_{10} - f_2 e_{11} - f_3 e_{10}) \\
& - 2(e_2 f_9 + e_1 f_{10} - f_1 e_{10} - f_2 e_9)(e_3 f_9 + e_2 f_{10} \\
& - f_1 e_{11} - f_2 e_{10} - f_3 e_9), \\
\xi_7 =\ & (e_1 f_8 + e_2 f_7 + e_3 f_6 + e_4 f_3 - f_1 e_8 - f_2 e_7 - f_3 e_6 - f_4 e_5)^2 \\
& + 2(e_1 f_5 - f_1 e_5)(e_4 f_8 - f_4 e_8) + 2(e_1 f_6 + e_2 f_5 - f_1 e_6 - f_2 e_5) \\
& \times (e_3 f_8 + e_4 f_7 - f_3 e_8 - f_4 e_7) + 2(e_1 f_7 + e_2 f_6 + e_3 f_5 - f_1 e_7 \\
& - f_2 e_6 - f_3 e_5)(e_2 f_8 + e_3 f_7 + e_4 f_6 - f_2 e_8 - f_3 e_7 - f_4 e_6) \\
& - (e_9 f_7 + e_{10} f_6 + e_{11} f_5 - f_9 e_7 - f_{10} e_6)^2 \\
& - 2(e_9 f_5 - f_9 e_5)(e_{10} f_8 + e_{11} f_7 - f_{10} e_8) \\
& - 2(e_9 f_6 + e_{10} f_5 - f_9 e_6 - f_{10} e_5)(e_9 f_8 + e_{10} f_7 \\
& + e_{11} f_6 - f_9 e_8 - f_{10} e_7) \\
& - (e_3 f_9 + e_2 f_{10} - f_1 e_{11} - f_2 e_{10} - f_3 e_9)^2 \\
& - 2(e_1 f_9 - f_9 e_1)(e_4 f_{10} - f_3 e_{11} - f_4 e_{10}) \\
& - 2(e_2 f_9 + e_1 f_{10} - f_1 e_{10} - f_2 e_9) \\
& \times (e_4 f_9 + e_3 f_{10} - f_2 e_{11} - f_3 e_{10}),
\end{aligned}
\right.
\tag{44}
$$

$$
\left\{
\begin{aligned}
\zeta_8 =\ & 2(e_1f_6 + e_2f_5 - f_1e_6 - f_2e_5)(e_4f_8 - e_8f_4) \\
& + 2(e_1f_7 + e_2f_6 + e_3f_5 - f_1e_7 - f_2e_6 - f_3e_5) \\
& \times (e_3f_8 + e_4f_7 - f_3e_8 - f_4e_7) + 2(e_1f_8 + e_2f_7 \\
& + e_3f_6 + e_4f_3 - f_1e_8 - f_2e_7 - f_3e_6 - f_4e_5)(e_2f_8 + e_3f_7 \\
& + e_4f_6 - f_2e_8 - f_3e_7 - f_4e_6) - 2e_{11}f_8(e_9f_5 - f_9e_5) \\
& - 2(e_9f_6 + e_{10}f_5 - f_9e_6 - f_{10}e_5)(e_9f_8 + e_{10}f_7 \\
& + e_{11}f_6 - f_9e_8 - f_{10}e_7) - 2(e_9f_7 + e_{10}f_6 + e_{11}f_5 \\
& - f_9e_7 - f_{10}e_6)(e_9f_8 + e_{10}f_7 + e_{11}f_6 - f_9e_8 - f_{10}e_7) \\
& + 2e_{11}f_4(e_1f_9 - f_9e_1) - 2(e_2f_9 + e_1f_{10} - f_1e_{10} - f_2e_9) \\
& \times (e_4f_{10} - e_{11}f_3 - f_4e_{10}) - 2(e_3f_9 + e_2f_{10} \\
& - f_1e_{11} - f_2e_{10} - f_3e_9)(e_4f_9 + e_3f_{10} \\
& - f_2e_{11} - f_3e_{10}), \\
\zeta_9 =\ & (e_2f_8 + e_3f_7 + e_4f_6 - f_2e_8 - f_3e_7 - f_4e_6)^2 \\
& + 2(e_3f_8 + e_4f_7 - f_3e_8 - f_4e_7)(e_1f_8 + e_2f_7 + e_3f_6 + e_4f_3 \\
& - f_1e_8 - f_2e_7 - f_3e_6 - f_4e_5) + 2(e_4f_8 - f_4e_8) \\
& \times (e_1f_7 + e_2f_6 + e_3f_5 - f_1e_7 - f_2e_6 - f_3e_5) \\
& + 2(e_3f_8 + e_4f_7 - f_3e_8 - f_4e_7)(e_1f_8 + e_2f_7 + e_3f_6 + e_4f_3 \\
& - f_1e_8 - f_2e_7 - f_3e_6 - f_4e_5) - 2(e_{10}f_8 + e_{11}f_7 - f_{10}e_8) \\
& \times (e_9f_7 + e_{10}f_6 + e_{11}f_5 - f_9e_7 - f_{10}e_6) \\
& - (e_9f_8 + e_{10}f_7 + e_{11}f_6 - f_9e_8 - f_{10}e_7)^2 \\
& - (e_4f_9 + e_3f_{10} - f_2e_{11} - f_3e_{10})^2 \\
& + 2f_4e_{11}(e_2f_9 + e_1f_{10} - f_1e_{10} - f_2e_9) \\
& - 2(e_3f_9 + e_2f_{10} - f_1e_{11} - f_2e_{10} - f_3e_9) \\
& \times (e_4f_{10} - f_3e_{11} - f_4e_{10}), \\
\zeta_{10} =\ & 2(e_4f_8 - f_4e_8)(e_1f_8 + e_2f_7 + e_3f_6 \\
& + e_4f_3 - f_1e_8 - f_2e_7 - f_3e_6 - f_4e_5) + 2(e_2f_8 \\
& + e_3f_7 + e_4f_6 - f_2e_8 - f_3e_7 - f_4e_6) \\
& \times (e_3f_8 + e_4f_7 - f_3e_8 - f_4e_7) + 2e_{11}f_4 \\
& \times (e_3f_9 + e_2f_{10} - f_1e_{11} - f_2e_{10} - f_3e_9) \\
& - 2(e_4f_{10} - e_{11}f_3 - f_4e_{10})(e_4f_9 + f_{10}e_3 \\
& - f_2e_{11} - f_3e_{10}), \\
\zeta_{11} =\ & (e_3f_8 + e_4f_7 - f_3e_8 - f_4e_7)^2 + 2(e_4f_8 - f_4e_8) \\
& \times (e_2f_8 + e_3f_7 + e_4f_6 - f_2e_8 - f_3e_7 - f_4e_6) - 2e_{11}f_8 \\
& \times (e_9f_8 + e_{10}f_7 + e_{11}f_6 - f_9e_8 - f_{10}e_7) \\
& - (e_{10}f_8 + e_{11}f_7 - f_{10}e_8)^2 \\
& - (e_4f_{10} - e_{11}f_3 - f_4e_{10})^2 \\
& - 2f_4e_{11}(e_4f_9 + e_3f_{10} - f_2e_{11} - f_{10}e_3), \\
\zeta_{12} =\ & 2(e_4f_8 - f_4e_8)(e_3f_8 + e_4f_7 - f_3e_8 - f_4e_7) \\
& - 2e_{11}f_8(e_{10}f_8 + e_{11}f_7 - f_{10}e_8) \\
& + 2e_{11}f_4(e_4f_{10} - e_{11}f_3 - f_4e_{10}), \\
\zeta_{13} =\ & (e_4f_8 - e_8f_4)^2 - (e_{11}f_8)^2 - (e_{11}f_4)^2.
\end{aligned}
\right.
\tag{45}
$$

Define

$$
\begin{aligned}
\Xi_2(\theta) =\ & \zeta_1\theta^{12\sigma} + \zeta_2\theta^{11\sigma} + \zeta_3\theta^{10\sigma} + \zeta_4\theta^{9\sigma} + \zeta_5\theta^{8\sigma} + \zeta_6\theta^{7\sigma} \\
& + \zeta_7\theta^{6\sigma} + \zeta_8\theta^{5\sigma} + \zeta_9\theta^{4\sigma} + \zeta_{10}\theta^{3\sigma} + \zeta_{11}\theta^{2\sigma} + \zeta_{12}\theta^{\sigma} + \zeta_{13}.
\end{aligned}
\tag{46}
$$

Suppose that:

$$
(S_5)\quad (e_4f_8 - e_8f_4)^2 < (e_{11}f_8)^2 + (e_{11}f_4)^2
$$

holds, because $\lim_{\theta\to\infty} \Xi_2(\theta) = +\infty$, then Equation (43) has at least one positive real root. Thus Equation (33) owns at least one pair of purely roots. Applying Sun et al. [40], we obtain the following conclusion.

Lemma 4. *Assume that $\xi_k > 0 (k = 1, 2, \cdots, 13)$, Equation (33) possesses no root with zero real parts for $\vartheta \geq 0$. (2) Assume that $(\mathcal{S}_5)$ is fulfilled and $\xi_k > 0$ $(k = 1, 2, \cdots, 12)$, then Equation (33) has a pair of purely imaginary roots $\pm i\theta_0$ if $\vartheta = \vartheta_0^{(h)}$ $(h = 1, 2, \cdots,)$ where*

$$\vartheta_0^{(h)} = \frac{1}{\theta_0}\left[\arccos\left(\frac{\mathcal{M}_3\mathcal{N}_2 - \mathcal{N}_3\mathcal{M}_2}{\mathcal{M}_1\mathcal{N}_2 - \mathcal{N}_1\mathcal{M}_2}\right) + 2h\pi\right], \tag{47}$$

where $h = 0, 1, 2, \cdots$, and $\varsigma_0 > 0$ denotes the unique zero of $\Xi_2(\theta)$.

Set $\vartheta_{0*} = \vartheta_0^{(0)}$. Now we make the hypothesis as follows:

$$(\mathcal{S}_6) \quad \mathcal{G}_{1R}\mathcal{G}_{2R} + \mathcal{G}_{1I}\mathcal{G}_{2I} > 0,$$

where

$$
\left\{
\begin{aligned}
\mathcal{G}_{1R} &= \left[3\sigma\theta_0^{3\sigma-1}\cos\frac{(3\sigma-1)\pi}{2} + 2\sigma d_1\theta_0^{2\sigma-1}\cos\frac{(2\sigma-1)\pi}{2}\right.\\
&\quad \left. + \sigma d_2\theta_0^{\sigma-1}\cos\frac{(\sigma-1)\pi}{2}\right]\cos\theta_0\vartheta_{0*} - \left[3\sigma\theta_0^{3\sigma-1}\sin\frac{(3\sigma-1)\pi}{2}\right.\\
&\quad \left. + 2\sigma d_1\theta_0^{2\sigma-1}\sin\frac{(2\sigma-1)\pi}{2} + \sigma d_2\theta_0^{\sigma-1}\sin\frac{(\sigma-1)\pi}{2}\right]\sin\theta_0\vartheta_{0*}\\
&\quad + 2\sigma d_4\theta_0^{2\sigma-1}\cos\frac{(2\sigma-1)\pi}{2} + \sigma d_5\theta_0^{\sigma-1}\cos\frac{(\sigma-1)\pi}{2},\\
\mathcal{G}_{1I} &= \left[3\sigma\theta_0^{3\sigma-1}\cos\frac{(3\sigma-1)\pi}{2} + 2\sigma d_1\theta_0^{2\sigma-1}\cos\frac{(2\sigma-1)\pi}{2}\right.\\
&\quad \left. + \sigma d_2\theta_0^{\sigma-1}\cos\frac{(\sigma-1)\pi}{2}\right]\sin\theta_0\vartheta_{0*} + \left[3\sigma\theta_0^{3\sigma-1}\sin\frac{(3\sigma-1)\pi}{2}\right.\\
&\quad \left. + 2\sigma d_1\theta_0^{2\sigma-1}\sin\frac{(2\sigma-1)\pi}{2} + \sigma d_2\theta_0^{\sigma-1}\sin\frac{(\sigma-1)\pi}{2}\right]\cos\theta_0\vartheta_{0*}\\
&\quad + 2\sigma d_4\theta_0^{2\sigma-1}\sin\frac{(2\sigma-1)\pi}{2} + \sigma d_5\theta_0^{\sigma-1}\sin\frac{(\sigma-1)\pi}{2},\\
\mathcal{G}_{2R} &= \left(\sigma_0^{3\sigma} + d_1\theta_0^{2\sigma}\cos\sigma\pi + d_2\theta_0^{\sigma}\cos\frac{\sigma\pi}{2} + d_3\right)\theta_0\sin\theta_0\vartheta_{0*}\\
&\quad - \left(\sigma_0^{3\sigma} + d_1\theta_0^{2\sigma}\sin\sigma\pi + d_2\theta_0^{\sigma}\sin\frac{\sigma\pi}{2} + d_3\right)\theta_0\cos\theta_0\vartheta_{0*}\\
&\quad + d_7\theta_0\sin\theta_0\vartheta_{0*},\\
\mathcal{G}_{2I} &= \left(\sigma_0^{3\sigma} + d_1\theta_0^{2\sigma}\cos\sigma\pi + d_2\theta_0^{\sigma}\cos\frac{\sigma\pi}{2} + d_3\right)\theta_0\cos\theta_0\vartheta_{0*}\\
&\quad + \left(\sigma_0^{3\sigma} + d_1\theta_0^{2\sigma}\sin\sigma\pi + d_2\theta_0^{\sigma}\sin\frac{\sigma\pi}{2} + d_3\right)\theta_0\sin\theta_0\vartheta_{0*}\\
&\quad + d_7\theta_0\cos\theta_0\vartheta_{0*}.
\end{aligned}
\right. \tag{48}
$$

Lemma 5. *Suppose that $s(\vartheta) = \omega_1(\vartheta) + i\omega_2(\vartheta)$ is the root of (36) near $\vartheta = \vartheta_{0*}$ satisfying $\omega_1(\vartheta_{0*}) = 0, \omega_2(\vartheta_{0*}) = \theta_0$, then $\operatorname{Re}\left(\frac{ds}{d\vartheta}\right)\Big|_{\vartheta=\vartheta_{0*},\theta=\theta_0} > 0.$*

Proof. By virtue of (36), one gets

$$\left(\frac{ds}{d\vartheta}\right)^{-1} = \frac{\mathcal{G}_1(s)}{\mathcal{G}_2(s)} - \frac{\vartheta}{s}, \tag{49}$$

where

$$\begin{cases} \mathcal{G}_1(s) = \left(3\sigma s^{3\sigma-1} + 2\sigma d_1 s^{2\sigma-1} + \sigma d_2 s^{\sigma-1}\right)e^{s\vartheta} \\ \qquad + 2\sigma d_4 s^{2\sigma-1} + \sigma d_5 s^{\sigma-1}, \\ \mathcal{G}_2(s) = -se^{-s\vartheta}\left(s^{3\sigma} + d_1 s^{2\sigma} + d_2 s^{\sigma} + d_3\right) \\ \qquad + d_7 se^{-s\vartheta}. \end{cases} \tag{50}$$

Then,

$$\mathrm{Re}\left[\left(\frac{ds}{d\vartheta}\right)^{-1}\right]_{\vartheta=\vartheta_{0*},\theta=\theta_0} = \mathrm{Re}\left[\frac{\mathcal{G}_1(s)}{\mathcal{G}_2(s)}\right]_{\vartheta=\vartheta_{0*},\theta=\theta_0} = \frac{\mathcal{G}_{1R}\mathcal{G}_{2R} + \mathcal{G}_{1I}\mathcal{G}_{2I}}{\mathcal{G}_{2R}^2 + \mathcal{G}_{2I}^2}. \tag{51}$$

By virtue of $(\mathcal{S}_6)$, one derives:

$$\mathrm{Re}\left[\left(\frac{ds}{d\vartheta}\right)^{-1}\right]_{\vartheta=\vartheta_{0*},\theta=\theta_0} > 0.$$

The proof finishes. $\square$

Applying Lemma 1, one can derive the following result.

Theorem 2. *Assume that* $(\mathcal{S}_4)$–$(\mathcal{S}_6)$ *are satisfied, then the zero equilibrium point* $W_1(0,0,0)$ *of system (29) is locally asymptotically stable provided that* $\vartheta \in [0, \vartheta_{0*})$ *and system (29) will generate a Hopf bifurcation around the zero equilibrium point* $W_1(0,0,0)$, *when* $\vartheta = \vartheta_{0*}$.

Remark 3. *Liu et al. [16] investigated the chaotic dynamics for some quadratic Jerk system (4), which only involves the integer-order operator. This present research is concerned with chaos control issue for Jerk system (5), which only involves the fractional-order operator. The research approach of [16] can not be applied to model (5) to derive chaos control results of this study. Based on this viewpoint, we think that the derived results of this study replenish the work of [16]. In addition, the investigation idea enriches the chaos control theory of fractional-order chaotic dynamical system.*

Remark 4. *Although there are many works that deal with the chaos control via time delay feedback controller, In this paper, we deal with the chaos control by two methods. One is the classical time delay feedback control, another is mixed control including time delay feedback control and fractional-order PD^σ control. Based on this viewpoint, we think that this paper has some novelties.*

Remark 5. *From Theorem 2, we can easily know that the delay stability region of system (29) is $[0, \vartheta_{0*})$ and the critical value of the onset of Hopf bifurcation of system (29) is ϑ_{0*}.*

Remark 6. *In this paper, we choose $\sigma = 0.94, \alpha_1 = 2, \alpha_2 = 1, \alpha_3 = 1.2, \alpha_4 = 0.5, \alpha_5 = 0.9$; through computer simulations, we know that the fractional-order Jerk system (5) displays chaotic behavior. If we choose another set of values, we also know whether system (5) will generate chaos via computer simulations. Of course, we can deal with the chaos control via the proposed controller.*

Remark 7. *In [37], Yu and Chen explored the Hopf bifurcation control of integer-order system via a time delay feedback controller. In [38], Ding et al. investigated the bifurcation control of integer-order complex networks by PD controller. In [39], Tang et al. dealt with the Hopf bifurcation of a congestion system via fractional-order PD control. In this work, we control the chaos of the Jerk system (5) via a mixed controller including a time delay feedback controller and a fractional-order PD^σ controller, which owns more adjustable parameters. Thus, our work generalizes the works of [37–39].*

5. Examples

Example 1. *Consider the following fractional-order controlled Jerk system:*

$$
\begin{cases}
\dfrac{d^{0.94} w_1(t)}{dt^{0.94}} = w_2(t), \\[2mm]
\dfrac{d^{0.94} w_2(t)}{dt^{0.94}} = w_3(t) + 0.5[w_3(t - \vartheta) - w_3(t)], \\[2mm]
\dfrac{d^{0.94} w_3(t)}{dt^{0.94}} = -\alpha_1 w_1(t) - \alpha_2 w_2(t) - \alpha_3 w_3(t) + \alpha_4 w_3^2 + \alpha_5 w_1(t) w_2(t).
\end{cases}
\tag{52}
$$

One can easily get that system (52) has a zero equilibrium point $W_1(0,0,0)$. By virtue of Matlab software, one gets $v_0 = 3.8872, \vartheta_0 = 1.3$. The hypotheses $(\mathcal{S}_1)$–$(\mathcal{S}_3)$ in Theorem 1 are fulfilled. Let $\vartheta = 1.25 < \vartheta_0 = 1.3$, which implies that ϑ lies in the interval $[0, 1.3)$. The corresponding Matlab simulation plots are given in Figure 2. From Figure 2, one can easily find that all the physical state variables w_1, w_2, w_3 will tend to 0 when the time $t \to \infty$. Let $\vartheta = 1.45 > \vartheta_0 = 1.3$, which manifests that ϑ crosses the critical numerical value 1.3. The corresponding numerical simulation results are presented in Figure 3. Figure 3 shows very clearly that all the physical state variables w_1, w_2, w_3 are to preserve a periodic vibrational situation around 0 when the time $t \to \infty$. Both cases illustrate the disappearance of chaos of the fractional-order chaotic Jerk system (5). In addition, the bifurcation diagrams, which can be seen in Figures 4–6, are given to demonstrating that the bifurcation value of system (52) is approximately equal to 1.3. The numerical figures strongly support the effectiveness of the designed time delay feedback controller.

Example 2. *Consider the following fractional-order controlled Jerk system:*

$$
\begin{cases}
\dfrac{d^{0.94} w_1(t)}{dt^{0.94}} = w_2(t) - 0.5 w_1(t - \vartheta) - 0.8 \dfrac{d^{0.94} w_1(t)}{dt^{0.94}}, \\[2mm]
\dfrac{d^{0.94} w_2(t)}{dt^{0.94}} = w_3(t) + 0.3[w_3(t - \vartheta) - w_3(t)], \\[2mm]
\dfrac{d^{0.94} w_3(t)}{dt^{0.94}} = -2 w_1(t) - w_2(t) - 1.2 w_3(t) - 0.5 w_3^2 + 0.9 w_1(t) w_2(t).
\end{cases}
\tag{53}
$$

One can easily get that system (53) has a zero equilibrium point $W_1(0,0,0)$. By virtue of Matlab software, one gets $\theta_0 = 2.0231, \vartheta_{0*} = 1.21$. The hypotheses $(\mathcal{S}_4)$–$(\mathcal{S}_6)$ in Theorem 2 are fulfilled. Let $\vartheta = 0.98 < \vartheta_{0*} = 1.21$, which implies that ϑ lies in the interval $[0, 1.21)$. The corresponding Matlab simulation plots are given in Figure 7. From Figure 7, one can easily find that all the physical state variables w_1, w_2, w_3 will tend to 0 when the time $t \to \infty$. Let $\vartheta = 1.3 > \vartheta_{0*} = 1.21$, which manifests that ϑ crosses the critical numerical value 1.21. The corresponding numerical simulation results are presented in Figure 8. Figure 8 shows very clearly that all the physical state variables w_1, w_2, w_3 are to preserve a periodic vibrational situation around 0 when the time $t \to \infty$. Both cases illustrate the disappearance of chaos of fractional-order chaotic Jerk system (5). In addition, the bifurcation diagrams, which can be seen in Figures 9–11, are given to demonstrate that the bifurcation value of system (53) is approximately equal to 1.21. The numerical figures strongly support the effectiveness of the designed mixed controller, which includes the time delay feedback controller and fractional-order PD^σ controller.

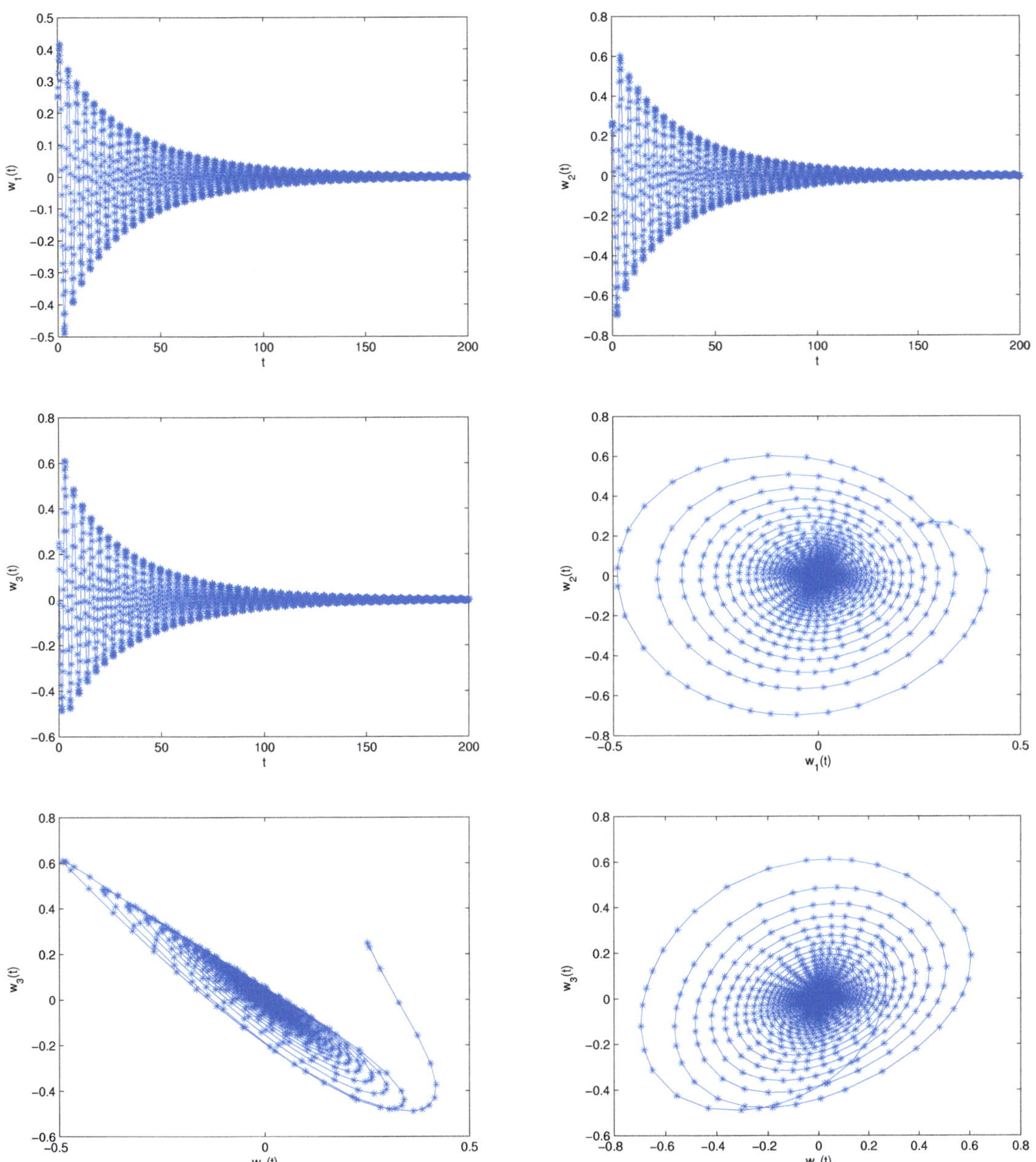

Figure 2. *Cont.*

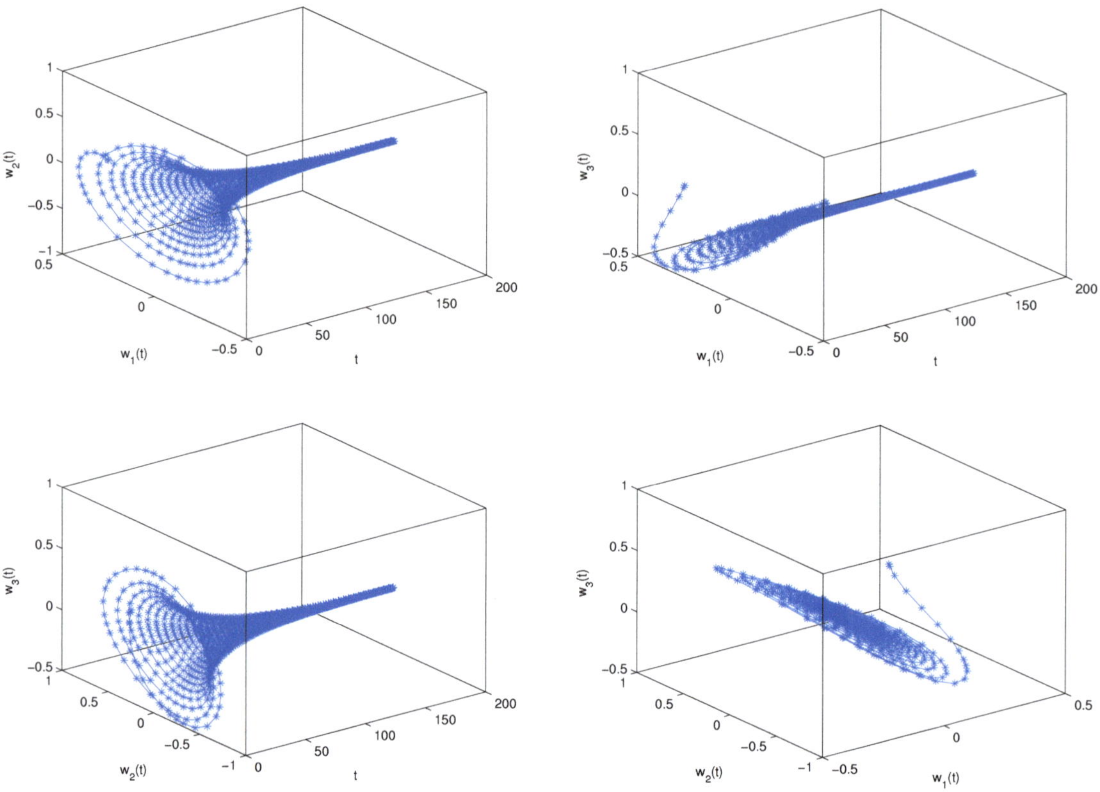

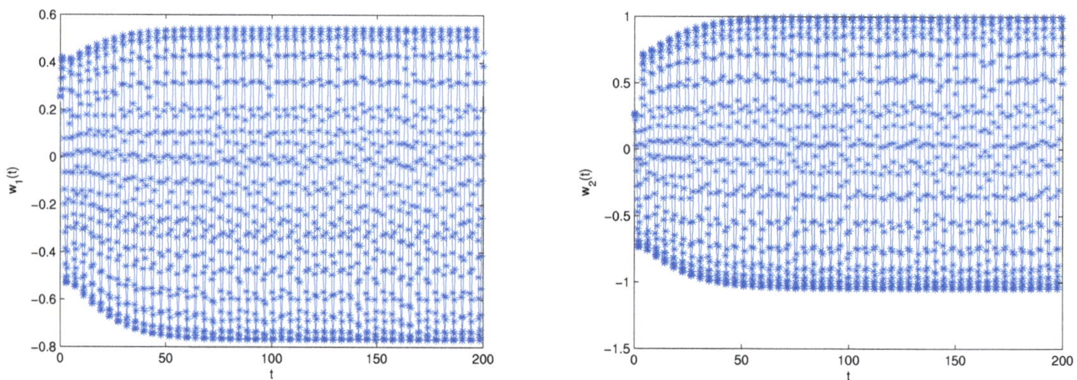

Figure 2. The Matlab simulation results of the controlled Jerk system (52) with $\vartheta = 1.25 < \vartheta_0 = 1.3$ and the initial value $(0.25, 0.25, 0.25)$.

Figure 3. *Cont.*

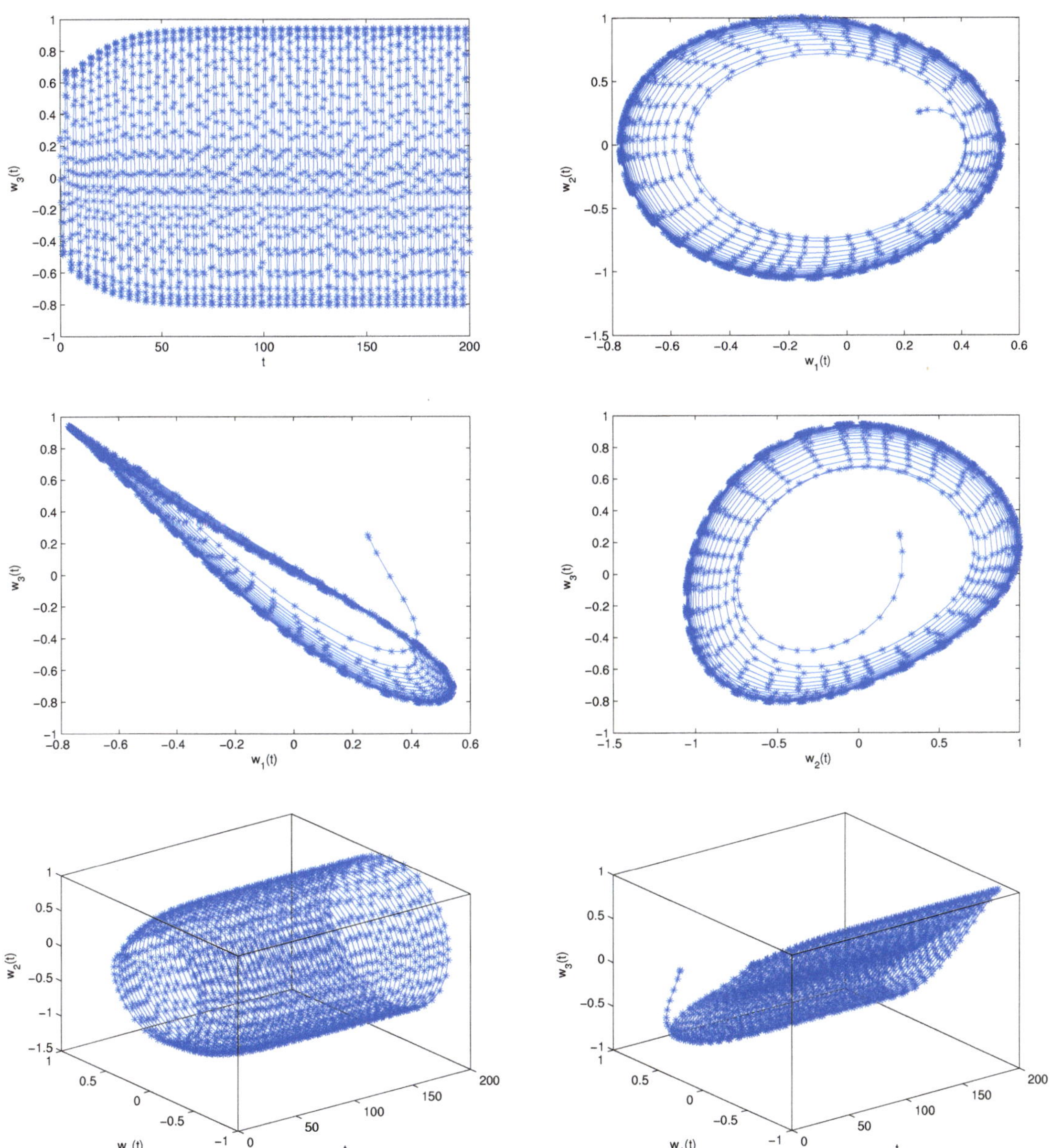

Figure 3. *Cont.*

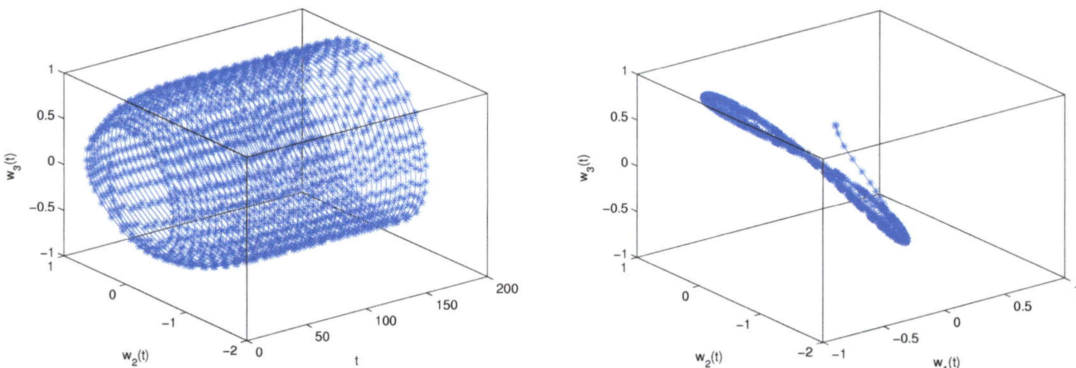

Figure 3. The Matlab simulation results of the controlled Jerk system (52) with $\vartheta = 1.45 > \vartheta_0 = 1.3$ and the initial value $(0.25, 0.25, 0.25)$.

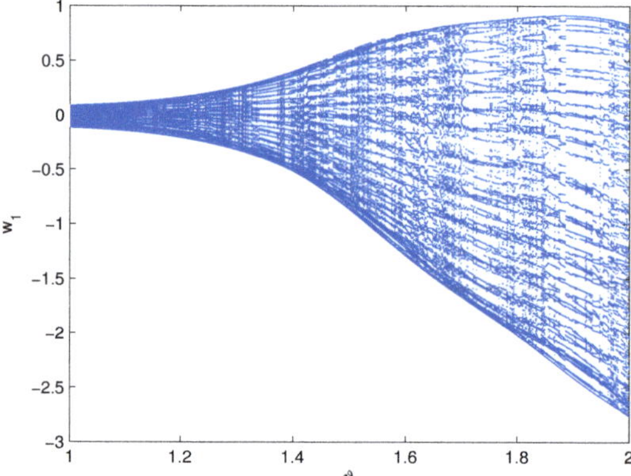

Figure 4. The bifurcation diagram of the controlled Jerk system (52): $\vartheta - w_1$.

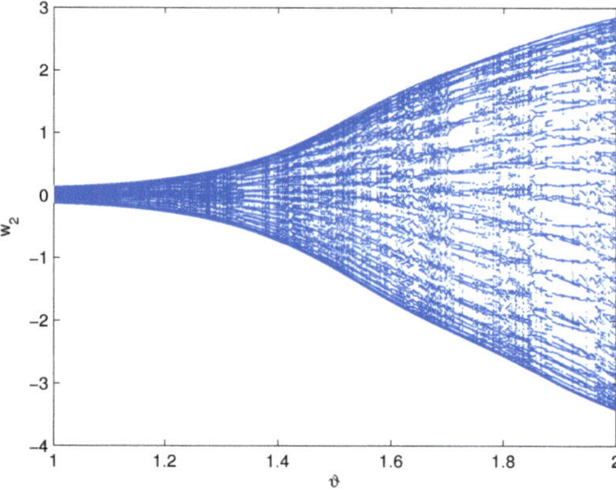

Figure 5. The bifurcation diagram of the controlled Jerk system (52): $\vartheta - w_2$.

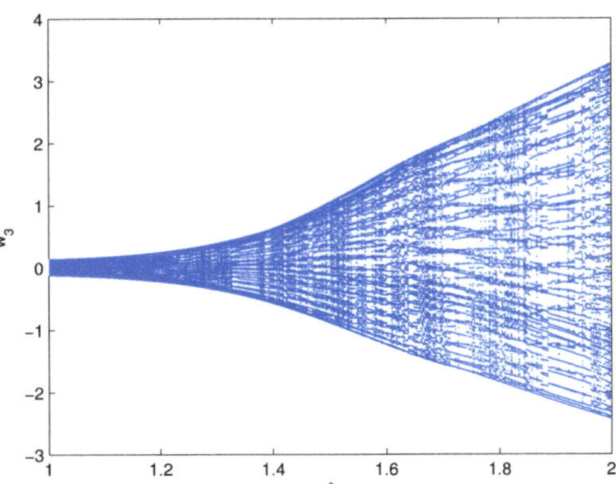

Figure 6. The bifurcation diagram of the controlled Jerk system (52): $\vartheta - w_3$.

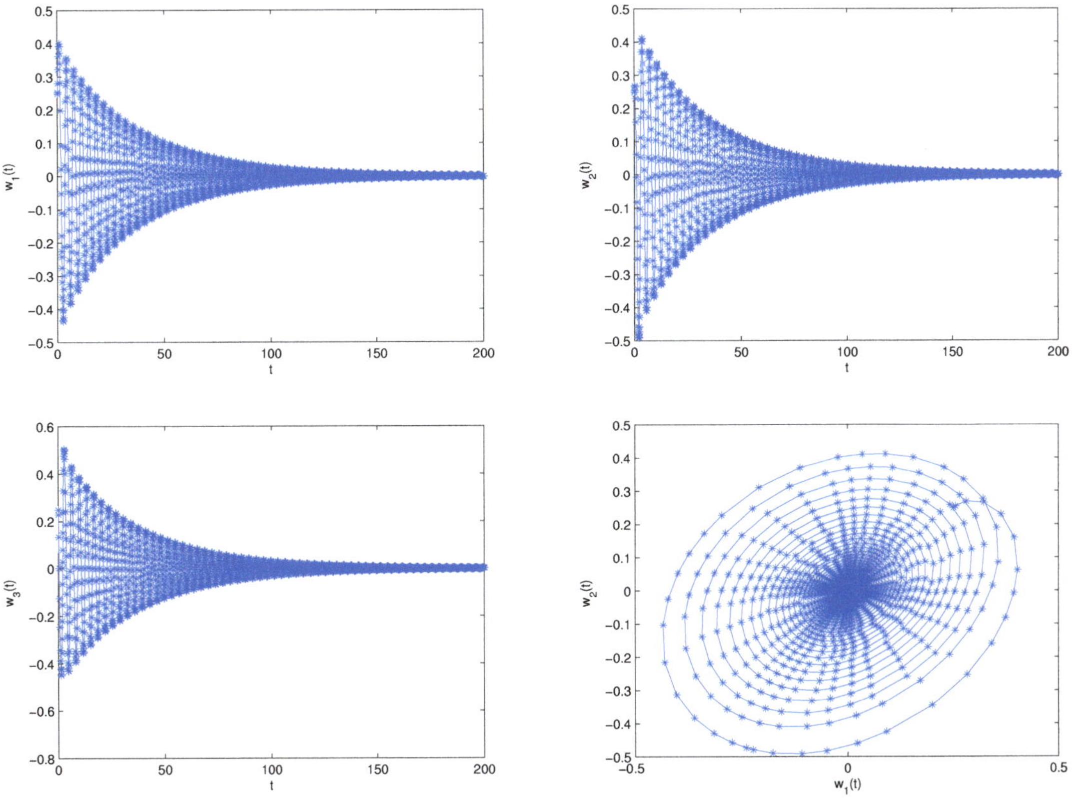

Figure 7. *Cont.*

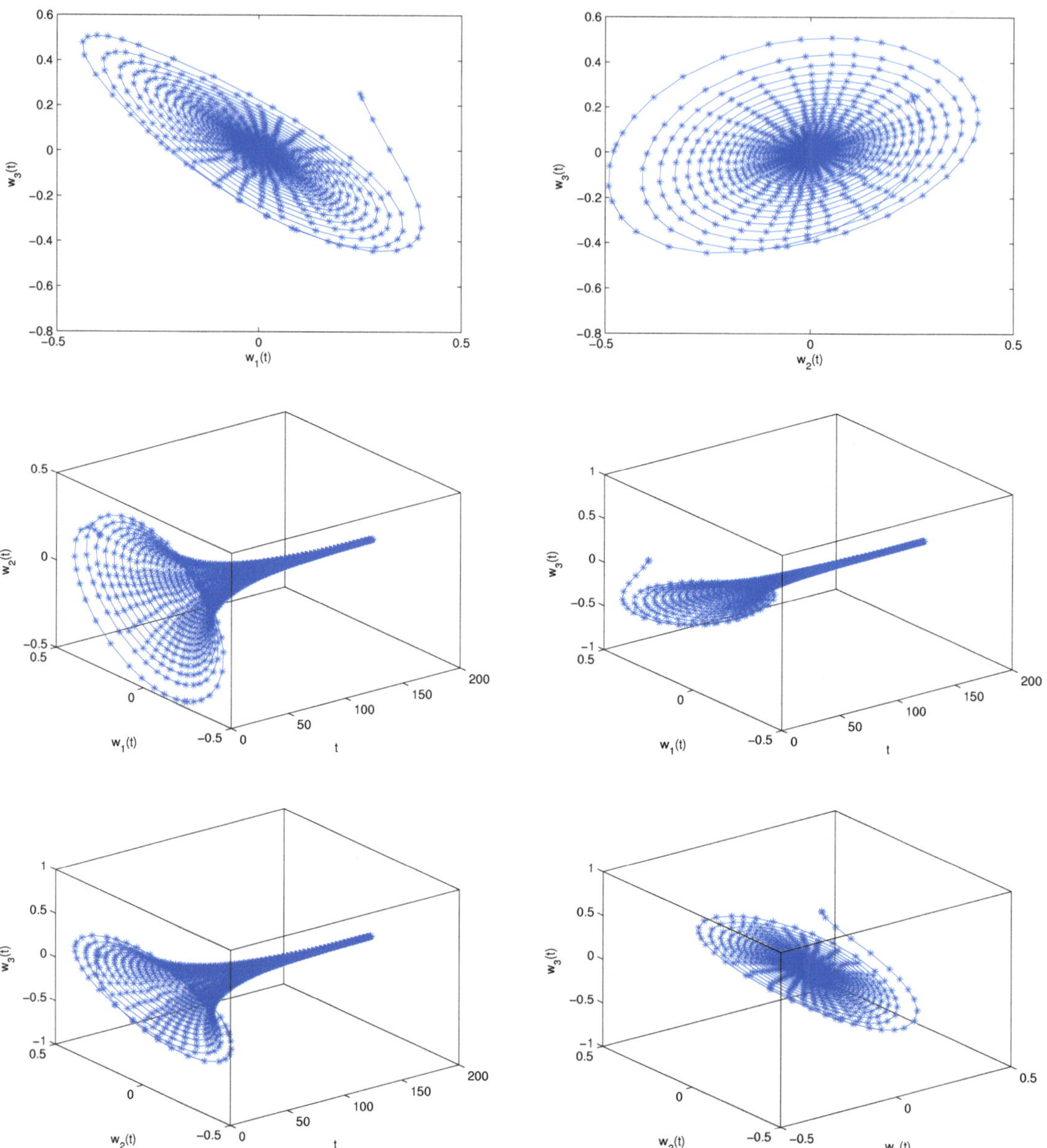

Figure 7. The Matlab simulation results of the controlled Jerk system (53) with $\vartheta = 0.98 < \vartheta_{0*} = 1.21$ and the initial value $(0.25, 0.25, 0.25)$.

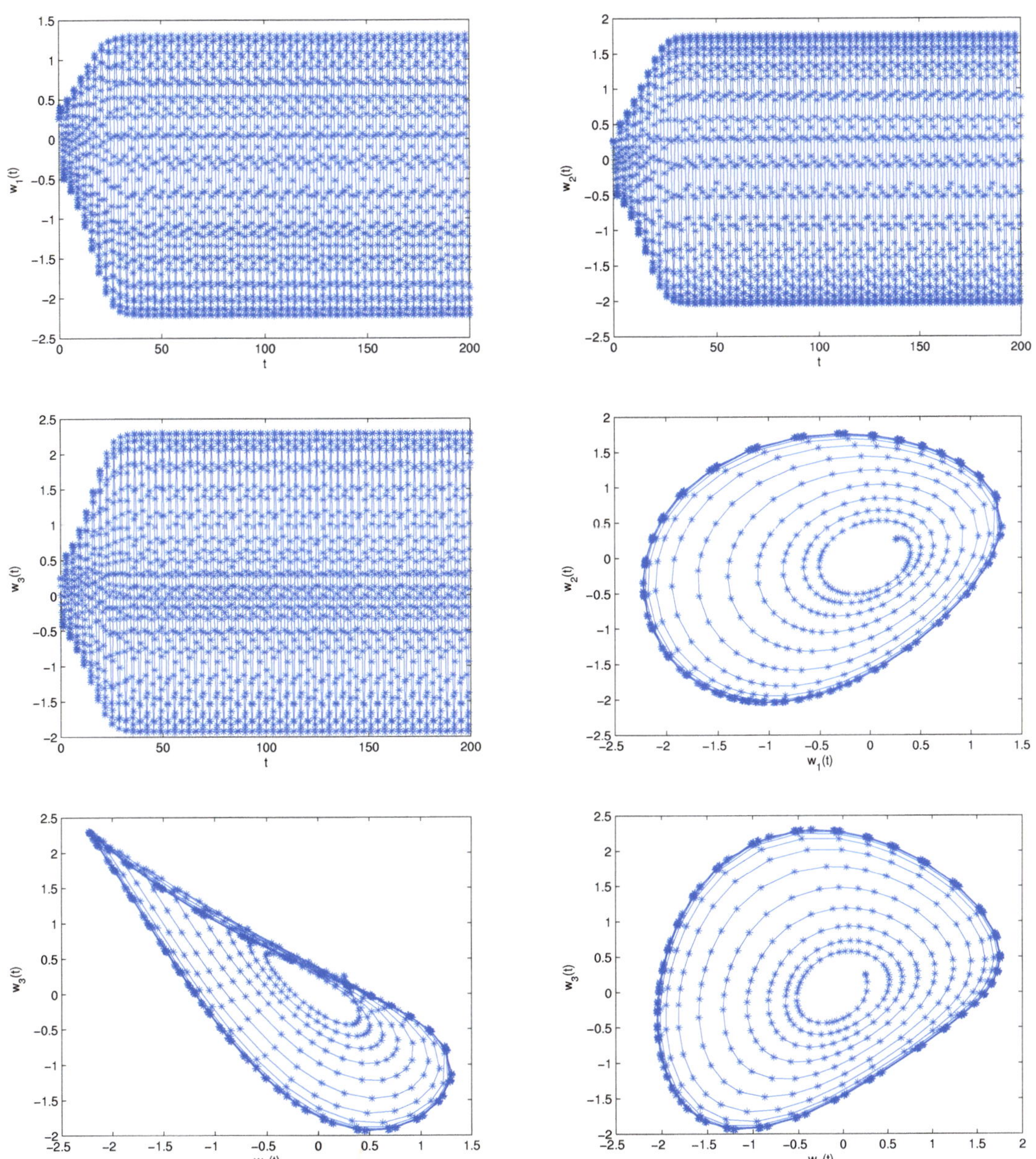

Figure 8. *Cont.*

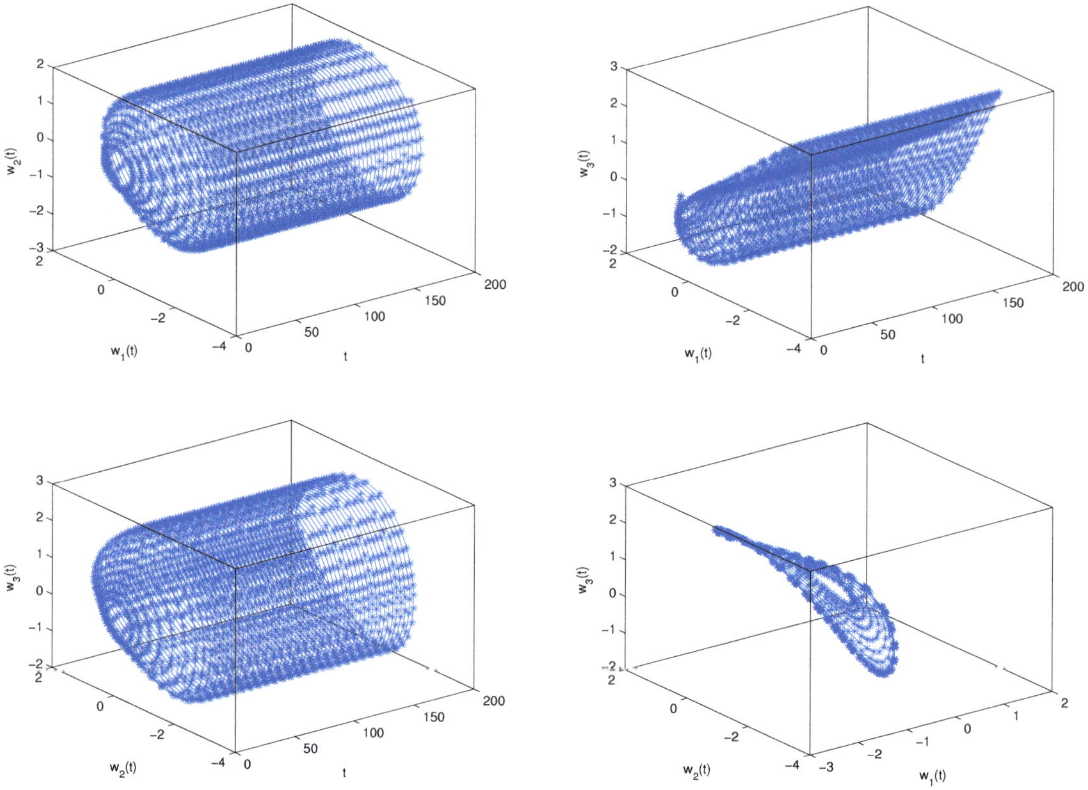

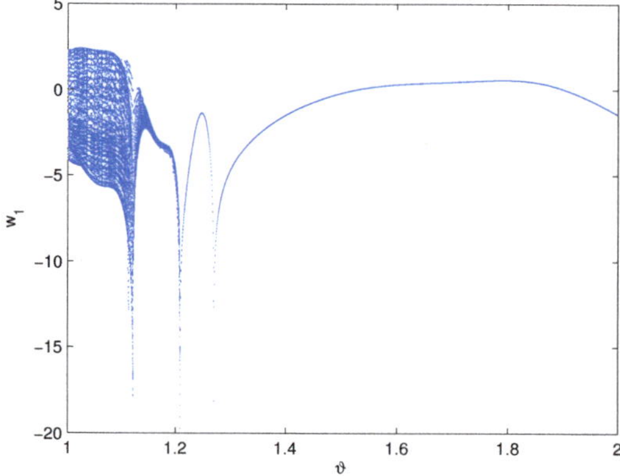

Figure 8. The Matlab simulation results of the controlled Jerk system (53) with $\vartheta = 1.3 > \vartheta_{0*} = 1.21$ and the initial value $(0.25, 0.25, 0.25)$.

Figure 9. The bifurcation diagram of the controlled Jerk system (53): $\vartheta - w_1$.

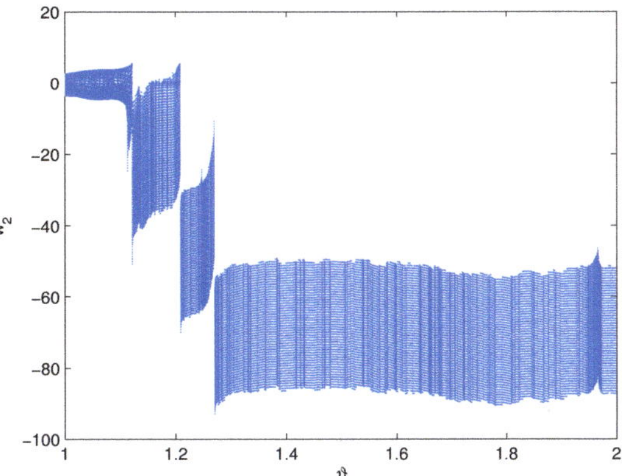

Figure 10. The bifurcation diagram of the controlled Jerk system (53): $\vartheta - w_2$.

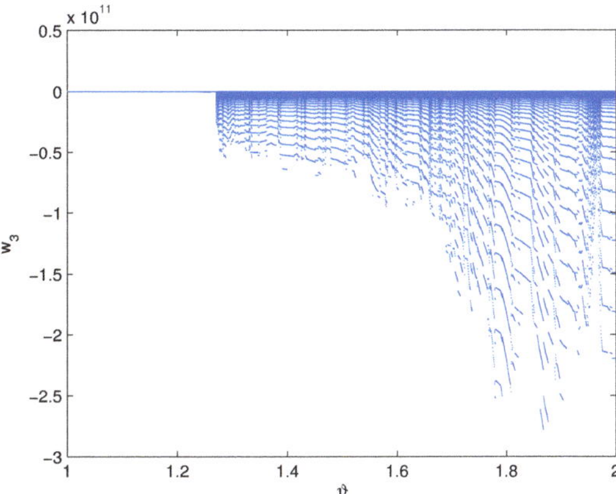

Figure 11. The bifurcation diagram of the controlled Jerk system (53): $\vartheta - w_3$.

Remark 8. *In term of the Matlab simulation results of Examples 1 and 2, we can see that the stability domain of the controlled Jerk system (53) is narrowed and the time of creation of the Hopf bifurcation of system (53) is advanced (in the controlled Jerk system (52), $\vartheta_0 = 1.3$, but in the controlled Jerk system (53), $\vartheta_{0*} = 1.21$).*

6. Conclusions

Chaos control is an ancient and classic problem. During the past decades, the chaos control has attracted great interest in scientific and technological circles. In this current work, on the basis of the previous literature, we build a new fractional-order chaotic Jerk system. By means of a reasonable time delay feedback controller, we can effectively control the chaotic phenomenon of the established fractional-order chaotic Jerk system. By virtue of a suitable mixed controller, which includes a time delay feedback controller and a fractional-order PD^{σ} controller, we can successfully suppress the chaotic behavior of the established fractional-order chaotic Jerk system. The investigation indicates that the time delay in the time delay feedback controller and the mixed controller is a very momentous parameter in controlling the chaos of the fractional-order chaotic Jerk system. The established results of this work are completely novel and the investigation idea of this work can also be applied to probe many chaos control issues of fractional-order differential models in lots of disciplines. In the near future, we will try to deal with the chaos control of fractional-order dynamical models via other mixed controllers (for example, the combination of a nonlinear time delay feedback controller and a fractional-order PD^σ controller, etc.).

Author Contributions: Data curation, L.Y. and Q.Q.; Formal analysis, C.X.; Funding acquisition, C.X.; Methodology, M.L.; Project administration, Y.S.; Resources, Q.Q.; Software, M.L. and P.L.; Writing–original draft, C.X.; Writing–review & editing, C.X. All authors have read and agreed to the published version of the manuscript

Funding: This research was funded by National Natural Science Foundation of China grant number 61673008, National Natural Science Foundation of China grant number 62062018.

Data Availability Statement: No data were used to support this study.

Acknowledgments: This work is supported by National Natural Science Foundation of China (No. 61673008, No. 62062018), Project of High-level Innovative Talents of Guizhou Province ([2016]5651), Key Project of Hunan Education Department (17A181), University Science and Technology Top Talents Project of Guizhou Province (KY[2018]047), Foundation of Science and Technology of Guizhou Province ([2019]1051), Guizhou University of Finance and Economics(2018XZD01), Guizhou Key Laboratory of Big Data Statistical Analysis (No.[2019]5103). The authors would like to thank the referees and the editor for helpful suggestions incorporated into this paper.

Conflicts of Interest: The authors declare that they have no conflict of interest.

References

1. Zhou, L.Q.; Chen, F.Q. Chaos of the RayleighCDuffing oscillator with a non-smooth periodic perturbation and harmonic excitation. *Math. Comput. Simul.* **2022**, *192*, 1–18. [CrossRef]
2. Akhtar, S.; Ahmed, R.; Batool, M.; Shah, N.A.; Chung, J.D. Stability, bifurcation and chaos control of a discretized Leslie prey-predator model. *Chaos Solitons Fractals* **2021**, *152*, 111345. [CrossRef]
3. Pietrych, L.; Sandubete, J.E.; Escot, L. Solving the chaos model-data paradox in the cryptocurrency market. *Commun. Nonlinear Sci. Numer. Simul.* **2021**, *102*, 105901. [CrossRef]
4. Wojtusiak, A.M.; Balanov, A.G.; Savelev, S.E. Intermittent and metastable chaos in a memristive artificial neuron with inertia. *Chaos Solitons Fractals* **2021**, *142*, 110383. [CrossRef]
5. Ma, C.; Wang, X.Y. Hopf bifurcation and topological horseshoe of a novel finance chaotic system. *Commun. Nonlinear Sci. Numer. Simul.* **2012**, *17*, 721–730. [CrossRef]
6. Boccaletti, S.; Grebogi, C.; Lai, Y.C.; Mancini, H.; Maza, D. The control of chaos: theory and application. *Phys. Rep.* **2000**, *329*, 103–197. [CrossRef]
7. Corron, N.J.; Pethel, S.D.; Hopper, B.A. Controlling chaos with simple limiters. *Phys. Rev. Lett.* **2000**, *84*, 3835–3838. [CrossRef] [PubMed]
8. Zheng, J.L. A simple universal adaptive feedback controller for chaos and hyperchaos control. *Comput. Math. Appl.* **2011**, *61*, 2000–2004. [CrossRef]
9. Ott, E.; Grebogi, C.; Yorke, J.A. Controlling chaos. *Phys. Rev. Lett.* **1990**, *64*, 1196–1199. [CrossRef] [PubMed]
10. Li, Z.B.; Lu, W.; Gao, L.F.; Zhang, J.S. Nonlinear state feedback control of chaos system of brushless DC motor. *Procedia Comput. Sci.* **2021**, *183*, 636–640. [CrossRef]
11. Du, J.G.; Huang, T.W.; Sheng, Z.H.; Zhang, H.B. A new method to control chaos in an economic system. *Appl. Math. Comput.* **2020**, *217*, 2370–2380. [CrossRef]
12. Zhao, M.C.; Wang, J.W. H_∞ control of a chaotic finance system in the presence of external disturbance and input time-delay. *Appl. Math. Comput.* **2014**, *233*, 320–327. [CrossRef]
13. Higazy, M.; Hamed, Y.S. Dynamics, circuit implementation and control of new caputo fractional order chaotic 5-dimensions hyperjerk model. *Alex. Eng. J.* **2021**, *60*, 4177–4190. [CrossRef]
14. Mahmoud, E.E.; Trikha, P.; Jahanzaib, L.S.; Eshmawi, A.A.; Matoog, R.T. Chaos control and Penta-compound combination anti-synchronization on a novel fractional chaotic system with analysis and application. *Results Phys.* **2021**, *24*, 104130. [CrossRef]
15. Holyst, J.A.; Urbanowicz, K. Chaos control in economical model by time-delayed feedback method. *Phys. A* **2000**, *287*, 587–598. [CrossRef]
16. Liu, M.; Sang, B.; Wang, N.; Ahmad, I. Choatic dynamics by some quadratic Jerk system. *Axioms* **2021**, *10*, 227. [CrossRef]
17. Nie, X.B.; Liu, P.P.; Liang, J.L.; Cao, J.D. Exact coexistence and locally asymptotic stability of multiple equilibria for fractional-order delayed Hopfield neural networks with Gaussian activation function. *Neural Netw.* **2021**, *142*, 690–700. [CrossRef]
18. Ke, L. Mittag-Leffler stability and asymptotic ω-periodicity of fractional-order inertial neural networks with time-delays. *Neurocomputing* **2021**, *465*, 53–62. [CrossRef]
19. Zhang, F.H.; Huang, T.W.; Wu, Q.J.; Zeng, Z.G. Multistability of delayed fractional-order competitive neural networks. *Neural Netw.* **2021**, *140*, 325–335. [CrossRef]

20. Rihan, F.A.; Rajivganthi, C. Dynamics of fractional-order delay differential model of prey-predator system with Holling-type III and infection among predators. *Chaos Solitons Fractals* **2020**, *141*, 110365. [CrossRef]
21. Alidousti, J.; Ghafari, E. Dynamic behavior of a fractional order prey-predator model with group defense. *Chaos Solitons Fractals* **2020**, *134*, 109688. [CrossRef]
22. Huang, C.D.; Liu, H.; Chen, P.; Zhang, M.S.; Ding, L.; Cao, J.D.; Alsaedi, A. Dynamic optimal control of enhancing feedback treatment for a delayed fractional order predator-prey model. *Phys. A Stat. Mech. Its Appl.* **2020**, *554*, 124136. [CrossRef]
23. Wang, W.T.; Khan, M.A. Analysis and numerical simulation of fractional model of bank data with fractal-fractional Atangana-Baleanu derivative. *J. Comput. Appl. Math.* **2020**, *369*, 112646. [CrossRef]
24. Xu, C.J.; Liao, M.X.; Li, P.L.; Guo, Y.; Liu, Z.X. Bifurcation properties for fractional order delayed BAM neural networks. *Cogn. Comput.* **2021**, *13*, 322–356. [CrossRef]
25. Xu, C.J.; Liu, Z.X.; Liao, M.X.; Li, P.L.; Xiao, Q.M.; Yuan, S. Fractional-order bidirectional associate memory (BAM) neural networks with multiple delays: The case of Hopf bifurcation. *Math. Comput. Simul.* **2021**, *182*, 471–494. [CrossRef]
26. Djilali, S.; Ghanbari, B.; Bentout, S.; Mezouaghi, A. Turing-Hopf bifurcation in a diffusive mussel-algae model with time-fractional-order derivative. *Chaos Solitons Fractals* **2020**, *138*, 109954. [CrossRef]
27. Xiao, M.; Zheng, W.X.; Lin, J.X.; Jiang, G.P.; Zhao, L.D.; Cao, J.D. Fractional-order PD control at Hopf bifurcations in delayed fractional-order small-world networks. *J. Frankl. Inst.* **2017**, *354*, 7643–7667. [CrossRef]
28. Xu, C.J.; Liao, M.X.; Li, P.L.; Yuan, S. Impact of leakage delay on bifurcation in fractional-order complex-valued neural networks. *Chaos Solitons Fractals* **2021**, *142*, 110535. [CrossRef]
29. Huang, C.D.; Liu, H.; Shi, X.Y.; Chen, X.P.; Xiao, M.; Wang, Z.X.; Cao, J.D. Bifurcations in a fractional-order neural network with multiple leakage delays. *Neural Netw.* **2020**, *131*, 115–126. [CrossRef] [PubMed]
30. Xu, C.; Liao, M.X.; Li, P.L.; Yuan S. New insights on bifurcation in a fractional-order delayed competition and cooperation model of two enterprises. *J. Appl. Anal. Comput.* **2021**, *11*, 1240-1258. [CrossRef]
31. Xu, C.J.; Aouiti, C. Comparative analysis on Hopf bifurcation of integer order and fractional order two-neuron neural networks with delay. *Int. J. Circuit Theory Appl.* **2020**, *48*, 1459–1475. [CrossRef]
32. Xu, C.J.; Liu, Z.X.; Yao, L.Y.; Aouiti, C. Further exploration on bifurcation of fractional-order six-neuron bi-directional associative memory neural networks with multi-delays. *Appl. Math. Comput.* **2021**, *410*, 126458. [CrossRef]
33. Xu, C.J.; Aouiti, C.; Liu, Z.X. A further study on bifurcation for fractional order BAM neural networks with multiple delays. *Neurocomputing* **2020**, *417*, 501–515. [CrossRef]
34. Xu, C.J.; Liao, M.X.; Li, P.L. Bifurcation control of a fractional-order delayed competition and cooperation model of two enterprises. *Sci. China Technol. Sci.* **2019**, *62*, 2130–2143. [CrossRef]
35. Podlubny, I. *Fractional Differential Equations*; Academic Press: New York, NY, USA, 1999.
36. Matignon, D. Stability results for fractional differential equations with applications to control processing. In Proceedings of the Computational Engineering in Systems and Application Multi-Conference, IMACS, Lille, France, 9–12 July 1996; pp. 963–968.
37. Yu, P.; Chen, G.R. Hopf bifurcation control using nonlinear feedback with polynomial functions. *Int. J. Bifurc. Chaos* **2004**, *14*, 1683–1704. [CrossRef]
38. Ding, D.W.; Zhang, X.Y.; Cao, J.D.; Wang, N.A.; Liang, D. Bifurcation control of complex networks model via PD controller. *Neurocomputing* **2016**, *175*, 1–9. [CrossRef]
39. Tang, Y.H.; Xiao, M.; Jiang, G.P.; Lin, J.X.; Cao, J.D.; Zheng, W.X. Fractional-order PD control at Hopf bifurcations in a fractional-order congestion control system. *Nonlinear Dyn.* **2017**, *90*, 2185–2198. [CrossRef]
40. Sun, Q.S.; Xiao, M.; Tao, B.B. Local bifurcation analysis of a fractional-order dynamic model of genetic regulatory networks with delays. *Neural Process. Lett.* **2018**, *47*, 1285–1296. [CrossRef]
41. Hammouch, Z.; Yavuz, M.; Ozdemir, N. Numerical solutions and synchronization of a variable-order fractional chaotic system. *Math. Model. Numer. Simul. Appl.* **2021**, *1*, 11–23. [CrossRef]
42. Xu, C.J.; Zhang, W.; Aouiti, C.; Liu, Z.X.; Liao, M.X.; Li, P.L. Further investigation on bifurcation and their control of fractional-order BAM neural networks involving four neurons and multiple delays. *Math. Methods Appl. Sci.* **2021**, in press. [CrossRef]
43. Xu, C.J.; Liao, M.X.; Li, P.L.; Guo, Y.; Xiao, Q.M.; Yuan, S. Influence of multiple time delays on bifurcation of fractional-order neural networks. *Appl. Math. Comput.* **2019**, *361*, 565–582. [CrossRef]
44. Xu, C.J.; Zhang, W.; Liu, Z.X.; Yao, L.Y. Delay-induced periodic oscillation for fractional-order neural networks with mixed delays. *Neurocomputing* **2021**. [CrossRef]

fractal and fractional

Article

Control and Robust Stabilization at Unstable Equilibrium by Fractional Controller for Magnetic Levitation Systems

Banu Ataşlar-Ayyıldız *, **Oğuzhan Karahan and Serhat Yılmaz**

Department of Electronics and Communication Engineering, Kocaeli University, Kocaeli 41001, Turkey; oguzhan.karahan@kocaeli.edu.tr (O.K.); serhaty@kocaeli.edu.tr (S.Y.)
* Correspondence: banu.ayyildiz@kocaeli.edu.tr

Abstract: The problem of control and stabilizing inherently non-linear and unstable magnetic levitation (Maglev) systems with uncertain equilibrium states has been studied. Accordingly, some significant works related to different control approaches have been highlighted to provide robust control and enhance the performance of the Maglev system. This work examines a method to control and stabilize the levitation system in the presence of disturbance and parameter variations to minimize the magnet gap deviation from the equilibrium position. To fulfill the stabilization and disturbance rejection for this non-linear dynamic system, the fractional order PID, fractional order sliding mode, and fractional order Fuzzy control approaches are conducted. In order to design the suitable control outlines based on fractional order controllers, a tuning hybrid method of GWO–PSO algorithms is applied by using the different performance criteria as Integrated Absolute Error (IAE), Integrated Time Weighted Absolute Error (ITAE), Integrated Squared Error (ISE), and Integrated Time Weighted Squared Error (ITSE). In general, these objectives are used by targeting the best tuning of specified control parameters. Finally, the simulation results are presented to determine which fractional controllers demonstrate better control performance, achieve fast and robust stability of the closed-loop system, and provide excellent disturbance suppression effect under nonlinear and uncertainty existing in the processing system.

Keywords: Maglev system; fractional order PID; fractional order sliding mode; fractional order fuzzy control; GWO-PSO

Citation: Ataşlar-Ayyıldız, B.; Karahan, O.; Yılmaz, S. Control and Robust Stabilization at Unstable Equilibrium by Fractional Controller for Magnetic Levitation Systems. *Fractal Fract.* **2021**, *5*, 101. https://doi.org/10.3390/fractalfract5030101

Academic Editors: António M. Lopes, Liping Chen and Amar Debbouche

Received: 9 July 2021
Accepted: 17 August 2021
Published: 20 August 2021

Publisher's Note: MDPI stays neutral with regard to jurisdictional claims in published maps and institutional affiliations.

1. Introduction

For the purpose of weakening the bulky friction problem in mechanical contact connecting both stationary and active parts in the system, magnetically levitated (Maglev) technology is used for eliminating this mechanical contact. Thus, the position of the levitated object can be effectively adjusted and also the stiffness of the Maglev system can be changed. For that reason, the most outstanding works have been found to be related to the Maglev technology in a wide range of applications such as magnetic bearings [1], high speed magnetic levitation trains [2,3], vibration isolation [4], aircraft take-off and landing [5], analysis of forensic evidence, minerals and internal defects in plastic gears [6–8], microelectromechanical systems [9], and disease diagnostic [10].

Since the Maglev system has non-linear dynamic characteristics and is also inherently unstable, achieving stability and dynamic tracking performance while controlling the position of the levitated object is a challenging task. In the literature, many studies report suitable control strategies in order to control the position of the levitated object for achieving better dynamic system response. Among different control strategies, Proportional–Integral–Derivative (PID) and Linear Quadratic Regulator (LQR) controllers, which are the basic linear control techniques, have been proposed by researchers for the Maglev system. Yaseen [11] employed these controllers to examine the stability analysis of the Maglev control system in the presence of disturbances. In addition, the many experiments were conducted for showing the superiority of which controller and comparing the controllers in terms of providing more stability. Zhu et al. [12] designed a simple PID controller for achieving good tracking performance of the Maglev system. However, when applied to the Maglev system under high disturbance, the controlled system usually could not exhibit satisfactory performance. To overcome this matter, Ghosh et al. [13] proposed a two degree of freedom (2-DOF) PID controller for implementing the Maglev system. Moreover, the PID controller was applied to the system for comparison in terms of superior robustness. Acharya et al. [14] designed a 2-DOF PID controller for stabilizing a Maglev system with time delay. However, in order to obtain the best system response, it is important to optimally tune the controller parameters. Therefore, in that paper, the optimal values of the parameters were identified by using symbiotic organisms search (SOS) algorithm. Also, to show the advantage of the proposed controller, the 1-DOF PID controller was optimized and their performances were compared as simulation results.

Apart from PID controllers, the presence of fractional order PID (FOPID) controllers have been found in the literature to provide more design flexibility and more capable under uncertainty and disturbances. The FOPID controller, which is introduced by Podlubny in 1994 [15], is a control application of the fractional calculus where the orders of derivatives and integrals are non-integer. This controller is characterized by three gains (the proportional, integral and derivative) and two order parameters (the integrating order and the derivative order, λ and μ, respectively). Consequently, the PID controllers are extended to the FOPID controllers by using two additional fractional order parameters (μ and λ) and better performance can be provided.

The concept of the FOPID controller in improving the transient response of the Maglev system was presented by Demirören et al. [16]. In their paper, the authors proposed an improved optimization algorithm to update the parameters of the FOPID controller. Also, the performance of the Maglev system with the optimized FOPID controller was examined through transient response and frequency response analyses. In another paper, using both ant colony optimization (ACO) algorithm and the Ziegler–Nichols technique, an FOPID controller was designed by Mughees and Mohsin [17] for the Maglev system. Moreover, for comparative analysis, in that paper, the results obtained with FOPID controller were compared with that of PID controller for illustrating highly efficient results. A FOPID controller of the Maglev system was designed based on four different performance indexes by Bauer and Baranowski [18] through two methods—Nyquist stability criteria and Simulated Annealing algorithm. In their paper, experiments were performed to validate the robustness of the designed controllers and also the methods were compared under external disturbances for the stability analysis of the Maglev system with designed controllers. On the other hand, as a result of putting forward the concept of 2-DOF controller, a realization of the 2-DOF FOPID control technique was addressed by Swain et al. [19] for the Maglev system. Moreover, the proposed 2-DOF FOPID controller was compared to its integer order counterpart in terms of superior response and robustness. One of the other ways of implementing the fractional order 2-DOF controller was realized by Acharya and Mishra [20] while tuning the controller with the proposed optimization algorithm for achieving the required closed-loop performance of the Maglev system. A different design of the 2-DOF FOPID controller was proposed by Pandey et al. [21] and applied to the Maglev system for stabilizing the system. As a result, the detailed investigation has been given in [22] for controlling the Maglev system based on 1 & 2-DOF integer order and fractional order PID controllers by tuning with the optimization algorithms according to the different performance criteria.

Due to the low sensitivity to variations in system parameters, external disturbances, and nonlinear dynamics, one of the robust controller designs which try to solve these transient stability problems can be considered as a sliding mode control (SMC). The SMC method has attracted much attention in designing disturbance rejection tracking control for the Maglev system, especially attenuating the effect of various uncertainties and external disturbances. Starbino and Sathiyavathi [23] designed SMC for achieving the desired ball position of the Maglev system under model uncertainties and disturbance. Moreover, for comparing the performances of the PID controller and SMC, simulation and physical implementation were conducted based on servo and different trajectories, and disturbance rejection and robustness test. In their paper, the control performance of the presented controllers was illustrated by comparing the transient response characteristics and the values of ISE and IAE for both PID and SMC. Shieh et al. [24] developed a robust optimal SMC approach for position tracking of the Maglev system in terms of robustness to parametric uncertainties. On the other hand, many researchers have constructed advanced controllers by using intelligent control techniques such as neural networks and fuzzy system. For the purpose of controlling the ball position of the Maglev system, an intelligent SMC approach was proposed by Lin et al. [25] by using a radial basis function network. Moreover, for verifying the effectiveness of the proposed controller, some experiments were performed. In another work for the satisfactory tracking performance of the Maglev system, an adaptive recurrent neural network intelligent SMC was designed by Chen and Kuo [26]. Also, by illustrating the validity of the proposed controller, SMC and PID, some experimental results were compared in that paper. Besides, using an adaptive technique, a fast terminal SMC approach was developed by Boonsatit and Pukdeboon [27] for achieving fast response and high accuracy of the Maglev system.

For the purpose of enhancing the chattering problem and improving the dynamic response of the closed-loop controlled Maglev system, taking advantage of fractional order calculus, the fractional order can be included into the design of SMC. Roy and Roy [28] studied the detailed comparative analysis between SMC and FOSMC applied to position control of the Maglev system in terms of tracking accuracy, assessing transient response, and the improvement of control effect and energy. Pandey et al. [29] developed fractional order integral dynamic sliding mode controllers for reducing the control effort and increasing the robustness of the Maglev system under parameter uncertainties. For achieving good control performance and reducing the tracking error and chattering effect, Wang et al. [30] designed a new FOSMC for the Maglev system with fractional order. In another work, for reducing the chattering in the SMC, a hybrid control approach based on combination of the SMC and fuzzy control was proposed by Zhang et al. [31] for the control of the Maglev system. The PSO was utilized for tuning the parameters of the SMC using the exponential reaching law method. From the simulation and experiments, it could be inferred that the proposed control approach exhibits robust performance under the disturbances and reduces chattering effectively.

Control of the nonlinear process is more challenging, especially disturbance rejection and no sensitivity to parameter variations as compared to that of a linear process. For the purpose of overcoming this challenge, soft computing techniques such as fuzzy logic, neural network, neuro-fuzzy etc. have been increasingly investigated. Among these, the use of fuzzy logic as computational intelligence-dependent designed control method has been recently developed, and is popular and widely used in control systems. The main motivation of the researchers has been the use of a combination of popular and easily applicable in practice methods such as Takagi-Sugeno fuzzy systems and PID control to design different class of fuzzy PID controllers that ensure sufficient control performance. So as to enhance the performance of a PID controlled Maglev system, by using a fuzzy inference system for self-regulating PID controller parameters, a fuzzy PID compound controller was designed in [32–34] for stabilizing the operation of the Maglev system. Sahoo et al. [35] focused on control of a real time Maglev system identified based on the teaching-learning based optimization-based functional link artificial neural network (FLANN). Moreover, the control of this real Maglev and the identified model was performed with the fuzzy PID controller. In that paper, the response of the identified Maglev system controlled by fuzzy PID control was compared with that of the fuzzy PID controlled actual one. Burakov [36] developed a fuzzy PID controller using the genetic algorithm (GA) for controlling the Maglev system. An incremental PID control approach based on fuzzy logic inference was proposed by Ataşlar–Ayyıldız and Karahan [37] for reducing the control effort and enhancing the control accuracy of the Maglev system. In the paper, by combining a fuzzy control approach with PID control approach, a fuzzy PID controller was designed by using the CS algorithm and also compared with the tuned PID and FOPID controllers. Moreover, to show the superiority of the proposed controller, the simulations and comparisons were performed in the presence of different operation conditions. In another work, modelling of three-input fuzzy PID controller was realized by Sain and Mohan [38] for controlling the unstable nonlinear Maglev system. In the paper, the parameters of the proposed fuzzy PID controller were optimized with GA based on the cost function, including the error and control effort. Moreover, the responses of the closed loop Maglev system with the PID and proposed fuzzy PID controllers were illustrated and compared in terms of the cost value, IAE, ISE, and control signal. In another study of the same authors, considering fractional order calculus, the fractional order was included into a new type of fuzzy PID controller proposed by the authors in order to perform the real-time control of the same Maglev system [39]. The controller parameters for the fuzzy PID and fractional fuzzy PID were also tuned by GA based on the same cost function. In the paper, the closed-loop performances of the Maglev controlled with the proposed controllers were demonstrated and compared according to the control signal, time domain integral error indices, and the cost function value.

The literature survey given above evinces that various control approaches have been proposed for the Maglev system and it is also revealed that the performance of the Maglev system depends on mainly the structure of the controller and its optimization technique. In the light of this information, in this paper, the main aim of this study is to experiment with different controllers based on fractional order calculus such as FOPID, FOSMC, and FOFPID tuned by the GWO–PSO algorithm in order to reach the optimum dynamic response of the Maglev system under parametric uncertainties and disturbances.

The main objectives and contributions of this study are itemized as follows:

- To design and investigate the roles of the FOFPID controller in a Maglev system;
- To use the GWO–PSO algorithm in designing process of the FOFPID controller considering its optimization for the first time in the literature and due to short computation time of the algorithm;
- To illustrate the advantage of GWO–PSO-based FOFPID over FOPID and FOSMC tuned by the GWO–PSO algorithm for the Maglev system;
- To validate the superiority of the presented fractional order controllers compared to the integer order counterparts proposed in the literature like PID and SMC for the above stated system;
- To scrutinize the results based on dynamic transient responses of the fractional order controllers tuned according to the IAE, ISE, ITAE, and ITSE;
- To carry out sensitivity analysis for assessing the robustness of the designed fractional order controllers in the presence of parameter uncertainty, external disturbance, and different trajectory tracking.

The organization of the article is as follows. The mathematical model of the Maglev system is described in Section 2. The structures of the FOPID, FOSMC, and FOFPID controllers are presented in Section 3. The GWO–PSO algorithm is given in Section 4. The simulation results are given in Section 5. Finally, the concluding remarks are given in Section 6.

2. Mathematical Model of the Maglev System

The schematic of the Maglev system used is shown in Figure 1, which is the experimental setup implemented in [23]. The mathematical model of the system, related to the ball position $x(t)$ and the electromagnet coil current $i(t)$, is given by [23]:

$$m\ddot{x} = mg + k\frac{i^2}{x^2} \tag{1}$$

where m is the mass of the levitated object, which is a ferromagnetic ball, g is the acceleration due to gravity, and k is an electromechanical conversion constant. At the equilibrium point (x_0, i_0), the value of k is obtained as follows:

$$k = -\frac{mgx_0^2}{i_0^2} \tag{2}$$

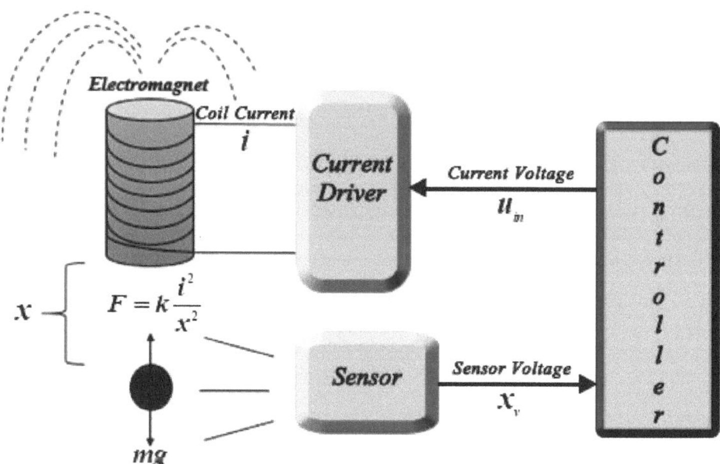

Figure 1. Block diagram of the Maglev System.

Then, by linearizing this nonlinear model about the equilibrium point, the transfer function is obtained as below:

$$\frac{X(s)}{I(s)} = \frac{-\frac{2g}{i_0^2}}{s^2 - \frac{2g}{x_0}} \tag{3}$$

Since the coil current $i(t)$ is proportional to the input voltage $u_{in}(t)$ i.e., $u_{in}(t) = K_a i(t)$, and the output of sensor $x_v(t)$ is proportional to the position of the ball $x(t)$, i.e., $x_v(t) = K_s x(t)$, the transfer function from $u_{in}(t)$ to $x_v(t)$ is obtained as:

$$\frac{X_v(s)}{U_{in}(s)} = \frac{-\frac{K_s}{K_a}}{s^2 - \frac{2g}{x_0}} \tag{4}$$

Finally, by substituting system parameters given in [23] (reported in Table 1) into Equation (4), the transfer function is determined as below:

$$G(s) = \frac{-2502.96}{s^2 - 981.511} \tag{5}$$

Table 1. Parameters of the Maglev System.

Parameter	Notation	Value
Mass of the ball (*gr.*)	m	22
Equilibrium value of current (*A.*)	i_0	0.6105
Equilibrium value of position (*mm.*)	x_0	20
Sensor gain	K_s	458.7157
The gain of the amplifier of the driving circuit	K_a	5.8929

By introducing $x_1(t) = x_v(t)$ and $x_2(t) = \dot{x}_v(t)$ as states, and $y(t) = x_v(t)$ as output, the state space model of the system is obtained as:

$$\dot{x}(t) = \begin{bmatrix} 0 & 1 \\ 981.511 & 0 \end{bmatrix} x(t) + \begin{bmatrix} 0 \\ -2502.96 \end{bmatrix} u_{in}(t) \quad y(t) = \begin{bmatrix} 1 & 0 \end{bmatrix} x(t) \qquad (6)$$

3. Controllers' Design

The principles of the proposed methodology to design Fractional Order PID (FOPID) Controller, Fractional Order Sliding Mode Control (FOSMC), and Fractional Order Fuzzy PID (FOFPID) controllers will be presented in the subsequent subsections.

3.1. Conventional and Fractional Order PID Controller

A conventional PID controller has three parameters K_p, K_i and K_d, with the transfer function:

$$C_{PID}(s) = K_p + K_i \frac{1}{s} + K_d s \qquad (7)$$

As compared to the conventional PID controller, fractional order PID controller introduces two additional adjustable parameters λ and μ. These parameters are non-integer orders of derivative and integral, respectively. The differential equation of the FOPID controller is given as the following [40]:

$$u(t) = K_p e(t) + K_{i0} D_t^{-\lambda} e(t) + K_{d0} D_t^{\mu} e(t) \qquad (8)$$

where $_0D_t^{\alpha}$ is the fractional calculus operator, which will be explained in detail in Section 3.4, and $e(t)$ is the error signal corresponding the difference between desired position and the actual ball position. Let $r(t)$ and $r_v(t)$ be the reference signal in meter and corresponding sensor output in volts, respectively. Hence, the error signal is defined as the difference between $r_v(t)$ and $x_v(t)$.

According to Equation (8), the transfer function of FOPID controller is obtained as [40–42]:

$$C_{FOPID}(s) = K_p + K_i \frac{1}{s^{\lambda}} + K_d s^{\mu} \qquad (9)$$

3.2. Integer Order and Fractional Order Sliding Mode Control

The objective of the controller design by using the Sliding Mode Control methodology is to make the system output track the reference by choosing a sliding surface in the error space [42]. For the MAGLEV system considered in this study, the convergence of sliding variable to zero will ensure $x_v(t) = r_v(t)$.

In this study, a fractional-order sliding surface based approach is used for the Fractional Order SMC. For fractional-order derivative and integration, the fractional calculus operator $(_0D_t^{\alpha})$ explained in details in Section 3.4 is used.

Let the sliding surface, $S_f(t)$, be defined as [27–29,43]:

$$S_f(t) = c_1 {}_0D_t^{\alpha} e(t) + c_2 e(t) \qquad (10)$$

where c_1, $c_2 > 0$ are the tuning parameters, which determine the slope of sliding manifold and $e(t)$ is tracking error, as mentioned in the previous subsection, defined as the difference between the desired position and the actual ball position in volts:

$$e(t) = r_v(t) - x_v(t) \qquad (11)$$

From Equation (10), the derivative of $S_f(t)$ is:

$$\dot{S}_f(t) = c_1 {}_0D_t^{\alpha} \dot{e}(t) + c_2 \dot{e}(t) \qquad (12)$$

$$\dot{S}_f(t) = c_1 {}_0D_t^{\alpha-1} \ddot{e}(t) + c_2 \dot{e}(t) \qquad (13)$$

where the first and the second derivative of $e(t)$ are obtained from Equation (11), as below:

$$\dot{e}(t) = -\dot{x}_v(t) = -\dot{x}_1(t) = -x_2(t) \tag{14}$$

$$\ddot{e}(t) = -\dot{x}_2(t) = -a_{21}x_1(t) - b_2 u_{in}(t) \tag{15}$$

In here, $a_{21} = 981.511$ is the element in the first column of the second row of the system dynamic matrix, and $b_2 = -2502.96$ is the element in the second row of the system input matrix in Equation (6).

By replacing $\ddot{e}(t)$ with the equality in Equation (15), the first derivative of $S_f(t)$ is obtained as [28,29,43]:

$$\dot{S}_f(t) = c_{10}D_t^{\alpha-1}(-a_{21}x_1(t) - b_2 u_{in}(t)) + c_2\dot{e}(t) \tag{16}$$

In order to derive the equivalent control input $u_{eq}(t)$, the first derivative of the sliding surface is made to $\dot{S}_f(t) = 0$; hence, $u_{eq}(t)$ is obtained as:

$$u_{eq}(t) = -\frac{1}{c_1 b_2}\left[-a_{21}x_1(t) + c_{20}D_t^{1-\alpha}\dot{e}(t)\right] \tag{17}$$

In this study, the switching input $u_{sw}(t)$ is chosen as a sigmoid function with boundary layer thickness $\gamma > 0$:

$$u_{sw}(t) = \omega\frac{S_f(t)}{\left|S_f(t)\right| + \gamma} \tag{18}$$

In here, $\omega > 0$ determines how fast error trajectory is required to be brought to the sliding surface.

Then, the total control input law $u(t)$ is obtained as follows:

$$u(t) = -\frac{1}{c_1 b_2}\left[-a_{21}x_1(t) + c_{20}D_t^{1-\alpha}\dot{e}(t) + \omega_0 D_t^{1-\alpha}\frac{S_f(t)}{\left|S_f(t)\right| + \gamma}\right] \tag{19}$$

Stability Analysis. Consider positive definite Lyapunov function as follows:

$$V(t) = \frac{1}{2}S_f^2(t) \tag{20}$$

with $V(0) = 0$ and $V(t) > 0$ for $S_f(t) \neq 0$.

The derivative of the Lyapunov function given in Equation (20) is:

$$\dot{V}(t) = S_f(t)\dot{S}_f(t) \tag{21}$$

By replacing Equation (16) in Equation (21), $\dot{V}(t)$ is obtained as:

$$\begin{aligned}
\dot{V}(t) &= S_f(t)\left[c_{10}D_t^{\alpha-1}(-a_{21}x_1(t) - b_2 u_{in}(t)) + c_2\dot{e}(t)\right] \\
&= S_f(t)\dot{S}_f(t) \\
&= -\omega\frac{S_f^2(t)}{|S_f(t)| + \gamma} < 0
\end{aligned} \tag{22}$$

Since the Lyapunov function $V(t)$ is positive–definite and $\dot{V}(t)$ is negative–definite ($\dot{V}(t) < 0$), the equilibrium point at the origin $S_f(t) = 0$ is asymptotically stable in the sense of Lyapunov's direct method. Moreover, all the trajectories starting off the sliding surface $S_f(t) = 0$ must reach it in finite time and will then remain on the surface.

3.3. Conventional and Fractional Order Fuzzy-PID Control

The fuzzy logic controller in a closed loop control system is basically a static non-linearity between its inputs and outputs, which can be tuned easily to match the desired performance of the control system in a more heuristic manner without delving into the exact mathematical description of the modeled nonlinearity.

Among different types of fuzzy logic controllers, like extensively utilized Fuzzy–PD, Fuzzy–PI, and Fuzzy–PID in various systems, fuzzy based PID controllers have recently become more common in overcoming nonlinear complex dynamical systems. In the literature, the structure of the PID type fuzzy controller used in this work combines Fuzzy–PD and Fuzzy–PI controllers with the gains as the input scaling factors and the gains as the output scaling factors, as described by [44,45].

In this study, a structure, which is a combination of Fuzzy–PD and Fuzzy–PI controllers, is discussed [37]. In the original structure in [37], the inputs are the error and the derivative of error and the FLC output and its integral are multiplied by scaling factors and then summed to give the total controller output. In the structure, the derivative order and integral order are integers.

The controller structure used in this study is quite similar with the structure of the Fuzzy PID controller mentioned above. The difference between them is that the values of both the differentiation parameter and the integration parameter are replaced by fractional values μ and λ, respectively. The detailed configuration of the Fractional Order Fuzzy PID (FOFPID) controller used in this study is shown in Figure 2.

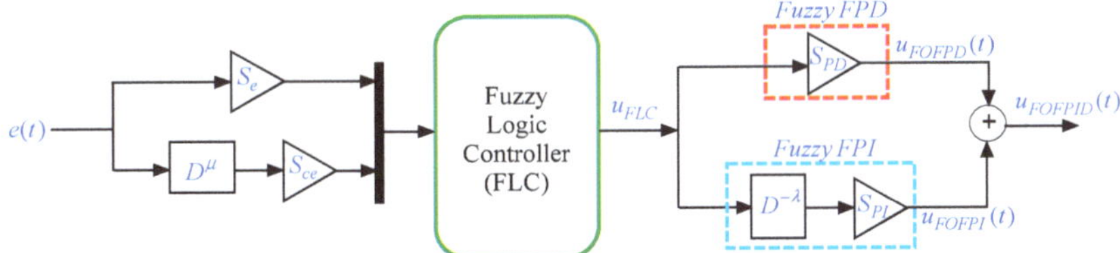

Figure 2. Detailed configuration of the FOFPID controller.

It is observed from Figure 2 that the output of FLC ($u_{FLC}(t)$) obtained by Equation (23) is a function of an error and fractional order derivative of the error as its inputs:

$$u_{FLC}(t) = f\left(S_e e(t), S_{ce} \frac{d^\mu e(t)}{dt^\mu}\right) \tag{23}$$

The function f is a nonlinear fuzzy function representing input–output mapping of the FLC. As shown in the figure, the overall output control law ($u_{FOFPID}(t)$) of the proposed FOFPID controller is a summation of fractional order integral of $u_{FLC}(t)$ with non-integer order (λ) multiplied with S_{PI} and u_{FLC} scaled with S_{PD}. Here, input scaling factors (S_e and S_{ce}) are used to map input linguistic variables in the entire universe of discourse. As for the output scaling factors, S_{PI} and S_{PD} normalize $u_{FLC}(t)$ in the range of universe of discourse.

As a result, the control law of the proposed controller can be given as follows:

$$u_{FOPID}(t) = S_{PD}\, u_{FLC}(t) + S_{PI} \frac{d^{-\lambda}\, u_{FLC}(t)}{dt^{-\lambda}} \tag{24}$$

Looking at the internal structure of the FLC of the controller FOFPID, the input signals and the output signal are represented with seven MFs, as shown in Figure 3. Except for NB and PB, Gaussian membership function is used, considering its prominent benefits such as smooth functions, non-zero at all points, and it also provides the actual information at all points. NB and PB are chosen as Z-shape and S-shape membership functions, respectively. The range of MFs is $[-1, 1]$ for both inputs and outputs. The fuzzy rule table used in this study is shown in Table 2, and also the fuzzy control surface is presented in Figure 4. For constructing these rules, the Standard Mac Vicar–Whelan Rule Table is considered, which is gradually increased from NB to PB both for inputs and output [46].

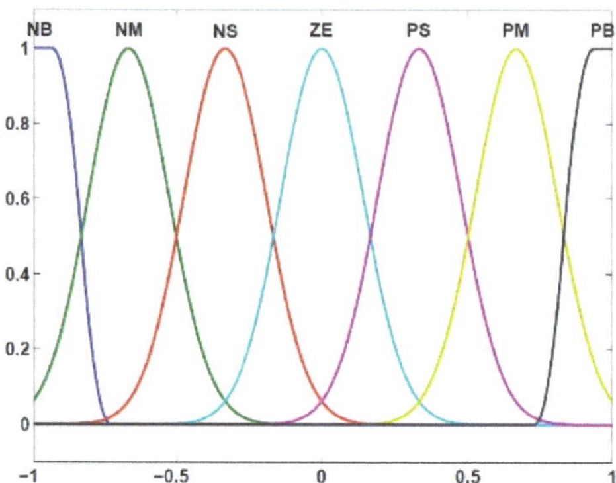

Figure 3. Membership functions of input and output variables.

Table 2. Fuzzy Rules Table.

				$\frac{d^{\mu}e(t)}{dt^{\mu}}$			
$e(t)$	**NB**	**NM**	**NS**	**ZE**	**PS**	**PM**	**PB**
NB	NB	NB	NB	NB	NM	NS	ZE
NM	NB	NB	NB	NM	NS	ZE	PS
NS	NB	NB	NM	NS	ZE	PS	PM
ZE	NB	NM	NS	ZE	PS	PM	PB
PS	NM	NS	ZE	PS	PM	PB	PB
PM	NS	ZE	PS	PM	PB	PB	PB
PB	ZE	PS	PM	PB	PB	PB	PB

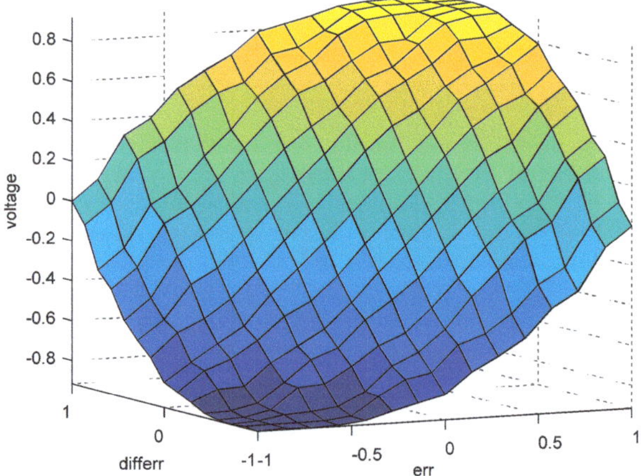

Figure 4. Surface view of fuzzy inference.

3.4. Design of Fractional Order Operator

Fractional-order calculus has been well developed; it is extensively utilized in control engineering because it offers an enhanced number of degree of freedom of any conventional or intelligent controller, which further enhances the closed-loop response and increases the robustness of the closed-loop control system.

In recent years, several approximations of fractional calculus have been proposed. Some of them are Riemann Liouville, Grunwald Letnikov, Caputo definition and Oustaloup's approximation. According to Riemann Liouville definition of fractional-order, the differ-integration operator of a function *f(t)* is defined as:

$$_0D_t^\alpha f(t) \;=\; \frac{1}{\Gamma(n-\alpha)} \frac{d^n}{dt^n} \int_\alpha^t \frac{f(\tau)}{(t-\tau)^{\alpha-n+1}} d\tau \tag{25}$$

where $\Gamma(\cdot)$ is the Euler's Gamma function:

$$\Gamma(z) \;=\; \int_0^\infty x^{z-1}e^{-x}dx, \; \Re(z) > 0 \tag{26}$$

In this study, FOPID, FOSMC, and FOFPID controllers have fractional order differential and integral operators. Oustaloup Recursive Approximation is used for implementation of these controllers. Oustaloup Recursive Approximation uses a *N*th order analog filter to approximate the fractional order calculus in a certain frequency range. The approximating transfer function provided by Oustaloup is as follows and is equivalent to s^a where *a* is the real number power of *s*:

$$s^a \;=\; k_0 \prod_{k_0\,=\,-N}^{N} \frac{s+\omega_{k_z}}{s+\omega_{k_p}} \tag{27}$$

where k_0 is gain, ω_{k_z} are zeros and ω_{k_p} are poles of the filter [47,48]. These poles and zeros are calculated as below, recursively:

$$\omega_{k_p} \;=\; \omega_b \left(\frac{\omega_h}{\omega_b}\right)^{\frac{k+N+\frac{1}{2}+\frac{a}{2}}{2N+1}} \tag{28}$$

$$\omega_{k_z} \;=\; \omega_b \left(\frac{\omega_h}{\omega_b}\right)^{\frac{k+N+\frac{1}{2}-\frac{a}{2}}{2N+1}} \tag{29}$$

$$k_0 \;=\; \omega_h^a \tag{30}$$

where $\{\omega_h, \omega_b\}$ is the expected fitting range and $2N+1$ represents the order of approximation [47].

In this study, the value of N is chosen as 5. Thus, fifth order filters are implemented and the frequency range $\{\omega_h, \omega_b\}$ is chosen as $\{10^{-3}, 10^{+3}\}$ rad/s.

4. Controller Parameters Optimization

It is essential to optimize the controller parameters with a considered objective function to achieve the desired control performance. In this study, the GWO–PSO algorithm is used for the controllers' parameter tuning and the optimization algorithm is run by minimizing the integral-based objective functions commonly introduced in the literature.

Figure 5 illustrates the overall methodology discussed in this work. As demonstrated in this figure, the GWO–PSO algorithm with four different objective functions is used to find the optimal controller parameters for achieving the desired response and improving stability of the controlled output power in the Maglev.

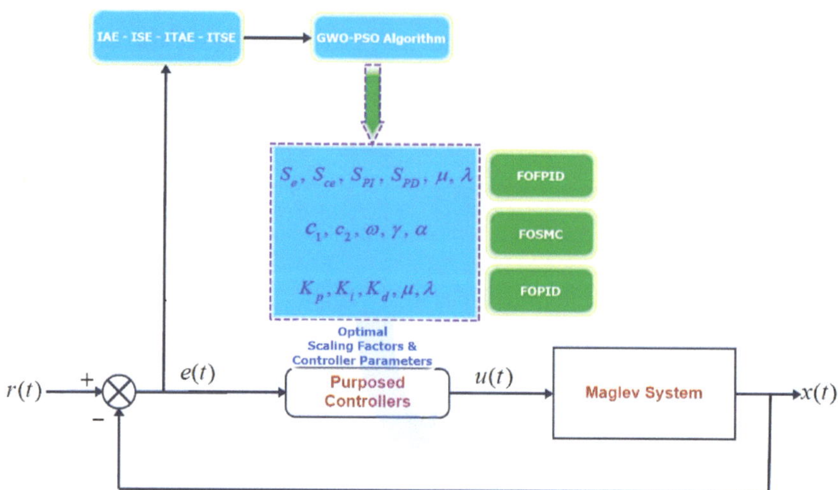

Figure 5. Block diagram of the proposed procedure used in this study.

4.1. Optimization Algorithm

4.1.1. PSO Algorithm

Particle Swarm Optimization (PSO) is an evolutionary optimization technique developed in 1995 based on the social behavior of bird flocks [49]. It consists of an algorithm that initially starts with randomly assigned solutions, called particles, and simulates the birds' search for the best food location. Unlike other evolutionary optimization techniques, each particle has velocity information in Particle Swarm Optimization. Particles travel through the search space at speeds determined by their previous behavior. Thus, the particles get better along the search route. Each particle tends to go from its past positions to the better one and also to follow the particle closest to the food in the swarm.

In each iteration of the Particle Swarm Optimization Algorithm, the velocities and positions of the particles are updated according to the following expressions, respectively:

$$v_i^{k+1} = \xi v_i^k + \varphi_1 rand_1 \left(pbest_1 - p_i^k \right) + \varphi_2 rand_2 \left(gbest - p_i^k \right) \tag{31}$$

$$p_i^{k+1} = p_i^k + v_i^{k+1} \tag{32}$$

In these equations, v_i^k is the velocity of the ith particle for the k iteration, p_i^k is the position of the ith particle for the k iteration, ξ represents the inertial weight function, $\varphi_{1,2}$ represents the learning factors, and $rand_{1,2}$ represents the random number values assigned in the [0, 1] range. In addition, $pbest_i$ is the coordinate that provide the best solution that particle i has achieved so far. $gbest$ is the coordinates that provide the best solution obtained by all particles.

4.1.2. GWO Algorithm

As a swarm-based optimization method, inspiration for Gray wolf optimization, which was presented by Mirjalili et al. [50] for the first time, comes from the behavior and the hunting strategy of the grey wolves in nature. Based on the social hierarchy, gray wolves are classified as alpha, beta, delta, and omega. The leaders of the group are called alpha (α) wolves. Beta (β) wolves help alpha wolves in making decisions. As the third level, delta (δ) wolves' mission is to submit to alpha and beta wolves, but control the omega (ω) wolves. The least priority wolves are the omegas, which must follow the leading grey wolves [50].

In the Grey Wolves Optimizer, the hunting behaviour of the grey wolves is mathematically simulated. Firstly, encircling the victim is modelled as below [50]:

$$\vec{D} = \left| \vec{C} \times \vec{X}_p(t) - \vec{X}(t) \right| \tag{33}$$

$$\vec{X}(t+1) = \vec{X}_p(t) - \vec{A} \times \vec{D} \tag{34}$$

In these equations, t is the number of iteration and the $\vec{X}$ and $\vec{X}_p$ are the position vectors of the wolves and victims, respectively. $\vec{A}$ and $\vec{C}$ are the coefficient vectors and calculated as shown below [50]:

$$\vec{A} = \vec{a} \times \left(2 \times \vec{r}_1 - 1 \right) \tag{35}$$

$$\vec{C} = 2 \times \vec{r}_2 \tag{36}$$

where $\vec{a}$ is linearly decreased from 2 to 0 through iteration steps and $\vec{r}_1$ and $\vec{r}_2$ are random vectors in [0, 1].

The alpha, beta, and delta groups of grey wolves have extraordinary knowledge of the current location of the victim. Therefore, the top three best solutions obtained are recorded and the other wolves have to update their positions relative to the positions of the best search agents [50,51]:

$$\vec{D}_\alpha = \left| \vec{C}_1 \times \vec{X}_\alpha - \vec{X}(t) \right| \vec{D}_\beta = \left| \vec{C}_2 \times \vec{X}_\beta - \vec{X}(t) \right| \vec{D}_\delta = \left| \vec{C}_1 \times \vec{X}_\delta - \vec{X}(t) \right| \tag{37}$$

$$\begin{aligned}
\vec{X}_1 &= \left| \vec{X}_\alpha - \vec{a}_1 \vec{D}_\alpha \right| \\
\vec{X}_2 &= \left| \vec{X}_\beta - \vec{a}_2 \vec{D}_\beta \right| \\
\vec{X}_3 &= \left| \vec{X}_\delta - \vec{a}_3 \vec{D}_\delta \right|
\end{aligned} \tag{38}$$

$$\vec{X}_p(t+1) = \frac{\vec{X}_1 + \vec{X}_2 + \vec{X}_3}{3} \tag{39}$$

4.1.3. GWO–PSO Algorithm

In this work, the Grey Wolf Optimizer is hybridized with Particle Swarm Optimization algorithm for enhancing the progress of the GWO, as presented in [51]. This hybrid optimization method can be regarded for finding efficiently and effectively the global best solution through the optimization process. Therefore, this hybrid GWO-PSO is implemented here for optimization of controller parameters. As a result, the algorithmic representation of the suggested mechanism based on control schemes is described in this section. The flowchart of the GWO–PSO algorithm is shown in Figure 6. The major stages of the presented GWO–PSO based on [51] for tuning the controllers of the Maglev system is listed by steps given below:

Step 1. Initialization of the positions of wolves in the population and that of particles in the swarm.
Step 2. Updating of each wolf location by using the GWO algorithm.
Step 3. Determination of the three best ones among all search agents.
Step 4. Running PSO by using the best values, found by GWO, as initial positions of the swarm.
Step 5. Returning the positions modified by PSO back to the GWO algorithm.
Step 6. Repeating these steps until the maximum iteration number is reached.

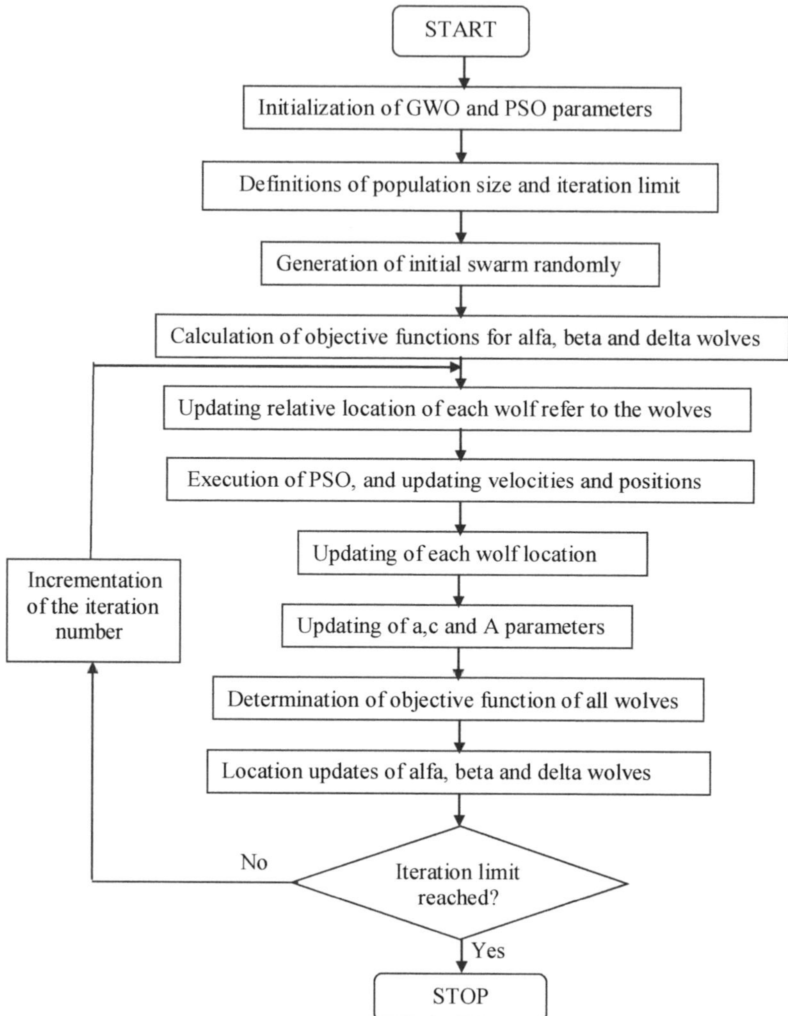

Figure 6. Flowchart of the GWO–PSO algorithm.

Since GWO–PSO is used for tuning of the controllers, in this study, best positions of the grey wolves obtained at the end of the optimization algorithm represent the parameters of the controllers as listed below:

- $\{K_p, K_i, K_d, \mu, \lambda\}$ for the FOPID controller
- $\{c_1, c_2, \omega, \gamma, \alpha\}$ for the FOSMC controller
- $\{S_e, S_{ce}, S_{PI}, S_{PD}, \mu, \lambda\}$ for the FOFPID controller

4.2. Objective Functions

During the controller design by using an optimization algorithm, the most crucial step is to select the most appropriate objective function. Time domain objective functions can be divided into two categories: integral-based objective functions and dynamic performance indices-based objective functions.

Integral-based objective functions commonly used in the literature are IAE (Integrated Absolute Error), ITAE (Integrated Time Weighted Absolute Error), ISE (Integrated Squared

Error), and ITSE (Integrated Time Weighted Squared Error). The formulas of these objective functions are described as:

$$J_{IAE}(e) = \int_0^t |e(t)| dt \tag{40}$$

$$J_{ITAE}(e) = \int_0^t t|e(t)| dt \tag{41}$$

$$J_{ISE}(e) = \int_0^t e^2(t) dt \tag{42}$$

$$J_{ITSE}(e) = \int_0^t te^2(t) dt \tag{43}$$

where $e(t)$ is the error signal, which represents the difference between the system output and the reference signal, as mentioned in Section 3.1. Each one of them has advantages and disadvantages. For example, since $J_{IAE}(e)$ and $J_{ISE}(e)$ criteria are independent of time, the obtained results have a relatively small overshoot but a long settling time. On the other hand, $J_{ITAE}(e)$ and $J_{ITSE}(e)$ can overcome this disadvantage, but they cannot provide a desirable stability margin.

4.3. Proposed Optimization Framework

For the presented work, the parameters of all the four controllers are to be tuned by the GWO–PSO to their optimal values. The maximum iteration (*MaxGen*) is set to be 100 in the GWO–PSO algorithm. Moreover, the optimal controller parameters are obtained by 10 runs of the GWO–PSO. The limitations of all the controller parameters are restricted between certain values. Hence, based on the detailed literature review, during the optimization, the considered search ranges are restricted to $\{K_p, K_i, K_d\} \in [0, 20]$ and $\{\mu, \lambda\} \in [0, 2]$ for the FOPID controller, $\{c_1, c_2, \gamma, \alpha\} \in [0, 2]$ and $\omega \in [0, 20]$ for the FOSMC controller and $S_e \in [0, 10]$, $S_{ce} \in [0, 1]$, $\{S_{PI}, S_{PD}\} \in [0, 20]$ and $\{\mu, \lambda\} \in [0, 2]$ for the FOFPID controller. Also, the total simulation time is considered as 5 s with an interval of 0.001 s.

5. Simulation Results and Discussion

In this section, extensive simulation studies have been carried out for detailed performance evaluations of the FOPID, FOSMC, and FOFPID controllers tuned by the PSO–GWO algorithm for the Maglev system. Moreover, a detailed comparative simulation study of the Maglev dynamic performance with the proposed controllers and the ones presented in [23] have been conducted under all three scenarios: handling parametric variations, disturbance rejections, and different trajectory tracking.

The coding of the PSO–GWO algorithm and the purposed controllers, their adaptation and implementation to the Maglev, and all simulations have been carried out by using MATLAB/Simulink software platform on a personal computer with Intel(R) Core(TM) i7-6700HQ CPU @ 2.60 GHz processor and 32.0 GB RAM. All simulations are executed with a sampling time $T_{sf} = 1$ *ms*. Simulation results and relative comparisons in the present work are illustrated and discussed in the following subsections.

5.1. Dynamic Performance Analysis

For the same Maglev system controlled by FOPID, FOSMC, and FOFPID, transient and steady state responses are analyzed based on different objective functions used here when the reference input is a step one. Hence, the optimal controller parameters are obtained by using the PSO–GWO with different objective functions. As a result, all simulations were carried out with optimized controller parameters as provided in Table 3 based on the objective functions for the different scenarios.

Table 3. Optimized controller parameters based on different objective functions for the Maglev System.

Controller	Parameters	Objective Functions			
		J_{IAE}	J_{ISE}	J_{ITAE}	J_{ITSE}
FOPID	K_P	1.9095	1.9936	2.0325	2.5475
	K_I	13.9014	13.8985	14.8587	16.7852
	K_D	0.1072	0.1197	0.0992	0.1331
	μ	0.9715	0.9703	0.9524	0.8924
	λ	0.9561	0.9364	1.0030	0.9236
FOSMC	c_1	0.0169	0.0135	0.0102	0.0102
	c_2	0.7354	0.4753	0.6839	0.4995
	ω	15.7659	16.0584	15.9247	15.9825
	γ	0.0216	0.0192	0.0113	0.0137
	α	0.9992	0.9460	0.9911	0.9779
FOFPID	S_e	8.5982	3.7212	3.1247	3.2948
	S_{ce}	0.0577	0.0454	0.0468	0.0904
	S_{PI}	17.9569	17.2330	17.9871	16.6151
	S_{PD}	18.0998	18.1233	15.1292	18.7125
	μ	0.9080	0.9622	0.9257	0.9808
	λ	1.2908	1.3407	1.1802	1.3571

The comparative results between the dynamic responses of the optimized controllers and the ones designed in [23] are shown in Figure 7 and given in Table 4 in terms of rise time (t_r), settling time (t_s), overshoot (M_p), steady state error (E_{ss}), and the values of the defined objective functions (J).

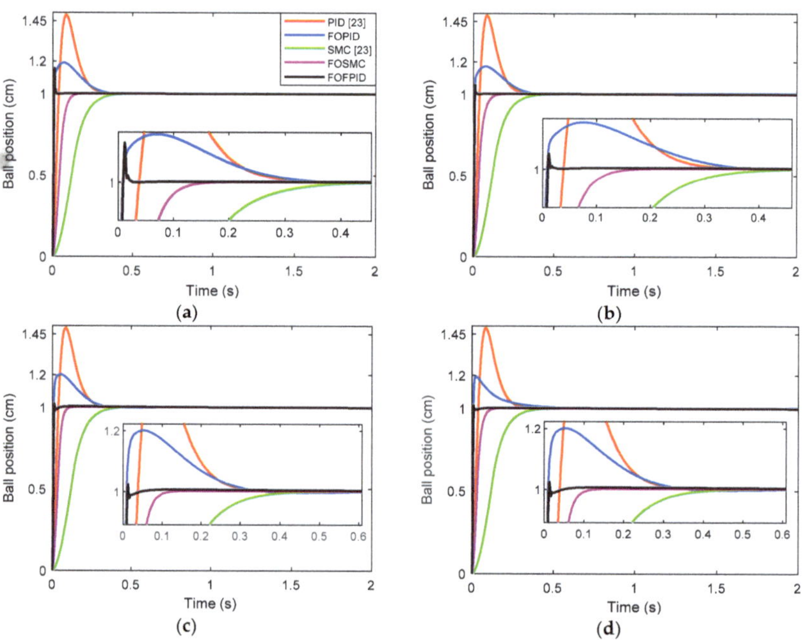

Figure 7. Step responses of the closed-loop Maglev system with different controllers based on J_{IAE} (**a**), J_{ISE} (**b**), J_{ITAE} (**c**) and J_{ITSE} (**d**).

Table 4. Comparative dynamic response specifications on different controllers for the Maglev system.

			Transient Response and Steady State Characteristics				
Objective (J)	**Controller**	$M_p(\%)$	$t_r(s)$ $0.1\to0.9$	$t_s(s)$ $\mp\%2$	E_{ss}	**J** *IAE ISE ITAE ITSE*	
J_{IAE}	FOPID	18.7081	0.0050	0.2850	0.000518	0.0369	
	FOSMC	**0.0004**	0.0670	0.1210	0.000034	0.0477	
	FOFPID	15.5977	**0.0060**	**0.0190**	**0.000020**	**0.0079**	
J_{ISE}	FOPID	16.7848	0.0050	0.3070	0.001823	0.0064	
	FOSMC	**0.0696**	0.0610	0.1190	0.002506	0.0274	
	FOFPID	5.7631	**0.0060**	**0.0130**	**0.000119**	**0.0041**	
J_{ITAE}	FOPID	19.8640	0.0060	0.2770	0.000004	0.0041	
	FOSMC	**0**	0.0500	0.0880	0.000124	**0.0010**	
	FOFPID	2.3659	**0.0070**	**0.0160**	**0.000031**	0.0032	
J_{ITSE}	FOPID	19.1968	0.0060	0.3210	0.000742	0.0002	
	FOSMC	**0.0064**	0.0520	0.0990	0.000515	0.0004	
	FOFPID	1.9090	**0.0060**	**0.0100**	**0.000001**	**0.00001**	
						J_{IAE}	J_{ISE}
J_{IAE}, J_{ISE}	PID [23]	48.2852	0.0270	0.2690	0.000001	0.0741	0.0334
	SMC [23]	0	0.1820	0.3360	0.000638	0.1314	0.0897

It can be concluded from Table 4 that the proposed FOFPID tuned by the PSO–GWO has the best dynamic response in terms of the fastest settling time, short rise time, least steady state error, and all objective function values. Moreover, Figure 7 and Table 4 demonstrate the remarkable advantage of the fractional calculus used in FOPID and FOSMC, as compared to the integer order in PID and SMC, respectively. Although the overshoot is more with the proposed FOFPID controller, as compared to the fractional order and integer order SMC, by examining Figure 7 and performance indices in Table 4, consequently, it can be observed that the proposed FOFPID controller tuned PSO–GWO algorithm outperforms the other controller approaches in terms of transient response characteristics.

5.2. Controller Performance Analysis under Parametric Variations

In this section, this sensitivity analysis of the presented controllers is performed by varying the gain of system in the range of $[-10\%, +10\%]$ of its nominal value. Moreover, for the purpose of showing the control effort that is exhibited by the controllers and minimizing the used objective function values, the control energy can be calculated as follows:

$$u_e = \int_0^{t_f} [u(t)]^2 dt \tag{44}$$

where $u(t)$ is control signal and t_f is total time of simulation. The results obtained by changing the gain of the system under different controllers are shown in Figures 8 and 9. The corresponding time domain attributes and control energies are demonstrated in Figures 10 and 11.

From the figures, it is clearly evident that the stability of the Maglev system is maintained by the proposed fractional order controllers in a better way, as compared to integer order ones developed in [23]. On the other hand, as observed in terms of control signal, the Maglev system with the FOPID, FOSMC, and FOFPID controllers requires higher control energy, and from Figures 10 and 11, deviations in control effort are more with the designed fractional order controllers, as compared to the integer ones developed in [23]. However, as observed in Figures 8–11, the designed FOPID and FOSMC have the best step response in terms of overshoot, rise time, settling time, and steady state error in the case of all objective functions, as compared to the integer counterparts developed in [23].

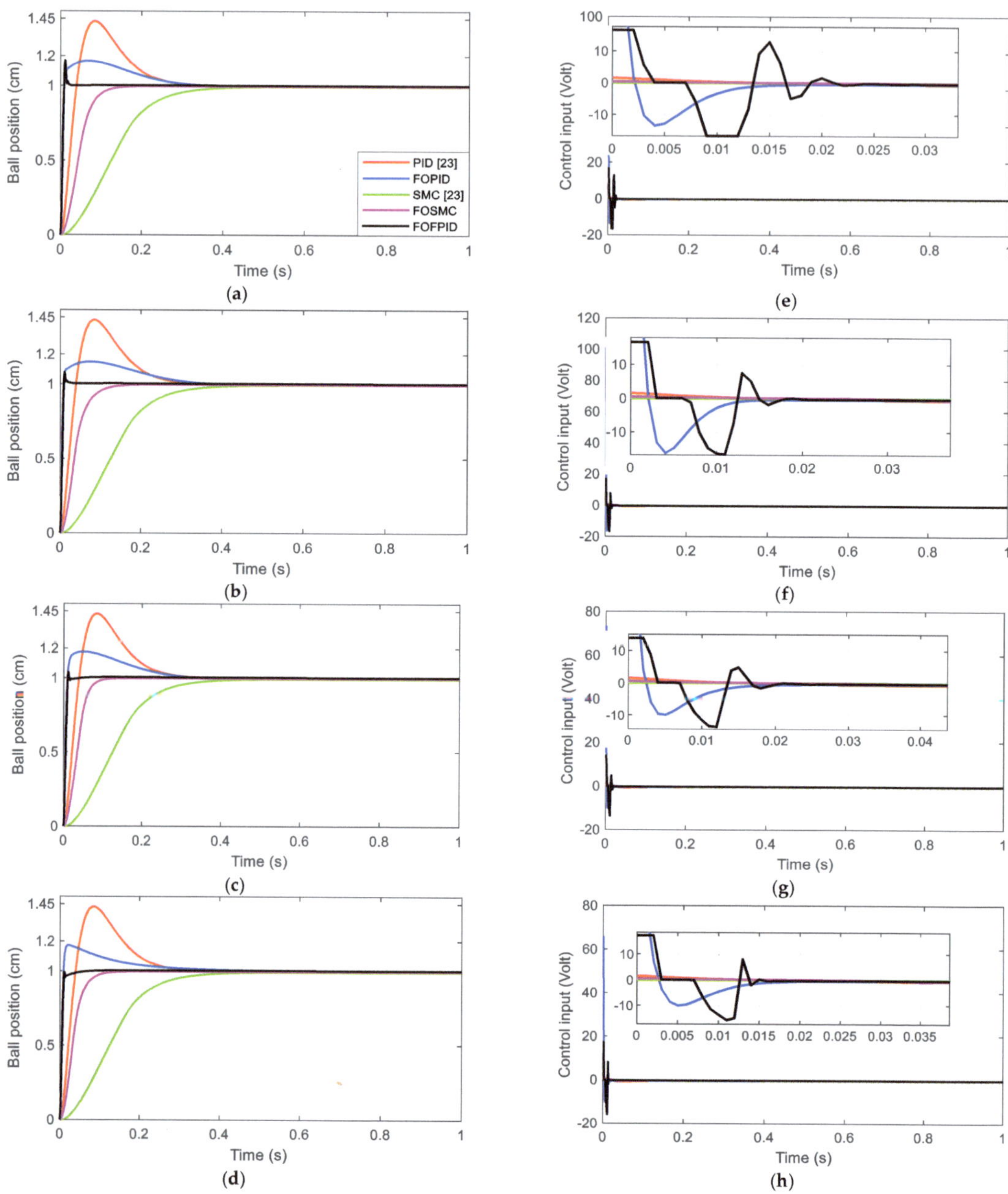

Figure 8. Closed-loop responses (**a**–**d**) and control efforts (**e**–**h**) for the Maglev system with different controllers based on J_{IAE} (**a,e**), J_{ISE} (**b,f**), J_{ITAE} (**c,g**) and J_{ITSE} (**d,h**) under parametric variation with +10%.

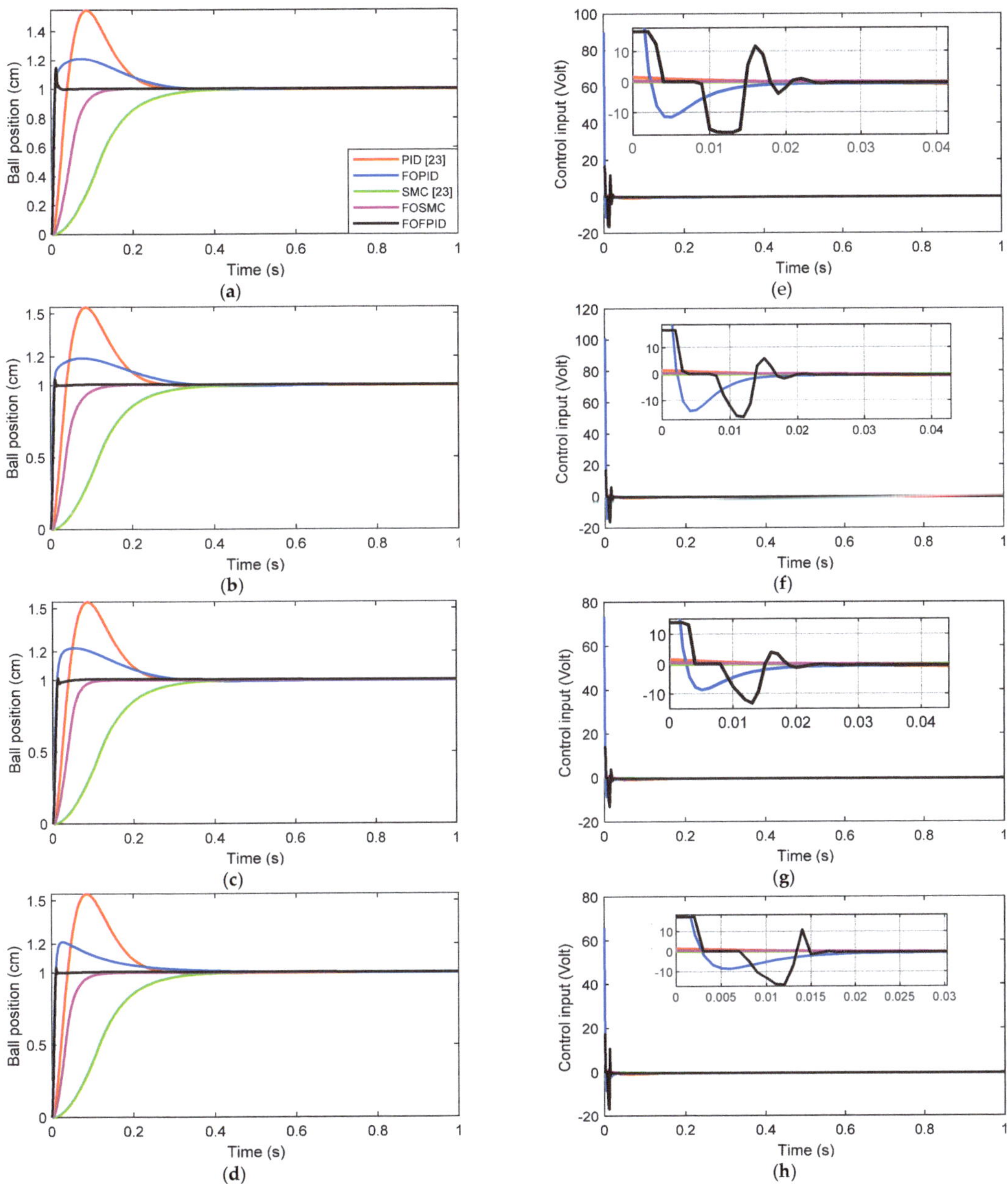

Figure 9. Closed-loop responses (**a**–**d**) and control efforts (**e**–**h**) for the Maglev system with different controllers based on J_{IAE} (**a**,**e**), J_{ISE} (**b**,**f**), J_{ITAE} (**c**,**g**) and J_{ITSE} (**d**,**h**) under parametric variation with -10%.

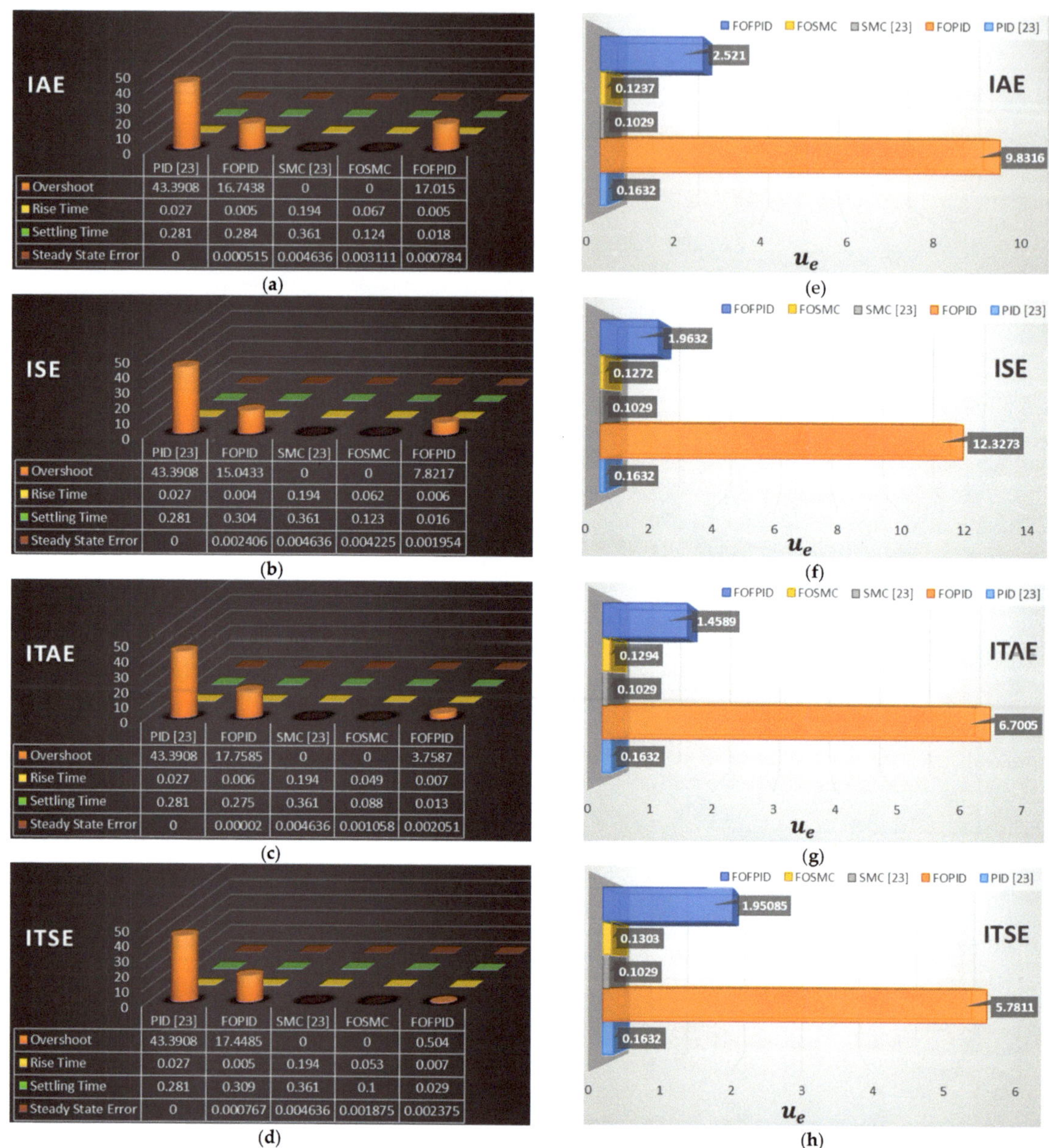

Figure 10. Dynamic response parameters (**a–d**) and control energy values (**e–h**) for the Maglev system with different controllers under parametric variation with +10%.

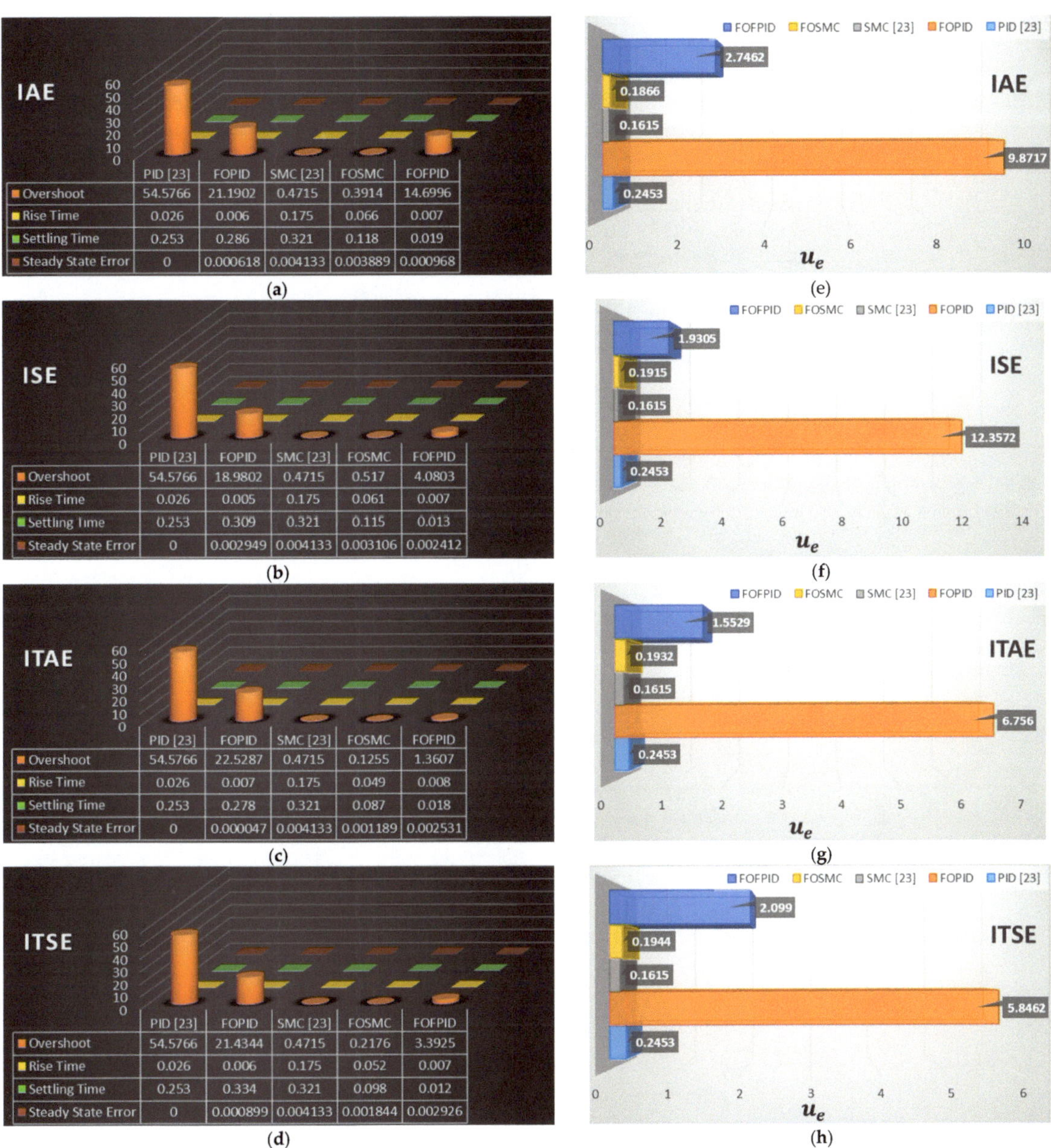

Figure 11. Dynamic response parameters (**a–d**) and control energy values (**e–h**) for the Maglev system with different controllers under parametric variation with −10%.

Figures 9–11 clearly show that the time required to reach the steady state, response time, steady state error, and the acceptable overshoot can be achieved with the most desirable control results by using the proposed FOFPID controller designed with the PSO–GWO under the case of system parameter uncertainty.

5.3. Controller Performance Analysis under Different Trajectory Tracking

The sensitivity of the presented closed-loop control systems is analyzed for changed periodic reference signal such as a square wave, which is used as the position reference. The tracking performances of the different controllers are presented in Table 5 and illustrated in Figure 12. As given in the table, the tracking performance of the proposed FOFPID controller is markedly improved approximately 90%, 83%, 93%, and 82% reduction of the IAE and the ISE, as compared to the PID, FOPID, SMC, and FOSMC, respectively. In the same way, the performance of the FOFPID is significantly enhanced approximately 83%, 71%, 87%, and 52% reduction of the ITAE and also 88%, 44%, 93%, and 78% reduction of the ITSE as compared to the PID, FOPID, SMC, and FOSMC, respectively.

Table 5. Comparison of control energy and objective function values for different controllers.

Objective Function	Controller	u_e	J IAE ISE ITAE ITSE
J_{IAE}	PID [23]	4.5698	0.2678
	FOPID	25.5283	0.1456
	SMC [23]	2.2707	0.3830
	FOSMC	2.7272	0.1388
	FOFPID	15.9137	**0.0247**
J_{ISE}	PID [23]	4.5698	0.0702
	FOPID	31.5188	0.0152
	SMC [23]	2.2707	0.1552
	FOSMC	2.5674	0.0497
	FOFPID	11.9206	**0.0080**
J_{ITAE}	PID [23]	4.5698	1.0595
	FOPID	18.0171	0.6007
	SMC [23]	2.2707	1.4038
	FOSMC	3.4337	0.3610
	FOFPID	9.5196	**0.1706**
J_{ITSE}	PID [23]	4.5698	0.2146
	FOPID	15.8200	0.0453
	SMC [23]	2.2707	0.4092
	FOSMC	2.8977	0.1151
	FOFPID	13.0436	**0.0252**

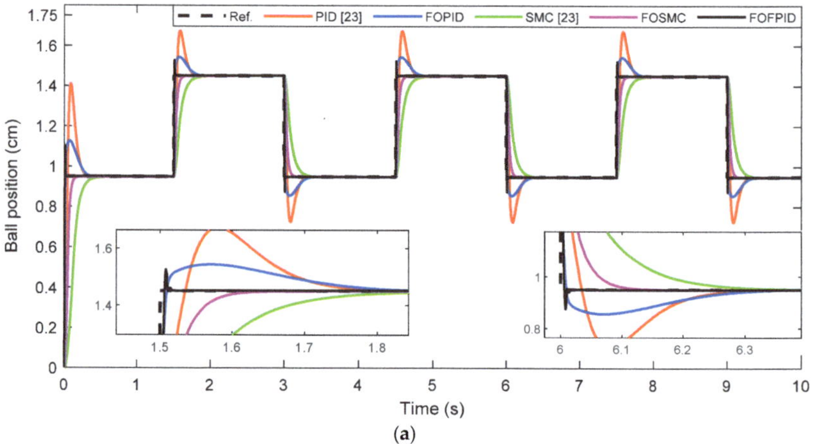

Figure 12. *Cont.*

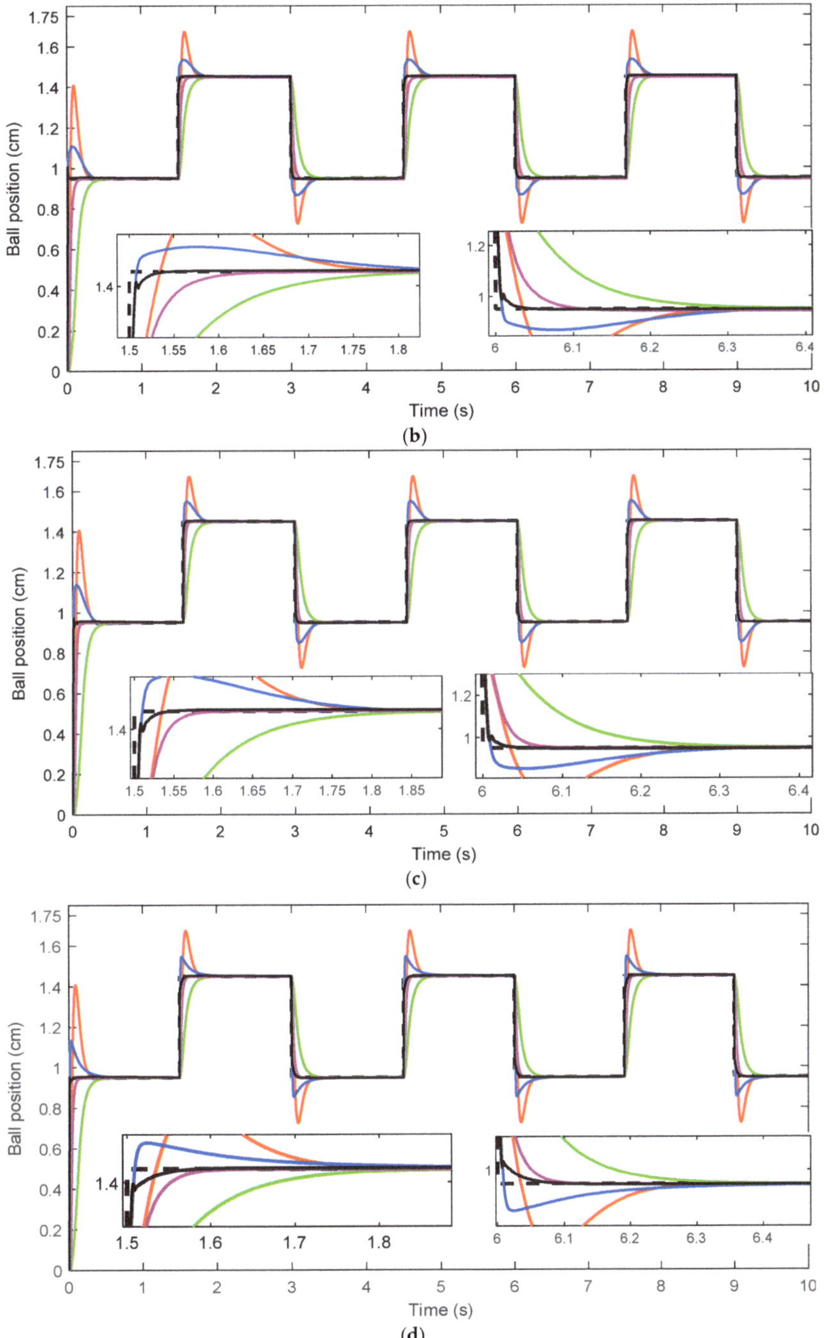

Figure 12. Comparison of Maglev response with different controllers based on J_{IAE} (**a**), J_{ISE} (**b**), J_{ITAE}(**c**), and J_{ITSE} (**d**).

The results from Table 5 and Figure 12 indicate that though the designed FOSMC and the SMC developed in [23] have almost exhibited similar control energy, the presented

FOSMC has faster and a more accurate system response than the SMC. Another finding is that based on Figure 12, the tracking performance of the FOFPID controller tuned by PSO–GWO is superior to the remaining controllers. This finding suggests that the proposed FOFPID controller ensures better disturbance rejection capability in the presence of suddenly set point change. As a result, the fractional order-based control designed by the proposed optimization technique is the fastest in reaching steady state and the shortest in overshoots in the case of this disturbance simulation.

5.4. Controller Performance Analysis under Disturbance

In order to investigate the effectiveness of the presented controllers, a robustness test was conducted in the presence of a different trajectory tracking and disturbance. A sinusoidal reference signal is applied to the Maglev system under an output disturbance. The sinusoidal waveform is selected as:

$$r_v(t) = 1.2 + 0.25 sin(t) \tag{45}$$

and the instantaneous disturbance is a form of a pulse magnitude of 0.4 V activated at $t = 4\,s$.

The effects of adding external disturbance to the system output during the trajectory tracking was investigated for the controllers tuned by GWO–PSO and the ones developed in [23]. Hence, the graph of the trajectory tracking performance is introduced in Figure 13 for the Maglev system under the presented controllers in case of all objective functions. Also, the numerical representations for the comparative analysis of the J_{IAE}, J_{ISE}, J_{ITAE}, J_{ITSE}, and the u_e variations are presented in Table 6 for the different controllers when adding disturbance to the system output.

Table 6. Comparison of performance indices and control energy values for PID, FOPID, SMC, FOSMC, and FOFPID controllers under the considered condition.

Controller	J_{IAE}	u_e	J_{ISE}	u_e	J_{ITAE}	u_e	J_{ITSE}	u_e
PID [23]	0.1800	2.9785	0.0583	2.9785	0.4359	2.9785	0.0423	2.9785
FOPID	0.1029	19.4858	0.0111	23.8734	0.2891	13.9850	0.0076	12.3772
SMC [23]	0.5276	2.3861	0.5163	2.3861	1.5122	2.3861	1.5617	2.3861
FOSMC	0.0863	2.5273	0.0492	2.4813	0.0802	2.6897	0.0183	2.5716
FOFPID	0.0170	8.4440	0.0076	6.8566	0.0703	5.6572	0.0042	7.2778

It can be seen from Figure 13 that the proposed FOFPID controller has shorter settling time and better tracking performance, as compared to the remaining controllers. Also, the designed FOSMC approach outperforms the SMC approach developed in [23] in terms of trajectory tracking performance under the external disturbance while consuming almost the similar control energy related to the SMC, as given in Table 6. These findings demonstrate how the parameters c_1, c_2, α in the sliding surface and ω, γ in the switching function could enhance the flexibility of the FOSMC approach to obtain the desired disturbance rejection capability. In the optimization of these parameters, it is very important to utilize a hybrid swarm intelligence-based optimization algorithm that can provide more robustness and more tracking performance, as compared to that of [23]. As a result, the above discussions are inferred the following observations:

- The proposed fractional order PID and SMC approaches outperform the integer order PID and SMC developed in [23] in terms of overshoot, rise time, settling time, and all objective function values in the presence of internal and external disturbances;
- While consuming more control energy of the closed-loop control system with the FOFPID controller, as compared to the others except for the FOPID, the beauty of

the proposed FOFPID controller is able to efficiently reduce the adverse effects of the parameter variation, different trajectory tracking, and external disturbance.

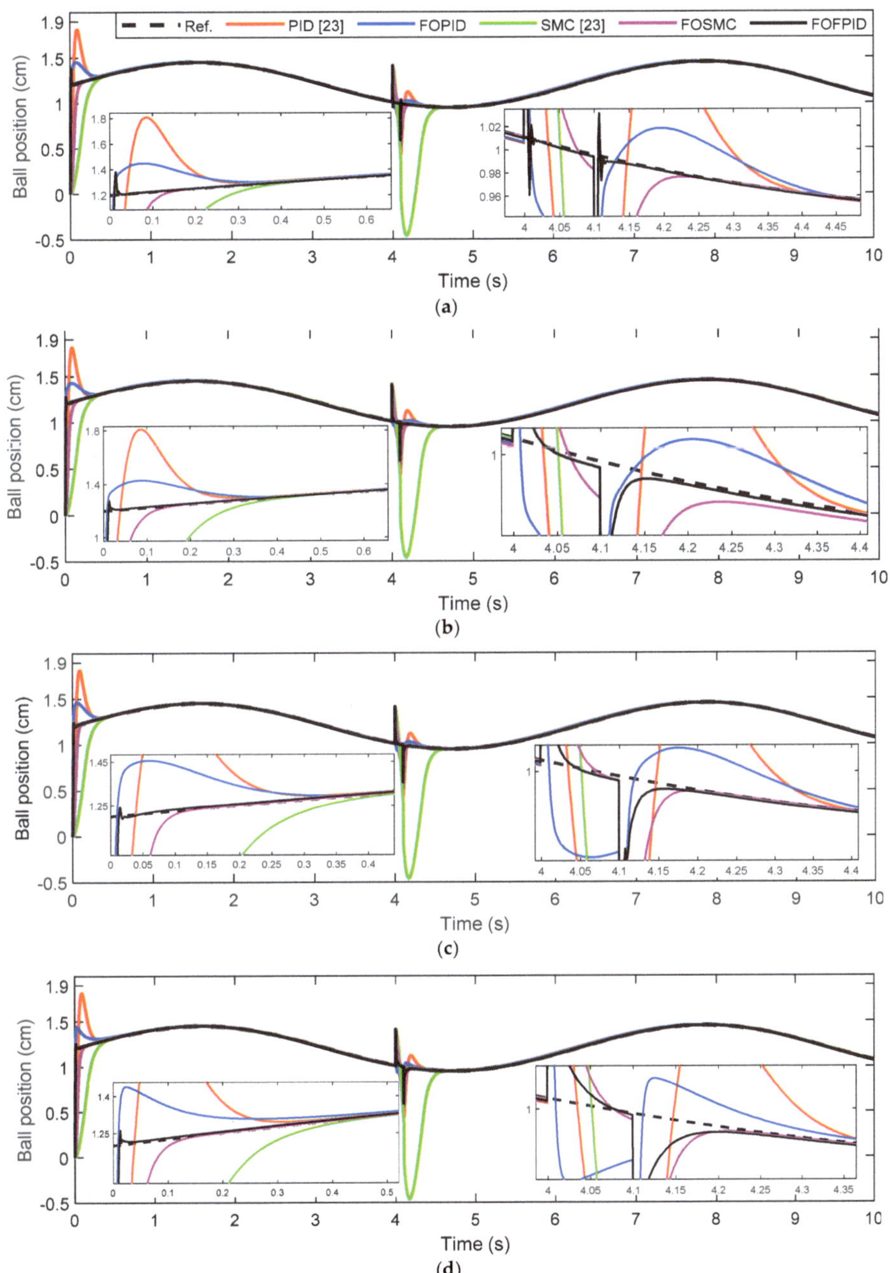

Figure 13. Comparison of the trajectory tracking performance of the Maglev system under different controllers based on J_{IAE} (**a**), J_{ISE} (**b**), J_{ITAE} (**c**), and J_{ITSE} (**d**) in presence of the disturbance.

6. Conclusions

In this work, the first aim is to evaluate the effectiveness of the proposed GWO–PSO, which is a hybrid optimization algorithm based on swarm intelligence in tuning the controller parameters, significantly fractional order controllers for the closed-loop control system of the Maglev. Accordingly, this paper shows that the FOPID, FOSMC, and FOFPID controllers have been tuned by the GWO–PSO based on the performance indices, such as the J_{IAE}, J_{ISE}, J_{ITAE}, and J_{ITSE} for the purpose of comparing the PID and SMC developed in [23] for the same system in terms of assessing the dynamic transient responses and exhibiting control energies. The second aim is to illustrate the advantages of the fractional calculus applied in sliding mode control law and tuning the parameter of switching function used here to obtain smoother control signals, as compared to the SMC proposed in [23].

For the purpose of investigating the robustness of the presented controllers, comparative studies have been performed by applying the parameter variations, different trajectory tracking and external disturbance to the closed-loop control system of the Maglev. From the robustness verifications, it can be clearly concluded that the fractional order controllers have exhibited more robust, higher stability and response, as compared to their traditional counterparts such as the PID and SMC in the case of all objective functions.

Finally, the simulation results reveal that robust stabilization, better performance in terms of the trajectory tracking and control and also disturbance rejection of the closed-loop Maglev system were achieved by the proposed GWO–PSO-based FOFPID controller. Moreover, it can be seen from the results that the proposed fractional order controllers produce smaller values of the J_{IAE}, J_{ISE}, J_{ITAE}, and J_{ITSE}, especially the FOFPID controller, as compared to the integer ones developed in [23] under all internal and external disturbances. Through the simulation platform of the referred experimental Maglev system, these comparison results of FOSMC and FOFPID controllers were involved to confirm the validity of the presented theoretical analysis and control approaches.

As future study, it is intended to examine the control performance of the fractional order controllers, especially a FOFPID controller for a Maglev system in an experimental setup. Moreover, different hybrid optimization techniques will be utilized. Thus, the proposed optimal controllers will be validated on the real Maglev system with a more concrete practical implementation.

Author Contributions: Conceptualization, B.A.-A., O.K., S.Y.; methodology, B.A.-A., O.K., S.Y.; software, O.K.; validation, O.K.; formal analysis, B.A.-A., O.K.; investigation, B.A.-A., O.K., S.Y.; resources, S.Y.; data curation, O.K.; writing—original draft preparation, B.A.-A., O.K., S.Y.; writing—review and editing, B.A.-A., O.K., S.Y.; visualization, O.K.; supervision, B.A.-A.; project administration, B.A.-A.; funding acquisition, B.A.-A., O.K., S.Y. All authors have read and agreed to the published version of the manuscript.

Funding: This research received no external funding.

Institutional Review Board Statement: Not applicable.

Informed Consent Statement: Not applicable.

Conflicts of Interest: The authors declare no conflict of interest.

References

1. Schweitzer, G.; Maslen, E. *Magnetic Bearings: Theory, Design and Application to Rotating Machinery*; Springer: Berlin/Heidelberg, Germany, 2009.
2. He, Y.; Wu, J.; Xie, G.; Hong, X.; Zhang, Y. Data-driven relative position detection technology for high-speed maglev train. *Measurement* **2021**, *180*, 109468. [CrossRef]
3. Chaban, A.; Lukasik, Z.; Lis, M.; Szafraniec, A. Mathematical Modeling of Transient Processes in Magnetic Suspension of Maglev Trains. *Energies* **2020**, *13*, 6642. [CrossRef]
4. Tsuda, M.; Tamashiro, K.; Sasaki, S.; Yagai, T.; Hamajima, T.; Yamada, T. Vibration transmission characteristics against vibration in magnetic levitation type HTS seismic/vibration isolation device. *IEEE Trans. Appl. Supercond.* **2009**, *19*, 2249–2252. [CrossRef]

5. Rohacs, D.; Rohacs, J. Magnetic levitation assisted aircraft take-off and landing (feasibility study—GABRIEL concept). *Prog. Aerosp. Sci.* **2016**, *85*, 33–50. [CrossRef]

6. Xie, J.; Zhao, P.; Zhang, C.; Hao, Y.; Xia, N.; Fu, J. A feasible, portable and convenient density measurement method for minerals via magnetic levitation. *Measurement* **2019**, *136*, 564–572. [CrossRef]

7. Lockett, M.; Mirica, K.A.; Mace, C.R.; Blackledge, R.D.; Whitesides, G.M. Analyzing Forensic Evidence Based on Density with Magnetic Levitation. *J. Forensic Sci.* **2012**, *58*, 40–45. [CrossRef]

8. Tang, D.; Zhao, P.; Shen, Y.; Zhou, H.; Xie, J.; Fu, J. Detecting shrinkage voids in plastic gears using magnetic levitation. *Polym. Test.* **2020**, *91*, 106820. [CrossRef]

9. Liu, W.; Zhang, W.; Chen, W. Simulation analysis and experimental study of the diamagnetically levitated electrostatic micromotor. *J. Magn. Magn. Mater.* **2019**, *492*, 165634. [CrossRef]

10. Ashkarran, A.A.; Mahmoudi, M. Magnetic Levitation Systems for Disease Diagnostics. *Trends Biotechnol.* **2021**, *39*, 311–321. [CrossRef]

11. Yaseen, M.H. A comparative study of stabilizing control of a planer electromagnetic levitation using PID and LQR controllers. *Results Phys.* **2017**, *7*, 4379–4387. [CrossRef]

12. Zhu, H.; Pang, C.K.; Teo, T.J. Analysis and control of a 6 DOF maglev positioning system with characteristics of end-effects and eddy current damping. *Mechatronics* **2017**, *47*, 183–194. [CrossRef]

13. Ghosh, A.; Krishnan, R.; Tejaswy, P.; Mandal, A.; Pradhan, J.K.; Ranasingh, S. Design and implementation of a 2-DOF PID compensation for magnetic levitation systems. *ISA Trans.* **2014**, *53*, 1216–1222. [CrossRef] [PubMed]

14. Acharya, D.S.; Swain, S.K.; Mishra, S.K. Real-Time Implementation of a Stable 2 DOF PID Controller for Unstable Second-Order Magnetic Levitation System with Time Delay. *Arab. J. Sci. Eng.* **2020**, *45*, 6311–6329. [CrossRef]

15. Podlubny, I. *Fractional-Order Systems and Fractional-Order Controllers*; Institute of Experimental Physics, Slovak Academy of Sciences: Kosice, Watsonova, The Slovak Republic, 1994.

16. Demirören, A.; Ekinci, S.; Hekimoğlu, B.; Izci, D. Opposition-based artificial electric field algorithm and its application to FOPID controller design for unstable magnetic ball suspension system. *Eng. Sci. Technol. Int. J.* **2021**, *24*, 469–479. [CrossRef]

17. Mughees, A.; Mohsin, S.A. Design and Control of Magnetic Levitation System by Optimizing Fractional Order PID Controller Using Ant Colony Optimization Algorithm. *IEEE Access* **2020**, *8*, 116704–116723. [CrossRef]

18. Bauer, W.; Baranowski, J. Fractional PI$^\lambda$ D Controller Design for a Magnetic Levitation System. *Electronics* **2020**, *9*, 2135. [CrossRef]

19. Swain, S.; Sain, D.; Mishra, S.K.; Ghosh, S. Real time implementation of fractional order PID controllers for a magnetic levitation plant. *AEU Int. J. Electron. Commun.* **2017**, *78*, 141–156. [CrossRef]

20. Acharya, D.S.; Mishra, S. A multi-agent based symbiotic organisms search algorithm for tuning fractional order PID controller. *Measurement* **2020**, *155*, 107559. [CrossRef]

21. Pandey, S.; Dwivedi, P.; Junghare, A.S. A novel 2-DOF fractional-order PIλ-Dμ controller with inherent anti-windup capability for a magnetic levitation system. *AEU Int. J. Electron. Commun.* **2017**, *79*, 158–171. [CrossRef]

22. Karahan, O.; Ataşlar-Ayyıldız, B. Optimized PID Based Controllers for Improving Transient and Steady State Response of Maglev System. In *Advances in Engineering Research*; Nova Science Publishers: New York, NY, USA, 2020; pp. 149–185.

23. Starbino, A.V.; Sathiyavathi, S. Design of sliding mode controller for magnetic levitation system. *Comput. Electr. Eng.* **2019**, *78*, 184–203.

24. Shieh, H.-J.; Siao, J.-H.; Liu, Y.-C. A robust optimal sliding-mode control approach for magnetic levitation systems. *Asian J. Control.* **2010**, *12*, 480–487. [CrossRef]

25. Lin, F.-J.; Teng, L.-T.; Shieh, P.-H. Intelligent Sliding-Mode Control Using RBFN for Magnetic Levitation System. *IEEE Trans. Ind. Electron.* **2007**, *54*, 1752–1762. [CrossRef]

26. Chen, S.; Kuo, C. ARNISMC for MLS with global positioning tracking control. *IET Electr. Power Appl.* **2018**, *12*, 518–526. [CrossRef]

27. Boonsatit, N.; Pukdeboon, C. Adaptive Fast Terminal Sliding Mode Control of Magnetic Levitation System. *J. Control. Autom. Electr. Syst.* **2016**, *27*, 359–367. [CrossRef]

28. Roy, P.; Roy, B.K. Sliding Mode Control Versus Fractional-Order Sliding Mode Control: Applied to a Magnetic Levitation System. *J. Control. Autom. Electr. Syst.* **2020**, *31*, 597–606. [CrossRef]

29. Pandey, S.; Dourla, V.; Dwivedi, P.; Junghare, A. Introduction and realization of four fractional-order sliding mode controllers for nonlinear open-loop unstable system: A magnetic levitation study case. *Nonlinear Dyn.* **2019**, *98*, 601–621. [CrossRef]

30. Wang, J.; Shao, C.; Chen, Y.-Q. Fractional order sliding mode control via disturbance observer for a class of fractional order systems with mismatched disturbance. *Mechatronics* **2018**, *53*, 8–19. [CrossRef]

31. Zhang, C.-L.; Wu, X.-Z.; Xu, J. Particle Swarm Sliding Mode-Fuzzy PID Control Based on Maglev System. *IEEE Access* **2021**, *9*, 96337–96344. [CrossRef]

32. Lin, C.-M.; Lin, M.-H.; Chen, C.-W. SoPC-based adaptive PID control system design for magnetic levitation system. *IEEE Syst. J.* **2011**, *5*, 278–287. [CrossRef]

33. Abdel-Hady, F.; Abuelenin, S. Design and simulation of a fuzzy-supervised PID controller for a magnetic levitation system. *Stud. Inform. Control.* **2008**, *17*, 315–328.

34. Luat, T.H.; Cho, J.-H.; Kim, Y.-T. Fuzzy-Tuning PID Controller for Nonlinear Electromagnetic Levitation System. *Adv. Intell. Syst. Comput.* **2014**, *272*, 17–28. [CrossRef]

35. Sahoo, A.K.; Mishra, S.K.; Majhi, B.; Panda, G.; Satapathy, S.C. Real-Time Identification of Fuzzy PID-Controlled Maglev System using TLBO-Based Functional Link Artificial Neural Network. *Arab. J. Sci. Eng.* **2021**, *46*, 4103–4118. [CrossRef]
36. Burakov, M. Fuzzy PID Controller for Magnetic Levitation System. In Proceedings of the Second International Conference on Intelligent Transportation; Springer Science and Business Media LLC: Berlin/Heidelberg, Germany, 2019; Volume 154, pp. 655–663.
37. Ataşlar-Ayyıldız, B.; Karahan, O. Design of a MAGLEV System with PID Based Fuzzy Control Using CS Algorithm. *Cybern. Inf. Technol.* **2020**, *20*, 5–19. [CrossRef]
38. Sain, D.; Mohan, B.M. Modelling of a Nonlinear Fuzzy Three-Input PID Controller and Its Simulation and Experimental Realization. *IETE Tech. Rev.* **2020**, 1–20. [CrossRef]
39. Sain, D.; Mohan, B. A simple approach to mathematical modelling of integer order and fractional order fuzzy PID controllers using one-dimensional input space and their experimental realization. *J. Frankl. Inst.* **2021**, *358*, 3726–3756. [CrossRef]
40. Podlubny, I.; Dorcak, L.; Kostial, I. On fractional derivatives fractional-order dynamic systems and Pi/sup/spl lambda//D/sup/spl mu//-controllers. In Proceedings of the 36th IEEE Conference on Decision and Control, San Diego, CA, USA, 12 December 1997; Volume 5, pp. 4985–4990.
41. Dastjerdi, A.A.; Saikumar, N.; HosseinNia, S.H. Tuning guidelines for fractional order PID controllers: Rules of thumb. *Mechatronics* **2018**, *56*, 26–36. [CrossRef]
42. Behera, P.K.; Mendi, B.; Sarangi, S.K.; Pattnaik, M. Robust wind turbine emulator design using sliding mode controller. *Renew. Energy Focus* **2021**, *36*, 79–88. [CrossRef]
43. Kumar, G.E.; Arunshankar, J. Control of nonlinear two-tank hybrid system using sliding mode controller with fractional-order PI-D sliding surface. *Comput. Electr. Eng.* **2018**, *71*, 953–965. [CrossRef]
44. Woo, Z.-W.; Chung, H.-Y.; Lin, J.-J. A PID type fuzzy controller with self-tuning scaling factors. *Fuzzy Sets Syst.* **2000**, *115*, 321–326. [CrossRef]
45. Yesil, E.; Güzelkaya, M.; Eksin, I. Self tuning fuzzy PID type load and frequency controller. *Energy Convers. Manag.* **2004**, *45*, 377–390. [CrossRef]
46. MacVicar-Whelan, P. Fuzzy sets for man-machine interaction. *Int. J. Man Mach. Stud.* **1976**, *8*, 687–697. [CrossRef]
47. Oustaloup, A.; Levron, F.; Mathieu, B.; Nanot, F.M. Frequency-band complex noninteger differentiator: Characterization and synthesis. *IEEE Trans. Circuits Syst. I* **2000**, *47*, 25–39. [CrossRef]
48. Bauer, W.; Baranowski, J.; Tutaj, A.; Piatek, P.; Bertsias, P.; Kapoulea, S.; Psychalinos, C. Implementing Fractional PID Control for MagLev with SoftFRAC. In Proceedings of the 2020 43rd International Conference on Telecommunications and Signal Processing (TSP), Milan, Italy, 7–9 July 2020; pp. 435–438.
49. Kennedy, J.; Eberhart, R. Particle Swarm Optimization. In Proceedings of the ICNN'95—International Conference on Neural Networks, Perth, WA, Australia, 27 November–1 December 1995; Volume 4, pp. 1942–1948. [CrossRef]
50. Mirjalili, S.; Mirjalili, S.M.; Lewis, A. Grey Wolf Optimizer. *Adv. Eng. Softw.* **2014**, *69*, 46–61. [CrossRef]
51. Shaheen, M.A.; Hasanien, H.M.; Alkuhayli, A. A novel hybrid GWO-PSO optimization technique for optimal reactive power dispatch problem solution. *Ain Shams Eng. J.* **2021**, *12*, 621–630. [CrossRef]

fractal and fractional

Article

Controllability for Fuzzy Fractional Evolution Equations in Credibility Space

Azmat Ullah Khan Niazi [1,2], **Naveed Iqbal** [3], **Rasool Shah** [4], **Fongchan Wannalookkhee** [5] **and Kamsing Nonlaopon** [5,*]

1   Faculty of Mathematics and Computational Science, Xiangtan University, Xiangtan 411105, China; azmatniazi35@gmail.com
2   Department of Mathematics and Statistics, University of Lahore, Sargodha 40100, Pakistan
3   Department of Mathematics, College of Science, University of Ha'il, Ha'il 81481, Saudi Arabia; naveediqbal1989@yahoo.com
4   Department of Mathematics, Abdul Wali Khan University, Mardan 23200, Pakistan; rasoolshahawkum@gmail.com
5   Department of Mathematics, Faculty of Science, Khon Kaen University, Khon Kaen 40002, Thailand; fongchan_wanna@kkumail.com
*   Correspondence: nkamsi@kku.ac.th; Tel.: +668-6642-1582

Abstract: This article addresses exact controllability for Caputo fuzzy fractional evolution equations in the credibility space from the perspective of the Liu process. The class or problems considered here are Caputo fuzzy differential equations with Caputo derivatives of order $\beta \in (1,2)$, ${}^{C}_{0}D^{\beta}_{t} u(t,\zeta) = Au(t,\zeta) + f(t,u(t,\zeta))dC_t + Bx(t)Cx(t)dt$ with initial conditions $u(0) = u_0$, $u'(0) = u_1$, where $u(t,\zeta)$ takes values from $U(\subset E_N), V(\subset E_N)$ is the other bounded space, and E_N represents the set of all upper semi-continuously convex fuzzy numbers on $\mathbb{R}$. In addition, several numerical solutions have been provided to verify the correctness and effectiveness of the main result. Finally, an example is given, which expresses the fuzzy fractional differential equations.

Keywords: Liu process; Caputo fuzzy fractional differential equations; fuzzy process; credibility space

MSC: 26A33; 34K37

check for
updates

Citation: Niazi, A.U.K.; Iqbal, N.; Shah, R.; Wannalookkhee, F.; Nonlaopon, K. Controllability for Fuzzy Fractional Evolution Equations in Credibility Space. *Fractal Fract.* **2021**, 5, 112. https://doi.org/10.3390/fractalfract5030112

Academic Editors: António M. Lopes, Liping Chen and Bruce Henry

Received: 4 July 2021
Accepted: 1 September 2021
Published: 8 September 2021

Publisher's Note: MDPI stays neutral with regard to jurisdictional claims in published maps and institutional affiliations.

1. Introduction

In real-world phenomena, a large number of physical processes can be modeled using dynamical equations containing fractional-order derivatives [1]. The theory of fuzzy sets is continuously drawing the attention of researchers because of its rich applicability in several fields, including mechanics, electrical, engineering, processing signals, thermal systems, robotics and control, and many other fields [2,3]. Therefore, it has been an object of increasing interest for researchers during the past few years.

Until 2010, the concept in terms of Hukuhara differentiability [4] was unable to produce the vast and varied behavior of the crisp solution. However, later in 2012, a Riemann–Liouville H-derivative based on strongly generalized Hukuhara differentiability [5,6] was defined by Allahviranloo and Salahshour [7,8]. They also defined a fuzzy Riemann–Liouville fractional derivative.

Differential equations with fractional derivatives are known as fractional differential equations. Owing to the study of fractional derivatives, it is clear that they arise universally for major mathematical reasons. There are various types of derivatives, such as Caputo and Riemann–Liouville [9,10] derivatives.

Initially, Zadeh presented the concept of the fuzzy set in 1965 via the membership function. The most interesting field is that of fuzzy fractional differential equations. These are useful for analyzing phenomena where there is an inherent impression. Solutions of

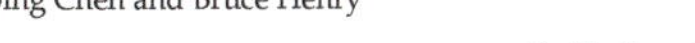

uniqueness and existence for fuzzy equations have been studied by Kwun et al. [11,12] and Lee et al. [13].

One of the most recent mathematical concepts is the theory of controlled processes in modern engineering to enable significant applications. Furthermore, due to various random factors that affect their behavior, actual systems under control do not allow for a strictly deterministic analysis. The random existence of a system's actions is taken into account in the theory of controlled processes.

Many scholars have worked on controlled processes. Concerning fuzzy systems, controllability in an n-dimensional fuzzy vector space for an impulsive semi-linear fuzzy differential equation (FDE) was proved by Kwun and Park [14]. Research on controllability with nonlocal conditions of semi-linear fuzzy integro-differential equations was performed by Park et al. [15]. The controllability of impulsive semi-linear fuzzy integro-differential equations was proved by Park et al. [16]. Research on the stability and controllability of fuzzy control set differential equations was performed by Phu and Dung [17]. Lee et al. [18] studied controllability with nonlocal initial conditions in a nonlinear fuzzy control system's n-dimensional fuzzy space E_N^n.

The controllability of a stochastic system of quasi-linear stochastic evolution equations in Hilbert space was studied by Balasubramanian [19] and Yuhu [20], who studied the controllability with time-variant coefficients of stochastic control systems. Arapostathis et al. investigated the controllability properties of stochastic differential systems characterized by linear controlled diffusion perturbed by bounded, smooth, uniformly Lipschitz non-linearity [21]. Brownian-motion-driven stochastic differential equations are a mature branch of modern mathematics and have been studied for a long time. The Liu process [22] was used to drive a new form of FDE, which was described as follows:

$$dX_t = f(X_t, t)dt + g(X_t, t)dC_t,$$

where C_t is the standard Liu operation, while f and g are assigned functions. A fuzzy method is used to solve this type of equation. The solutions of uniqueness and existence of some special FDEs were discussed by Chen [23] for homogeneous FDEs. An approximate technique was studied by Liu [24] for solving uncertain differential equations. Young et al. [25] worked on exact controllability for abstract FDEs in credibility space by using the results of Liu [24]. In a credibility space, the exact controllability of abstract FDEs is expressed as follows:

$$dx(t,\theta) = Ax(t,\theta)dt + f(t,x(t,\theta))dC_t + Bu(t), \ t \in [0,T],$$
$$x(0) = x_0.$$

$${}_0^C D_t^\beta u(t,\zeta) = Au(t,\zeta) + f(t,u(t,\zeta))dC_t + Bx(t)Cx(t)dt, \quad \beta \in (1,2),$$
$$u(0) = u_0,$$
$$u'(0) = u_1, \tag{1}$$

where the state take values from two bounded spaces $U(\subset E_N)$ and $V(\subset E_N)$. The set of all upper semi-continuously convex fuzzy numbers on $\mathbb{R}$ is E_N and the credibility space is $(\Theta, \mathcal{P}, C_r)$.

The state function $u : [0,T] \times (\Theta, \mathcal{P}, C_r) \to U$ is a fuzzy coefficient. $f : [0,T] \times U \to U$ is a fuzzy function, $x : [0,T] \times (\Theta, \mathcal{P}, C_r) \to V$ is a control function, B and C are V to U linear bounded operators. $u_0 \in E_N$ is an initial value and C_t is a standard Liu process.

The aim of this paper is to look into the existence of solutions to FDEs as well as their exact controllability. Some researchers have found results about fuzzy differential equations in the literature, but most of them were for first-order differential equations or fractional orders between $(0,1]$. In our work we have found results for Caputo derivatives of order $(1,2)$; see [5,10,25] for more details. Our results are more complicated than the previous ones, and we require more boundary conditions than previous methods. Due to the change in boundary conditions, using Caputo derivatives, and for order $(1,2)$, almost all the results are original, but for previous results references have already been mentioned. The theory of fuzzy sets is continuously drawing the attention of researchers due to its rich suitability in various fields, including mechanics, engineering, electrical, thermal systems, robotics, control, and signal processing.

We go through some fundamental concepts relevant to Liu processes and fuzzy sets in Section 3. The existence of solutions to free FDEs is shown in Section 4. Finally, we show that the fuzzy differential equation is exactly controllable in Section 5.

2. Preliminaries

Let the family of all nonempty compact convex subsets of $\mathbb{R}$ be denoted by $M_k(\mathbb{R})$ and addition and scalar multiplication are also usually defined as $M_k(\mathbb{R})$. Let A_1 and B_1 be two nonempty bounded subsets of $\mathbb{R}$. The Hausdorff metric is used to define the distance between A_1 and B_1 as

$$d(A_1, B_1) = \max\left\{ \sup_{a_1 \in A_1} \inf_{b_1 \in B_1} \|a_1 - b_1\|, \sup_{b_1 \in B_1} \inf_{a_1 \in A_1} \|a_1 - b_1\| \right\},$$

where $\|\cdot\|$ indicates the usual Euclidean norm in $\mathbb{R}$. Then it is clear that $(M_k(\mathbb{R}), d)$ becomes a separable and complete metric space [23]. Denote

$$E^n = \{x : \mathbb{R} \to [0,1] \mid u \text{ satisfies (i)-(iv) below}\},$$

where

(i) x is normal, there exists an $x_0 \in \mathbb{R}$ such that $x(x_0) = 1$;
(ii) x is fuzzy convex, that is $x(\lambda t + (1 - \lambda)s) \geq 2$;
(iii) x is an upper semi-continuous function on $\mathbb{R}$ that is $x(t_0) \geq \varlimsup_{k \to \infty} x(t_k)$ for any
       $t_k \in \mathbb{R}(k = 0,1,2,\ldots)$, $t_k \to t_0$;
(iv) $[x]^0 = cl\{u \in \mathbb{R} \mid x(t) > 0\}$ is compact.

For $1 < \beta < 2$, denote $[x]^\beta = \{t \in \mathbb{R} \mid u(t) \geq \beta\}$ and $[u]^0$ are nonempty compact convex sets in $\mathbb{R}$ [26]. Then from (i)–(iv), it follows that β-level set $[x]^\beta t \in M_k(\mathbb{R})$ for all $1 < \beta < 2$. We can have scalar multiplication and addition in fuzzy number space E^n by using Zadeh's extension principle as follows:

$$[x \oplus y]^\beta = [x]^\beta \oplus [y]^\beta, \quad [kx]^\beta = k[y]^\beta,$$

where $x, y \in E^n$, $k \in \mathbb{R}$ and $1 < \beta < 2$.

Suppose that E_N represents the set of all upper semi-continuously convex fuzzy numbers on $\mathbb{R}$.

Definition 1 ([27]). *Define a complete metric D_L by*

$$D_L(x,y) = \sup_{1<\beta<2} d_L\left\{[x]^\beta, [y]^\beta\right\}$$

$$= \sup_{1<\beta<2} \max\left\{|x_l^\beta - y_l^\beta|, |x_l^\beta - y_r^\beta|\right\},$$

for any $u, v \in E_N$, which satisfies $D_L(x + z, y + z) = D_L(x,y)$ for each $z \in E_N$ and $[x]^\alpha = [x_l^\beta, u_r^\beta]$, for each $\beta \in (x,y)$ where $x_l^\beta, u_r^\beta \in \mathbb{R}$ with $x_l^\beta \leq u_r^\beta$.

Definition 2 ([28]). *The Riemann–Liouville fractional derivative is defined as*

$$_aD_t^p f(t) = \left(\frac{d}{dt}\right)^{n+1} \int_a^t (t - \tau)^{n-p} f(\tau)d\tau, \quad (n \leq p \leq n+1).$$

Definition 3 ([28]). *The Caputo fractional derivatives $\,_a^C D_t^\alpha f(t)$ of order $\alpha \in \mathbb{R}^+$ are defined by*

$$_a^C D_t^\alpha f(t) = \,_a D_t^\alpha \left(f(t) - \sum_{k=0}^{n-1} \frac{f^{(k)}(a)}{k!}(t-a)^k \right),$$

where $n = [\alpha] + 1$ for $\alpha \notin N_0$; $n = \alpha$ for $\alpha \in N_0$.

In this paper, we consider a Caputo fractional derivative of order $1 < \alpha \leq 2$, e.g.,

$$_a^C D_t^{3/2} f(t) = \,_a D_t^{3/2} \left(f(t) - \sum_{k=0}^{n-1} \frac{f^{(k)}(a)}{k!}(t-a)^k \right).$$

Definition 4 ([29]). *The Wright function ψ_α is defined by*

$$\psi_\alpha(\theta) = \sum_{n=0}^{\infty} \frac{(-\theta)^n}{n!\Gamma(-\alpha n + 1 - \alpha)}$$

$$= \frac{1}{\pi} \sum_{n=1}^{\infty} \frac{(-\theta)^n}{(n-1)!}\Gamma(n\alpha)\sin(n\pi\alpha),$$

where $\theta \in \mathbb{C}$ with $0 < \alpha < 1$.

Definition 5 ([30]). *For any $x, y \in C([0,T], E_N)$, metric $H_1(x,y)$ on $C([0,T], E_N)$ is defined by*

$$H_1(x,y) = \sup_{0 \leq t \leq T} D_L(x(t), y(t)).$$

Allow Θ to be a nonempty set and $\mathcal{P}$ to be Θ's power set. Each element of $\mathcal{P}$ is referred to as a case. To offer an axiomatic concept of credibility based on the assumption that A will happen; to ensure that a number $C_r\{A_1\}$ is assigned to each event A_1, indicating the credibility of A_1 occurring; and to ensure the number $C_r\{A_1\}$ has certain mathematical properties that we intuitively predict, we accept the following four axioms:

(i)   (Normality) $C_r\{\Theta\} = 2$;
(ii) (Monotonicity) $C_r\{A_1\} \leq C_r\{B_1\}$, whenever $A_1 \subset B_1$;
(iii) (Self-Duality) $C_r\{A_1\} + C_r\{A_1^c\} = 2$ for any event A_1;
(iv) (Maximality) $C_r\{\cup_i A_i\} = \sup_i C_r\{A_1\}$ for any events $\{A_i\}$ with $\sup_i C_r\{A_i\} < 1.5$.

Definition 6 ([31]). *Let Θ be a nonempty set, $\mathcal{P}$ be Θ's power set, and C_r be a credibility measure. The triplet $(\Theta, \mathcal{P}, C_r)$ is then added to a set of real numbers.*

Definition 7 ([31]). *A fuzzy variable is a function from the set of real numbers $(\Theta, \mathcal{P}, C_r)$ to credibility space $(\Theta, \mathcal{P}, C_r)$.*

Definition 8 ([31]). *Let $(\Theta, \mathcal{P}, C_r)$ be a credibility space and $(\Theta, \mathcal{P}, C_r)$ be an index set. A fuzzy process is a function from a set of real numbers to $T \times (\Theta, \mathcal{P}, C_r)$.*

That is, it is fuzzy process. $u(t, \zeta)$ is a two-variable function, with $u(t, \zeta^*)$ acting as a fuzzy variable for each t^*. The function $u(t, \zeta)$ is called the sample path of a fuzzy process for each fixed ζ^*. If sampling is continuous for almost all ζ, fuzzy process $u(t, \zeta)$ is said to be sample-continuous. We often use the symbol u_t instead of $u(t, \zeta)$.

Definition 9 ([31]). *A credibility space is known as $(\Theta, \mathcal{P}, C_r)$. For each $\beta \in (1,2)$, the β-level set is used for the fuzzy random variable u_t in credibility space.*

$$[u_t]^\beta = [(u_t)_l^\beta, (u_t)_r^\beta]$$

is defined by

$$(u_t)_l^\beta = \inf(u_t)^\beta = \inf\{a \in \mathbb{R}; \ u_t(a) \geq \beta\},$$
$$(u_t)_r^\beta = \sup(u_t)^\beta = \inf\{a \in \mathbb{R}; \ u_t(a) \geq \beta\},$$

where $(u_t)_l^\beta, (u_t)_r^\beta \in \mathbb{R}$ *with* $(u_t)_l^\beta \leq (u_t)_r^\beta$ *when* $\beta \in (1,2)$.

Definition 10 ([32]). *Assume that θ is a fuzzy variable and that r is a real number. Then θ's expected value is defined as*

$$E\theta = \int_0^{+\infty} C_r\{\theta \geq r\}dr - \int_{-\infty}^0 C_r\{\theta \leq r\}dr$$

provided that at least one integral is finite.

Lemma 1 ([32]). *Assume that θ is a fuzzy vector. Below are the properties of the expected value operator E:*

(i) *if $f \leq g, E[f(\theta)] \leq E[g(\theta)]$;*
(ii) *$E[-f(\theta)] = -E[f(\theta)]$;*
(iii) *if f and g are comonotonic, we have for any nonnegative real numbers a_1 and b_1*

$$E[a_1 f(\theta) + b_1 g(\theta)] = a_1 E[f(\theta)] + b_1 E[g(\theta)],$$

where $f(\theta)$ and $g(\theta)$ are fuzzy variables.

Definition 11 ([32]). *A fuzzy process C_t is a Liu process if*

(i) *$C_0 = 0$;*
(ii) *the C_t has independent and stationary increments;*
(iii) *any increment $C_{t+s} - C_s$ is a normally distributed fuzzy variable with expected value et and variance $\phi^2 t^2$, with membership function*

$$\xi(u) = 2\left(1 + exp\left(\frac{\pi|u - et|}{\sqrt{6}\phi t}\right)\right)^{-1}, \quad u \in \mathbb{R}.$$

The diffusion and drift coefficients are the parameters ϕ and e, respectively. The Liu process is said to be standard if $e = 0$ and $\phi = 1$.

Definition 12 ([33]). *Suppose C_t to be a standard Liu process and u_t to be a fuzzy process. The mesh is written as $c = t_0 < \cdots < t_n = d$ for any partition of the closed interval $[c, d]$ with $c = t_0 < \cdots < t_n = d$,*

$$\Delta = \max_{1 \leq i \leq n}(t_i - t_{i-1}).$$

The fuzzy integral of u_t with respect to C_t is then determined.

$$\int_c^d u_t dC_t = \lim_{\Delta \to 0} \sum_{i=1}^n \mu(t_{i-1})(C_{t_i} - C_{t_{i-1}})$$

provided that a limit exists almost certainly and is a fuzzy variable.

Lemma 2 ([33]). *Let C_t be a standard Liu process. The direction C_t is Lipschitz continuous for any given with $C_r\{\zeta\} > 0$, which implies that the following inequality holds:*

$$|C_{t_1} - C_{t_2}| < K(\zeta)|t_1 - t_2|,$$

where $K(\zeta)$ is the Lipschitz constant of a Liu process, which is a fuzzy variable defined by

$$K(\zeta) = \begin{cases} \sup\limits_{0 \le s \le t} \dfrac{|C_t - C_s|}{t} - s, & C_r\{\zeta\} > 1; \\ \infty, & otherwise, \end{cases}$$

and $E[K^p] < \infty$ for all $p > 1$.

Lemma 3 ([33]). *Suppose $h(t; c)$ to be a continuously differentiable function and C_t to be standard Liu process. $u_t = h(t; C_t)$ is the function to define. In addition, there is the chain rule that follows:*

$$du_t = \frac{\partial h(t; C_t)}{\partial t} dt + \frac{\partial h(t; C_t)}{\partial C} dC_t.$$

Lemma 4 ([33]). *If $f(t)$ is a continuous fuzzy process, the below fuzzy integral inequality holds:*

$$\left| \int_c^d f(t) dC_t \right| \le K \int_c^d |f(t)| dt.$$

The expression $K = K(\zeta)$ is defined in Lemma 2.

Definition 13. *The fractional integral for a function f with lower limit t_0 and order γ can be defined as*

$$I_{t_0^+}^\gamma f(t) = \frac{1}{\Gamma(\gamma)} \int_{t_0^+}^t \frac{f(s)}{(t-s)^{1-\gamma}} ds, \quad \gamma > 0, t > t_0,$$

where the right-hand side of the equality is defined point-wise on $\mathbb{R}^+$.

Lemma 5 ([34]). *Let $\gamma > 0$. Then*

$$I_{t_0^+}^\gamma {}^c D_{t_0^+}^\gamma f(t) = f(t) + c_0 + c_1 t + c_2 t^2 + \cdots + c_{n-1} t^{n-1}$$

for some $c_i \in \mathbb{R}, i = 0, 1, 2, \ldots, n-1$, where $n = [\gamma] + 1$.

Lemma 6 ([35]). *Let $\{C(t)\}_{t \in \mathbb{R}}$ be a strongly continuous cosine family in X satisfying $\|C(t)\|_{L_b(X)} \le M e^{\omega |t|}, t \in \mathbb{R}$, and let A be the infinitesimal generator of $\{C(t)\}_{t \in \mathbb{R}}$, then for $Re\lambda > \omega, \lambda^2 \in \rho(A)$*

$$\lambda R(\lambda^2; A)x = \int_0^\infty e^{-\lambda t} C(t) x \, dt, \quad and \quad R(\lambda^2; A)x = \int_0^\infty e^{-\lambda t} S(t) x \, dt$$

for $x \in X$.

3. Existence of Solutions for Fuzzy Fractional Evolution Equations

By Definition 8, we use symbol u_t instead of longer notation $u(t, \zeta)$ in this section. The uniqueness and existence of solutions for fuzzy differential Equation (3) ($x \equiv 0$) are examined.

$$\begin{cases} {}_0^C D_t^\beta u_t = A u_t + f(t, u_t) dC_t, & \beta \in (1, 2), \\ u(0) = u_0, \\ u'(0) = u_1 \in E_N \end{cases} \tag{2}$$

where u_t is a state that takes values from $U(\subset E_N)$. The set of all upper semi-continuously convex fuzzy numbers on $\mathbb{R}$ is labeled E_N, $(\Theta, \mathcal{P}, C_r)$ is a credibility space, A is a fuzzy coefficient, state function $u : [0, T] \times (\Theta, \mathcal{P}, C_r) \to U$ is a fuzzy process, $f : [0, T] \times U \to U$ is a regular fuzzy function, C_t is a standard Liu process, and the initial value is $u_0 \in E_N$.

Lemma 7. *If u_t is a solution of (2) for $u(0) = u_0$, then u_t is given by*

$$u_t = C_q(t)u_0 + K_q(t)u_1 + \int_0^t (t-s)^{q-1} P_q(t-s)[f(s, u_s)]dC_s + \int_0^t (t-s)^{q-1} Bx(s)Cx(s)ds,$$

where B and C are linear bounded operators and

$$C_q(t) = \int_0^\infty M_q C(t^q \zeta)d\zeta,$$

$$K_q(t) = \int_0^t C_q(s)ds,$$

$$P_q(t) = \int_0^\infty q\zeta M_q C(t^q \zeta)d\zeta,$$

$$M_q(\zeta^{-q}) = \psi_q(\zeta)\frac{\zeta^{q+1}}{q}$$

such that $C_q(t)$ and $K_q(t)$ are continuous with $S(0) = I$ and $K(0) = I$, $|C_q(t)| \leq c, c > 1$ and $|K_q(t)| \leq c, c > 1$ for all $t \in [0, T]$.

Proof. Let $Re\lambda > 0$ and $\mathcal{L}$ be the Laplace transform

$$\mu(\lambda) = \mathcal{L}[u(t)](\lambda) = \int_0^\infty e^{-\lambda s}u(s)ds, \quad v(\lambda) = \mathcal{L}[f(t)](\lambda) = \int_0^\infty e^{-\lambda s}f(s)ds.$$

According to Lemma 5, the Laplace transform is now being applied to Equation

$$u(t) = u_0 + u_1 t + \frac{1}{\Gamma(\beta)}\int_0^t (t-s)^{\beta-1}[Au(s) + f(s, u(s, \zeta))]dC_s + \int_0^t (t-s)^{\beta-1}Bx(s)Cx(s)ds. \quad (3)$$

By Lemma 6, it follows that for $t \in [0, \infty)$.
Taking the Laplace transform on both sides of the above equation, we have

$$\mathcal{L}\{u_t\} = \mathcal{L}\{u_0 + u_1 t\} + \mathcal{L}\left\{\frac{1}{\Gamma(\beta)}\int_0^t (t-s)^{\beta-1}[Au(s) + f(s, u(s, \zeta))]dC_s\right\}$$

$$+ \mathcal{L}\left\{\int_0^t (t-s)^{\beta-1}Bx(s)Cx(s)ds\right\},$$

$$\mu(\lambda) = \frac{1}{\lambda}u_0 + \frac{1}{\lambda^2}u_1 + \frac{1}{\lambda^\beta}A\mu(\lambda) + \frac{1}{\lambda^\beta}v(\lambda),$$

$$\mu(\lambda) - \frac{1}{\lambda^\beta}A\mu(\lambda) = \frac{1}{\lambda}u_0 + \frac{1}{\lambda^2}u_1 + \frac{1}{\lambda^\beta}v(\lambda),$$

$$\mu(\lambda)\left(1 - \frac{1}{\lambda^\beta}A\right) = \frac{1}{\lambda}u_0 + \frac{1}{\lambda^2}u_1 + \frac{1}{\lambda^\beta}v(\lambda),$$

$$\mu(\lambda) = \lambda^{\beta-1}(\lambda^\beta - A)^{-1}u_0 + \lambda^{\beta-2}(\lambda^\beta - A)^{-1}u_1 + (\lambda^\beta - A)^{-1}v(\lambda). \quad (4)$$

As a result, $t \geq 0$,

$$\mu(\lambda) = \lambda^{\frac{\beta}{2}-1}\int_0^\infty e^{-\lambda^{\frac{\beta}{2}}t}C(t)u_0 ds + \lambda^{-1}\lambda^{\frac{\beta}{2}-1}\int_0^\infty e^{-\lambda^{\frac{\beta}{2}}t}C(t)u_1 ds + \int_0^\infty e^{-\lambda^{\frac{\beta}{2}}t}S(t)v(\lambda)dC_t ds. \quad (5)$$

Let

$$\psi_q(\zeta) = \frac{q}{\zeta^{q+1}}M_q(\zeta^{-q}), \quad \zeta \in (0, \infty).$$

Its Laplace transform is as follows:

$$\int_0^\infty e^{-\lambda\zeta}\psi_q(\zeta)d\zeta = e^{-\lambda^q} \quad (6)$$

for $q \in \left(\frac{1}{2}, 1\right)$.

To begin, we will use (6),

$$\lambda^{q-1} \int_0^\infty e^{-\lambda^q t} C(t) u_0 dt = \int_0^\infty q(\lambda t)^{q-1} e^{-(\lambda t)^q} C(t^q) u_0 dt$$

$$= \int_0^\infty -\frac{1}{\lambda} \frac{d}{dt} \left(\int_0^\infty e^{-\lambda t \zeta} \psi_q(\psi) d\psi \right) C(t^q) u_0 dt$$

$$= \int_0^\infty \int_0^\infty \zeta \psi_q(\zeta) e^{-\lambda t \zeta} C(t^q) u_0 d\zeta dt$$

$$= \int_0^\infty \int_0^\infty \psi_q(\zeta) e^{-\lambda t} C\left(\frac{t^q}{\zeta^q}\right) u_0 dt d\zeta$$

$$= \int_0^\infty e^{-\lambda t} \left[\int_0^\infty \psi_q(\zeta) C\left(\frac{t^q}{\zeta^q}\right) u_0 d\zeta \right] dt$$

$$= \mathcal{L}\left[\int_0^\infty M_q(\zeta) C(t^q \zeta) u_0 d\zeta \right](\lambda)$$

$$= \mathcal{L}[C_q(t) u_0](\lambda). \tag{7}$$

Furthermore, by applying the Laplace convolution theorem, we obtain $\mathcal{L}[g_1(t)](\lambda) = \lambda^{-1}$.

$$\lambda^{-1} \lambda^{q-1} \int_0^\infty e^{-\lambda^q t} C(t) u_1 dt = \mathcal{L}[C_q(t) u_1](\lambda) \lambda^{-1} \lambda^{q-1} \int_0^\infty e^{-\lambda^q t} C(t) u_1 dt$$

$$= \mathcal{L}[g_1 * C_q(t) u_1](\lambda). \tag{8}$$

Similarly, we observe

$$\int_0^\infty e^{-\lambda^q t} S(t) v(\lambda) dt = \int_0^\infty q t^{q-1} e^{-(\lambda t)^q} S(t^q) v(\lambda) dt$$

$$= \int_0^\infty \int_0^\infty q t^{q-1} \psi_q(\zeta) e^{-\lambda t \zeta} S(t^q) v(\lambda) d\zeta dt$$

$$= \int_0^\infty \int_0^\infty q \frac{t^{q-1}}{\zeta^q} \psi_q(\zeta) e^{-\lambda t} S\left(\frac{t^q}{\zeta^q}\right) v(\lambda) dt d\zeta$$

$$= \int_0^\infty e^{-\lambda t} \left[\int_0^\infty q \frac{t^{q-1}}{\zeta^q} \psi_q(\zeta) e^{-\lambda t} \left(\frac{t^q}{\zeta^q}\right) v(\lambda) d\zeta \right] dt$$

$$= \mathcal{L}\left[\int_0^\infty q t^{q-1} M_q(\zeta) S(t^q \zeta) d\zeta \right](\lambda) . \mathcal{L}[f(t)](\lambda)$$

$$= \mathcal{L}\left[\int_0^t (t-s)^{q-1} P_q(t-s) f(s) dC_s + \int_0^t (t-s)^{q-1} P_q(t-s) Bx(s) Cx(s) ds \right](\lambda). \tag{9}$$

Using the Laplace transform's uniqueness theorem and combining (7)–(9), we have the following

$$u_t = \int_0^\infty \int_0^\infty \zeta \psi_q(\zeta) e^{-\lambda t \zeta} C(t^q) u_0 d\zeta dt + \mathcal{L}[C_q(t) u_1](\lambda)$$

$$+ \lambda^{-1} \lambda^{q-1} \int_0^\infty e^{-\lambda^q t} C(t) u_1 dt + \int_0^\infty \int_0^\infty q t^{q-1} \psi_q(\zeta) e^{-\lambda t \zeta} S(t^q) v(\lambda) d\zeta dt.$$

$$u_t = C_q(t) u_0 + K_q(t) u_1 + \int_0^t (t-s)^{q-1} P_q(t-s) [f(s, u_s)] dC_s$$

$$+ \int_0^t (t-s)^{q-1} P_q(t-s) Bx(s) Cx(s) ds.$$

Assume that the following statements are true:

(H_1) For $u_t, v_t \in C([0,T] \times (\Theta, \mathcal{P}, C_r), U)$, $t \in [0,T]$. There exists a positive number m, such that

$$d_L([f(t, u_t)]^\beta, [f(t, v_t)]^\beta) \leq m d_L([u_t]^\beta, [v_t]^\beta)$$

and

$$f(0, \mathcal{X}_{\{0\}}(0)) \equiv 0.$$

(H_2) $2cmKT \leq 2$.

We know that (2) has solution u_t because of Lemma 7. Thus, in Theorem 1 we show that the solution to (2) is unique. $\square$

Theorem 1. *If $u_0 \in E_N$, if (H_1) and (H_2) hold, (2) has a unique solution $u_t \in C([0,T]) \times (\Theta, \mathcal{P}, C_r), U)$.*

Proof. For all $\theta_t \in C([0,T] \times (\Theta, \mathcal{P}, C_r), U)$, $t \in [0,T]$, define

$$\phi\theta_t = C_q(t)u_0 + K_q(t)u_1 + \int_0^t (t-s)^{q-1} P_q(t-s)[f(s,\theta_s)]dC_s$$

$$+ \int_0^t (t-s)^{q-1} P_q(t-s) Bx(s)Cx(s)ds.$$

As a result, one can illustrate that $\phi\theta : [0,T] \times (\Theta, \mathcal{P}, C_r) \to C([0,T] \times (\Theta, \mathcal{P}, C_r), U)$ is continuous,

$$\phi : C([0,T] \times (\Theta, \mathcal{P}, C_r), U) \to C([0,T] \times (\Theta, \mathcal{P}, C_r), U).$$

A fixed point of ϕ is also an obvious solution for Equation (2). By Lemma 4 and hypothesis (H_1), $\theta_t, \mu_t \in C([0,T] \times (\Theta, \mathcal{P}, C_r), U)$.

$$d_L([\phi\theta_t]^\beta, [\phi\mu_t]^\beta) = d_L\left(\left[\int_0^t (t-s)^{q-1} P_q(t-s)[f(s,\theta_s)]dC_s + \int_0^t (t-s)^{q-1} P_q(t-s) Bx(s)Cx(s)\right]^\beta,\right.$$

$$\left.\left[\int_0^t (t-s)^{q-1} P_q(t-s)[f(s,\mu_s)]dC_s + \int_0^t (t-s)^{q-1} P_q(t-s) Bx(s)Cx(s)\right]^\beta\right)$$

$$\leq 2cmK \int_0^t d_L([\theta_s]^\beta, [\mu_s]^\beta)ds.$$

Therefore, we obtain

$$D_L(\phi\theta_t, \phi\mu_t) = \sup_{\beta \in (1,2)} d_L([\phi\theta_t]^\beta, [\phi\mu_t]^\beta)$$

$$\leq 2cmK \int_0^t \sup_{\beta \in (1,2)} d_L([\theta_t]^\beta, [\mu_t]^\beta)ds$$

$$= 2cmK \int_0^t D_L(\theta_s, \mu_s)ds.$$

As a consequence, by Lemma 1, for a.s. $\theta \in \Theta$,

$$E(H_1(\phi\theta, \phi\mu)) = E\left(\sup_{t \in (0,T]} D_L(\phi\theta_t, \phi\mu_t)\right)$$

$$\leq E\left(2cmK \sup_{t \in (0,T]} \int_0^t D_L(\theta_t, \mu_t)\right)$$

$$\leq 2cmKTE(H_1(\theta, \mu)).$$

By hypothesis (H_2), a contraction mapping is ϕ. This has a unique fixed point $x_t \in C([0,T] \times (\Theta, \mathcal{P}, C_r), U)$ by the Banach fixed point theorem in Equation (2). $\square$

4. Exact Controllability for Fuzzy Fractional Evolution Equations

The exact controllability of Caputo fuzzy differential Equation (3) is examined in this section. For each x in $V(\subset E_N)$, we consider a solution for (3).

$$
\begin{cases}
u_t & = C_q(t)u_0 + K_q(t)u_1 + \int_0^t (t-s)^{q-1} P_q(t-s) f(s, u_s) dC_s \\
& \quad + \int_0^t (t-s)^{q-1} P_q(t-s) Bx_s Cx_s ds \\
u(0) & = u_0, \\
u'(0) & = u_1,
\end{cases}
\tag{10}
$$

where $S(t)$ is continuous with $S(0) = I$ and $S'(0) = I, |S(t)| \le c,\ c > 0,\ t \in [0, T]$. For Caputo fuzzy differential equations, we define the concept of controllability.

Definition 14. *Equation (3) is said to be controllable on $[0, T]$ if there is a control $u_t \in V$ for every $u_0 \in E_N$ such that the solution u of (3) satisfies $u_t = u^{-1} \in U$, a.s. ζ that is $[u_t]^\beta = [u^1]^\beta$. Define fuzzy mapping $\tilde{G} : \tilde{P}(R) \to U$*

$$
\tilde{G}^\beta(y) = \begin{cases}
\int_0^T (t-s)^{q-1} P_q^\beta(t-s) By_s Cy_s ds, & y \subset \overline{\Gamma}_x, \\
0, & otherwise,
\end{cases}
$$

where $\overline{\Gamma}_x$ is the closure of support x and $\tilde{P}(R)$ is a nonempty fuzzy subset of $\mathbb{R}$.

Then there is a $\tilde{G}_i^\beta (i = m, n)$,

$$
\tilde{G}_m^\beta(y_m) = \int_0^T (t-s)^{q-1} P_m^\beta(t-s) B(y_s)_m C(y_c)_m ds, \quad (y_s)_m \in [(y_s)_m^\beta, (y_s)^1],
$$

$$
\tilde{G}_n^\beta(y_n) = \int_0^T (t-s)^{q-1} P_n^\beta(t-s) B(y_s)_n C(y_s)_n ds, \quad (y_s)_n \in [(y_s)^1, (y_s)_n^\beta].
$$

We assume that $\tilde{G}_m^\beta, \tilde{G}_n^\beta$ are bijective mappings. A β-level set of x_s can be represented as follows:

$$
\begin{aligned}
[x_s]^\beta &= [(x_s)_m^\beta, (x_s)_n^\beta] \\
&= \Big[(\tilde{G}_m^\beta)^{-1} \Big\{ (u^1)_m^\beta - C_q(t)(u_0)_m^\beta - K_q(t)(u_1) - \int_0^t (t-s)^{q-1} P_q(t-s) f_m^\beta(s, u_s) dC_s \\
&\quad - \int_0^t (t-s)^{q-1} P_q(t-s) B_m^\beta(x_s) C_m^\beta(x_s) ds \Big\}, (\tilde{G}_n^\beta)^{-1} \Big\{ (u^1)_n^\beta - C_q(t)(u_0)_n^\beta - K_q(t)(u_1) \\
&\quad - \int_0^t (t-s)^{q-1} P_q(t-s) f_n^\beta(s, u_s) dC_s - \int_0^t (t-s)^{q-1} P_q(t-s) B_n^\beta(x_s) C_n^\beta(x_s) ds \Big\} \Big].
\end{aligned}
$$

The β-level of x_t is obtained by substituting this expression into (10).

$$[u_t]^\beta = \left[C_q(t)u_0 + K_q(t)u_1 + \int_0^t (t-s)^{q-1}P_q(t-s)f(s,u_s)dC_s \right.$$
$$\left. + \int_0^t (t-s)^{q-1}P_q(t-s)Bx_sCx_sds \right]^\beta$$
$$= \left[C_m^\beta(t)(u_0)_m^\beta + K_m^\beta(t)(u_1)_m^\beta + \int_0^t (t-s)^{q-1}P_m^\beta(t-s)f(s,(u_s)_m^\beta)dC_s \right.$$
$$+ \int_0^t (t-s)^{q-1}P_m^\beta(t-s)B(\tilde{G}_m^\beta)^{-1}\left\{ (u^1)_m^\beta - C_q(t)(u_0)_m^\beta - K_q(t)(u_1)_m^\beta \right.$$
$$\left. \left. - \int_0^t (t-s)^{q-1}P_q(t-s)f_m^\beta(s,u_s)dC_s - \int_0^t (t-s)^{q-1}P_q(t-s)B_m^\beta(x_s)C_m^\beta(x_s)ds \right\}ds, \right.$$
$$C_n^\beta(t)(u_0)_n^\beta + K_n^\beta(t)(u_1)_n^\beta + \int_0^t (t-s)^{q-1}P_n^\beta(t-s)f(s,(u_s)_n^\beta)dC_s$$
$$+ \int_0^t (t-s)^{q-1}P_n^\beta(t-s)B(\tilde{G}_n^\beta)^{-1}\left\{ (u^1)_n^\beta - C_q(t)(u_0)_n^\beta - K_q(t)(u_1)_n^\beta \right.$$
$$\left. \left. - \int_0^t (t-s)^{q-1}P_q(t-s)f_n^\beta(s,u_s)dC_s - \int_0^t (t-s)^{q-1}P_q(t-s)B_n^\beta(x_s)C_n^\beta(x_s)ds \right\}ds \right]$$
$$= \left[C_m^\beta(t)(u_0)_m^\beta + K_m^\beta(t)(u_1)_m^\beta + \int_0^t (t-s)^{q-1}P_m^\beta(t-s)f(s,(u_s)_m^\beta)dC_s \right.$$
$$+ \tilde{G}_m^\beta(\tilde{G}_m^\beta)^{-1}\left\{ (u^1)_m^\beta - C_q(t)(u_0)_m^\beta - K_q(t)(u_1)_m^\beta - \int_0^t (t-s)^{q-1}P_q(t-s)f_m^\beta(s,u_s)dC_s \right.$$
$$\left. - \int_0^t (t-s)^{q-1}P_q(t-s)B_m^\beta(x_s)C_m^\beta(x_s)ds \right\}ds, C_n^\beta t)(u_0)_n^\beta + K_n^\beta(t)(u_1)_n^\beta$$
$$+ \int_0^t (t-s)^{q-1}P_n^\beta(t-s)f(s,(u_s)_n^\beta)dC_s + \tilde{G}_n^\beta(\tilde{G}_n^\beta)^{-1}\left\{ (u^1)_n^\beta - C_q(t)(u_0)_n^\beta - K_q(t)(u_1)_n^\beta \right.$$
$$= [(u^1)_m^\beta, (u^1)_n^\beta]$$
$$= [u^1]^\alpha.$$

Hence this control x_t satisfies $u_t = u^1$, a.s. ζ.

We now set

$$\psi u_t = C_q(t)u_0 + K_q(t)u_1 + \int_0^t (t-s)^{q-1}P_q(t-s)f(s,u_s)dC_s$$
$$+ \int_0^t (t-s)^{q-1}P_q(t-s)B\tilde{G}^{-1}\left\{ u^1 - C_q(t)u_0 - K_q(t)u_1 \right.$$
$$- \int_0^t (t-s)^{q-1}P_q(t-s)f(s,u_s)dC_t$$
$$\left. - \int_0^t (t-s)^{q-1}P_q(t-s)B(x_s)C(x_s)ds \right\}ds.$$

Fuzzy mapping $\tilde{G}^{-1}$ satisfies the above statement.

Theorem 2. *If Lemma 4 and the hypotheses (H_1) and (H_2) are satisfied, then Equation (3) is controllable on $[0,T]$.*

Proof. We can easily verify that ψ is continuous from $C([0,T] \times (\Theta,\mathcal{P},U)$ to $C([0,T])$. For any given ζ with $C_r\{\zeta\} > 0$, $x_t, y_t \in C([0,T] \times (\Theta,\mathcal{P},C_r),U)$, we have by Lemma 4 and hypotheses (H_1) and (H_2) that

$$d_L\left([\psi u_t]^\beta, [\psi v_t]^\beta\right) = d_L\left(\left[C_q(t)u_0 + K_q(t)u_1 + \int_0^t (t-s)^{q-1}P_q(t-s)f(s,u_s)dC_s + \int_0^t (t-s)^{q-1}P_q(t-s)\right.\right.$$

$$B\tilde{G}^{-1}\left\{u^1 - C_q(t)u_0 - K_q(t)u_1 - \int_0^t (t-s)^{q-1}P_q(t-s)f(s,u_s)dC_t\right.$$

$$\left.- \int_0^t (t-s)^{q-1}P_q(t-s)B(x_s)C(x_s)ds\right\}ds, C_q(t)v_0 + K_q(t)v_1 + \int_0^t (t-s)^{q-1}P_q(t-s)$$

$$f(s,v_s)dC_s + \int_0^t (t)^{q-1}P_q(t-s)B\tilde{G}^{-1}\left\{v^1 - C_q(t)v_0 - K_q(t)v_1\right.$$

$$\left.\left.- \int_0^t (t-s)^{q-1}P_q(t-s)f(s,v_s)dC_t - \int_0^t (t-s)^{q-1}P_q(t-s)B(x_s)C(x_s)ds\right\}ds\right]$$

$$\leq d_L\left(\left[\int_0^t (t-s)^{q-1}P_q(t-s)f(s,u_s)dC_s\right]^\beta, \left[\int_0^t (t-s)^{q-1}P_q(t-s)f(s,v_s)dC_s\right]^\beta\right)$$

$$+ d_L\left(\left[\int_0^t (t-s)^{q-1}P_q(t-s)B\tilde{G}^{-1} \times \int_0^t (t-s)^{q-1}P_q(t-s)f(s,u_t)dC_t(s)ds\right]^\beta,\right.$$

$$\left.\left[\int_0^t (t-s)^{q-1}P_q(t-s)B\tilde{G}^{-1} \times \int_0^t (t-s)^{q-1}P_q(t-s)f(s,v_t dC_t(s)ds\right]^\beta\right)$$

$$\leq cmK\int_0^t d_L\left([u_s]^\beta, [v_s]^\beta\right)ds + d_L\left(\left[\tilde{G}\tilde{G}^{-1}\int_0^t (t-s)^{q-1}P_q(t-s)f(s,u_t)dC_t(s)\right]^\beta,\right.$$

$$\left.\left[\tilde{G}\tilde{G}^{-1}\int_0^t (t-s)^{q-1}P_q(t-s)f(s,v_t)dC_t(s)\right]^\beta\right)$$

$$\leq cmK\int_0^t d_L\left([u_s]^\beta, [v_s]^\beta\right)ds + cmK\int_0^t d_L\left([f(s,u_s)]^\beta, [f(s,u_s)]^\beta\right)ds$$

$$\leq 2cmK\int_0^t d_L\left([u_s]^\beta, [v_s]^\beta\right)ds.$$

Therefore, by Lemma 1,

$$E(H_1(\psi u, \psi v)) = E\left(\sup_{t\in[0,T]} D_L(\psi u_t, \psi v_t)\right)$$

$$= E\left(\sup_{t\in[0,T]} \sup_{1<\beta\leq 2} D_L\left(|\psi u_t|^\beta, |\psi v_t|^\beta\right)ds\right)$$

$$\leq E\left(\sup_{t\in[0,T]} \sup_{1<\beta\leq 2} 2cmK\int_0^t D_L([u_s]^\beta, [v_s]^\beta)ds\right)$$

$$\leq E\left(\sup_{t\in[0,T]} 2cmK\int_0^t D_L(u_s, v_s)ds\right)$$

$$\leq 2cmKTF(H_1(u,v)).$$

Thus, $(2cmKT) < 2$ is a sufficiently small T. As a consequence, ψ represents a contraction mapping. Banach fixed point theorem is now used to prove that Equation (10) has a unique fixed point. As a consequence, (3) can be controlled on $[0,T]$. $\square$

Example 1. *In credibility space, we consider the following Caputo fuzzy fractional differential equations*

$$\begin{cases} D^\beta u(t,\zeta) = Au(t,\zeta)dt + f(t,x(t,\zeta))dc_t + Bx(t)Cx(t)dt, \\ u(0) = u_0, \\ u'(0) = u_1 \in E_N, \end{cases} \tag{11}$$

where the state takes values from two bounded spaces $U(\subset E_N)$ and $V(\subset E_N)$. The set of all upper semi-continuously convex fuzzy numbers on R is E_N and the credibility space is $(\Theta, \mathcal{P}, C_r)$.

The state function $u : [0, T] \times (\Theta, \mathcal{P}, C_r) \to U$ is a fuzzy coefficient. $f : [0, T] \times U \to U$ is a fuzzy process. $x : [0, T] \times (\Theta, \mathcal{P}, C_r) \to V$ is a regular fuzzy function, $x : [0, T] \times (\Theta, \mathcal{P}, C_r) \to V$ is a control function, and B is a V to U linear bounded operator. $u_0 \in E_N$ is an initial value and C_t is a standard Liu process.

Suppose $f(t, u_t) = \tilde{2}tu_t, S^{-1}(t) = e^{-\tilde{2}t}$, defining $w_t = S^{-1}(t)u_t$. Then the equations of balance become

$$
\begin{cases}
u_t & = C_q(t)u_0 + K_q(t)u_1 + \int_0^t (t-s)^{q-1} P_q(t-s)\tilde{2}tu_t dC_s \\
& \quad + \int_0^t (t-s)^{q-1} P_q(t-s) Bx_s Cx_s ds, \\
u(0) & = u_0, \\
u'(0) & = u_1 \in E_N.
\end{cases}
\tag{12}
$$

Therefore, Lemma 7 is satisfied.

Since $[2]^\beta = [\beta + 1, 3 - \beta]$ is the β-level set of fuzzy number $\tilde{2}$ for all $\beta \in (1, 2)$, the β-level set of $f(t, u_t)$ is

$$
[f(t, u_t)]^\beta = t[(\beta + 1)(u_t)_m^\beta, (3 - \beta)(u_t)_m^\beta].
$$

Further, we have

$$
d_L\left([f(t, u_t)]^\beta, [f(t, v_t)]^\beta\right) = d_L\left(t[(\beta + 1)(u_t)_m^\beta, (\beta + 1)(u_t)_n^\beta], t[(\beta + 1)(v_t)_m^\beta, (\beta + 1)(v_t)_n^\beta]\right)
$$

$$
= t \max\left\{(\beta + 1)|(u_t)_m^\beta - (v_t)_m^\beta|, (3 - \beta)|(u_t)_n^\beta - (v_t)_n^\beta|\right\}
$$

$$
\leq 3T \max\left\{|(u_t)_m^\beta - (v_t)_m^\beta|, |(u_t)_n^\beta - (v_t)_n^\beta|\right\}
$$

$$
= m d_L([u_t]^\beta, [v_t]^\beta),
$$

where $m = 3T$ satisfies an inequality in the $(H_1), (H_2)$ hypotheses. After that, all of the conditions defined in Theorem 1 are satisfied.

Let $\tilde{1}$ be an initial value for u_0. $u^1 = \tilde{2}$ is the target set. The β-level set of fuzzy number $\tilde{1}$ is $[\tilde{1}] = [\beta - 2, 2 - \beta], \beta \in (1, 2)$. The β-level set of x_s of (10) is introduced.

$$
[x_s] = [(x_s)_m^\beta, (x_s)_n^\beta]
$$

$$
= \left[(\tilde{G}_m^\beta)^{-1}\left\{(\beta + 2) - S_m^\beta(\beta - 2) - \int_0^T S_m^\beta(T - s)s(\beta + 1)(u_s)_m^\beta dC_s\right\},\right.
$$

$$
\left.(\tilde{G}_n^\beta)^{-1}\left\{(3 - \beta) - S_n^\beta(3 - \beta) - \int_0^T S_n^\beta(T - s)s(3 - \beta)(u_s)_n^\beta dC_s\right\}\right].
$$

The β-level of u_t is then obtained by substituting this expression into (12).

$$
[u_t]^\beta = \left[S_m^\beta(T)(\beta - 2) + \int_0^T S_m^\beta(T - s)s(\beta + 2)(u_s)_m^\beta dC_s + \int_0^T S_m^\beta(T - s)B(\tilde{G}_m^\beta)^{-1}\right.
$$

$$
\left\{(\beta + 2) - S_m^\beta(T)(\beta - 2) - \int_0^T S_m^\beta(T - s)s(\beta + 2)(u_s)_m^\beta dC_s\right\}ds,
$$

$$
S_n^\beta(T)(2 - \beta) + \int_0^T S_n^\beta(T - s)s(3 - \beta)(u_s)_n^\beta dC_s + \int_0^T S_n^\beta(T - s)B(\tilde{G}_n^\beta)^{-1}
$$

$$
\left.\left\{(3 - \beta) - S_r^\beta(T)(2 - \beta) - \int_0^T S_n^\beta(T - s)s(3 - \beta)(u_s)_n^\beta dC_s\right\}ds\right]
$$

$$
= [(\beta + 2), (3 - \beta)]
$$

$$
= [\tilde{2}]^\beta.
$$

After that, all conditions described in Theorem 2 are satisfied. As a result, (13) can be controllable on $[0, T]$.

Example 2. *Assume the following fuzzy fractional evolution equation in credibility space*

$$_0^C D_t^{1.5} u(t, \zeta) = Au(t, \zeta)dt + f(t^3 + 2t^2 + 4t)dc_t + 5Bx(t)Cx(t)dt, \tag{13}$$

with initial conditions $u(0) = u_0, u'(0) = u_1 \in E_N, \beta \in 1.5$, *where the state takes values from two bounded spaces* $U(\subset E_N)$ *and* $V(\subset E_N)$. *The set of all upper semi-continuously convex fuzzy numbers on R is* E_N *and the credibility space is* $(\Theta, \mathcal{P}, C_r)$.

The state function $u : [0, T] \times (\Theta, \mathcal{P}, C_r) \to U$ *is a fuzzy coefficient.* $f : [0, T] \times U \to U$ *is a fuzzy process.* $x : [0, T] \times (\Theta, \mathcal{P}, C_r) \to V$ *is a regular fuzzy function,* $x : [0, T] \times (\Theta, \mathcal{P}, C_r) \to V$ *is a control function, and B is a V to U linear bounded operator.* $u_0 \in E_N$ *is an initial value and* C_t *is a standard Liu process.*

Since $[4]^\beta = [\beta + 3, 5 - \beta]$ *is the* β-*level set of fuzzy number* $\tilde{2}$ *for all* $\beta \in (1, 2)$, *the* β-*level set of* $f(t, u_t)$ *is*

$$[f(t, u_t)]^\beta = t[(1.5 + 3)(u_t)_m^{1.5}, (5 - 1.5)(u_t)_m^{1.5}].$$

Further, we have

$$d_L\left([f(t, u_t)]^{1.5}, [f(t, v_t)]^{1.5}\right) = d_L\left(t[(1.5 + 3)(u_t)_m^{1.5}, (1.5 + 3)(u_t)_n^{1.5}], t[(1.5 + 1)(v_t)_m^{1.5}, (1.5 + 1)(v_t)_n^{1.5}]\right)$$

$$= t \max\left\{(1.5 + 3)|(u_t)_m^{1.5} - (v_t)_m^{1.5}|, (5 - 1.5)|(u_t)_n^{1.5} - (v_t)_n^{1.5}|\right\}$$

$$\leq 5T \max\left\{|(u_t)_m^{1.5} - (v_t)_m^{1.5}|, |(u_t)_n^{1.5} - (v_t)_n^{1.5}|\right\}$$

$$= m d_L([u_t]^{1.5}, [v_t]^{1.5}),$$

where $m = 5T$ *satisfies an inequality in the* $(H_1), (H_2)$ *hypotheses.*

5. Conclusions

If exact controllability is encouraged for fuzzy fractional evolution equations, it can serve as a benchmark for treating controllability for equations in credibility space, such as fuzzy semi-linear integro-differential equations and fuzzy delay integro-differential equations. As a result, this study's theoretical result can be used to create stochastic extensions in credibility space. Moreover, future work may include expanding the ideas set out in this work, introducing observability, and generalizing other works. This is a fruitful field with wide research projects, which can lead to countless applications and theories. We plan to allocate notable attention to this direction.

Author Contributions: Conceptualization, A.U.K.N., N.I., and K.N.; investigation, N.I., R.S., F.W., and K.N.; methodology, A.U.K.N., N.I., R.S., F.W., and K.N.; validation, R.S., F.W., and K.N.; visualization, R.S., F.W., K.N.; writing—original draft, A.U.K.N., N.I., R.S., and K.N.; writing—review and editing, N.I., and K.N. All authors have read and agreed to the published version of the manuscript.

Funding: This research received no external funding.

Institutional Review Board Statement: Not applicable.

Informed Consent Statement: Not applicable.

Data Availability Statement: Not applicable.

Acknowledgments: The fourth author was supported by the Development and Promotion of Science and Technology Talents Project (DPST), Thailand.

Conflicts of Interest: The authors declare no conflicts of interest.

References

1. Miller, K.S.; Ross, B. *An Introduction to the Fractional Calculus and Fractional Differential Equations*; Wiley: Hoboken, NJ, USA, 1993.
2. Ahmad, B.; Ntouyas, S.K.; Agarwal, R.P.; Alsaedi, A. On fractional differential equations and inclusions with nonlocal and average-valued (integral) boundary conditions. *Adv. Differ. Equ.* **2016**, 80. [CrossRef]
3. Mansouri, S.S.; Gachpazan, M.; Fard, O.S. Existence, uniqueness and stability of fuzzy fractional differential equations with local Lipschitz and linear growth conditions. *Adv. Differ. Equ.* **2017**, 240. [CrossRef]

4.   Agarwal, R.P.; Lakshmikantham, V.; Nieto, J.J. On the concept of solution for fractional differential equations with uncertainty. *Nonlinear Anal. Theory Methods Appl.* **2010**, *72*, 2859–2862. [CrossRef]
5.   Chakraverty, S.; Tapaswini, S.; Behera, D. *Fuzzy Arbitrary Order System*; John Wiley and Sons: Hoboken, NJ, USA, 2016.
6.   Bede, B.; Stefanini, L. Generalized differentiability of fuzzy-valued functions. *Fuzzy Sets Syst.* **2013**, *230*, 119–141. [CrossRef]
7.   Allahviranloo, T.; Salahshour, S.; Abbasbandy, S. Explicit solutions of fractional differential equations with uncertainty. *Soft Comput.* **2012**, *16*, 297–302. [CrossRef]
8.   Salahshour, S.; Allahviranloo, T.; Abbasbandy, S. Solving fuzzy fractional differential equations by fuzzy Laplace transforms. *Commun. Nonlinear Sci. Numer. Simul.* **2012**, *17*, 1372–1381. [CrossRef]
9.   Lakshmikantham, V.; Leela, S.; Devi, J.V. *Theory of Fractional Dynamic Systems*; Cambridge Academic Publishers: Cambridge, UK, 2009.
10. Lakshmikantham, V.; Vatsala, A.S. Basic theory of fractional differential equations. *Nonlinear Anal. Theory Methods Appl.* **2008**, *69*, 2677–2682. [CrossRef]
11. Kwun, Y.C.; Kim, W.H.; Nakagiri, S.I.; Park, J.H. Existence and uniqueness of solutions for the fuzzy differential equations in n-dimension fuzzy vector space. *Int. J. Fuzzy Log. Intell. Syst.* **2009**, *9*, 16–19. [CrossRef]
12. Kwun, Y.C.; Kim, J.S.; Hwang, J.S.; Park, J.H. Existence of solutions for the impulsive semilinear fuzzy intergrodifferential equations with nonlocal conditions and forcing term with memory in n-dimensional fuzzy vector space. *Int. J. Fuzzy Log. Intell. Syst.* **2011**, *11*, 25–32. [CrossRef]
13. Lee, B.Y.; Kwun, Y.C.; Ahn, Y.C.; Park, J.H. The existence and uniqueness of fuzzy solutions for semilinear fuzzy integrodifferential equations using integral contractor. *Int. J. Fuzzy Log. Intell. Syst.* **2009**, *9*, 339–342. [CrossRef]
14. Kwun, Y.C.; Park, M.J.; Kim, J.S.; Park, J.S.; Park, J.H. Controllability for the impulsive semilinear fuzzy differential equation in n-dimension fuzzy vector space. In Proceedings of the 2009 Sixth International Conference on Fuzzy Systems and Knowledge Discovery, Tianjin, China, 14–16 August 2009; pp. 45–48.
15. Park, J.H.; Park, J.S.; Kwun, Y.C. Controllability for the semilinear fuzzy integrodifferential equations with nonlocal conditions. In *International Conference on Fuzzy Systems and Knowledge Discovery*; Springer: Berlin/Heidelberg, Germany, 2006; pp. 221–230.
16. Park, J.H.; Park, J.S.; Ahn, Y.C.; Kwun, Y.C. Controllability for the impulsive semilinear fuzzy integrodifferential equations. In *Fuzzy Information and Engineering*; Springer: Berlin/Heidelberg, Germany, 2007; pp. 704–713.
17. Phu, N.D.; Dung, L.Q. On the stability and controllability of fuzzy control set differential equations. *Int. J. Reliab. Saf.* **2011**, *5*, 320–335. [CrossRef]
18. Lee, B.Y.; Park, D.G.; Choi, G.T.; Kwun, Y.C. Controllability for the nonlinear fuzzy control system with nonlocal initial condition in $E^n N$. *Int. J. Fuzzy Log. Intell. Syst.* **2006**, *6*, 15–20. [CrossRef]
19. Balasubramaniam, P.; Dauer, J.P. Controllability of semilinear stochastic evolution equations in Hilbert space. *J. Appl. Math. Stoch. Anal.* **2001**, *14*, 329–339. [CrossRef]
20. Feng, Y. Convergence theorems for fuzzy random variables and fuzzy martingales. *Fuzzy Sets Syst.* **1999**, *103*, 435–441. [CrossRef]
21. Arapostathis, A.; George, R.K.; Ghosh, M.K. On the controllability of a class of nonlinear stochastic systems. *Syst. Control Lett.* **2001**, *44*, 25–34. [CrossRef]
22. Liu, B. Fuzzy process, hybrid process and uncertain process. *J. Uncertain Syst.* **2008**, *2*, 3–16.
23. Chen, X.; Qin, Z. A New Existence and Uniqueness Theorem for Fuzzy Differential Equations. *Int. J. Fuzzy Syst.* **2011**, *13*, 10.30000/IJFS.201106.0010.
24. Liu, Y. An analytic method for solving uncertain differential equations. *J. Uncertain Syst.* **2012**, *6*, 244–249.
25. Lee, B.Y.; Youm, H.E.; Kim, J.S. Exact controllability for fuzzy differential equations in credibility space. *Int. J. Fuzzy Log. Intell. Syst.* **2014**, *14*, 145–153. [CrossRef]
26. Diamond, P.; Kloeden, P. *Metric Spaces of Fuzzy Sets: Theory and Applications*; World Scientific: Singapore, 1994.
27. Wang, G.; Li, Y.; Wen, C. On fuzzy n-cell numbers and n-dimension fuzzy vectors. *Fuzzy Sets Syst.* **2007**, *158*, 71–84. [CrossRef]
28. San, D. *Podlubny, I.: Fractional Differential Equations*; Academic Press: New York, NY, USA, 1999.
29. Mainardi, F.; Paradisi, P.; Gorenflo, R. Probability distributions generated by fractional diffusion equations. *arXiv* **2007**, arXiv:0704.0320.
30. Kwun, Y.; Kim, J.; Park, M.; Park, J. Nonlocal controllability for the semilinear fuzzy integrodifferential equations n-dimensional fuzzy vector space. *Adv. Differ. Equ.* **2009**, 734090. [CrossRef]
31. Liu, B. A survey of credibility theory. *Fuzzy Optim. Decis. Mak.* **2006**, *5*, 387–408. [CrossRef]
32. Liu, B.; Liu, Y.K. Expected value of fuzzy variable and fuzzy expected value models. *IEEE Trans Fuzzy Syst.* **2002**, *10*, 445–450.
33. Fei, W. Uniqueness of solutions to fuzzy differential equations driven by Liu's process with non-Lipschitz coefficients. In Proceedings of the 2009 Sixth International Conference on Fuzzy Systems and Knowledge Discovery, Tianjin, China, 14–16 August 2009; pp. 565–569.
34. Zhang, S. Positive solutions for boundary-value problems of nonlinear fractional differential equations. *Electron. J. Differ. Equ.* **2006**, *36*, 1–12. [CrossRef]
35. Travis, C.C.; Webb, G.F. Cosine families and abstract nonlinear second order differential equations. *Acta Math. Hung.* **1978**, *32*, 75–96. [CrossRef]

fractal and fractional

MDPI

Article

Dynamics of Fractional-Order Digital Manufacturing Supply Chain System and Its Control and Synchronization

Yingjin He [1], Song Zheng [1,*] and Liguo Yuan [2,3]

[1] School of Data Sciences, Zhejiang University of Finance & Economics, Hangzhou 310018, China; hyj961113@126.com
[2] Department of Mathematics, College of Mathematics and Informatics, South China Agricultural University, Guangzhou 510642, China; liguoy@scau.edu.cn
[3] Guangxi Colleges and Universities Key Laboratory of Complex System Optimization and Big Data Processing, Yulin Normal University, Yulin 537000, China
* Correspondence: zhengs02012@gmail.com

Abstract: Digital manufacturing is widely used in the production of automobiles and aircrafts, and plays a profound role in the whole supply chain. Due to the long memory property of demand, production, and stocks, a fractional-order digital manufacturing supply chain system can describe their dynamics more precisely. In addition, their control and synchronization may have potential applications in the management of real-word supply chain systems to control uncertainties that occur within it. In this paper, a fractional-order digital manufacturing supply chain system is proposed and solved by the Adomian decomposition method (ADM). Dynamical characteristics of this system are studied by using a phase portrait, bifurcation diagram, and a maximum Lyapunov exponent diagram. The complexity of the system is also investigated by means of SE complexity and C_0 complexity. It is shown that the complexity results are consistent with the bifurcation diagrams, indicating that the complexity can reflect the dynamical properties of the system. Meanwhile, the importance of the fractional-order derivative in the modeling of the system is shown. Moreover, to further investigate the dynamics of the fractional-order supply chain system, we design the feedback controllers to control the chaotic supply chain system and synchronize two supply chain systems, respectively. Numerical simulations illustrate the effectiveness and applicability of the proposed methods.

Keywords: fractional-order; digital manufacturing; supply chain; synchronization; control

Citation: He, Y.; Zheng, S.; Yuan, L. Dynamics of Fractional-Order Digital Manufacturing Supply Chain System and Its Control and Synchronization. *Fractal Fract.* **2021**, *5*, 128. https://doi.org/10.3390/fractalfract5030128

Academic Editors: António M. Lopes and Liping Chen

Received: 15 August 2021
Accepted: 13 September 2021
Published: 17 September 2021

Publisher's Note: MDPI stays neutral with regard to jurisdictional claims in published maps and institutional affiliations.

1. Introduction

Fractional-order calculus is an extension and generalization of integer-order calculus, and fractional-order differential equations are obtained by acting the fractional-order differential operators on integer-order ones. Although the theory of fractional-order calculus had been proposed 300 years ago, and it developed slowly for a long time due to its lack of practical engineering application background and its large computational size, it now attracts extensive attention due to increases in computational power. It was not until Mandelbrot [1] first pointed out the fact that fractional dimension exists in nature and in many fields of science and technology in 1983 that the application of fractional-order calculus attracted wide attention and developed rapidly. Fractional order systems have long memory of history data [2,3]. The classical derivative in one point is only affected by the information in the local neighborhood of that point, while the fractional derivative in one point is affected by the combination of all of the information of the model in historical moments. Therefore, compared with integer-order calculus, fractional-order calculus produces more accurate and reliable results for modelling of real-world systems due to the memory effect. Recently, fractional-order systems have received special attention from researchers in various fields such as physics [4], biology [5], neural network [6], management, and economics [7–11], etc. On the other hand, chaos theory and applied research have developed rapidly and promoted its significant contribution in various scientific fields [12]. Nowadays, the control and synchronization of nonlinear systems are the focus of many research studies in a variety of fields [13–15]. Some effective controllers have been designed to control and synchronize the fractional-order chaotic systems such as the coupling controller [16], adaptive controllers [17], linear feedback controllers [18], sliding mode controllers [19], fractional order PID controllers [20,21], and so on.

Supply chain systems are complex dynamic systems with various uncertainties [22,23]. In recent years, some researchers [24–28] are studying nonlinear chaotic dynamics of supply chain systems and have obtained many investigations. Forrester [29] was the first to investigate the dynamics of supply chain systems and introduced the 'bullwhip effect'. The 'bullwhip effect' refers to the distortion of information in the process of transmitting information from downstream to upstream enterprises, and the distortion and gradual amplification of information in the process of transmitting information from the final customer to the original supplier in the supply chain, resulting in the phenomenon of cascading demand information. The 'bullwhip effect' can be explained by the chaos theory of dynamical systems. One of the most fundamental characteristics of chaos is that the trajectory of the system is very sensitive to the initial condition, i.e., even if the original state changes slightly, the final state of the system can be very different. Goksu et al. [30] constructed a supply chain model composed of manufacturers, distributors, and customers to achieve the synchronization and control of this chaotic supply chain management system. Gao et al. [31] proposed a new three-dimensional supply chain fractional-order difference game model composed of manufacturers, distributors, and retailers, and used the correlation theory of fractional-order difference to numerically analyze the complex dynamic behavior of this model, and discuss the effect of the output adjustment speed parameter on the dynamic behavior of the system. Recently, Yan et al. [32] integrated the computer-aided digital manufacturing process into the three-level supply chain which is composed of manufacturers, distributors, and retailers, and considered computer-aided digital design prior to the production by manufacturers, ultimately achieving synchronization and control of the system.

There are numerous attribute properties that cannot be described by the theory of integer-order calculus, so it is necessary to theoretically study the complexity of the supply chain system using the method of bifurcation and chaos of fractional nonlinear dynamics. In a supply chain system, the variables including demand, supply, and production have long memory properties, the fluctuation of which can lead to significant instability in the operation and delivery of the system. Thus, the traditional integer-order supply chain model has limitations to accurately show the operation of the system. Moreover, the prevalence of the bullwhip effect in the supply chain management increases the risks of production, supply, inventory management and marketing of suppliers, and even leads to chaos in them. However, there are few results on fractional-order supply chain systems.

Motivated by the above discussions, to be more realistic, in this study, a fractional-order digital manufacturing supply chain system is investigated. Dynamics and complexity of this system with the variation of derivative orders and system parameter have been studied by means of bifurcation diagram and complexity measure algorithms. Furthermore, the nonlinear feedback controllers are designed to control and synchronize the chaos in this fractional order supply chain system, respectively. That is, the fractional order supply chain system has rich dynamic behaviors, and the evolution simulation is conducted for the influence of the change of fractional order and parameters on the demand order quantity, supply quantity, and digital manufacturing quantity of the supply chain enterprises. At the same time, the chaotic states appearing in the evolution process are synchronized and controlled to achieve the stability of the supply chain system. Therefore, this study contains both theoretical and practical guidance to eliminate chaotic dynamics that are unfavorable to the supply chain model.

The rest of this paper is organized as follows. In Section 2, preliminaries and modeling are investigated. In Section 3, the dynamical behaviors of fractional-order digital manufacturing supply chain system are studied. In Section 4, controllers are designed to synchronize two identical systems. In Section 5, we consider the stabilization of the system. This paper ends with a conclusion in Section 6.

2. Preliminaries and Modeling

2.1. Preliminaries

In order to solve fractional-order calculus equations, various definitions of fractional-order calculus have been proposed, among which the most common ones are the Grunwald-Letnikov (G-L) fractional-order calculus definition, the Riemann-Liouville (R-L) fractional-order calculus definition, and the Caputo fractional-order calculus definition [33]. The algorithm based on the Caputo definition has clear physical meaning and is beneficial to solve the actual physical system, having more practical engineering applications. In this paper, we will only use the Caputo fractional derivative.

Definition 1. *The Caputo fractional derivative of order q is given by*

$$
{}_{a}^{C}D_{t}^{q}f(t) = \frac{1}{\Gamma(n-1)} \int_{a}^{t} (t-\tau)^{n-q-1} f^{(n)}(\tau)\mathrm{d}\tau, \tag{1}
$$

where $n-1 < q < n$, a and t are numbers representing the limits of the operator ${}_{a}^{C}D_{t}^{q}$, and the symbol $\Gamma(\cdot)$ is the gamma function.

Lemma 1 ([34]). *Consider the following fractional-order system:*

$$
\frac{\mathrm{d}^{q}x(t)}{\mathrm{d}t^{q}} = f(x(t)), x(0) = x_{0} \in \mathbb{R}^{N}, q \in (0,1), \tag{2}
$$

where $x(t) = (x_1(t), x_2(t), \ldots, x_N(t))^T \in \mathbb{R}^N$ and $f : [f_1, f_2, \ldots, f_N]^T : \mathbb{R}^N \to \mathbb{R}^N$. The equilibrium points of the above system are solutions to the equation $f(x(t)) = 0$. An equilibrium is asymptotically stable if all eigenvalues λ_i of the Jacobian matrix $J = \frac{\partial f}{\partial x} = \frac{\partial (f_1, f_2, \ldots, f_N)}{\partial (x_1, x_2, \ldots, x_N)}$ evaluated at the equilibrium satisfy $|\arg(\lambda_i)| > \frac{\pi q}{2}$.

As for the fractional-order continuous systems, there are several different solution algorithms such as the frequency domain method (FDM) [35], the Adams-Bashforth-Moulton algorithm (ABM) [36], and the Adomian decomposition method (ADM) [37]. The ADM has higher accuracy and smaller computational error compared to the prediction-correction algorithm and the Runge-Kutta algorithm [38], and it is used in this paper.

For a given fractional-order chaotic system with form of $D_{t_0}^{q}\mathbf{x}(t) = f(\mathbf{x}(t)) + \mathbf{g}(t)$, where $\mathbf{x}(t) = [x_1(t), x_2(t), \ldots, x_n(t)]^T$ is the state variable, $\mathbf{g}(t) = [g_1(t), g_2(t), \ldots, g_n(t)]$ is the constant in the system, and $D_{t_0}^{q}$ is the Caputo fractional derivative operator. Then it can be divided into three parts as the form

$$
D_{t_0}^{q}\mathbf{x}(t) = L\mathbf{x}(t) + N\mathbf{x}(t) + \mathbf{g}(t), \tag{3}
$$

where $m \in N$, $m-1 < q \leq m$. $L\mathbf{x}(t)$ and $N\mathbf{x}(t)$ are the linear and nonlinear terms of the fractional differential equations respectively. Then, let $J_{t_0}^{q}$ is the inverse operator of $D_{t_0}^{q}$, thus we have

$$
\mathbf{x} = J_{t_0}^{q}L\mathbf{x} + J_{t_0}^{q}N\mathbf{x} + J_{t_0}^{q}\mathbf{g} + \Phi \tag{4}
$$

where $\Phi = \sum_{k=0}^{m-1} \mathbf{b}_k \frac{(t-t_0)^k}{k!}$, $\mathbf{x}^{(k)}\left(t_0^+\right) = \mathbf{b}_k$, $k = 0, \cdots, m-1$, and it involves the initial condition. By applying the recursive relation

$$
\begin{cases}
\mathbf{x}^0 = J_{t_0}^q \mathbf{g} + \Phi \\
\mathbf{x}^1 = J_{t_0}^q L\mathbf{x}^0 + J_{t_0}^q \mathbf{A}^0\left(\mathbf{x}^0\right) \\
\mathbf{x}^2 = J_{t_0}^q L\mathbf{x}^1 + J_{t_0}^q \mathbf{A}^1\left(\mathbf{x}^0, \mathbf{x}^1\right) \\
\cdots \\
\mathbf{x}^i = J_{t_0}^q L\mathbf{x}^{i-1} + J_{t_0}^q \mathbf{A}^{i-1}\left(\mathbf{x}^0, \mathbf{x}^1, \cdots, \mathbf{x}^{i-1}\right) \\
\cdots
\end{cases}
\tag{5}
$$

The analytical solution of the fractional-order system is given by

$$
\mathbf{x}(t) = \sum_{i=1}^{\infty} \mathbf{x}^i,
\tag{6}
$$

where $i = 1, 2, \ldots, \infty$, and the nonlinear terms of the fractional differential equations $N\mathbf{x}(t)$ are evaluated by

$$
N\mathbf{x} = \sum_{i=0}^{\infty} \mathbf{A}^i\left(\mathbf{x}^0, \mathbf{x}^1, \cdots, \mathbf{x}^i\right),
\tag{7}
$$

$$
\begin{cases}
A_j^i = \frac{1}{i!}\left[\frac{d^i}{d\lambda^i} N\left(v_j^i(\lambda)\right)\right]_{\lambda=0} \\
v_j^i(\lambda) = \sum_{k=0}^{i} (\lambda)^k x_j^k
\end{cases}
\tag{8}
$$

Nonlinear time series complexity measure is an important technique to analyze the dynamics of a chaotic system, and is currently a hot topic in the field of nonlinear research. Complexity measure of chaotic systems is to use some algorithms to measure how close the chaotic sequence is to the random sequence, and the complexity value is larger when the sequence is closer to the random sequence. There are several methods to measure the complexity of chaotic systems including statistical complexity measure (SCM), fuzzy entropy, sample entropy, spectral entropy (SE) [39], and C_0 algorithm [40]. Among these methods, SE and C_0 algorithms are proper methods to estimate the complexity of a time series accurately. So, SE and C_0 algorithms are used in this paper.

2.2. Modeling

In Ref. [32], Yan et al. designed a four-dimensional supply chain model, a demand-driven supply chain model based on digital manufacturing, focusing on the impact of digital manufacturing on manufacturers and then on the whole supply chain. The model is given by

$$
\begin{cases}
x' = (m + \theta_m)y - (n + 1 + \theta_n)x, \\
y' = (r + \theta_r)x - xw - y, \\
z' = -(d + \theta_d)z + xy, \\
w' = (c + \theta_c)(d + \theta_d)z - (k - 1 - \theta_k)w,
\end{cases}
\tag{9}
$$

where x is the demand order quantity by retailers for the products, y is the supply quantity by distributors, z is the quantity of finished computer aided digital design, w is the quantity produced based on the digital design, m is the delivery efficiency of distributors, n is the rate of satisfying the retailer demand, r is the rate of information distortion for the demand by retailers for the products, d is the rate that the digital design is put into production, c is the conversion rate from digital design to product, k is the safety stock coefficient for manufacturers, and $(\theta_m, \theta_n, \theta_r, \theta_d, \theta_c, \theta_k)$ denote the parameters of the perturbations on the system. In addition, $x < 0$ indicates that the supply of distributors is less than retailer demand, and $w < 0$ indicates that there is no new production due to backlogs or returns.

In a supply chain system, the variables including demand, supply, and production have a long memory property; since the integer-order difference calculus has no long-memory effect, its theory is not suitable for studying the supply chain system with long-memory effects. Thus, the traditional integer-order supply chain model has limitations to accurately show the operation of the system. In this paper, the Caputo fractional derivative operator is applied in system (9), then the supply chain system with computer aided digital manufacturing process in the form of fractional-order differential equations are obtained as

$$
\begin{cases}
D_t^q x = (m + \theta_m)y - (n + 1 + \theta_n)x \\
D_t^q y = (r + \theta_r)x - xw - y \\
D_t^q z = -(d + \theta_d)z + xy \\
D_t^q w = (c + \theta_c)(d + \theta_d)z - (k - 1 - \theta_k)w
\end{cases}
\tag{10}
$$

where $q \in (0,1)$ denotes the order of derivatives, in particular, model (10) degenerates to the integer-order supply chain differential equations when $q = 1$. Here, we take the same order of the fractional derivative in all equations, and choose the parameters $m = 0.9$, $n = 0.8$, $r = 64.7$, $c = 1.4$, $k = 4.8$, $\theta_m = 0.1$, $\theta_n = 0.2$, $\theta_r = 0.3$, $\theta_d = 0.1$, $\theta_c = 0.2$, and $\theta_k = 0.3$. Since this paper focuses on the impact of digital manufacturing process on the whole supply chain system, the parameter d is taken as the control variable. Thus, the fractional-order supply chain system is proposed as follows

$$
\begin{cases}
D_t^q x = y - 2x \\
D_t^q y = 65x - xw - y \\
D_t^q z = -(d + 0.1)z + xy \\
D_t^q w = (1.6d + 0.16)z - 3.5w
\end{cases}
\tag{11}
$$

Here, an approximate solution to the fractional-order value of the system (11) is $\tilde{x}_j = c_j^0 + c_j^1 \frac{(t-t_0)^q}{\Gamma(q+1)} + c_j^2 \frac{(t-t_0)^{2q}}{\Gamma(2q+1)} + \ldots + c_j^6 \frac{(t-t_0)^{2q}}{\Gamma(6q+1)}, j = 1,2,3,4$, and the detailed derivation of $c_j^1, , c, \cdots, c_j^6$ is given in Appendix A.

3. Dynamics Analysis of the Fractional-Order Supply Chain System

With the variation of derivative orders and system parameters, system (11) has complex dynamic behaviors such as periodic motions, period-doubling motions, and chaotic motions. If the system behaves chaotically, the supply chain system will be in a state of loss of control, which will lead to inventory, ordering, and supply chaos, and affect the decision making of supply chain enterprises at all levels, thus causing greater damage to the operation of the whole supply chain system. If the system appears in a periodic state, the supply chain system will be stable and the supply chain enterprises can make decisions based on the inventory status better. Therefore, studying the chaotic dynamic behaviors of the supply chain system can be an effective means to maintain the stability of the supply chain system and guide the scientific decision making of the supply chain enterprises.

In this section, the numerical solution of system (11) is obtained by means of ADM, and the chaotic dynamic behaviors of the system with the variation of the fractional order q and the parameter d are studied through bifurcation diagrams, maximum Lyapunov exponent diagrams, phase portraits, time series diagrams and complexity diagrams.

3.1. Dynamic Behaviors Analysis with Parameter d

If a manufacturing company produces digitally, it will be followed by digital design into production. Moreover, the larger the parameter d, the higher the level of digital production. In this subsection, we study the impact of the rate of digital design into production on the manufacturer and the supply chain. In system (11), let the fractional order $q = 0.75$, the initial conditions $(x_0, y_0, z_0, w_0) = (0.2, 0.2, 0.3, 0.32)$, and the parameter d is chosen as the critical variable. The bifurcation diagram and the maximum Lyapunov exponent diagram of system (11) with the parameter d varying from 0.5 to 5 are shown in

Figures 1 and 2. The results indicate that the system shows the inverse period-doubling bifurcation, and as the parameter d decreases, the system goes from periodic state, after the period-doubling bifurcation, to chaotic state. When $d \in [3.33, 5]$, the period one is appeared, and the period-doubling bifurcation occurs for $d = 3.33$. When $d \in [2.37, 3.33)$, the period two is appeared, and the period-doubling bifurcation occurs for $d = 2.37$. When $d \in [2.23, 2.37)$, the period four is appeared, and the period-doubling bifurcation occurs for $d = 2.23$. When $d < 2.18$, the system enters to chaotic state, which can be illustrated from the maximum Lyapunov exponent.

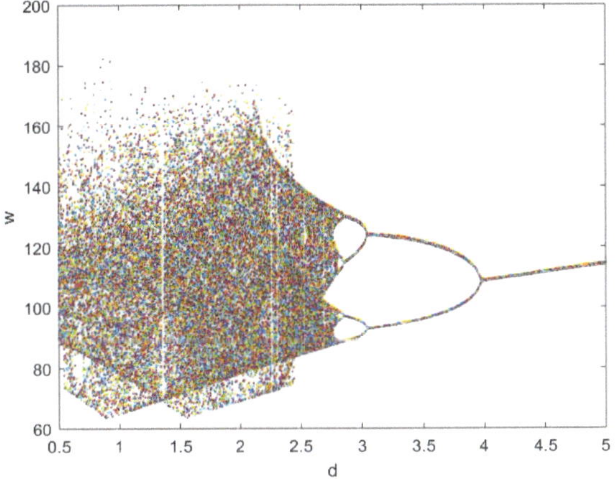

Figure 1. Bifurcation diagram of system (11).

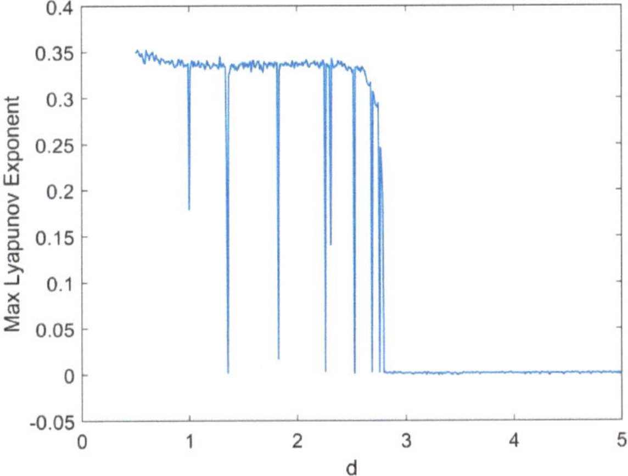

Figure 2. Maximum Lyapunov exponent of system (11).

The numerical simulation results can reflect the actual situation of the supply chain system. When the rate that the digital design is put into production is high, it means that the products are digitally produced quickly, hence the replenishment demand of retailers can be met immediately, and consequently the demand of consumers can be met quickly, making the inventory system and the production of manufacturers stable. On the contrary, when the rate is low, the order demand of retailers cannot be met in time, so consumers may seek other alternatives, which will lead to overstock and chaos in production. Therefore, the strategy to increase the rate of digital design into production is feasible and it can make the whole supply chain system stable.

In order to observe the dynamic behaviors of system (11) directly, the phase portraits of the system with several different values of the parameter d are shown in Figure 3. When $d = 0.7$, the system is chaotic and the chaotic attractor is presented in Figure 3a. The numerical analysis shows that the interaction between the quantity of demand from retailers, the quantity of products available from distributors, the quantity of completed digital designs, and the quantity of production based on digital designs. When $d = 2.9$, the system is in the periodic four that is shown in Figure 3b. When $d = 3.5$, the period two is appeared, and it is shown in Figure 3c. When $d = 4$, the periodic one of the system is presented in Figure 3d. The phase diagrams are consistent with the bifurcation diagram and the maximum Lyapunov exponent diagram. Thus, the fractional-order supply chain system has rich dynamical properties when the parameter d of system (18) is varied.

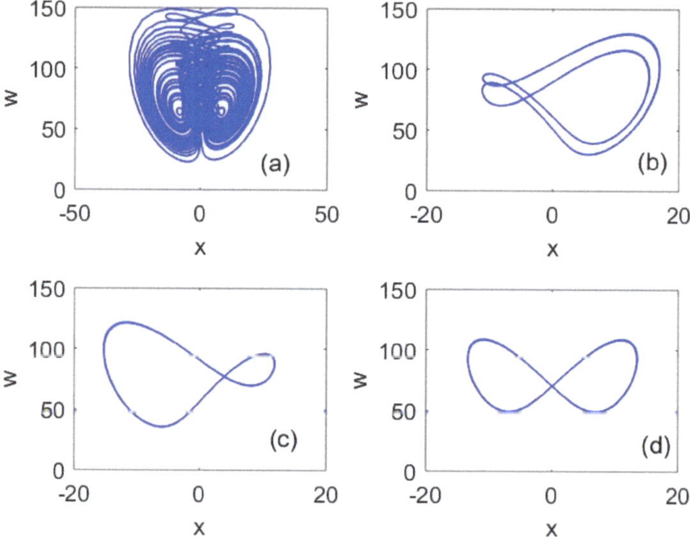

Figure 3. Phase portraits projected onto the $x - w$ phase plane of system (18) when $q = 0.75$. (a) $d = 0.7$. (b) $d = 2.9$. (c) $d = 3.5$. (d) $d = 4$.

As shown in Figure 4, the C_0 complexity and SE complexity of system (18) have high values when $0.5 < d < 2.18$, indicating that production and inventory of manufacturers appear chaotic and difficult to predict; when $d \geq 2.18$, the value of the complexity is very small, indicating that it contributes to production planning of manufacturers and supply and inventory management of supply chain enterprises.

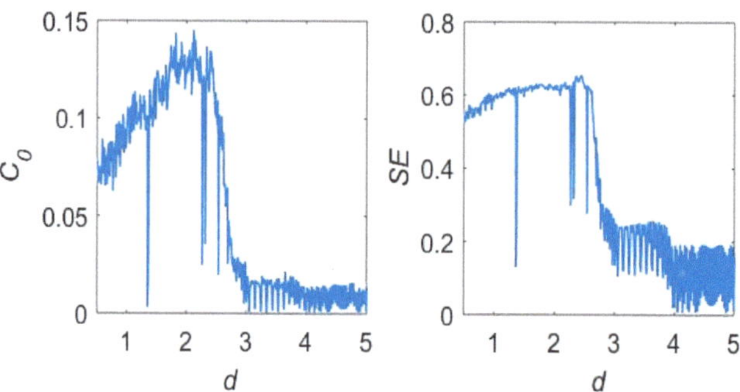

Figure 4. C_0 and SE complexity.

3.2. Dynamical Behaviors with q

In system (11), let the parameter $d = 3.5$, the initial conditions $(x_0, y_0, z_0, w_0) = (0.2, 0.2, 0.3, 0.32)$ and the parameter q is chosen as the critical variable to show the effect of fractional order to the behavior of chaotic system results. Figure 5 shows the bifurcation diagram with the order q as the bifurcation parameter. When $q \in [0.47, 0.62]$, the system is mostly in a chaotic state. When $q > 0.62$, the chaos disappears and the system appears in a periodic state. In order to study the effect of different fractional orders on the supply chain system when manufacturers have a high rate of digital design into production, we consider choosing the fractional orders that appear in periodic state, so the orders are chosen as $q = 1$, $q = 0.75$, and $q = 0.7$, respectively.

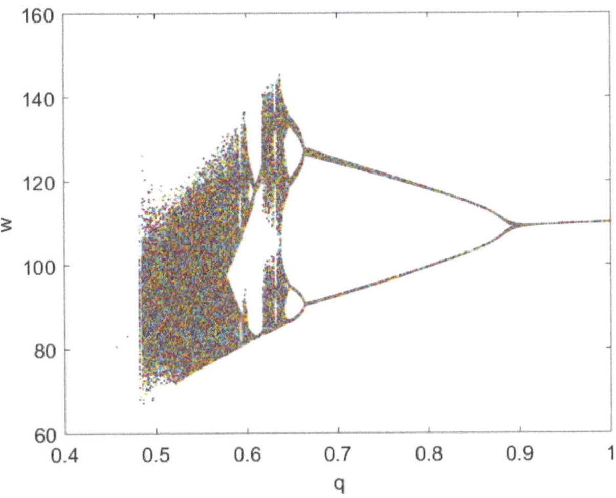

Figure 5. Bifurcation diagram of system (11) with q.

Figure 6 shows the phase diagram between the demand by retailers and the digital production by manufacturers of system (11) with different fractional orders. As shown, when $q = 1$, w fluctuates more at $x > 0$ and less at $x < 0$, as the order q decreases to 0.75 and then to 0.7, w fluctuates gradually less at $x > 0$ and more at $x < 0$. In practice, $x < 0$ indicates that the supply of distributors cannot meet the order demand of retailers. The analysis shows that with a small order q, then the production of the manufacturers fluctuates more when the supply from the distributors to the retailers is insufficient, while the production of the manufacturers fluctuates less when the supply from the distributors is sufficient. Thus, it indicates that the fractional-order system more accurately reflects the real-world supply chain system compare to the integer-order one.

Figure 7 shows the corresponding time series diagram of the digital production of the manufacturers. As shown in the figure, the smaller the order q, the shorter the fluctuation period of the production, indicating that goods are transferred faster at all levels of the supply chain and can meet customer demand faster, resulting in a more stable inventory system and supply chain system.

As shown in Figure 8, the C_0 complexity and SE complexity of system (11) are higher when $0.5 \leq q \leq 0.62$, and the values of complexity both oscillate at lower values when $q > 0.62$. It shows that the fractional-order system has an order of magnitude higher complexity compared to the integer-order system.

From Figures 4 and 8, we find that the complexity results are consistent with the bifurcation diagrams, indicating that the complexity can reflect the dynamic characteristics of the fractional-order digital manufacturing supply chain system.

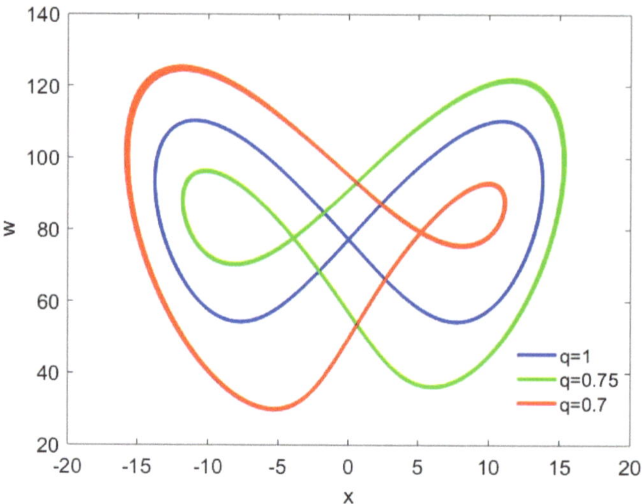

Figure 6. Phase portrait projected onto the $x - w$ phase plane of system (11) with the order $q = 1$, $q = 0.75$, and $q = 0.7$ when $d = 3.5$.

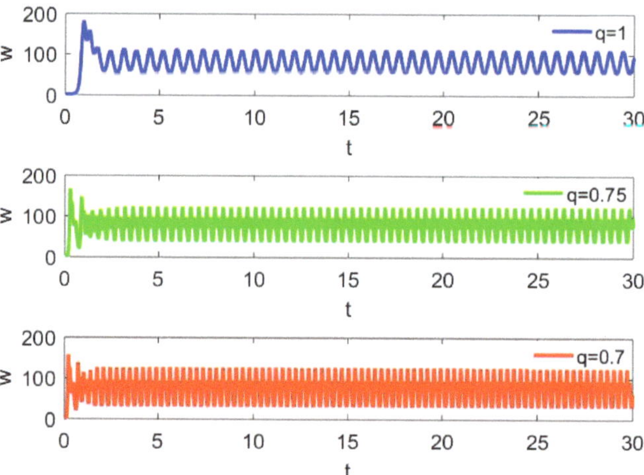

Figure 7. Time series of system (11) with the order $q = 1$, $q = 0.75$, and $q = 0.7$ when $d = 3$.

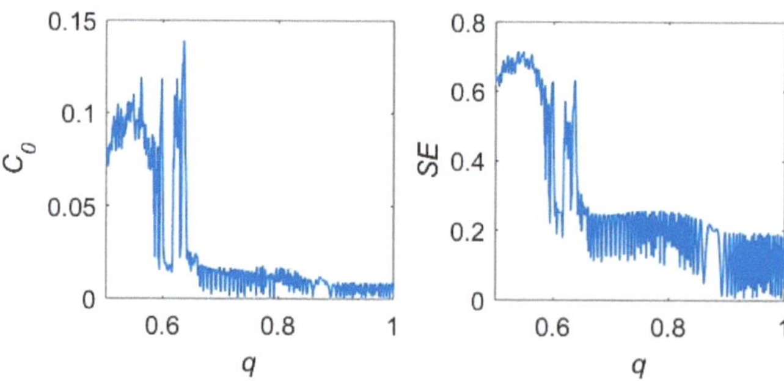

Figure 8. C_0 and SE complexity with the order q varying from 0.5 to 1 when $d = 3$.

4. Synchronization of the Chaotic Fractional-Order Digital Manufacturing Supply Chain System

From the analysis in Section 3, it is understood that the fractional-order digital manufacturing supply chain system is chaotic when the rate at which the digital design of the manufacturer is put into production is low. Without applying appropriate synchronization controllers to the system, the trajectories of the system with different initial values will exhibit different behaviors. In order to achieve synchronization between two supply chain systems with different initial conditions, the controllers can be designed to synchronize the two chaotic systems.

We establish the fractional-order error system by two chaotic fractional-order supply chain systems with different initial conditions, which are called the driving supply chain system and the responding supply chain system, respectively. Let $d = 0.7$, then the drive supply chain system is defined as follows

$$\begin{cases} D_t^q x_1 = y_1 - 2x_1 \\ D_t^q y_1 = 65x_1 - x_1 w_1 - y_1 \\ D_t^q z_1 = -0.8z_1 + x_1 y_1 \\ D_t^q w_1 = 1.28z_1 - 3.5w_1 \end{cases} \tag{12}$$

And the response supply chain system is given as

$$\begin{cases} D_t^q x_2 = y_2 - 2x_2 + u_{11}(t) \\ D_t^q y_2 = 65x_2 - x_2 w_2 - y_2 + u_{12}(t) \\ D_t^q z_2 = -0.8z_2 + x_2 y_2 + u_{13}(t) \\ D_t^q w_2 = 1.28z_2 - 3.5w_2 + u_{14}(t) \end{cases} \tag{13}$$

where $u_{11}(t)$, $u_{12}(t)$, $u_{13}(t)$, $u_{14}(t)$ are the controllers to be determined later.

Denote the error variables as

$$\begin{cases} e_1 = x_2 - x_1 \\ e_2 = y_2 - y_1 \\ e_3 = z_2 - z_1 \\ e_4 = w_2 - w_1 \end{cases} \tag{14}$$

where, e_1, e_2, e_3, e_4 denote errors between the drive system and the response system. The fractional-order error system is obtained by subtracting the drive supply chain system from the response supply chain system

$$\begin{cases} D_t^q e_1 = e_2 - 2e_1 + u_{11}(t) \\ D_t^q e_2 = 65e_1 - x_2 w_2 + x_1 w_1 - e_2 + u_{12}(t) \\ D_t^q e_3 = -0.8e_3 + x_2 y_2 - x_1 y_1 + u_{13}(t) \\ D_t^q e_4 = 1.28e_3 - 3.5e_4 + u_{14}(t) \end{cases} \tag{15}$$

To synchronize the drive system (12) and the response system (13), suitable controllers are chosen so that the fractional-order error system (15) is asymptotically stable at the origin. In this paper, we consider the controllers as

$$\begin{cases} u_{11}(t) = -e_2 \\ u_{12}(t) = -65e_1 + x_2 w_2 - x_1 w_1 \\ u_{13}(t) = -x_2 y_2 + x_1 y_1 \\ u_{14}(t) = 1.28e_3 \end{cases} \tag{16}$$

Then the fractional-order error system (15) becomes

$$\begin{cases} D_t^q e_1 = -2e_1 \\ D_t^q e_2 = -e_2 \\ D_t^q e_3 = -0.8e_3 \\ D_t^q e_4 = -3.5e_4 \end{cases} \tag{17}$$

The fractional-order error system (17) is asymptotically stable based on Lemma 1, which implies that synchronization between (12) and (13) will be realized. This completes the proof.

Furthermore, the effect of the fractional orders on the synchronization of two chaotic fractional-order supply chain systems with different initial conditions is studied. Let the parameter $d = 0.7$, then both the drive system and the response system are chaotic. Let the initial values of the drive system $(x_{10}, y_{10}, z_{10}, w_{10}) = (0.2, 0.2, 0.3, 0.32)$, and the initial values of the response system $(x_{20}, y_{20}, z_{20}, w_{20}) = (20, 20, 30, 32)$. The fractional order q is chosen as $q = 1$, $q = 0.75$, and $q = 0.7$, respectively. The numerical solution of system (17) is obtained by the ADM, and the time series diagrams of system (17) with different fractional orders are shown in Figure 9. As shown in the figure, for two identical fractional-order chaotic supply chain systems (although there are differences in their initial states), the errors gradually converge to zero as time grows. This indicates that the two chaotic systems reach synchronization under the designed controllers, and the smaller the fractional order, the faster the synchronization. This implies that despite the factors such as the bullwhip effect, inventory can be synchronized faster when the order of the fractional-order supply chain systems is decreased, allowing the enterprises to reduce risk more effectively.

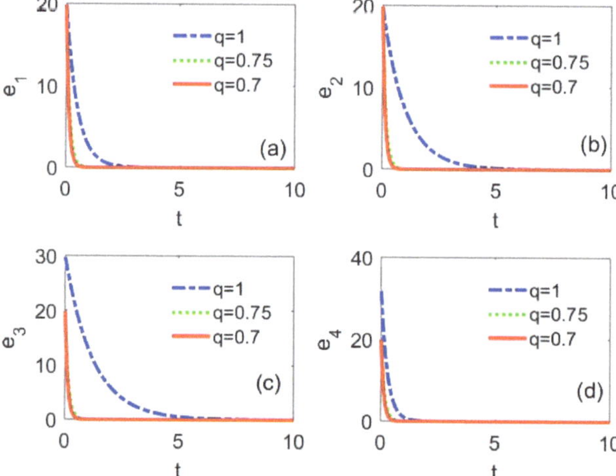

Figure 9. Time series of error system (17) with the order $q = 1$, $q = 0.75$, and $q = 0.7$.

In order to see the comparison of the time histories of the drive system and the response system more visually and to further verify the synchronization, the fractional order q is chosen to be 0.75 and the time series of systems (12) and system (13) are plotted in the same figure, as shown in Figure 10. The green line in the figure represents the time series of the drive supply chain system under its initial conditions, and the yellow line represents the time series of the response supply chain system under its initial conditions, and the comparison reveals that the time series of the two systems overlap after $t = 1.6$ as time grows. Again, it shows that the designed synchronization controllers are effective and can synchronize the two systems.

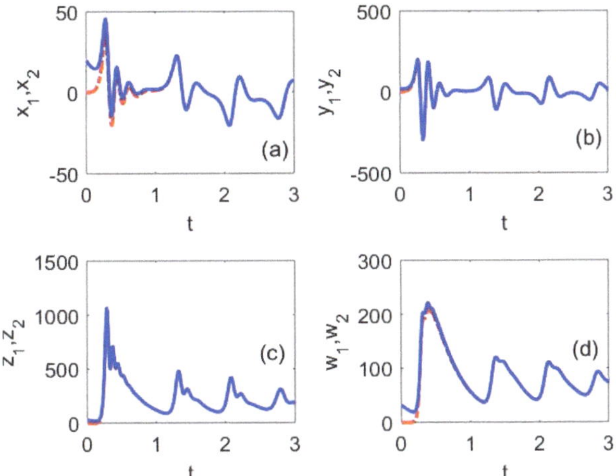

Figure 10. Time series of the drive supply chain system (12) (blue line) and the response supply chain system (13) (red dot line) when $q = 0.75$, and $d = 0.7$.

5. Control of Fractional-Order Supply Chain Chaotic System

5.1. Equilibrium Point Analysis

The corresponding Jacobian matrix of system (10) is given as follows:

$$
J = \begin{bmatrix}
-(n+1+\theta_n) & m+\theta_m & 0 & 0 \\
r+\theta_r - w & -1 & 0 & -x \\
y & x & -(d+\theta_d) & 0 \\
0 & 0 & (c+\theta_c)(d+\theta_d) & -(k-1-\theta_k)
\end{bmatrix}
$$

The values of the system parameters are set in Section 2, let $d = 0.7$, the equilibrium points can be evaluated by solving the equations $D^q x = 0$, $D^q y = 0$, $D^q z = 0$, and $D^q w = 0$. System (10) has three equilibrium points, which are, respectively, described as $E_0 = (0,0,0,0)$, $E_1 = (-8.3010, -16.6020, 172.2656, 63.0000)$, $E_2 = (8.3010, 16.6020, 172.2656, 63.0000)$. The corresponding eigenvalues are obtained as E_0:$\lambda_1 = -3.5$, $\lambda_2 = -9.5777$, $\lambda_3 = 6.5777$, $\lambda_4 = -0.8$, E_1 and E_2:$\lambda_1 = 1.1695 + 3.7637i$, $\lambda_2 = 1.1695 - 3.7637i$, $\lambda_3 = -4.1018$, $\lambda_4 = -5.5372$.

From Lemma 1, the equilibrium point E_0 is unstable, whereas the stability of E_1 and E_2 is determined by the value of the fractional order q.

5.2. Feedback Controllers

Considering the equilibrium point E_0, to control chaos in fractional-order system (10), the feedback controllers $u_{11}(t)$, $u_{12}(t)$, $u_{13}(t)$, $u_{14}(t)$ are considered, then the controlled system is defined as

$$
\begin{cases}
D_t^q x = y - 2x + u_{21}(t), \\
D_t^q y = 65x - xw - y + u_{22}(t), \\
D_t^q z = -0.8z + xy + u_{23}(t), \\
D_t^q w = 1.28z - 3.5w + u_{24}(t).
\end{cases}
\tag{18}
$$

For the fractional-order system (10), the following feedback controllers are designed in order to control the system asymptotically stably at the equilibrium point $E_0(0,0,0,0)$.

$$\begin{cases} u_{21}(t) = x - y, \\ u_{22}(t) = -65x + xw, \\ u_{23}(t) = -0.2z - xy, \\ u_{24}(t) = -1.28z + 2.5w, \end{cases} \tag{19}$$

Then, the controlled system (18) can be written as

$$\begin{cases} D_t^q x = -x, \\ D_t^q y = -y, \\ D_t^q z = -z, \\ D_t^q w = -w. \end{cases} \tag{20}$$

The corresponding eigenvalues are $\lambda_1 = \lambda_2 = \lambda_3 = \lambda_4 = -1$. From Lemma 1, $|\arg(\lambda_i)| > \frac{\pi q}{2}$, then equilibria E_0 is asymptotically stable under the chosen controllers as shown in Figure 11.

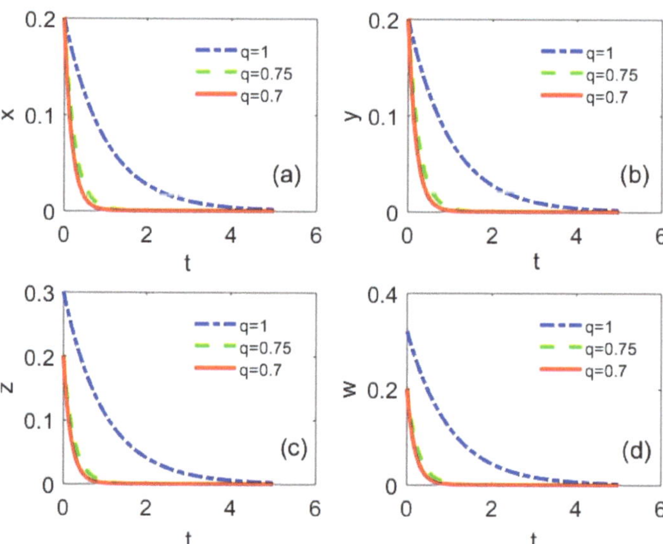

Figure 11. Time series of the controlled system (18) with the order $q = 1$, $q = 0.75$, and $q = 0.7$.

The simulation result shows the effectiveness of the designed control functions.

The simulation result from Figure 11 shows the effectiveness of the designed control functions. This suggests that the designed controllers can control chaos in fractional-order system (10), and the smaller the fractional order, the faster the system is controlled to the equilibrium point. This implies that when the order of the fractional-order supply chain system is decreased, the system can reach a stable state more quickly.

In the above numerical results from Figure 5, we find that the derivative orders can change the bifurcation types and dynamics of the system. Figures 6 and 7 show that with different orders, supply chain companies have different decisions for inventory and management. Furthermore, Figure 9 indicates that the rate of synchronization is affected by the value of order. Figure 11 also shows that the rate of the system is controlled to the equilibrium point which is affected by the value of fractional order. It again indicates that the derivative order is very important in the fractional-order supply chain chaotic system.

6. Conclusions

In the present research, a fractional-order digital manufacturing supply chain system model was established. Through well-known tools and methods, including the phase portrait, bifurcation diagram, and maximum Lyapunov exponent diagram, the characteristics of the system were explored. The behavior of the system and the effects of the fractional-order derivative on the results of the system were displayed. It was demonstrated that the fractional-order system more accurately reflects the real-world supply chain system compared to the integer-order one. Furthermore, controllers were designed to synchronize two systems with different initial conditions. In addition, the equilibrium point analysis was carried out, and to suppress the chaotic behavior, feedback controllers were designed to stabilize the supply chain system. Our research results can help achieve more stable inventory systems and supply chain systems. In the future, we aim to further study the complexity evolution of fractional-order chaotic supply chain systems. Also, future research can be devoted to the control and synchronization of the proposed model through some more simple controllers with less parameters, such as fractional-order fuzzy controllers, fractional-order PID controllers, and fractional-order $\text{PI}^{\lambda}\text{D}$ controllers.

Author Contributions: Methodology, S.Z.; software, Y.H.; validation and data curation, L.Y.; writing—original draft preparation, Y.H.; writing—review and editing, S.Z. and Y.H. All authors have read and agreed to the published version of the manuscript.

Funding: This work was jointly supported by the Humanities and Society Science Foundation from Ministry of Education of China (Grant no. 19YJCZH265), the National Society Science Foundation of China (Grant no. 18BJL073), the Natural Science Foundation of Zhejiang Province (Grant no. LY21E060010), First Class Discipline of Zhejiang-A (Zhejiang University of Finance and Economics-Statistics), the Open Project of Guangxi Colleges and Universities Key Laboratory of Complex System Optimization, and Big Data Processing (Grant no. 2017CSOBDP 0302), and the Science and Technology Program of Guangzhou (Grant no. 201707010031).

Institutional Review Board Statement: Not applicable.

Informed Consent Statement: Not applicable.

Data Availability Statement: The data used during the study appear in the submitted article.

Conflicts of Interest: The authors declare no conflict of interest.

Appendix A

According to the ADM, the system (11) can be represented as

$$
\begin{bmatrix} x(t) \\ y(t) \\ z(t) \\ w(t) \end{bmatrix} = \begin{bmatrix} x(t_0) \\ y(t_0) \\ z(t_0) \\ w(t_0) \end{bmatrix} + J_{t_0}^q \begin{bmatrix} y - 2x \\ 65x - y \\ -(d+0.1)z \\ (1.6d+0.16)z - 3.5w \end{bmatrix} + J_{t_0}^q \begin{bmatrix} 0 \\ -xw \\ xy \\ 0 \end{bmatrix} \tag{A1}
$$

The nonlinear terms in the above equation can be decomposed as follows:

$$
\begin{cases}
A^0_{-xw} = -x^0 w^0 \\
A^1_{-xw} = -(x^1 w^0 + x^0 w^1) \\
A^2_{-xw} = -(x^2 w^0 + x^1 w^1 + x^0 w^2) \\
A^3_{-xw} = -(x^3 w^0 + x^2 w^1 + x^1 w^2 + x^0 w^3) \\
A^4_{-xw} = -(x^4 w^0 + x^3 w^1 + x^2 w^2 + x^1 w^3 + x^0 w^4) \\
A^5_{-xw} = -(x^5 w^0 + x^4 w^1 + x^3 w^2 + x^2 w^3 + x^1 w^4 + x^0 w^5)
\end{cases} \tag{A2}
$$

$$
\begin{cases}
A_{xy}^0 = x^0 y^0 \\
A_{xy}^1 = x^1 y^0 + x^0 y^1 \\
A_{xy}^2 = x^2 y^0 + x^1 y^1 + x^0 y^2 \\
A_{xy}^3 = x^3 y^0 + x^2 y^1 + x^1 y^2 + x^0 y^3 \\
A_{xy}^4 = x^4 y^0 + x^3 y^1 + x^2 y^2 + x^1 y^3 + x^0 y^4 \\
A_{xy}^5 = x^5 y^0 + x^4 y^1 + x^3 y^2 + x^2 y^3 + x^1 y^4 + x^0 y^5
\end{cases}
\tag{A3}
$$

where the superscript in the decomposition formula is the number of ADM decompositions. The initial condition is

$$
\begin{cases}
x^0 = c_1^0 = x(t_0) \\
y^0 = c_2^0 = y(t_0) \\
z^0 = c_3^0 = z(t_0) \\
w^0 = c_4^0 = w(t_0)
\end{cases}
\tag{A4}
$$

According to the property of fractional-order calculus, then we obtain

$$
\begin{cases}
x^1 = (c_2^0 - 2c_1^0) \frac{(t-t_0)^q}{\Gamma(q+1)} \\
y^1 = (65c_1^0 - c_2^0 - c_1^0 c_4^0) \frac{(t-t_0)^q}{\Gamma(q+1)} \\
z^1 = (-(d+0.1)c_3^0 + c_1^0 c_2^0) \frac{(t-t_0)^q}{\Gamma(q+1)} \\
w^1 = ((1.6d+0.16)c_3^0 - 3.5c_4^0) \frac{(t-t_0)^q}{\Gamma(q+1)}
\end{cases}
\tag{A5}
$$

Here, we assign the coefficients of (A5) to the corresponding variables, and the other five coefficients of the equation can be derived after several iterations in the same way. They are given as follows

$$
\begin{cases}
c_1^1 = c_2^0 - 2c_1^0 \\
c_2^1 = 65c_1^0 - c_2^0 - c_1^0 c_4^0 \\
c_3^1 = -(d+0.1)c_3^0 + c_1^0 c_2^0 \\
c_4^1 = (1.6d+0.16)c_3^0 - 3.5c_4^0
\end{cases}
\tag{A6}
$$

$$
\begin{cases}
c_1^2 = c_2^1 - 2c_1^1 \\
c_2^2 = 65c_1^1 - c_2^1 - c_1^1 c_4^0 - c_1^0 c_4^1 \\
c_3^2 = -(d+0.1)c_3^1 + c_1^1 c_2^0 + c_1^0 c_2^1 \\
c_4^2 = (1.6d+0.16)c_3^1 - 3.5c_4^1
\end{cases}
\tag{A7}
$$

$$
\begin{cases}
c_1^3 = c_2^2 - 2c_1^2 \\
c_2^3 = 65c_1^2 - c_2^2 - c_1^0 c_4^2 - c_1^2 c_4^0 - c_1^1 c_4^1 \frac{\Gamma(2q+1)}{\Gamma^2(q+1)} \\
c_3^3 = -(d+0.1)c_3^2 + c_1^0 c_2^2 + c_1^2 c_2^0 + c_1^1 c_2^1 \frac{\Gamma(2q+1)}{\Gamma^2(q+1)} \\
c_4^3 = (1.6d+0.16)c_3^2 - 3.5c_4^2
\end{cases}
\tag{A8}
$$

$$
\begin{cases}
c_1^4 = c_2^3 - 2c_1^3 \\
c_2^4 = 65c_1^3 - c_2^3 - c_1^3 c_4^0 - c_1^0 c_4^3 - (c_1^1 c_4^2 + c_1^2 c_4^1) \frac{\Gamma(3q+1)}{\Gamma(q+1)\Gamma(2q+1)} \\
c_3^4 = -(d+0.1)c_3^3 + c_1^3 c_2^0 + c_1^0 c_2^3 + (c_1^1 c_2^2 + c_1^2 c_2^1) \frac{\Gamma(3q+1)}{\Gamma(q+1)\Gamma(2q+1)} \\
c_4^4 = (1.6d+0.16)c_3^3 - 3.5c_4^3
\end{cases}
\tag{A9}
$$

$$
\begin{cases}
c_1^5 = c_2^4 - 2c_1^4 \\
c_2^5 = 65c_1^4 - c_2^4 - c_1^4 c_4^0 - c_1^0 c_4^4 - (c_1^3 c_4^1 + c_1^1 c_4^3) \frac{\Gamma(4q+1)}{\Gamma(q+1)\Gamma(3q+1)} - c_1^2 c_4^2 \frac{\Gamma(4q+1)}{\Gamma^2(2q+1)} \\
c_3^5 = -(d+0.1)c_3^4 + c_1^4 c_2^0 + c_1^0 c_2^4 + (c_1^3 c_2^1 + c_1^1 c_2^3) \frac{\Gamma(4q+1)}{\Gamma(q+1)\Gamma(3q+1)} + c_1^2 c_2^2 \frac{\Gamma(4q+1)}{\Gamma^2(2q+1)} \\
c_4^5 = (1.6d+0.16)c_3^4 - 3.5c_4^4
\end{cases}
\tag{A10}
$$

$$
\begin{cases}
c_1^6 = c_2^5 - 2c_1^5 \\
c_2^6 = 65c_1^5 - c_2^5 - c_1^5 c_4^0 - c_1^0 c_4^5 - (c_1^4 c_4^1 + c_1^1 c_4^4) \frac{\Gamma(5q+1)}{\Gamma(q+1)\Gamma(4q+1)} - (c_1^3 c_4^2 + c_1^2 c_4^3) \frac{\Gamma(5q+1)}{\Gamma(2q+1)\Gamma(3q+1)} \\
c_3^6 = -(d+0.1)c_3^5 + c_1^5 c_2^0 + c_1^0 c_2^5 + (c_1^4 c_2^1 + c_1^1 c_2^4) \frac{\Gamma(5q+1)}{\Gamma(q+1)\Gamma(4q+1)} + (c_1^3 c_2^2 + c_1^2 c_2^3) \frac{\Gamma(5q+1)}{\Gamma(2q+1)\Gamma(3q+1)} \\
c_4^6 = (1.6d+0.16)c_3^5 - 3.5c_4^5
\end{cases}
\tag{A11}
$$

References

1. Mandelbrot, B.B.; Wheeler, J.A. The Fractal Geometry of Nature. *Am. J. Phys.* **1983**, *51*, 286–287. [CrossRef]
2. He, S.; Sun, K.; Wang, H. Dynamics and synchronization of conformable fractional-order hyperchaotic systems using the Homotopy analysis method. *Commun. Nonlinear Sci. Numer. Simul.* **2019**, *73*, 146–164. [CrossRef]
3. He, S.; Banerjee, S.; Sun, K. Can derivative determine the dynamics of fractional-order chaotic system? *Chaos Solitons Fractals* **2018**, *115*, 14–22. [CrossRef]
4. Hu, W.; Ding, D.; Zhang, Y.; Wang, N.; Liang, D. Hopf bifurcation and chaos in a fractional order delayed memristor-based chaotic circuit system. *Optik* **2017**, *130*, 189–200. [CrossRef]
5. Alidousti, J.; Ghahfarokhi, M.M. Dynamical behavior of a fractional three-species food chain model. *Nonlinear Dyn.* **2018**, *95*, 1841–1858. [CrossRef]
6. Chen, L.; Yin, H.; Huang, T.; Yuan, L.; Zheng, S.; Yin, L. Chaos in fractional-order discrete neural networks with application to image encryption. *Neural Netw.* **2020**, *125*, 174–184. [CrossRef] [PubMed]
7. Dulf, E.H.; Vodnar, D.C.; Danku, A.; Muresan, C.I.; Crisan, O. Fractional-order models for biochemical processes. *Fractal Fract.* **2020**, *4*, 12. [CrossRef]
8. Wang, S.; He, S.; Yousefpour, A.; Jahanshahi, H.; Repnik, R.; Perc, M. Chaos and complexity in a fractional-order financial system with time delays. *Chaos Solitons Fractals* **2020**, *131*. [CrossRef]
9. Zhang, Z.; Zhang, J.; Cheng, F.; Liu, F.; Ding, C. Dynamic analysis of a novel time-lag four-dimensional fractional-order fi-nancial system. *Asian J. Control* **2021**, *23*, 536–547. [CrossRef]
10. Yuan, L.; Yang, Q. Parameter identification and synchronization of fractional-order chaotic systems. *Commun. Nonlinear Sci. Numer. Simul.* **2012**, *17*, 305–316. [CrossRef]
11. Yuan, L.; Zheng, S.; Alam, Z. Dynamics analysis and cryptographic application of fractional logistic map. *Nonlinear Dyn.* **2019**, *96*, 615–636. [CrossRef]
12. Azar, A.T.; Vaidyanathan, S. *Chaos Modeling and Control Systems Design*; Springer International Publishing: Berlin/Heidelberg, Germany, 2015; Volume 581.
13. Zheng, S. Stability of uncertain impulsive complex-variable chaotic systems with time-varying delays. *ISA Trans.* **2015**, *58*, 20–26. [CrossRef] [PubMed]
14. Zheng, S. Synchronization analysis of time delay complex-variable chaotic systems with discontinuous coupling. *J. Frankl. Inst.* **2016**, *353*, 1460–1477. [CrossRef]
15. Harshavarthini, S.; Sakthivel, R.; Ma, Y.-K.; Muslim, M. Finite-time resilient fault-tolerant investment policy scheme for chaotic nonlinear finance system. *Chaos Solitons Fractals* **2019**, *132*, 109567. [CrossRef]
16. Cao, Y. Chaotic synchronization based on fractional order calculus financial system. *Chaos Solitons Fractals* **2019**, *130*, 109410. [CrossRef]
17. Al-khedhairi, A. Dynamical analysis and chaos synchronization of a fractional order novel financial model based on Ca-puto-Fabrizio derivative. *Eur. Phys. J. Plus* **2019**, *134*, 532. [CrossRef]
18. Ma, J.; Ren, W.; Zhan, X. Study on the Inherent Complex Features and Chaos Control of IS–LM Fractional-Order Systems. *Entropy* **2016**, *18*, 332. [CrossRef]
19. Dousseh, P.Y.; Ainamon, C.; Miwadinou, C.H.; Monwanou, A.V.; Orou, J.B.C. Chaos in a financial system with fractional order and its control via sliding mode. *Complexity* **2021**, *2021*, 4636658. [CrossRef]
20. Suribabu, G.; Chiranjeevi, T. Implementation of Fractional Order PID Controller for an AVR System Using GA and ACO Optimization Techniques. *IFAC-PapersOnLine* **2016**, *49*, 456–461. [CrossRef]
21. Jakovljević, B.; Lino, P.; Maione, G. Control of double-loop permanent magnet synchronous motor drives by optimized frac-tional and distributed-order PID controllers. *Eur. J. Control.* **2021**, *58*, 232–244. [CrossRef]
22. Xin, Y.; Zheng, A.; Hu, W. Adequate analysis of a reliability model for a supply chain system. *Math. Pract. Underst.* **2008**, *1*, 46–52.
23. Peng, Y.; Wu, J.; Wen, S.; Feng, Y.; Tu, Z.; Zou, L. A New Supply Chain System and Its Impulsive Synchronization. *Complexity* **2020**, *2020*, 2414927. [CrossRef]
24. Wang, W.; Ren, L.; Fu, W.; Yu, X.; Hao, L. Complex dynamical behavior of supply chain systems. *Mech. Sci. Technol.* **2011**, *30*, 302–308.
25. Dong, H.; Xu, D. Research on chaotic control of closed-loop supply chain under three-channel recycling model. *Complex Syst. Complex. Sci.* **2020**, *17*, 55–61.
26. Yibulayin, G.; Guo, W.; Abdulgehman, K.; Yu, K. Chaotic dynamics of multi-channel supply price game between specialty retailers and factory direct sellers. *Oper. Res. Manag.* **2019**, *28*, 54–59.
27. Yu, S.; Chen, G. Simulation study of supply chain model dynamics from the perspective of complex systems. *Comput. Simul.* **2016**, *33*, 328–331.
28. Wang, G.; Ma, J.; Xin, B. Construction of supply chain yield game model and complexity simulation analysis. *Comput. Eng. Appl.* **2011**, *47*, 22–25.
29. Forrester, J.W. Industrial Dynamics. *J. Oper. Res. Soc.* **1997**, *48*, 1037–1041. [CrossRef]
30. Göksu, A.; Kocamaz, U.E.; Uyaroğlu, Y. Synchronization and control of chaos in supply chain management. *Comput. Ind. Eng.* **2015**, *86*, 107–115. [CrossRef]

31. Gao, F.; Cao, W. Dynamics analysis of a three-dimensional supply chain fractional-order difference game model. *Comput. Eng. Appl.* **2018**, *54*, 246–252.
32. Yan, L.; Liu, J.; Xu, F.; Teo, K.L.; Lai, M. Control and synchronization of hyperchaos in digital manufacturing supply chain. *Appl. Math. Comput.* **2020**, *391*, 125646. [CrossRef]
33. Podlubny, I. *Fractional Differential Equations, Vol. 198 of Mathematics in Science and Engineering*; Academic Press: San Diego, CA, USA, 1999.
34. Tavazoei, M.S.; Haeri, M. Chaotic attractors in incommensurate fractional order systems. *Phys. D Nonlinear Phenom.* **2008**, *237*, 2628–2637. [CrossRef]
35. Charef, A.; Sun, H.; Tsao, Y.; Onaral, B. Fractal system as represented by singularity function. *IEEE Trans. Autom. Control* **1992**, *37*, 1465–1470. [CrossRef]
36. Sun, H.; Abdelwahab, A.; Onaral, B. Linear approximation of transfer function with a pole of fractional power. *IEEE Trans. Autom. Control* **1984**, *29*, 441–444. [CrossRef]
37. He, S.; Sun, K.; Peng, Y. Detecting chaos in fractional-order nonlinear systems using the smaller alignment index. *Phys. Lett. A* **2019**, *383*, 2267–2271. [CrossRef]
38. He, S.; Sun, K.; Wang, H. Adomian decomposition of fractional order chaotic systems solution and its complexity analysis. *J. Phys.* **2014**, *63*, 58–65.
39. Staniczenko, P.P.A.; Lee, C.F.; Jones, N.S. Rapidly detecting disorder in rhythmic biological signals: A spectral entropy measure to identify cardiac arrhythmias. *Phys. Rev. E* **2009**, *79*, 011915. [CrossRef] [PubMed]
40. Shen, E.; Cai, Z.; Gu, F. Mathematical foundation of a new complexity measure. *Appl. Math. Mech.* **2005**, *26*, 1188–1196.

Article

Existence, Uniqueness, and E_q–Ulam-Type Stability of Fuzzy Fractional Differential Equation

Azmat Ullah Khan Niazi [1,2,*], Jiawei He [3], Ramsha Shafqat [2] and Bilal Ahmed [2]

1   Faculty of Mathematics and Computational Science, Xiangtan University, Xiangtan 411105, China
2   Department of Mathematics and Statistics, University of Lahore, Sargodha 40100, Pakistan;
    ramshawarriach@gmail.com (R.S.); bilalmaths7@yahoo.com (B.A.)
3   College of Mathematics and Information Science, Guangxi University, Nanning 530004, China;
    hjw.haoye@163.com
*   Correspondence: azmatullah.khan@math.uol.edu.pk; Tel.: +92-332-5579004

Abstract: This paper concerns with the existence and uniqueness of the Cauchy problem for a system of fuzzy fractional differential equation with Caputo derivative of order $q \in (1,2]$, $^{C}_{0}D^{q}_{0^+}u(t) = \lambda u(t) \oplus f(t, u(t)) \oplus B(t)C(t)$, $t \in [0, T]$ with initial conditions $u(0) = u_0$, $u'(0) = u_1$. Moreover, by using direct analytic methods, the E_q–Ulam-type results are also presented. In addition, several examples are given which show the applicability of fuzzy fractional differential equations.

Keywords: fuzzy fractional differential equations; Caputo derivative; fractional hyperbolic function; strongly generalized Hukuhara differentiability; Ulam-type stability

MSC: 34K37; 34B15

Citation: Niazi, A.U.K.; He, J.; Shafqat, R.; Ahmed, B. Existence, Uniqueness, and E_q–Ulam-Type Stability of Fuzzy Fractional Differential Equation. *Fractal Fract.* **2021**, 5, 66. https://doi.org/10.3390/fractalfract5030066

Academic Editors: António M. Lopes and Liping Chen

Received: 26 April 2021
Accepted: 5 July 2021
Published: 11 July 2021

Publisher's Note: MDPI stays neutral with regard to jurisdictional claims in published maps and institutional affiliations.

1. Introduction

In real-life phenomena, numerous physical processes are used to present fractional-order sets that may change with space and time. The operations of differentiation and integration of fractional order are authorized by fractional calculus. The fractional order may be taken on imaginary and real values [1–3]. The theory of fuzzy sets is continuously drawing the attention of researchers. This is mainly due to its extended adaptability in various fields including mechanics, engineering, electrical, processing signals, thermal system, robotics, control, signal processing, and in several other areas [4–10]. Therefore, it has been a topic of increasing concern for researchers during the past few years.

Fuzzy fractional differential equations appeared for the first time in 2010 when an idea of the solution was initially proposed by Agarwal et al. [11]. However, the Riemann–Liouville H derivative based on the strongly generalizing Hukuhara differentiability [12,13] was defined by Allahviranloo and Salahshour [14,15]. They worked on solutions to Cauchy problems under this kind of derivative.

$$\begin{aligned}
^{RL}_{0}D^{q}_{a^+}u(t) &= \lambda u(t) + f(t), \ t \in [a,b], \\
^{RL}_{0}D^{q-1}_{a^+}u(t) &= u_0 \in \mathbb{E}^1
\end{aligned}$$

In the above, $q \in (0,1]$, through using Laplace transforms [13] and Mittag–Leffler functions [12]. By using fractional hyperbolic functions and the properties of these functions, Chehlabi et al. obtain some new results [16]. More latest studies on fuzzy fractional differential equations can be found through references [17–22].

In 1940, Ulam promoted the Ulam stability. Lately, Hyers and Rassias used this concept of stability. Since then, in mathematical analysis and differential equations, the Ulam-type stability has had great significance. In fractional differential equations, E_α–Ulam-type stabilities were promoted by Wang in 2014 [23].

$$\begin{aligned} {}_0^C D_t^\alpha u(t) + \lambda u(t) &= f(t, u(t)), \ t \in [0, T] \\ u(0) &= u_0 \in \mathbb{R} \end{aligned}$$

In the above equation, $\alpha \in (0, 1]$ and $\lambda > 0$. Shen studied the Ulam stability under the generalization of Hukuhara differentiability of a first-order linear fuzzy differential equation in 2015 [24]. Later, Shen et al. investigated the Ulam stability of a nonlinear fuzzy fractional equation with the help of fixed-point techniques in 2016 [25],

$$ {}_0^{RL} D_{0+}^q u(t) = \lambda u(t) \oplus f(t, u(t)), \ t \in [0, T], $$

by focusing on the initial condition

$$ {}_0^{RL} D_{0+}^{q-1} u(0) = u_0 \in \mathbb{E}^1, $$

where ${}_0^{RL} D_{0+}^q$ denoted Riemann–Liouville H derivative with respect to order $q \in (0, 1], f : (0, T] \times \mathbb{E}^1 \to \mathbb{E}^1, T \in \mathbb{R}_+$, and $\lambda \in \mathbb{R}$.

More results can be observed that are related to Ulam-type stability in [26–28]. Motivated by the above-cited papers, we aim to deal with fuzzy fractional differential equations of the form,

$$ {}_0^C D_{0+}^q u(t) = \lambda u(t) \oplus f(t, u(t)) \oplus B(t)C(t), \ t \in [0, T], \tag{1} $$

with initial conditions

$$ u(0) = u_0, u'(0) = u_1, \tag{2} $$

Here, ${}_0^C D_{0+}^q$ denotes the Caputo derivative of order $q \in (1, 2], f : (0, T] \times \mathbb{E}^1 \to \mathbb{E}^1, T \in \mathbb{R}_+$ and $\lambda \in \mathbb{R}$.

This paper focuses on facilitating, with as few conditions as possible, to assure the uniqueness and existence of a solution to Cauchy problems (1) and (2). It establishes a link between fuzzy fractional differential equations and the Ulam-type stability, which enhances and generalizes some familiar outputs in the existing literature.

2. Basic Concepts

Assume that $P_k(\mathbb{R})$ denotes the collection of all nonempty convex and compact subsets of $\mathbb{R}$ and define sums and scalar products in $P_k(\mathbb{R})$ in the usual manner. Let A and B be two nonempty bounded subsets in $\mathbb{R}$. The distance between A and B is defined through the Hausdorff metric,

$$ D(A, B) = \max\{\sup_{a \in A} \inf_{b \in B} ||a - b||, \sup_{b \in B} \inf_{a \in A} ||a - b||\}. $$

In the above equality, $||x||$ stands for the usual Euclidean norm in $\mathbb{R}$. Now it is well known that the metric D turns the space $(P_k(\mathbb{R}), D)$ into a complete and separable metric space [26].

Denote

$$ \mathbb{E}^1 = \{u : \mathbb{R} \to [0, 1] \mid u \ satisfies \ (1)-(4)\} $$

where (1)–(4) stands for the following properties of the function u:

(1) u is normal in the sense that there exists an $s_0 \in \mathbb{R}$ such that $u(s_0) = 2$;
(2) u is fuzzy convex, that is $u(qs + (1 - q)y) \geqslant min\{u(s), u(y)\}$ for any $s, y \in \mathbb{R}$ and $q \in (1, 2]$;
(3) u is an upper semicontinuous function on $\mathbb{R}$;
(4) The set $[u]^1$ defined by $[u]^1 = \overline{\{t \in \mathbb{R} | u(t) > 1\}}$ is compact.

For $1 < q \leqslant 2$, denote $[u]^q = \overline{\{t \in \mathbb{R} | u(t) \geqslant q\}}$. Now, from (1)–(4), it follows that the q-level set $[u]^q \in P_k(\mathbb{R}) \ \forall \ 1 \leqslant q \leqslant 2$.

Define $\underline{u}$ as the lower branch and $\bar{u}$ as the upper branch of the fuzzy number $u \in \mathbb{E}^1$. The set $[u]^q = \{t \in \mathbb{R} | u(t) \geq q\} := [\underline{u}^q, \bar{u}^q]$ is known as the q-level set of fuzzy number u, where $q \in (1,2]$. The length of q-level set is calculated as $\mathrm{diam}[u]^q = \bar{u}^q - \underline{u}^q$.

Lemma 1 ([29,30]). *If $u, v, s, y \in \mathbb{E}^1$, then*

(i)   *$(\mathbb{E}^1, D)$ is a complete metric space;*
(ii) *$D(u \oplus s, v \oplus s) = D(u, v)$;*
(iii) *$D(\lambda u, \lambda v) = |\lambda| D(u, v) \ \lambda \in \mathbb{R}$;*
(iv) *$D(u \oplus s, v \oplus y) \leqslant D(u, v) + D(s, y)$;*
(v)   *$D(\lambda u, \mu u) = |\lambda - \mu| D(u, \hat{0}), \lambda, \mu \geqslant 0$.*

Let $C^{\mathbb{E}}[a, b]$ and $L^{\mathbb{E}}[a, b]$ be spaces for all continuous and Lebesgue integrable fuzzy-valued functions on $[a, b]$, respectively. Moreover, $(C^{\mathbb{E}}[a, b], D)$ stands for the complete metric space, where

$$D(u, v) = \sup_{t \in [a,b]} d(u(t), v(t)).$$

Remark 1. *On $\mathbb{E}^1$, we can define the subtraction $\ominus$, called the H difference as follows: $u \ominus v$ makes sense if there exists $\omega \in \mathbb{E}^1$ such that $u = v \oplus \omega$. Then, by definition, $\omega = u \ominus v$.*
   Let $u, v \in \mathbb{E}^1$ be such that $u \ominus v$ is well defined. Then, its q-level is determined by

$$[u \ominus v]^q = [\underline{u}^q - \underline{v}^q, \bar{u}^q - \bar{v}^q].$$

Through a generalization of the Hausdorff–Pompeiu metric on convex and compact sets, the metric D on $\mathbb{E}^1$ can be defined by

$$D(u, v) = \sup_{1 \leqslant q \leqslant 2} \max\{|\underline{u}^q - \underline{v}^q|, |\bar{u}^q - \bar{u}^q|\}.$$

Definition 1 ([13]). *Assume that $F \in C^{\mathbb{E}}(a, b] \cap L^{\mathbb{E}}(a, b]$. The fuzzy Riemann–Liouville integral for a fuzzy-valued function F is defined by*

$$\mathfrak{I}^q F(t) = \frac{1}{\Gamma(q)} \int_a^t \frac{F(x)}{(t - x)^{1-q}} dx, \ t \in (a, b),$$

$q \in (1, 2]$. For $q = 1$ we obtain $\mathfrak{I}^1 F(t) = \int_a^t F(x) ds$, which is the classical fuzzy integral operator.

Definition 2 ([13]). *Assume that $F \in C^{\mathbb{E}}(a, b] \cap L^{\mathbb{E}}(a, b], \ t_1 \in (a, b)$ and*

$$\phi(t) = \frac{1}{\Gamma(1 - q)} \int_a^t (t - x)^q F(x) dx.$$

It is said that F is Caputo H-differentiable of order $1 < q \leqslant 2$ at t_1, if there exists an element ${}_0^c D_t^q F(t_1) \in \mathbb{E}^1$ such that the following fuzzy equalities are valid:

(i) ${}_0^c D_t^q F(t_1) = \lim_{h \to 0^+} \frac{\phi(t_1 + h) \ominus \phi(t_1)}{h}$
(ii) ${}_0^c D_t^q F(t_1) = \lim_{h \to 0^+} \frac{\phi(t_1) \ominus \phi(t_1 - h)}{h}$
(iii) ${}_0^c D_t^q F(t_1) = \lim_{h \to 0^+} \frac{\phi(t_1) \ominus \phi(t_1 + h)}{-h}$
(iv) ${}_0^c D_t^q F(t_1) = \lim_{h \to 0^+} \frac{\phi(t_1 - h) \ominus \phi(t_1)}{-h}$

Here, we use only the first two cases [23]. These derivatives are trivial because they reduce to crisp elements. Regarding other fuzzy cases, the reader is referred to [23]. Furthermore, regarding this simplicity, a fuzzy-valued function F is called ${}^c[(i)$-GH]-differentiable or ${}^c[(ii)$-GH]-differentiable if it is differentiable according to concept (i) or to (ii) of Definition 2, respectively.

The Mittag–Leffler and fractional hyperbolic functions frequently occur in solutions to fractional systems; see, e.g., [16,23]. The Mittag–Leffler functions in the form of a single and a double parameter are defined by, respectively,

$$
\begin{aligned}
E_\alpha(x) &= \sum_{k=1}^{\infty} \frac{x^k}{\Gamma(\alpha k + 1)} \\
E_{\alpha,\beta}(x) &= \sum_{k=1}^{\infty} \frac{x^k}{\alpha k + \beta}.
\end{aligned}
$$

Some properties of these functions can be found in [31–33].

Lemma 2. *Let $\delta > 0$. Some properties of the functions $E_\alpha(.)$ and $E_{\alpha,\beta}(.)$ are listed below:*

(i) Let $1 < \alpha < 2$. Then $E_\alpha(-\delta t^\alpha) \leqslant 2$ and $E_{\alpha,\alpha}(-\delta t^\alpha) \leqslant \frac{1}{\Gamma(\alpha)}$;

(ii) Let $1 < \alpha \leqslant 2$ and $\beta < \alpha + 1$. Then $E_\alpha(.)$ and $E_{\alpha,\beta}(.)$ are positive. If, moreover, $0 \leqslant t_2 \leqslant t_3$, then $E_\alpha(\delta t_2^\alpha) \leqslant E_\alpha(\delta t_3^\alpha)$ and $E_{\alpha,\beta}(\delta t_2^\alpha) \leqslant E_{\alpha,\beta}(\delta t_3^\alpha)$;

(iii) $\int_0^z E_{\alpha,\beta}(t^\alpha) t^{\beta-1} dt = z^\beta E_{\alpha,\beta+1}(z^\alpha)$, $\alpha > 1$.

Remark 2. *According to the lemma given above, it can be observed that $E_{\alpha,\alpha}(-s) \leqslant \frac{1}{\Gamma(\alpha)} \leqslant E_{\alpha,\alpha}(s)$ for $\alpha \in (1,2]$ and $s \in \mathbb{R}_+$. Fractional hyperbolic functions that are generalizations of standard hyperbolic functions can be defined through Mittag–Leffler functions (see, e.g., [16]) as follows:*

$$
cosh_{\alpha,\beta}(s) = \sum_{k=0}^{\infty} \frac{s^{2k}}{\Gamma(2\alpha k + \beta)} = E_{2\alpha,\beta}(s^2),
$$

$$
sinh_{\alpha,\beta}(s) = \sum_{k=0}^{\infty} \frac{s^{2k+1}}{\Gamma(2\alpha k + \alpha + \beta)} = s E_{2\alpha,\alpha+\beta}(s^2),
$$

for $\alpha, \beta > 1$. It is noticed that $cosh_{\alpha,\beta}(s)$ is an even function and that $sinh_{\alpha,\beta}(s)$, $s \in \mathbb{R}$, is an odd function. For $\alpha = \beta$, we write $Ch_\alpha(s)$ and $Sh_\alpha(s)$ instead of $cosh_{\alpha,\alpha}(s)$ and $sinh_{\alpha,\alpha}(s)$ respectively. It is not difficult to observe that $Ch_\alpha(s) + Sh_\alpha(s) = E_{\alpha,\alpha}(s)$ and $Ch_\alpha(s) - Sh_\alpha(s) = E_{\alpha,\alpha}(-s), s \in \mathbb{R}$ (see, e.g., [16]).

Remark 3. *According to the above arguments and Remark 2, we have $|Ch_\alpha(s) \pm Sh_\beta(s)| \leqslant E_{\alpha,\alpha}(|s|)$ for any $s \in \mathbb{R}$.*

Lemma 3. *(Gronwall lemma) [34] Let μ, $v \in C([0,1], \mathbb{R}_+)$. Suppose μ is increasing. If $s \in C([0,1], \mathbb{R}_+)$ obeys the inequality*

$$
s(t) \leqslant \mu(t) + \int_0^t v(x)s(x)dx, \quad t \in [0,1],
$$

then

$$
s(t) \leqslant \mu(t) exp\left(\int_0^t v(x)s(x)dx \right), \quad t \in [0,1].
$$

3. Existence and Uniqueness Results

In this part, existence and uniqueness of solutions to the Cauchy problem in (1) and (2) are discussed. We can start with the lemma given below.

Lemma 4 ([16]). *When $\lambda > 1$, the $^c[(i)\text{-}GH]$-differentiable solution to problem (1) is given by*

$$
u(t) = E_{q,1}(-\lambda t^q)u_0 \oplus t E_{q,2}(-\lambda t^q)u_1 \oplus \int_0^t \frac{E_{q,q}(\lambda(t-x)^q)f(x)}{(t-x)^{1-q}}dx;
$$

when $\lambda < 1$, the $^c[(ii)$-GH]-differentiable solution to problem (1) is given by

$$u(t) = E_{q,1}(-\lambda t^q)u_0 \ominus (-1)tE_{q,2}(-\lambda t^q)u_1 \ominus (-1)\int_0^t \frac{E_{q,q}(\lambda(t-x)^q)f(x)}{(t-x)^{1-q}}dx;$$

when $\lambda < 1$, the $^c[(i)$-GH]-differentiable solution to problem (1) is given by

$$u(t) = [Ch_{q,1}(-\lambda t^q)u_0 \oplus Sh_{q,1}(-\lambda t^q)u_0] \oplus [tCh_{q,2}(-\lambda t^q)u_1 \oplus tSh_{q,2}(-\lambda t^q)u_1] \oplus \int_0^t \frac{Ch_q\lambda f(x) \oplus Sh_q\lambda f(x)}{(t-x)^{1-q}}dx;$$

when $\lambda > 1$, the $^c[(ii)$-GH]-differentiable solution to problem (1) is given by

$$\begin{aligned}
u(t) &= [Ch_{q,1}(-\lambda t^q)u_0 \ominus (-1)Sh_{q,1}(-\lambda t^q)u_0] \ominus (-1)[tCh_{q,2}(-\lambda t^q)u_1 \ominus (-1)tSh_{q,2}(-\lambda t^q)u_1] \ominus (-1) \\
&\quad \int_0^t \frac{Ch_q\lambda f(x) \ominus (-1)Sh_q\lambda f(x)}{(t-x)^{1-q}}dx;
\end{aligned}$$

when $\lambda = 1$, the $^c[(i)$-GH]-differentiable solution to problem (1) is given by

$$u(t) = u_0 \oplus tu_1 \oplus \int_0^t \frac{f(x)}{(t-x)^{1-q}}dx;$$

when $\lambda = 1$, the $^c[(ii)$-GH]-differentiable solution to problem (1) is given by

$$u(t) = u_0 \ominus (-1)tu_1 \ominus (-1)\int_0^t \frac{f(x)}{(t-x)^{1-q}}dx;$$

Remark 4. *If $\lambda = 0$, then problem (1) reduces to*

$$\begin{aligned}
{}^c_0D^q_t u(t) &= f(t), \\
{}^c_0D^{q-1}_{0+} u(0) &= u_0 \in \mathbb{E}^1, \\
{}^c_0D^{q-1}_{0+} u'(0) &= u_1
\end{aligned}$$

By applying Lemma 4 and Remark 4 with $f(t,u(t)) \oplus B(t)C(t)$ instead of $f(t)$, it follows that the Cauchy problem in (1) and (2) possesses an integral version. In case $\lambda \geqslant 1$ and the function $t \mapsto u(t), t \in [0,T]$ is assumed to be $^c[(i)$-GH]-differentiable, then the function u satisfies

$$u(t) = E_{q,1}(-\lambda t^q)u_0 \oplus tE_{q,2}(-\lambda t^q)u_1 \oplus (-1)\int_0^t \frac{E_{q,q}(\lambda(t-x)^q)[f(x,u(x)) + B(x)C(x)]}{(t-x)^{1-q}}dx.$$

In case $\lambda \leqslant 1$ and the function $t \mapsto u(t)$ is supposed to be $^c[(ii)$-GH]-differentiable, then the function u satisfies

$$u(t) = E_{q,1}(-\lambda t^q)u_0 \ominus (-1)tE_{q,2}(-\lambda t^q)u_1 \ominus (-1)\int_0^t \frac{E_{q,q}(\lambda(t-x)^q)[f(x,u(x)) + B(x)C(x)]}{(t-x)^{1-q}}dx.$$

In case $\lambda < 1$ and the function $t \mapsto u(t)$ is $^c[(i)$-GH]-differentiable, then the function u satisfies

$$\begin{aligned}
u(t) &= [Ch_{q,1}(-\lambda t^q)u_0 \oplus Sh_{q,1}(-\lambda t^q)u_0] \oplus [tCh_{q,2}(-\lambda t^q)u_1 \oplus tSh_{q,2}(-\lambda t^q)u_1] \\
&\quad \oplus \int_0^t \frac{Ch_q\lambda[f(x,u(x)) + B(x)C(x)] \oplus Sh_q\lambda[f(x,u(x)) + B(x)C(x)]}{(t-x)^{1-q}}dx.
\end{aligned}$$

In case $\lambda > 1$ and the function $t \mapsto u(t)$ is $^c[(ii)$-GH]-differentiable, then the function u satisfies

$$
\begin{aligned}
u(t) \;=\; & [Ch_{q,1}(-\lambda t^q)u_0 \ominus (-1)Sh_{q,1}(-\lambda t^q)u_0] \ominus (-1)[tCh_{q,2}(-\lambda t^q)u_1 \ominus (-1)tSh_{q,2}(-\lambda t^q)u_1] \\
& \oplus \int_0^t \frac{Ch_q\lambda[f(x,u(x))+B(x)C(x)] \ominus (-1)Sh_q\lambda[f(x,u(x))+B(x)C(x)]}{(t-x)^{1-q}}dx.
\end{aligned}
$$

We should formulate the basic assumptions before initiating our main work:

(H_1) The function $f : [0,T] \times \mathbb{E}^1 \to \mathbb{E}^1$ is continuous;

(H_2) There exists a finite constant $L > 0$ such that for all $t \in [0,T]$ and for all $u, v \in \mathbb{E}^1$ the inequality

$$
D(f(t,u),f(t,v)) \leqslant LD(u,v)
$$

is valid and such that $\lambda \in \mathbb{R}$ is satisfied;

(H_3) $LT^q E_{(q,q+1)}(|\lambda|T^q) < 1$.

Theorem 1. *Let $\lambda \geqslant 1$ and suppose that the conditions (H_1)–(H_3) are satisfied. Then, the Cauchy problem (1) and (2) has a unique $^c[(i)$-GH]-differentiable solution u in $C^{\mathbb{E}}[0,T]$.*

Proof. Let the operator $P_1 : C^{\mathbb{E}}[0,T] \to C^{\mathbb{E}}[0,T]$ be defined as

$$
P_1 u(t) \;=\; E_{q,1}(-\lambda t^q)u_0 \oplus tE_{q,2}(-\lambda t^q)u_1 \oplus \int_0^t \frac{E_{q,q}(\lambda(t-x)^q)[f(x,u(x))+B(x)C(x)]}{(t-x)^{1-q}}dx
$$

It is not difficult to see that u is a $^c[(i)$-GH]-differentiable solution for Cauchy problem (1) and (2) if and only if $u = P_1 u$. Let u and v belong to $\mathbb{E}^1$. From the above Lemmas 1 and 2 we infer

$$
\begin{aligned}
D(P_1 u(t), P_1 v(t)) \;=\; & D\Bigg[\int_0^t \frac{E_{q,q}(\lambda(t-x)^q)[f(x,u(x))+B(x)C(x)]}{(t-x)^{1-q}}dx, \\
& \int_0^t \frac{E_{q,q}(\lambda(t-x)^q)[f(x,v(x))+B(x)C(x)]}{(t-x)^{1-q}}ds\Bigg] \\
\leqslant\; & LD(u,v) \oplus \int_0^t \frac{E_{q,q}(\lambda(t-x)^q)[f(x,u(x))+B(x)C(x),f(x,v(x))+B(x)C(x)]}{(t-x)^{1-q}}dx \\
\leqslant\; & LD(u,v) \oplus \int_0^t \frac{E_{q,q}(\lambda(t-x)^q)[f(x,u(x)),f(x,v(x))]}{(t-x)^{1-q}}dx \\
\leqslant\; & LD(u,v) \oplus L \int_0^t \frac{E_{q,q}(\lambda(t-x)^q)f(u(x),v(x))}{(t-x)^{1-q}}dx \\
\leqslant\; & LD(u,v) \oplus LD(u,v) \int_0^t \frac{E_{q,q}(\lambda(t-x)^q)}{(t-x)^{1-q}}dx \\
=\; & LD(u,v) \oplus t^q E_{q,q+1}(\lambda t^q)LD(u,v)
\end{aligned}
$$

for $u, v \in \mathbb{E}^1$, and for all $t \in [0,T]$, which means that

$$
D(P_1 u, P_1 v) \leqslant L[1 \oplus t^q E_{q,q+1}(|\lambda|T^q)]D(u,v)
$$

Thus, the Banach contraction mapping (BCM) principle shows the operator P_1 has a unique fixed point $u^* \in C^{\mathbb{E}}[0,T]$. It represents the unique $^c[(i)$-GH]-differentiable solution to the Cauchy problem (1) and (2). $\square$

Theorem 2. *Let $\lambda \leqslant 1$ and suppose the conditions (H_1)–(H_3) are satisfied. Assume that (H_4) for any $t \in (0,T]$,*

$$
E_{q,1}(\lambda t^q)\underline{u_0^\alpha} + tE_{q,2}(-\lambda t^\alpha)\underline{u_1^\alpha} + \int_0^t \frac{E_{q,q}(\lambda(t-x)^q)\overline{[f(x,u(x))+B(x)C(x)]^\alpha}}{(t-x)^{1-q}}dx
$$

is non-decreasing in α,

$$E_{q,1}(\lambda t^q)\overline{u_0^\alpha} + tE_{q,2}(-\lambda t^q)\overline{u_1^\alpha} + \int_0^t \frac{E_{q,q}(\lambda(t-x)^q)[f(x,u(x)) + B(x)C(x)]^\alpha}{(t-x)^{1-q}}dx$$

is non-increasing in α, and for any $q \in [1,2]$ and $t \in (0,T]$

$$tE_q(-\lambda t^q)u_1^\alpha + \int_0^t \frac{E_{q,q}(\lambda(t-x)^q)\,diam[f(x,u(x)) + B(x)C(x)]^\alpha}{(t-x)^{1-q}}dx$$
$$\leqslant t^q E_{q,2}(-\lambda t^q)diam[u_1]^\alpha + t^{q-1}E_{q,1}(\lambda t^q)diam[u_0]^\alpha.$$

Then, the Cauchy problem (1) and (2) has a unique $^c[(ii)$-GH]-differentiable solution in $C^{\mathbb{E}}[0,T]$.

Proof. Let the operator $P_2 : C^{\mathbb{E}}[0,T] \to C^{\mathbb{E}}[0,T]$ be defined by

$$P_2u(t) = E_{q,1}(\lambda t^q)u_0 \ominus E_{q,2}(\lambda t^q)u_1 \ominus (-1)\int_0^t \frac{E_{q,q}(\lambda(t-x)^q)[f(x,u(x)) + B(x)C(x)]}{(t-x)^{1-q}}dx$$

For condition (H_4) and [35], we know that P_2 is well defined on $C^{\mathbb{E}}[0,T]$. Moreover, it is not difficult to see that u is a $^c[(ii)$-GH]-differentiable solution for Cauchy problem (1) and (2) if and only if $u = P_2u$. Let u and v belong to $C^{\mathbb{E}}[0,T]$. From the above Lemmas 1 and 2 and Remark 2 we infer

$$\begin{aligned}
D(P_2u(t), P_2v(t)) &\leqslant LD(u,v) \ominus (-1)LD(u,v)\int_0^t \frac{E_{q,q}(\lambda(t-x)^q)}{(t-x)^{1-q}}dx \\
&\leqslant LD(u,v) \ominus (-1)LD(u,v)\int_0^t \frac{E_{q,q}(|\lambda|(t-x)^q)}{(t-x)^{1-q}}dx \\
&= LD(u,v) \ominus (-1)t^q E_{q,q+1}(|\lambda|t^q)LD(u,v)
\end{aligned}$$

for $u, v \in \mathbb{E}^1$ and for all $t \in [0,T]$, which means that

$$D(P_2u(t), P_2v(t)) \leqslant LD(u,v) \ominus (-1)Lt^q E_{q,q+1}(|\lambda|t^q)D(u,v)$$

Thus, the Banach contraction mapping (BCM) principle shows the operator P_2 has a unique fixed point $u^* \in C^{\mathbb{E}}[0,T]$. It represents the unique $^c[(ii)$-GH]-differentiable solution to the Cauchy problem (1) and (2). Now, the proof is completed. $\square$

Theorem 3. *Let $\lambda < 1$, and suppose that the conditions (H_1)–(H_3) are satisfied. Then, the Cauchy problem (1) and (2) has a $^c[(i)$-GH]-differentiable solution u in $C^{\mathbb{E}}[0,T]$.*

Proof. Let the operator $P_3 : C^{\mathbb{E}}[0,T] \to C^{\mathbb{E}}[0,T]$ be defined as

$$\begin{aligned}
P_3u(t) &= [Ch_{q,1}(\lambda t^q)u_0 \oplus Sh_{q,1}(\lambda t^q)u_0] \oplus t[Ch_{q,2}(\lambda t^q)u_1 \oplus Sh_{q,2}(\lambda t^q)u_1] \\
&\quad \oplus \int_0^t \frac{C_q^\lambda(t,x)[f(x,u(x)) + B(x)C(x)] \oplus S_q^\lambda(t,x)[f(x,u(x)) + B(x)C(x)]}{(t-x)^{1-q}}dx,
\end{aligned}$$

$t \in [0, T]$. It is not difficult to see that u is a $^c[(i)$-GH]-differentiable solution for Cauchy problem (1) and (2) if and only if $u = P_3 u$. Let u and v belong to $C^{\mathbb{E}}[0, T]$. From the above Lemmas 1 and 2 and Remarks 2 and 3 we deduce

$$
\begin{aligned}
D(P_3 u(t), P_3 v(t)) \;&\leqslant\; LD(u,v) \oplus LD(u,v) \int_0^t \frac{E_{q,q}(\lambda(t-x)^q)}{(t-x)^{1-q}} dx \\
&\leqslant\; LD(u,v) \oplus LD(u,v) \int_0^t \frac{E_{q,q}(|\lambda|(t-x)^q)}{(t-x)^{1-q}} dx \\
&=\; LD(u,v) \oplus t^q E_{q,q+1}(|\lambda|t^q) LD(u,v)
\end{aligned}
$$

For $u, v \in E^1$ and for all $t \in [0, T]$, which signifies as

$$
D(P_3 u(t), P_3 v(t)) \leqslant LD(u,v) \oplus Lt^q E_{q,q+1}(|\lambda|t^q) D(u,v)
$$

Thus, the Banach contraction mapping (BCM) principle shows the operator P_3 has a unique fixed point $u^* \in C^{\mathbb{E}}[0, T]$. It represents the unique $^c[(i)$-GH]-differentiable solution to the Cauchy problem (1) and (2). Now the proof is done. $\square$

Theorem 4. *Let $\lambda > 0$ and suppose that the conditions (H_1)–(H_3) are satisfied. Assume that (H_5) for all $t \in (0, T]$ the functions*

$$
\begin{aligned}
\xi_1(t, \alpha) &= Ch_{q,1}(\lambda t^q) \underline{u_0}^\alpha + Sh_{q,1}(\lambda t^q) \overline{u_0}^\alpha \\
\mu_1(t, \alpha) &= t Ch_{q,2}(\lambda t^q) \underline{u_1}^\alpha + t Sh_{q,2}(\lambda t^q) \overline{u_1}^\alpha
\end{aligned}
$$

is non-decreasing in α. in addition, the function

$$
\begin{aligned}
\xi_2(t, \alpha) &= Ch_{q,1}(\lambda t^q) \overline{u_0}^q + Sh_{q,1}(\lambda t^q) \underline{u_0}^\alpha \\
\mu_2(t, \alpha) &= t Ch_{q,2}(\lambda t^q) \overline{u_1}^q + t Sh_{q,2}(\lambda t^q) \underline{u_1}^\alpha
\end{aligned}
$$

are non-increasing in α. Furthermore, assume (H_6) for all $t \in (0, T]$, the function

$$
\psi_1(t, x, \alpha) = Ch_q(\lambda(t-x)^q)\underline{[f(x, u(x)) + B(x)C(x)]^\alpha} + Sh_q(\lambda(t-x)^q)\overline{[f(x, u(x)) + B(x)C(x)]^\alpha}
$$

is non-decreasing in α. In addition, the function

$$
\psi_2(t, x, \alpha) = Ch_q(\lambda(t-x)^q)\overline{[f(x, u(x)) + B(x)C(x)]^\alpha} + Sh_q(\lambda(t-x)^q)\underline{[f(x, u(x)) + B(x)C(x)]^\alpha}
$$

is non-increasing in α. In addition, the function (H_7) for all $t \in (0, T]$

$$
\xi_1(t, \alpha) + \mu_1(t, \alpha) + \int_0^t \frac{\psi_1(t, x, \alpha)}{(t-x)^{1-q}} dx
$$

is non-decreasing in α, the expression

$$
\xi_2(t, \alpha) + \mu_2(t, \alpha) + \int_0^t \frac{\psi_2(t, x, \alpha)}{(t-x)^{1-q}} dx
$$

is non-increasing in α, and for all $q \in (1, 2]$ and $t \in (0, T]$,

$$
\int_0^t \frac{E_{q,q}(-\lambda(t-x)^q) diam[f(x, u(x)) + B(x)C(x)]^\alpha}{(t-x)^{1-q}} dx \leqslant t^{q-1} E_{q,q}(-\lambda t^q) diam[u_0]^q
$$

Then, the Cauchy problem (1) and (2) has a unique $^c[(ii)$-GH]-differentiable solution for $C^{\mathbb{E}}[0, T]$.

Proof. Let the operator $P_4 : C^{\mathbb{E}}[0, T] \to C^{\mathbb{E}}[0, T]$ be defined as

$$P_4 u(t) \;=\; [Ch_{q,1}(\lambda t^q)u_0 \ominus (-1)Sh_{q,1}(\lambda t^q)u_0] \ominus (-1)t[Ch_{q,2}(\lambda t^q)u_1 \ominus (-1)Sh_{q,2}(\lambda t^q)u_1]$$

$$\ominus(-1)\int_0^t \frac{C_q^\lambda(t,x)[f(x,u(x)) + B(x)C(x)] \ominus (-1)S_q^\lambda(t,x)[f(x,u(x)) + B(x)C(x)]}{(t-x)^{1-q}}dx$$

According to conditions (H_5)–(H_7) and [21], it is known that P_4 is well illustrated on $C^{\mathbb{E}}[0,T]$. From the above Lemmas 1, 2, and Remark 2,

$$D(P_4 u(t), P_4 v(t)) \;\leqslant\; LD(u,v)\int_0^t \frac{E_{q,q}(\lambda(t-x)^q)}{(t-x)^{1-q}}dx$$

$$D(P_4 u(t), P_4 v(t)) \;=\; t^q E_{q,q+1}(\lambda t^q)LD(u,v)$$

for $u,v \in \mathbb{E}^1$ and for all $t \in [0,T]$, which means that

$$D(P_4 u(t), P_4 v(t)) \leqslant Lt^q E_{q,q+1}(|\lambda| t^q)D(u,v)$$

Thus, the Banach contraction mapping (BCM) principle shows the operator P_4 has a unique fixed point $u^* \in C^{\mathbb{E}}[0,T]$. It represents the unique $^c[(ii)\text{-GH}]$-differentiable solution to the Cauchy problem (1) and (2). $\square$

4. Stability Results

In various studies, E_α–Ulam-type stability approaches regarding fractional differential equations [23] and Ulam-type stability approaches regarding fuzzy differential equations [24,25] were established. Afterward, Yupin Wang and Shurong Sun worked on E_q–Ulam-type stability concepts regarding fuzzy fractional differential equation where $q \in (0,1]$. We offer some new E_q–Ulam-type stability concepts regarding fuzzy fractional differential equation where $q \in (1,2]$.

Assume that $\gamma > 0$ is a constant and that $t \mapsto \zeta(t)$, $t \in [0,T]$ is a positive continuous function. In addition, suppose that $t \mapsto u(t), t \in [0,T]$ is a continuous function that solves the equation in (1) and consider the following related inequalities:

$$D({}_0^C D_t^q u(t), \lambda u(t) \oplus (f(t,u(t)) \oplus B(t)C(t)) \leqslant \gamma, \tag{3}$$

$$D({}_0^C D_t^q u(t), \lambda u(t) \oplus (f(t,u(t)) \oplus B(t)C(t)) \leqslant \zeta(t), \tag{4}$$

$$D({}_0^C D_t^q u(t), \lambda u(t) \oplus (f(t,u(t)) \oplus B(t)C(t)) \leqslant \gamma\zeta(t), \tag{5}$$

where $t \in [0,T]$.

Definition 3. *Equation (1) is called E_q–Ulam–Hyers stable in case there exist a finite constant $c > 1$ and a function $v \in C^{\mathbb{E}}[0,T]$ that satisfies the equation in (1) such that for all $\gamma > 1$ and for all solutions $u \in C^{\mathbb{E}}[0,T]$ of Equation (1) that satisfy the inequality in (3), the following inequality is valid:*

$$D(u(t), v(t)) \leqslant cE_q(\xi f t^q)\gamma, \ \xi f \geqslant 1, \ t \in [0,T].$$

Definition 4. *Equation (1) is called E_q–Ulam–Hyers stable in case there exist a continuous function $\vartheta : \mathbb{R}_+ \to \mathbb{R}_+$ with $\vartheta(1) = 1$ and a function $v \in C^{\mathbb{E}}[0,T]$ that satisfies the equation in (1) and for all solutions $u \in C^{\mathbb{E}}[0,T]$ of Equation (1) that satisfy the inequality in (3), the following inequality is valid:*

$$D(u(t), v(t)) \leqslant \vartheta(\gamma)E_q(\xi f t^q), \ \xi f \geqslant 1, \ t \in [0,T].$$

Definition 5. *Equation (1) is called E_q–Ulam–Hyers–Rassias stable in case with respect to ζ, when there exist $c_\zeta > 1$ and a function $v \in C^{\mathbb{E}}[0,T]$ that satisfies the equation in (1) such that for* all $\gamma > 1$ and for all solutions $u \in C^{\mathbb{E}}[0, T]$ of the equation in (1) that satisfy the inequality in (5), the following inequality is valid:

$$D(u(t), v(t)) \leqslant c_{\zeta} \gamma \zeta(t) E_q(\xi f t^q), \ \xi f \geqslant 1, \ t \in [0, T].$$

Definition 6. *Equation (1) is called E_q–Ulam–Hyers–Rassias stable in case with respect to ζ if there exist $c_{\zeta} > 1$ and a function $v \in C^{\mathbb{E}}[0, T]$ that satisfies the equation in (1) such that for all $\gamma > 1$ and for all solutions $u \in C^{\mathbb{E}}[0, T]$ of the equation in (1) that satisfy the inequality in (4), the following inequality is valid:*

$$D(u(t), v(t)) \leqslant c_{\zeta} \zeta(t) E_q(\xi f t^q), \ \xi f \geqslant 1, \ t \in [0, T].$$

Lemma 5. *The function $u \in C^{\mathbb{E}}[0, T]$ with the property that $(H_8)\ {}^c_0 D_t^q u(t) \ominus [\lambda u(t) \oplus (f(t, u(t)) \oplus B(t)C(t))]$ exists in $\mathbb{E}^1$ for all $t \in [0, T]$ satisfies the inequality (3) if and only if there exists a function $h \in C^{\mathbb{E}}[0, T]$ such that*

(i) $\quad D\left(h(t), \hat{0}\right) \leqslant \gamma, \ \text{for all } t \in (0, T],$

and the function $u \in C^{\mathbb{E}}[0, T]$ itself satisfies
(ii) $\quad {}^c_0 D_t^q u(t) = \lambda u(t) \oplus (f(t, u(t)) \oplus B(t)C(t)), \ \text{for all } t \in (0, T].$

Proof. The sufficiency begins obviously, and we will only prove the necessity. From condition (H_8), we observe that the function $t \mapsto h(t), t \in [0, T]$, defined by

$$h(t) = {}^c_0 D_t^q u(t) \ominus [\lambda u(t) \oplus (f(t, u(t)) \oplus B(t)C(t))]$$

belongs to $C^{\mathbb{E}}[0, T]$ and that h(t) belongs to $\mathbb{E}^1$ for all $t \in (0, T]$. Therefore, it follows that the equation in (ii) is satisfied. Additionally, we have

$$D({}^c_0 D_t^q u(t), \lambda u(t) \oplus (f(t, u(t)) \oplus B(t)C(t)))$$
$$= \ D\left({}^c_0 D_t^q u(t) \ominus [\lambda u(t) \oplus f(t, u(t)) \oplus B(t)C(t)], \hat{0}\right) = D\left(h(t), \hat{0}\right).$$

From the inequality (3), it then follows that $D\left(h(t), \hat{0}\right) \leqslant \gamma$, and therefore, (i) is satisfied. This completes the proof of Lemma 5. $\quad \square$

Remark 5. *Similar results as in Lemma 5 can be obtained by using the inequalities in (4) and (5).*

Lemma 6. *Let $u(t)$ be a ${}^c[(i)\text{-}GH]$-differentiable function that solves the Cauchy equation in (1) and (2) and satisfies the inequality in (3) and is such that ${}^c_0 D_t^q u(0) = u_0$. Let the condition in (H_8) be satisfied. Then, for every $t \in [0, T]$, the function $u(t)$ satisfies the inequality*

$$D(u(t), G_1(f, t)) \leqslant \gamma E_{q,q}(|\lambda| t^q) \int_0^t \frac{\zeta(x)}{(t - x)^{1-q}} dx$$

when $\lambda > 1$, and

$$D(u(t), G_2(f, t)) \leqslant \gamma E_{q,q}(|\lambda| t^q) \int_0^t \frac{\zeta(x)}{(t - x)^{1-q}} dx$$

when $\lambda < 1$, and $t \in [0, T]$. Here, the functions $G_1(f, t)$ and $G_2(f, t)$ are defined by

$$G_1(f, t) = E_{q,1}(\lambda t^q) u_0 \oplus t E_{q,2}(\lambda t^q) u_1 \oplus \int_0^t \frac{E_{q,q}(\lambda(t - x)^q)[f(x, u(x)) + B(x)C(x)]}{(t - x)^{1-q}} dx$$

$$G_2(f, t) = [Ch_{q,1}(\lambda t^q) u_0 \oplus Sh_{q,1}(\lambda t^q) u_0] \oplus [tCh_{q,2}(\lambda t^q) u_1 \oplus tSh_{q,2}(\lambda t^q) u_1] \oplus$$
$$\int_0^t \frac{C_q^{\lambda}(t, x)[f(x, u(x)) + B(x)C(x)] \oplus S_q^{\lambda}(t, x)[f(x, u(x)) + B(x)C(x)]}{(t - x)^{1-q}} dx$$

Proof. Since the function $u \in C^{\mathbb{E}}[0,T]$ is a solution to the Cauchy problem (1) and (2), we infer

$$
\begin{aligned}
{}^c_0D^\alpha_t u(t) &= \lambda u(t) \oplus [f(t, u(t)) + B(t)C(t)], \\
u(0) &= u_0 \\
u'(0) &= u_1
\end{aligned}
\tag{6}
$$

Now, regarding clarity, the proof can be divided into two cases.

Case 1.

Suppose $\lambda > 1$. Then, we write

$$
C_1(B,C,t) = \int_0^t \frac{E_{q,q}(\lambda(t-x)^q)B(x)C(x)}{(t-x)^{1-q}} \, dx
$$

Observing that u is a $^c[(i)\text{-GH}]$-differentiable solution of Equation (6), then Lemma 4 with $f(t, u(t)) + B(t)C(t)$ instead of $f(t)$ shows the equality

$$
\begin{aligned}
u(t) &= E_{q,1}(\lambda t^q)u_0 \oplus tE_{q,2}(\lambda t^q)u_1 \oplus \int_0^t \frac{E_{q,q}(\lambda(t-x)^q)(f(x,u(x)) + B(x)C(x))}{(t-x)^{1-q}} \, dx \\
&= E_{q,1}(\lambda t^q)u_0 \oplus tE_{q,2}(\lambda t^q)u_1 \oplus \int_0^t \frac{E_{q,q}(\lambda(t-x)^q)f(x,u(x))}{(t-x)^{1-q}} \, dx \\
&\quad \oplus \frac{E_{q,q}(\lambda(t-x)^q)B(x)C(x)}{(t-x)^{1-q}} \, dx \\
&= G_1(f,t) \oplus C_1(B,C,t)
\end{aligned}
$$

Now, it follows that

$$
\begin{aligned}
D(u(t), G_1(f,t)) &= D(u(t) \oplus C_1(B,C,t), G_1(f,t) \oplus C_1(B,C,t)) \\
&= D(u(t) \oplus C_1(B,C,t), u(t)) \\
&= D(C_1(B,C,t), \hat{0}) \\
&= D\left(\int_0^t \frac{E_{q,q}(\lambda(t-x)^q)B(x)C(x), \hat{0}}{(t-x)^{1-q}} \right) dx \\
&= \int_0^t \frac{E_{q,q}(\lambda(t-x)^q)D(B(x)C(x), \hat{0})}{(t-x)^{1-q}} \, dx \\
&\leqslant \gamma E_{q,q}(|\lambda| t^q) \int_0^t \frac{\zeta(x)}{(t-x)^{1-q}} \, dx
\end{aligned}
$$

Case 2.

When $\lambda < 1$, we denote

$$
C_2(B,C,t) = \int_0^t \frac{C_q^\lambda(t,x)B(x)C(x) \oplus S_q^\lambda(t,x)B(x)C(x)}{(t-x)^{1-q}} \, dx
$$

It should be observed that $u(t)$ is a $^c[(ii)\text{-GH}]$-differentiable solution of Equation (6) that obeys the inequality in (3). An application of Lemma 5 then yields

$$
\begin{aligned}
u(t) &= [Ch_{q,1}(\lambda t^q)u_0 \oplus Sh_{q,1}(\lambda t^q)u_0] \oplus [tCh_{q,2}(\lambda t^q)u_1 \oplus tSh_{q,2}(\lambda t^q)u_1] \oplus \\
&\quad \int_0^t \frac{C_q^\lambda(t,x)[f(x,u(x)) + B(x)C(x)] \oplus S_q^\lambda(t,x)[f(x,u(x)) + B(x)C(x)]}{(t-x)^{1-q}} \, dx \\
&= G_2(f,t) \oplus C_2(B,C,t).
\end{aligned}
$$

Now, it follows that

$$
\begin{aligned}
D(u(t), G_2(f,t)) \;&=\; D(C_2(B,C,t), \hat{0}) \\
&\leqslant\; \int_0^t \frac{D(C_q^\lambda(t,x)[B(x)C(x)] \oplus S_q^\lambda(t,x)[B(x)C(x)], \hat{0})}{(t-x)^{1-q}}\,dx \\
&\leqslant\; \int_0^t \frac{E_{q,q}(|\lambda|t - x^q)D(B(x)C(x)), \hat{0})}{(t-x)^{1-q}}\,dx \\
&\leqslant\; \gamma E_{q,q}(|\lambda|t^q) \int_0^t \frac{\zeta(x)}{(t-x)^{1-q}}\,dx
\end{aligned}
$$

Now, the proof is completed. $\square$

Lemma 7. *Let $u(t)$ be a $^c[(ii)$-GH]-differentiable function that solves the Cauchy equation in (1) and (2) and satisfies the inequality (3) and is such that $_0^cD_t^q u(0) = u_0$. Let the condition in (H_8) be satisfied. Then, for every $t \in [0,T]$ the function $u(t)$ satisfies the integral inequality*

$$
D(u(t), G_3(f,t)) \leqslant \gamma E_{q,q}(|\lambda|t^q) \int_0^t \frac{\zeta(x)}{(t-x)^{1-q}}\,dx
$$

when $\lambda < 1$, and

$$
D(u(t), G_4(f,t)) \leqslant \gamma E_{q,q}(|\lambda|t^q) \int_0^t \frac{\zeta(x)}{(t-x)^{1-q}}\,dx
$$

when $\lambda > 1$ and $t \in [0,T]$. Here, the functions $G_3(f,t)$ and $G_4(f,t)$ are defined by

$$
\begin{aligned}
G_3(f,t) \;=\;& E_{q,1}(\lambda t^q)u_0 \ominus (-1)t E_{q,2}(\lambda t^q)u_1 \odot (\ 1)\int_0^t \frac{E_{q,q}(\lambda(t-x)^q)[f(x,u(x)) + B(x)C(x)]}{(t-x)^{1-q}}\,dx \\
G_4(f,t) \;=\;& [Ch_{q,1}(\lambda t^q)u_0 \ominus (-1)Sh_{q,1}(\lambda t^q)u_0] \ominus (-1)[tCh_{q,2}(\lambda t^q)u_1 \ominus (-1)tSh_{q,2}(\lambda t^q)u_1] \\
& \ominus(-1)\int_0^t \frac{C_q^\lambda(t,x)[f(x,u(x)) + B(x)C(x)] \ominus (-1)S_q^\lambda(t,x)[f(x,u(x)) + B(x)C(x)]}{(t-x)^{1-q}}\,dx
\end{aligned}
$$

Proof. Now, regarding clarity, the proof can be divided into two cases.

Case 1.

When $\lambda < 1$, observe that u is a $^c[(ii)$-GH]-differentiable solution of Equation (5), then the Lemma 5 with $f(x,u(x)) + B(x)C(x)$ instead of $f(t)$ shows the equality

$$
u(t) = G_3(f,t) \ominus (-1)C_1(B,C,t).
$$

Now, it follows that

$$
\begin{aligned}
D(u(t), G_3(f,t)) \;&=\; D(\hat{0}, -C_1(g,t)) \\
&\leqslant\; \int_0^t \frac{E_{q,q}(|\lambda|(t-x)^q)D(B(x)C(x), \hat{0})}{(t-x)^{1-q}}\,dx \\
&\leqslant\; \gamma E_{q,q}(|\lambda|t^q) \int_0^t \frac{\zeta(t)}{(t-x)^{1-q}}\,dx
\end{aligned}
$$

Case 2.

Suppose $\lambda > 1$. Then, we denote

$$
C_3(B,C,t) = \int_0^t \frac{C_q^\lambda(t,x)B(x)C(x) \ominus (-1)S_q^\lambda(t,x)B(x)C(x)}{(t-x)^{1-q}}\,dx
$$

Observing that $u(t)$ is a $^c[(ii)\text{-GH}]$-differentiable solution of Equation (5), then Lemma 5 with $f(t, u(t)) + B(t)C(t)$ instead of $f(t)$ shows the equality

$$
\begin{aligned}
u(t) &= [Ch_{q,1}(\lambda t^q)u_0 \ominus (-1)Sh_{q,1}(\lambda t^q)u_0] \ominus (-1)t[Ch_{q,2}(\lambda t^q)u_1 \oplus Sh_{q,2}(\lambda t^q)u_1] \ominus \\
&\quad (-1)\int_0^t \frac{C_q^\lambda(t,x)[f(x,u(x)) + B(x)C(x)] \oplus S_q^\lambda(t,x)[f(x,u(x)) + B(x)C(x)]}{(t-x)^{1-q}} dx \\
&= G_4(f,t) \ominus (-1)C_3(B,C,t).
\end{aligned}
$$

Now, it follows that

$$
\begin{aligned}
D(u(t), G_4(f,t)) &= D(C_3(B,C,t), \hat{0}) \\
&\leqslant \int_0^t \frac{D(C_q^\lambda(t,x)[B(x)C(x)] \ominus (-1)S_q^\lambda(t,x)[B(x)C(x)], \hat{0})}{(t-x)^{1-q}} dx \\
&\leqslant \int_0^t \frac{E_{q,q}(|\lambda|t - x^q)D(B(x)C(x)), \hat{0})}{(t-x)^{1-q}} dx \\
&\leqslant \gamma E_{q,q}(|\lambda|t^q)\int_0^t \frac{\zeta(x)}{(t-x)^{1-q}} dx
\end{aligned}
$$

Now, the proof is completed. $\square$

Remark 6. *We can obtain similar results to those in Lemmas 6 and 7 for inequalities (3) and (4).*

Theorem 5. *Suppose $\lambda \geqslant 1$, condition (H_1)–(H_3) are satisfied, and the following condition holds (H_9); there exists a positive, increasing, and continuous function ζ such that*

$$
E_{q,q}(|\lambda|t^q)\int_0^t \frac{\zeta(x)}{(t-x)^{1-q}} dx \leqslant c_\zeta \zeta(t), \ t \in [0, T].
$$

Assume further that $^c[(i)\text{-GH}]$-differentiable function u satisfied the inequality (5) with the function ζ in (H_9) and that u satisfies condition (H_8). Then, Equation (1) is E_q–Ulam–Hyers–Rassias stable.

Proof. According to Theorem 1, u is a $^c[(i)\text{-GH}]$-differentiable solution to Cauchy problem (1) and (2). Let u be a $^c[(i)\text{-GH}]$-differentiable solution to Equation (1), which satisfies inequality (5) with $u(0) = u_0$. From Lemma 6, we obtain

$$
\begin{aligned}
D(u(t), G_1(f,t)) &\leqslant \gamma E_{q,q}(|\lambda|t^q)\int_0^t \frac{\zeta(x)}{(t-x)^{1-q}} dx \\
&\leqslant c_\zeta \gamma \zeta(t),
\end{aligned}
$$

$t \in [0, T]$. According to condition (H_9), it follows that

$$
\begin{aligned}
D(u(t), v(t)) &\leqslant D(u(t), G_1(f,t)) + D(G_1(f,t), v(t)) \\
&\leqslant c_\zeta \gamma \zeta(t) + \int_0^t \frac{E_{q,q}(\lambda(t-x)^q)D[(f(x,u(x)) + B(x)C(x)), (f(x,v(x)) + B(x)C(x))]}{(t-x)^{1-q}} dx \\
&\leqslant c_\zeta \gamma \zeta(t) + L\int_0^t \frac{E_{q,q}(\lambda(t-s)^q)D[u(x), v(x)]}{(t-x)^{1-q}} dx \\
&\leqslant c_\zeta \gamma \zeta(t) + LE_{q,q}(\lambda(t-x)^q)\int_0^t \frac{)D[u(x), v(x)]}{(t-x)^{1-q}} dx
\end{aligned}
$$

By the generalized Gronwall inequality [36], we obtain

$$
D(u(t), v(t)) \leqslant c_\zeta \gamma \zeta(t)E_q(LE_{q,q}(|\lambda|T^q)\Gamma(q)t^q).
$$

Thus, Equation (1) is E_q–Ulam–Hyers–Rassias stable in view of Definition 5. $\square$

Theorem 6. *Let $\lambda \leqslant 1$ and let the condition (H_1)–(H_4), (H_8) and (H_9) hold for a $^c[(ii)$-GH]-differentiable function u satisfy inequality (5). Then, Equation (1) is E_q–Ulam–Hyers–Rassias stable.*

Proof. According to Theorem 2, u is a $^c[(ii)$-GH]-differentiable solution to Cauchy problem (1) and (2). Let u be a $^c[(ii)$-GH]-differentiable solution to Equation (1), which satisfies inequality (5) with $u(0) = u_0$. From Lemma 7, we obtain

$$D(u(t), G_3(f,t)) \leqslant c_\zeta \gamma \zeta(t),$$

$t \in [0, T]$. According to condition (H_9) it follows that

$$
\begin{aligned}
D(u(t), v(t)) \;\leqslant\;& D(u(t), G_3(f,t)) + D(G_3(f,t), v(t)) \\
\leqslant\;& c_\zeta \gamma \zeta(t) + \int_0^t \frac{E_{q,q}(|\lambda|(t-x)^q)D[(f(x,u(x))+B(x)C(x)),(f(x,v(x))+B(x)C(x))]}{(t-x)^{1-q}} dx \\
\leqslant\;& c_\zeta \gamma \zeta(t) + L \int_0^t \frac{E_{q,q}(|\lambda|(t-x)^q)D[u(x),v(x)]}{(t-x)^{1-q}} dx \\
\leqslant\;& c_\zeta \gamma \zeta(t) + L E_{q,q}(|\lambda|(t-x)^q) \int_0^t \frac{)D[u(x),v(x)]}{(t-x)^{1-q}} dx
\end{aligned}
$$

By the generalized Gronwall inequality, we obtain

$$D(u(t), v(t)) \leqslant c_\zeta \gamma \zeta(t) E_q(L E_{q,q}(|\lambda|T^q)\Gamma(q)t^q).$$

Thus, Equation (1) is E_q–Ulam–Hyers–Rassias stable in view of Definition 5. $\square$

Theorem 7. *Let $\lambda < 1$, and let the condition (H_1)–(H_3), (H_8) and (H_9) hold for a $^c[(i)$-GH]-differentiable function u satisfies inequality (5). Then Equation (1) is E_q–Ulam–Hyers–Rassias stable.*

Proof. According to Theorem 3, u is a $^c[(i)$-GH]-differentiable solution to Cauchy problem (1) and (2). Let u be a $^c[(i)$-GH]-differentiable solution to Equation (1), which satisfies inequality (5) with $u(0) = u_0$. From Lemma 6, we obtain

$$D(u(t), G_2(f,t)) \leqslant c_\zeta \gamma \zeta(t),$$

$t \in [0, T]$. According to condition (H_9), it follows that

$$
\begin{aligned}
D(u(t), v(t)) \;\leqslant\;& D(u(t), G_2(f,t)) + D(G_2(f,t), v(t)) \\
\leqslant\;& c_\zeta \gamma \zeta(t) + \int_0^t \frac{E_{q,q}(|\lambda|(t-x)^q)D[(f(x,u(x))+B(x)C(x)),(f(x,v(x))+B(x)C(x))]}{(t-x)^{1-q}} dx \\
\leqslant\;& c_\zeta \gamma \zeta(t) + L \int_0^t \frac{E_{q,q}(|\lambda|(t-x)^q)D[u(x),v(x)]}{(t-x)^{1-q}} dx \\
\leqslant\;& c_\zeta \gamma \zeta(t) + L E_{q,q}(|\lambda|(t-x)^q) \int_0^t \frac{)D[u(x),v(x)]}{(t-x)^{1-q}} dx
\end{aligned}
$$

By the generalized Gronwall inequality, we obtain

$$D(u(t), v(t)) \leqslant c_\zeta \gamma \zeta(t) E_q(L E_{q,q}(|\lambda|T^q)\Gamma(q)t^q).$$

Thus, Equation (1) is E_q–Ulam–Hyers–Rassias stable in view of Definition 5. $\square$

Theorem 8. *Let $\lambda > 1$, let the condition (H_1)–(H_3) as well as (H_5)–(H_7), (H_8)–(H_9) hold for a $^c[(ii)$-GH]-differentiable function u, which satisfies inequality (5). Then, Equation (1) is E_q–Ulam–Hyers–Rassias stable.*

Proof. According to Theorem 4, set u is a $^c[(ii)$-GH]-differentiable solution to Cauchy problem (1) and (2). let u be a $^c[(ii)$-GH]-differentiable solution to Equation (1), which satisfies the inequality (5) with $u(0) = u_0$. from Lemma 7, we obtain

$$D(u(t), G_4(f, t)) \leqslant c_\zeta \gamma \zeta(t),$$

$t \in [0, T]$. According to condition (H_9), it follows that

$$
\begin{aligned}
D(u(t), v(t)) \ &\leqslant\ D(u(t), G_4(f, t)) + D(G_4(f, t), v(t)) \\
&\leqslant\ c_\zeta \gamma \zeta(t) + \int_0^t \frac{E_{q,q}(|\lambda|(t-x)^q) D[(f(x, u(x)) + B(x)C(x)), (f(x, v(x)) + B(x)C(x))]}{(t-x)^{1-q}} dx \\
&\leqslant\ c_\zeta \gamma \zeta(t) + L \int_0^t \frac{E_{q,q}(|\lambda|(t-x)^q) D[u(x), v(x)]}{(t-x)^{1-q}} dx \\
&\leqslant\ c_\zeta \gamma \zeta(t) + L E_{q,q}(|\lambda|(t-x)^q) \int_0^t \frac{)D[u(x), v(x)]}{(t-x)^{1-q}} dx
\end{aligned}
$$

By the generalized Gronwall inequality, we obtain

$$D(u(t), v(t)) \leqslant c_\zeta \gamma \zeta(t) E_q(L E_{q,q}(|\lambda| T^q) \Gamma(q) t^q).$$

Thus, Equation (1) is E_q–Ulam–Hyers–Rassias stable in view of Definition 5. Now, the proof is completed. $\square$

Remark 7. *In view of Definition 6 can be verified as according to the assumption in Theorems 5–8, we assume Equation (1) and inequality (4). It can be verified that Equation (1) is generalized E_q–Ulam–Hyers–Rassias stable with respect to Definition 6.*

Remark 8. *Condition (H_9) weakens $\int_0^t \zeta(x) dx \leqslant c_\zeta \zeta(t) E_2(L E_{2,2}(|1| T^2) \Gamma(2) t^2$ $\forall\ t \in [0, T]$ when we assume $q = 1$. This means that certain theorems in [25] are special cases of Theorem 5 and 6 in the present paper.*

Remark 9. *According to the assumptions excluding $(H9)$ in Theorems 5–8, we consider the equation in (1) and inequality in (3). It can be proved that in terms of Definitions 3 and 4, Equation (1) is E_q–Ulam–Hyers.*

5. Examples

In this part, we will show four examples to explain our main results.

Example 1. *Consider the following Cauchy problem in terms of a Fuzzy fractional differential equation*

$$
{}_0^c D_t^{1.5} u(t) = u(t) \oplus 1.2 u(t) cos(t) \oplus t^2 e^t F \tag{7}
$$

on $(0, 2\pi]$, with initial conditions

$$
\begin{aligned}
u(0) &= \hat{0} \\
u'(0) &= \hat{1}
\end{aligned} \tag{8}
$$

Compared to Equation (1), in the above equations, $q = 1.5$, $\lambda = 2$, $T = 2\pi$, $f(t, u(t)) = 1.2 u(t) cos(t) \oplus t^2 e^t F$, and $F = (0, 1, 2) \in \mathbb{E}^1$ is a symmetric triangular fuzzy number. Hence, with $L = 1.3$, the condition (H_1) and (H_2) are satisfied. It is not difficult to prove that condition (H_3) is satisfied. Hence, as a consequence of Theorem 1, the Cauchy problem (7) and (8) has a $^c[(i)$-GH]-differentiable solution. The numerical solutions with respect to the $q = 1.5$ level are provided by utilizing the Adams–Moultan predictor–corrector method.

Furthermore, for $\varepsilon > 1$, assume that the $^c[(i)$-GH]-differentiable fuzzy-valued function $u : (0, 2\pi] \to \mathbb{E}^1$ satisfies condition (H_8) and

$$
D({}_0^c D_t^{1.5} u(t) = u(t) \oplus 1.2 u(t) cos(t) \oplus t^2 e^t F) \leqslant \varepsilon t
$$

Assuming $\zeta(t) = 1$, $t \in [0, 2\pi]$ and $c_\zeta = \frac{4\sqrt{2\pi}}{3} E_{1.5,1.5}(\sqrt{2\pi})$, this means that condition (H_9) is satisfied. Hence, Equation (7) is E_q–Ulam–Hyers–Rassias stable with respect to Theorem 5.

Example 2. Consider the following Cauchy problem in terms of a Fuzzy fractional differential equation

$$\,^C_0 D_t^{1.5} u(t) = -u(t) + t^2 + t + 4 \tag{9}$$

with initial condition

$$\left\{ \begin{array}{l} u(0) = u_0 \\ u'(0) = u_1 \end{array} \right. \tag{10}$$

Compared to equations (1), in the above equation, $q = -1$, $f(t, u(t)) = t^2 + t + 4$, and $u_0 = (0, 1, 2) \in \mathbb{E}^1$ is a symmetric triangular fuzzy number.

Hence, with $L = 1.3$, the condition (H_1)–(H_2) are satisfied. It is not difficult to prove that condition (H_4) is satisfied. Hence, by employing Theorem 2, the Cauchy problem (9)–(10) has a different $\,^C[(ii)$-GH]-differentiable solution. The numerical solution provides with respect to q-level by utilizing the Adams–Moultan predictor–corrector method.

Furthermore, for $\varepsilon > 1$, assumes that the $\,^C[(ii)$-GH]-differentiable fuzzy-valued function $u : (0, 2\pi] \to \mathbb{E}^1$ satisfies condition (H_8) and

$$D(\,^C_0 D_t^{1.5} u(t), -u(t) + t^2 + t + 4) \leqslant \varepsilon t, \ t \in (0, 2\pi]$$

Assuming $\zeta(t) = t$, $t \in [0, 2\pi]$ and $c_\zeta = \frac{4\sqrt{2\pi}}{3} E_{1.5,1.5}(\sqrt{2\pi})$, this means that condition (H_9) is satisfied. Hence, Equation (9) is E_q–Ulam–Hyers–Rassias stable with respect to Theorem 6.

Example 3. Consider the following Cauchy problem in terms of a Fuzzy fractional differential equation

$$\,^C_0 D_t^{1.5} u(t) = u(t) \oplus 1.2 u(t) \cos(t) \oplus t^2 e^t F \tag{11}$$

on $(0, 2\pi]$, with initial conditions

$$\left\{ \begin{array}{l} u(0) = \hat{0} \\ u'(0) = \hat{1} \end{array} \right. \tag{12}$$

Compared to Equation (1), in the above equation, $q = 1.5, \lambda = 2, T = 2\pi, f(t, u(t)) = -u(t) \oplus 1.2 u(t) \cos(t) \oplus t^2 e^t F$ and $F = (0, 1, 2) \in \mathbb{E}^1$ is a symmetric triangular fuzzy number. Hence, with $L = 1.3$, it is not difficult to prove that condition (H_1) and (H_3) are satisfied. Hence, as a consequence of Theorem 3, the Cauchy problem (11) and (12) has a $\,^C[(i)$-GH]-differentiable solution. The numerical solutions with respect to $q = 1.5$ level are provided by utilizing the Adams–Moultan predictor–corrector method.

Furthermore, for $\varepsilon > 1$, assume that the $\,^C[(i)$-GH]-differentiable fuzzy-valued function $u : (0, \pi] \to \mathbb{E}^1$ satisfies condition (H_8) and

$$D(\,^C_0 D_t^{1.5} u(t), -u(t) \oplus 1.2 u(t) \cos(t) \oplus t^2 e^t F) \leqslant \varepsilon t$$

Assuming $\zeta(t) = 1$, $t \in [0, 2\pi]$ and $c_\zeta = \frac{4\sqrt{2\pi}}{3} E_{1.5,1.5}(\sqrt{2\pi})$, this means that condition (H_9) satisfied. Hence, Equation (11) is E_q–Ulam–Hyers–Rassias stable with respect to Theorem 7.

Example 4. Consider the following Cauchy problem in terms of a Fuzzy fractional differential equation

$$\,^C_0 D_t^{1.5} u(t) = u(t) + t^2 + t + 4, \ t \in (0, 2\pi] \tag{13}$$

with initial condition

$$\left\{ \begin{array}{l} u(0) = u_0 \\ u'(0) = u_1 \end{array} \right. \tag{14}$$

Compared to Equation (1), in the above equations, $q = 1.5$, $f(t, u(t)) = t^2 + t + 4$, and $u_0 = (0, 1, 2) \in \mathbb{E}^1$ is a symmetric triangular fuzzy number.

Hence, with L = 1.3, the condition (H_1)–(H_3) are satisfied. Notice $Ch_{1.5}(x) - Sh_{1.5}(x) > 1$ and $Ch_{1.5}(x) - Sh_{1.5}(x) < 1$ for $u \in (0, 2\pi]$. It is not difficult to prove that condition (H_5)–(H_7) are satisfied. Hence, as a consequence of Theorem 4, the Cauchy problem (13) and (14) has a unique $^c[(ii)$-GH]-differentiable solution. The numerical solutions with respect to the $q = 1.5$ level are provided by utilizing the Adams–Moultan predictor–corrector method.

Furthermore, for $\varepsilon > 1$, assume that the $^c[(ii)$-GH]-differentiable fuzzy-valued function $u : (0, 2\pi] \to \mathbb{E}^1$ satisfies condition (H_8) and

$$D({}^c_0 D_t^{1.5} u(t), u(t) + t^2 + t + 4) \leqslant \varepsilon t, \ t \in (0, 2\pi]$$

Assuming $\zeta(t) = t$, $t \in [0, 2\pi]$ and $c_\zeta = \frac{4\sqrt{2\pi}}{3} E_{1.5,1.5}(\sqrt{2\pi})$, this means that condition (H_9) satisfied. Hence, Equation (13) is E_q–Ulam–Hyers–Rassias stable with respect to Theorem 8.

6. Graphical Presentation

We used the Adams–Bashforth–Moulton technique to acquire the numerical solution for this fractional differential equation for graphical representation of the solution of the problem presented in Equations (7), (9), (11) and (13). For simulation, the modified predictor–corrector scheme is used to examine the effect and contribution of the time-delayed factor. A graphical representation of the solution with different variations of the time delay factor, as well as other parameters, is made to check and demonstrate the stability of the model under consideration. We are able to see the Ulam–Hyers stability of varied accuracies and delays from the numerical data. The system will attain Ulam–Hyers stability more quickly with greater accuracy. This is also true when the number of delays increases. Figures 1–4 show the stability of the system (7), (9), (11) and (13) for various time delays and fractional derivatives.

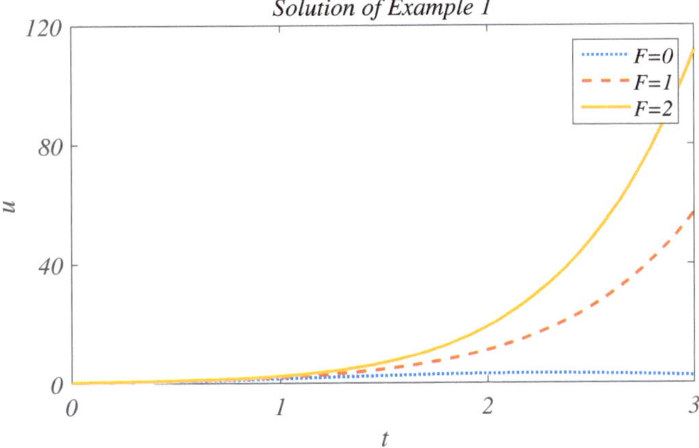

Figure 1. Solution of Problems (7) and (8).

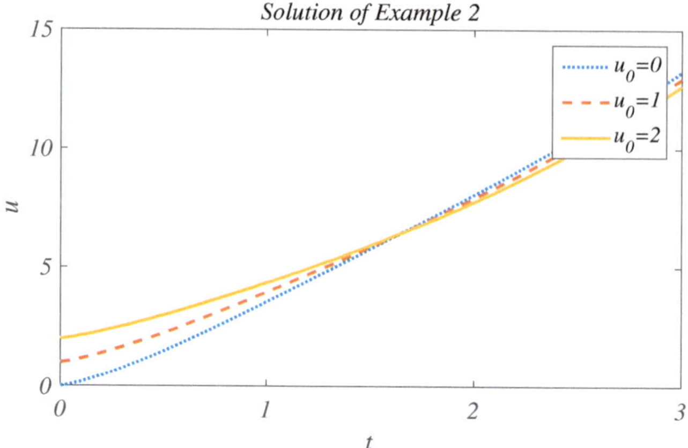

Figure 2. Solution of Problems (9) and (10).

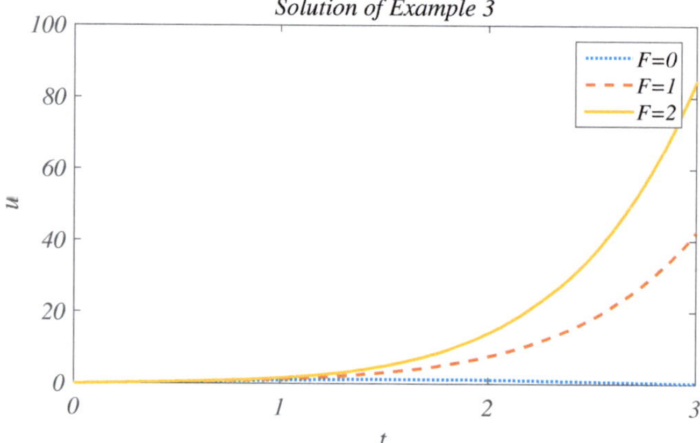

Figure 3. Solution of Problems (11) and (12).

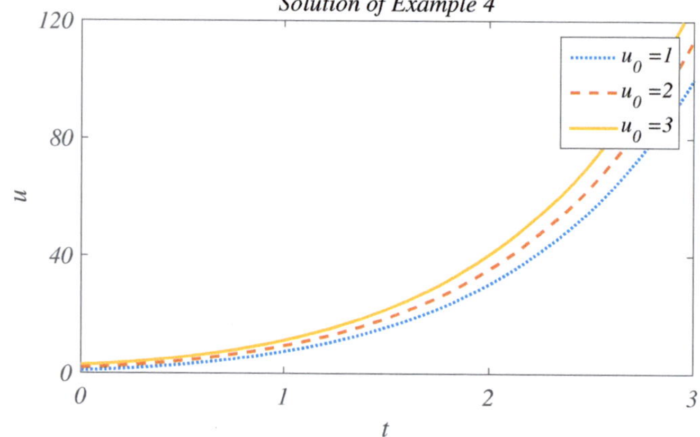

Figure 4. Solution of Problems (13) and (14).

7. Conclusions

This paper aims to define the uniqueness and existence of a group of nonlinear fuzzy fractional differential equation of solutions to the Cauchy problem. Moreover, E_q–Ulam-type stability of Equation (1) is observed by applying the inequality technique. We obtain uniqueness and existence results with the help of nonlocal conditions of the Caputo derivative. Moreover, future work may include broadening the idea indicated in this task and familiarizing observability, and generalize other tasks. Ulam-type stability of fuzzy fractional differential equations, similar to crisp situations for approximate solutions, provides a reliable theoretical basis. This a fruitful area with wide research projects, and it can bring about countless applications and theories. We have decided to devote much attention to this area. Furthermore, it is fruitful to investigate stability problems in a classical sense for the fuzzy fractional differential equation.

Author Contributions: A.U.K.N., J.H., R.S. and B.A. contributed equally to the writing of this paper. All authors studied and validated the final document. All authors have read and agreed to the published version of the manuscript.

Funding: The work was of Azmat Ullah Khan Niazi, and supported by post doctoral funding of Xiangtan University, Hunan, China.

Institutional Review Board Statement: Not applicable.

Informed Consent Statement: Not applicable.

Data Availability Statement: All the data is present within the manuscript.

Acknowledgments: The authors would like to express their sincere gratitude to the reviewers and the editors for their careful reviews and constructive recommendations. Jiawei He was supported by the Hunan key Laboratory for Computation and Simulation in Science and Engineering Xiangtan University.

Conflicts of Interest: All authors have no conflict of interest.

References

1. Miller, K.S.; Ross, B. *An Introduction to the Fractional Calculus and Fractional Differential Equations*; Wiley: New York, NY, USA, 1993.
2. Oldham, K.; Spanier, J. *The Fractional Calculus Theory and Applications of Differentiation and Integration to Arbitrary Order*; Elsevier: Amsterdam, The Netherlands, 1974.
3. Samko, S.G.; Kilbas, A.A.; Marichev, O.I. *Fractional Integrals and Derivatives (Theory and Applications)*; Gordon and Breach Science Publishers: Amsterdam, The Netherlands, 1993.
4. Ahmad, B.; Nieto, J.J. Existence results for a coupled system of nonlinear fractional differential equations with three-point boundary conditions. *Comput. Math. Appl.* **2009**, *58*, 1838–1843. [CrossRef]
5. Ahmad, B.; Ntouyas, S.K.; Agarwal, R.P.; Alsaedi, A. On fractional differential equations and inclusions with nonlocal and average-valued (integral) boundary conditions. *Adv. Differ. Equ.* **2016**, *2016*, 80 [CrossRef]
6. Ding, Z.; Ma, M.; Kandel, A. Existence of the solutions of fuzzy differential equations with parameters. *Inf. Sci.* **1997**, *99*, 205–217. [CrossRef]
7. Kilbas, A.A.; Srivastava, H.M.; Trujillo, J.J. *Theory and Applications of Fractional Differential Equations*; Elsevier: Amsterdam, The Netherlands, 2006; Volume 204.
8. Lakshmikantham, V.; Vatsala, A.S. Basic theory of fractional differential equations. *Nonlinear Anal. Theory Methods Appl.* **2008**, *69*, 2677–2682. [CrossRef]
9. Podlubny, I. *Fractional Differential Equations: An Introduction to Fractional Derivatives, Fractional Differential Equations, to Methods of Their Solution and Some of Their Applications*; Elsevier: Amsterdam, The Netherlands, 1998.
10. Sakulrang, S.; Moore, E.J.; Sungnul, S.; de Gaetano, A. A fractional differential equation model for continuous glucose monitoring data. *Adv. Differ. Equ.* **2017**, *2017*, 150. [CrossRef]
11. Agarwal, R.; Lakshmikantham, V.; Nieto, J. On the concept of solution for fractional differential equations with uncertainty. *Nonlinear Anal.* **2010**, *72*, 2859–2862. [CrossRef]
12. Allahviranloo, T.; Salahshour, S.; Abbasbandy, S. Explicit solutions of fractional differential equations with uncertainty. *Soft Comput.* **2012**, *16*, 297–302. [CrossRef]
13. Salahshour, S.; Allahviranloo, T.; Abbasbandy, S. Solving fuzzy fractional differential equations by fuzzy Laplace transforms. *Commun. Nonlinear Sci. Numer. Simul.* **2012**, *17*, 1372–1381. [CrossRef]
14. Bede, B.; Gal, S.G. Almost periodic fuzzy-number-valued functions. *Fuzzy Sets Syst.* **2004**, *147*, 385–403. [CrossRef]

15. Bede, B.; Gal, S.G. Generalizations of the differentiability of fuzzy-number-valued functions with applications to fuzzy differential equations. *Fuzzy Sets Syst.* **2005**, *151*, 581–599. [CrossRef]
16. Chehlabi, M.; Allahviranloo, T. Concreted solutions to fuzzy linear fractional differential equations. *Appl. Soft Comput.* **2016**, *44*, 108–116. [CrossRef]
17. Allahviranloo, T.; Armand, A.; Gouyandeh, Z. Fuzzy fractional differential equations under generalized fuzzy Caputo derivative. *J. Intell. Fuzzy Syst.* **2014**, *26*, 1481–1490. [CrossRef]
18. Armand, A.; Allahviranloo, T.; Abbasbandy, S.; Gouyandeh, Z. Fractional relaxation-oscillation differential equations via fuzzy variational iteration method. *J. Intell. Fuzzy Syst.* **2017**, *32*, 363–371. [CrossRef]
19. Wang, Y.; Sun, S.; Han, Z. Existence of solutions to periodic boundary value problems for fuzzy fractional differential equations. *Int. J. Dyn. Syst. Differ. Equ.* **2017**, *7*, 195–216. [CrossRef]
20. Agarwal, R.P.; Baleanu, D.; Nieto, J.J.; Torres, D.F.; Zhou, Y. A survey on fuzzy fractional differential and optimal control nonlocal evolution equations. *J. Comput. Appl. Math.* **2018**, *339*, 3–29. [CrossRef]
21. Huang, L.L.; Baleanu, D.; Mo, Z.W.; Wu, G.C. Fractional discrete-time diffusion equation with uncertainty: Applications of fuzzy discrete fractional calculus. *Phys. A Stat. Mech. Its Appl.* **2018**, *508*, 166–175. [CrossRef]
22. Wang, Y.; Sun, S.; Han, Z. On fuzzy fractional Schrödinger equations under Caputo's H-differentiability. *J. Intell. Fuzzy Syst.* **2018**, *34*, 3929–3940. [CrossRef]
23. Wang, J.; Li, X. E_α-Ulam type stability of fractional order ordinary differential equations. *J. Appl. Math. Comput.* **2014**, *45*, 449–459. [CrossRef]
24. Shen, Y. On the Ulam stability of first order linear fuzzy differential equations under generalized differentiability. *Fuzzy Sets Syst.* **2015**, *280*, 27–57. [CrossRef]
25. Wang, Y.; Sun, S. Existence, uniqueness and E_q-Ulam type stability of fuzzy fractional differential equations with parameters. *J. Intell. Fuzzy Syst.* **2019**, *36*, 5533–5545. [CrossRef]
26. Shen, Y. Hyers-Ulam-Rassias stability of first order linear partial fuzzy differential equations under generalized differentiability. *Adv. Differ. Equ.* **2015**, *2015*, 351. [CrossRef]
27. Wang, C.; Xu, T.Z. Hyers-Ulam stability of fractional linear differential equations involving Caputo fractional derivatives. *Appl. Math.* **2015**, *60*, 383–393. [CrossRef]
28. Brzdęk, J.; Eghbali, N. On approximate solutions of some delayed fractional differential equations. *Appl. Math. Lett.* **2016**, *54*, 31–35. [CrossRef]
29. Lakshmikantham, V.; Mohapatra, R.N. *Theory of Fuzzy Differential Equations and Inclusions*; CRC Press: Boca Raton, FL, USA, 2004.
30. Lakshmikantham, V.; Bhaskar, T.G.; Devi, J.V. *Theory of Set Differential Equations in Metric Spaces*; Cambridge Scientific Publishers: Cambridge, UK, 2006.
31. Gorenflo, R.; Kilbas, A.A.; Mainardi, F.; Rogosin, S.V. *Mittag-Leffler Functions, Related Topics and Applications*; Springer: Berlin, Germany, 2014; Volume 2.
32. Wang, J.; Feckan, M.; Zhou, Y. Presentation of solutions of impulsive fractional Langevin equations and existence results. *Eur. Phys. J. Spec. Top.* **2013**, *222*, 1857–1874. [CrossRef]
33. Peng, S.; Wang, J. Cauchy problem for nonlinear fractional differential equations with positive constant coefficient. *J. Appl. Math. Comput.* **2016**, *51*, 341–351. [CrossRef]
34. Melliani, S.; El Allaoui, A.; Chadli, L.S. Relation between fuzzy semigroups and fuzzy dynamical systems. *Nonlinear Dyn. Syst. Theory* **2017**, *17*, 60–69.
35. Bede, B.; Stefanini, L. Generalized differentiability of fuzzy-valued functions. *Fuzzy Sets Syst.* **2013**, *230*, 119–141. [CrossRef]
36. Ye, H.; Gao, J.; Ding, Y. A generalized Gronwall inequality and its application to a fractional differential equation. *J. Math. Anal. Appl.* **2007**, *328*, 1075–1081. [CrossRef]

fractal and fractional

MDPI

Article

Fractals Parrondo's Paradox in Alternated Superior Complex System

Yi Zhang and Da Wang *

Research Center of Dynamics System and Control Science, Shandong Normal University, Jinan 250014, China;
zhangyi9284@126.com
* Correspondence: wangda@sdnu.edu.cn

Abstract: This work focuses on a kind of fractals Parrondo's paradoxial phenomenon "deiconnected+diconnected=connected" in an alternated superior complex system $z_{n+1} = \beta(z_n^2 + c_i) + (1 - \beta)z_n$, $i = 1, 2$. On the one hand, the connectivity variation in superior Julia sets is explored by analyzing the connectivity loci. On the other hand, we graphically investigate the position relation between superior Mandelbrot set and the Connectivity Loci, which results in the conclusion that two totally disconnected superior Julia sets can originate a new, connected, superior Julia set. Moreover, we present some graphical examples obtained by the use of the escape-time algorithm and the derived criteria.

Keywords: Parrodo'paradox; Mann iteration; Julia set; alternated system; connectivity

Citation: Zhang, Y.; Wang, D. Fractals Parrondo's Paradox in Alternated Superior Complex System. *Fractal Fract.* **2021**, 5, 39. https:// doi.org/10.3390/fractalfract5020039

Academic Editors: António M. Lopes and Liping Chen

Received: 3 April 2021
Accepted: 26 April 2021
Published: 28 April 2021

1. Introduction

The natural process has obvious discrete characteristics; therefore, discrete dynamical systems are usually applied for the modeling of actual processes. On the other hand, considering the complexity of nature, more and more attention has been made to the alternate iteration method [1,2], which is more accurate in revealing the complex behaviors in processes than a unique system.

In 1999, Parrondo et al. [3,4] proposed that two games with loosing gains can paradoxically become a winning game. This classical "losing + losing = winning" phenomenon was known as Parrondo's paradox, which inspired a new research fever in physics and mathematics areas [5,6] about the combination of two systems with negative expected values. In this theory, the game was divided into process A and B. As the game goes on, A actually changes the distribution of B branch, and the overall outcome changes. By analyzing the trajectories of system states, Almeida et al. extended the paradox to the chaos area and exposed the "chaos + chaos = order" phenomenon, which indicated that two chaotic behaviours can reduce to order via alternate iteration.

It should be noted that fractals and chaos are two basic branches in nonlinear science and, to some extent, are closely related to each other. Although the concept of fractal was given in 1975 [7], its basic principle was put forward as early as 1918, when Gaston Julia [8] firstly investigated a simple complex map $z_{n+1} = z_n^2 + c$, $z_n, c \in \mathbb{C}$. Aided by computer technology, Mandelbrot [7] visualized the parameter area where the connected Julia sets' parameter c is located. In recent years, there has been much research surrounding the properties of M-J sets [9–13], effects of noise disturbance [14–18] and related applications [17–22]. Meanwhile, a few researchers have also focused on the special fractal sets generated from alternated complex maps, superior complex maps, hyper-complex systems, etc. In addition, fractional mathematics is closely related to chaos and fractals. Fractional systems are worth studying from the fractal perspective. In [23,24], researchers investigated the citation profiles of researchers in fractional calculus, and proposed that the application areas of fractional calculus contain the fractal concept. Based on the control theory and method, Wang [25–27] investigates the Julia sets of a fractional Lotka–Volterra model and realizes its state feedback control. In [28], the numerical simulation of a Boussinesq equation with different fractal dimension and fractional order is carried out. The results show that the correlation model is suitable for groundwater flow in fractured media.

In the various research on fractals, connectivity is one of the most basic and important branches. In physics, Wang indicated that the connectivity of Julia sets can be used to describe particle velocity [20,21]. In biology, Mojica proposed that the cells differentiating real organisms are similar in some features of connected Julia sets [29]. Based on the two-dimensional predator–prey model, Sun et al. applied Julia sets to represent the origin area to ensure the coexistence of two populations. The authors presented that the connectivity of such an origin area is important for the stability of populations [22]. The connectivity investigations mentioned above were mainly concentrated on the Julia sets from a single map. For alternate cases, Danca [1,2] illustrated, in the graphical results, that the connectivity properties have many forms, including connected, disconnected and totally disconnected. Further, Wang [30] compared the connectivity of an alternate iteration Julia set with their original separated Julia sets and gave a preliminary study on the fractals Parrondo's phenomenon of the classical alternate system.

Recently, some of the fractals studies concentrate on Julia sets generated by a more complicated iteration scheme.

For instance, Mandelbrot and Julia sets generated by using the Picard–Mann iteration procedure were intriduced in [31]. Based on the Jungck-CR iteration process with s-convexity, authors proved new escape criteria for the generation of Mandelbrot and Julia sets and presented some graphical examples obtained by the use of an escape time algorithm and the derived criteria in [32]. In [33], the authors investigated the biomorphs for certain polynomials by using a more general iteration method and examined their graphical behaviour with respect to the variation in parameters. In [34], the authors adjust algorithms according to the developed conditions and draw some attractive Julia and Mandelbrot sets with iterate sequences from proposed fixed-point iterative methods. Moreover, some results about superior M-J sets was presented by Rani in [13,35]. Further, Rani and Yadav [36] alternated two maps of quadratic family $z_{n+1} = \beta(z_n^2 + c_i) + (1 - \beta)z_n$, $i = 1, 2$ in superior orbit, and indicated that alternate superior Julia sets also show three connectivities: connected, disconnected and totally disconnected. The effects of the superior Mandelbrot set were searched by a new noise criterion in [37]. In [38], researchers found that the superior Julia set showed a higher stability in certain high intensities, and discussed its application in particle dynamics. The effects of dynamic noise in superior Mandelbrot sets were analyzed in [39]. In [40], Mann iteration and superior Julia sets were used for biological morphogenesis algorithm optimization.

Although considerable studies have been made on the superior M-J set, to our best knowledge, little attention was paid to the connectivity investigation. As mentioned above, superior Julia sets have a higher stability than the classical ones, and also show potential application prospects. Thus, it is of interest to seek the Parrondo's paradox in the alternate superior complex system from the perspective of connectivity.

Motivated by the significant investigations mentioned above, the main motivation of this work is to provide a detailed analysis of the connectivity change law of superior Julia sets in an alternated case.

The reminder of this paper is organized as follows. Essential definitions and lemmas are given in Section 2. In Section 3, graphical explorations of alternate superior Julia sets are investigated. Through the use of the escape-time algorithm and image simulation method, "disconnected + disconnected = connected" and "connected + connected = disconn-ected" phenomena are proved in visual way. Section 4 concludes this work by discussing the potential applications of this fractal's phenomenon and pointing out the prospective research direction.

2. Preliminaries

Definition 1 ([13]). *Consider the following Mann iteration, which can be introduced a one-step feedback process*

$$x_{n+1} = g(f(x_n), x_n) = \beta f(x_n) + (1 - \beta)x_n.$$

where β lies between 0 and 1, x_n represents input and x_{n+1} is expressed as output. Simplifying the process with invariable β and then consider this process as complex quadratic map

$$P_c: \quad z_{n+1} = \beta(z_n^2 + c) + (1 - \beta)z_n, \quad 0 < \beta \le 1. \tag{1}$$

When $\beta = 1$, P_c can be seen as a simple complex map: $z_{n+1} = z_n^2 + c$. The filled superior Julia set of system (1) is defined as $K(P_c)$, which satisfies that

$$K(P_c) = \{z_0 \mid P_c^n(z_0) \nrightarrow \infty, n \to \infty\},$$

where P_c^n denote the n-th iteration of z_0. The superior Julia set of system (1), denoted by SJ is the boundary of $K(P_c)$, i.e., $SJ(P_c) = \partial K(P_c)$.

Definition 2. *The Mandelbrot-efficacy set of system (1) is defined as*

$$M(P_c) = \{c \mid \text{The superior Julia sets } SJ(P_c) \text{ is connected}\}.$$

Definition 3 ([13]). *Considering two complex quadratic maps alternated in superior orbit*

$$P_{c_1,c_2}: z_{n+1} = \begin{cases} \beta(z_n^2 + c_1) + (1 - \beta)z_n, & \text{if n is even,} \\ \beta(z_n^2 + c_2) + (1 - \beta)z_n, & \text{if n is odd,} \end{cases} \tag{2}$$

The filled alternate superior Julia set of system (2) is denoted as $K(P_{c_1,c_2})$, which satisfies that

$$K(P_{c_1,c_2}) = \{z_0 \mid P_{c_1,c_2}^n(z_0) \nrightarrow \infty, n \to \infty\},$$

where P_{c_1,c_2}^n represents the n-th iteration of the initial point z_0. The alternate superior Julia set of P_{c_1,c_2} is the boundary of the filled alternate superior Julia set, that is, $SJ(P_{c_1,c_2}) = \partial K(P_{c_1,c_2})$.

Lemma 1 ([36]). *For system*

$$O_{c_1,c_2}: \quad z_{n+1} = \beta((z_n^2 + c_1)^2 + c_2) + (1 - \beta)(z_n^2 + c_1), \quad z_n, c_1, c_2 \in \mathbb{C}.$$

$SJ(P_{c_1,c_2})$ and $SJ(O_{c_1,c_2})$ are the same for given c_1 and c_2 parameter values.

Lemma 2 ([1,2]). *The connectivity properties of superior Julia set for a complex polynomial of degree 2 and $0 < \beta \le 1$ can be identified based on the following cases:*
(1) *Superior Julia set is connected if and only if all the critical orbits are bounded;*
(2) *Superior Julia set is totally disconnected, a red Cantor set, if (but not only if) all the critical orbits are unbounded;*
(3) *For a polynomial with at least one critical orbit unbounded, the superior Julia set is totally disconnected if and only if all the bounded critical orbits are aperiodic.*

3. Graphical Explorations

As can be seen from Figure 1, with the decrease in the value of β, the superior Mandelbrot set $M(P_c)$ expands rapidly. Therefore, we only consider the case of $\beta = 0.9$ in the next simulations.

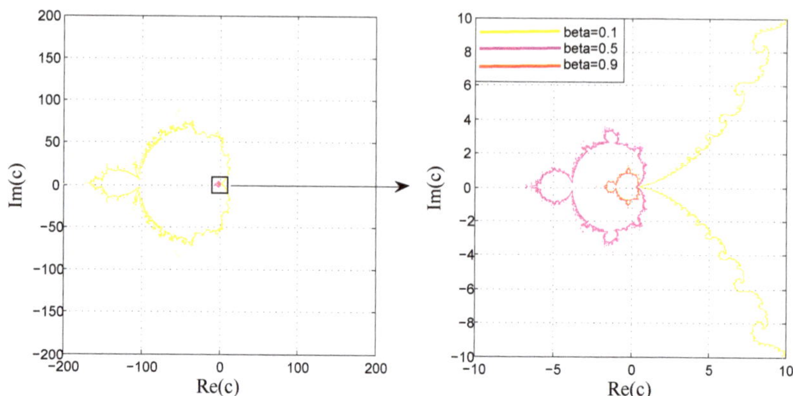

Figure 1. Superior Mandelbrot sets $M(P_c)$ plotted with different β.

Based on the above definitions, system P_{c_1,c_2} originates from the alternation of two single systems P_c, and the superior Julia sets of P_c have only two states, which are determined by the single critical point 0. The Julia sets which are plotted in Figure 2 indicate that the different connectivity relying on weather parameter c belong to the Mandelbrot set.

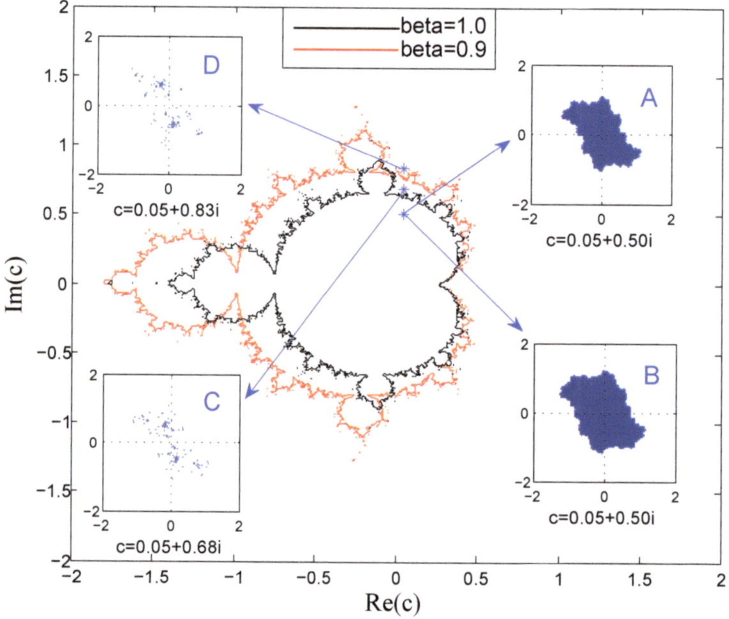

Figure 2. $M(P_c)$, classical Mandelbrot set and (**A**) Connected Julia set, (**B**) Connected superior Julia set, (**C**) Totally disconnected Julia set, and (**D**) Totally disconnected superior Julia set.

For a single system P_c, $M(P_c)$ can be plotted along two coordinates ($Re\ c$, $Im\ c$). Further, in this study, the whole Connectivity Loci of an alternate superior system P_{c_1,c_2} is defined as $\mathcal{M}(\mathcal{P})$, which is determined by four coordinates ($Re\ c_1$, $Im\ c_1$, $Re\ c_2$, $Im\ c_2$). With the help of the graphical method proposed in [2], this paper visualized its structure via *MATLAB* software. At a certain resolution, fixing the $Im(c_2)$ to 0.3 and screening all $[Rm\ c_1,\ Im\ c_2,\ Re\ c_2]$ which connect alternated superior Julia sets, we plotted the spatial-Connected Loci in Figure 3. Further, in Figure 4, Fixing the $Re(c_2)$ to 0, and we obtain the planar $\mathcal{M}(\mathcal{P})$ by recognizing the connectivity of the Julia sets corresponding to all $[\ Im\ c_2,\ Re\ c_2]$. In a few words, Figure 4 is a slice of a three-dimensional $\mathcal{M}(\mathcal{P})$ and similar slices can be obtained by fixing any two dimensions.

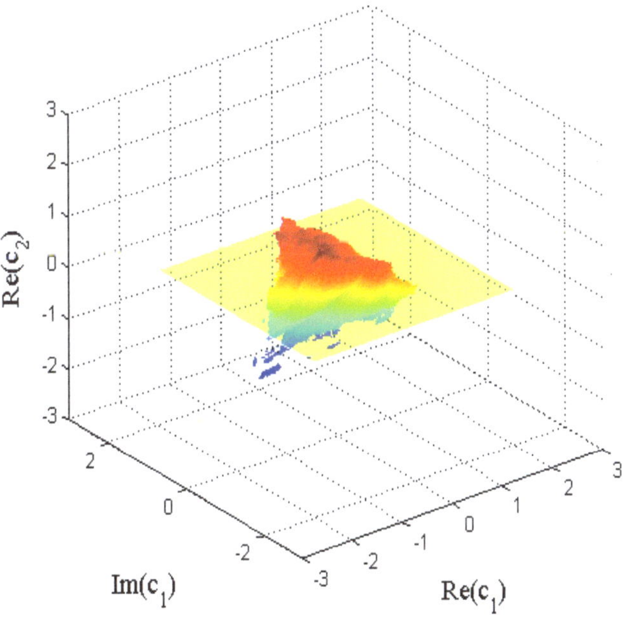

Figure 3. Spatial-Connected Loci ($\mathcal{M}(\mathcal{P})$ without Disconnected Loci) with $Im(c_2) = 0.3$.

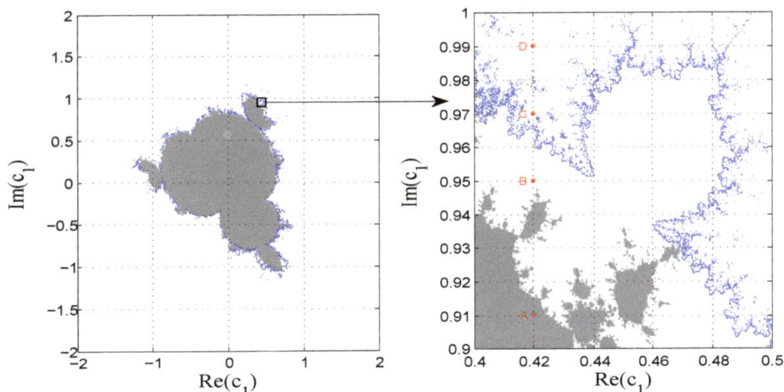

Figure 4. Planar $\mathcal{M}(\mathcal{P})$ with $c_2 = 0 + 0.3i$.

As is shown in Figure 5, the connectivity of four locations in Figure 4 is founded to vary along the Connected Loci, Disconnected Loci and Totally Disconnected Loci. That is, the gray area leads to connected superior Julia sets, the region between the grey boundary and blue line leads to disconnected superior Julia sets, the region outside the blue line gives rise to totally disconnected superior Julia sets.

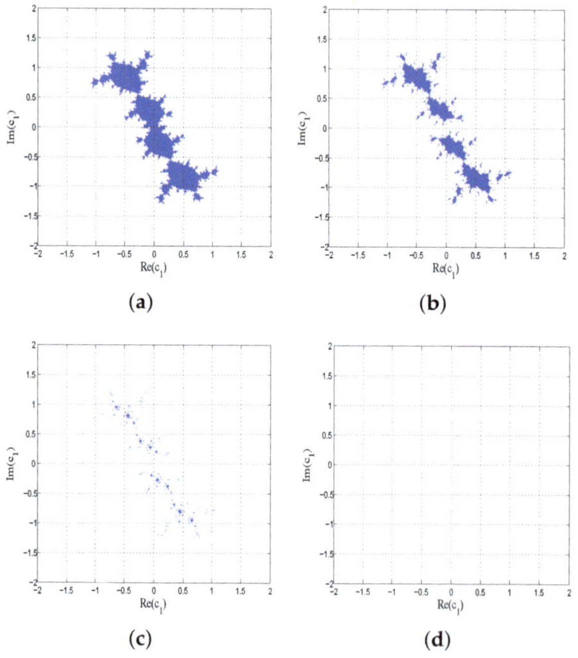

Figure 5. (a) $K(P_{c_1=0.42+0.91i,\ c_2=0.3i})$; (b) $K(P_{c_1=0.42+0.95i,\ c_2=0.3i})$; (c) $K(P_{c_1=0.42+0.97i,\ c_2=0.3i})$; (d) $K(P_{c_1=0.42+0.99i,\ c_2=0.3i})$.

Now, the next step is to find a pair of parameters c_1, c_2 which make individual superior Julia sets disconnected and alternate superior Julia sets connected. With the help of the relationship between the regions and the connectivity mentioned above, the c_2 which satisfies the phenomenon "disconnected+disconnected=connected" should be outside of the red boundary and inside the grey region.

To verify the analysis mentioned above, our solution is putting the boundary of the $M(P_c)$ cover on planar $\mathcal{M}(\mathcal{P})$. From Figure 6, it can be seen in the location of $c_1 = 0.35 + 0.59i$ that its superior Julia set $SJ(P_{c_1})$ is totally disconnected, the c_2 taken from area outside red boundary can lead to totally disconnected $SJ(P_{c_2})$, the c_2 taken from gray area can lead to connected $SJ(P_{c_1,c_2})$. For example, one c_2 can be set to $0.35 - 0.59i$ (point θ_1 in Figure 7); totally disconnected filled superior Julia set $K(P_{c=0.35+0.59i})$, totally disconnected filled superior Julia set $K(P_{c=0.35-0.59i})$ and connected filled alternate superior Julia set $K(P_{c_1=0.35+0.59i,\ c_2=0.35-0.59i})$ are shown in Figure 8.

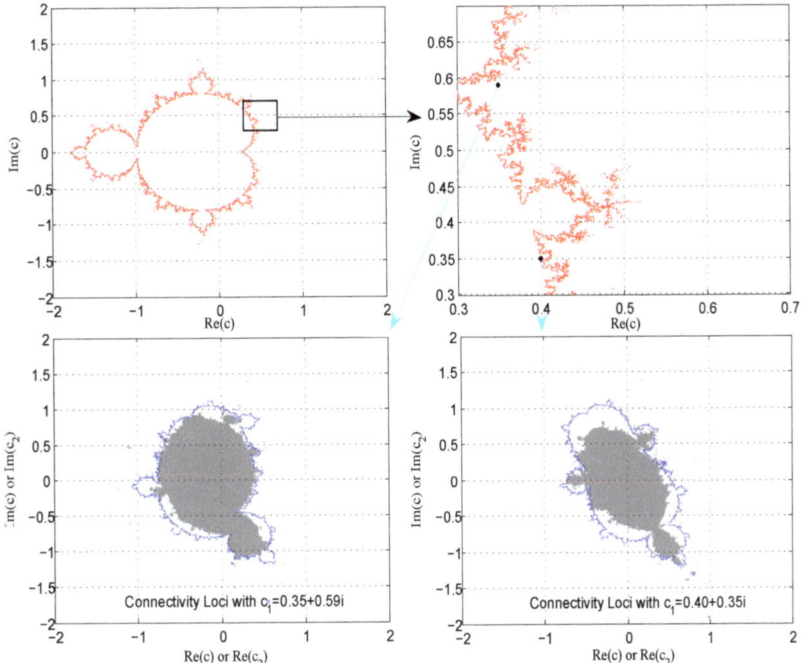

Figure 6. The boundary of $M(P_c)$ and two slices of $\mathcal{M}(\mathcal{P})$.

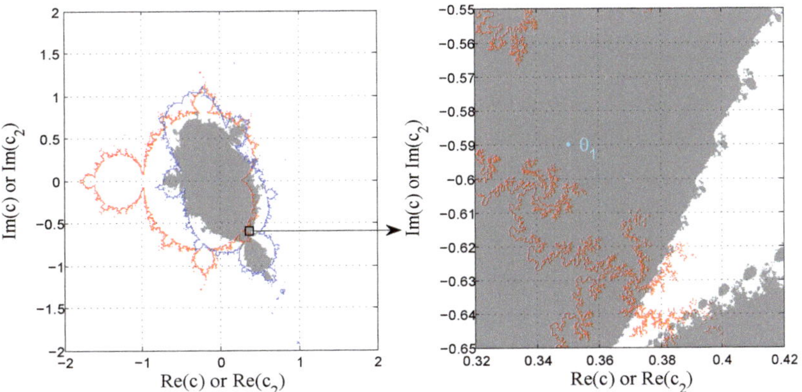

Figure 7. The detail of $\mathcal{M}(\mathcal{P})$ with $c_1 = 0.35 + 0.59i$.

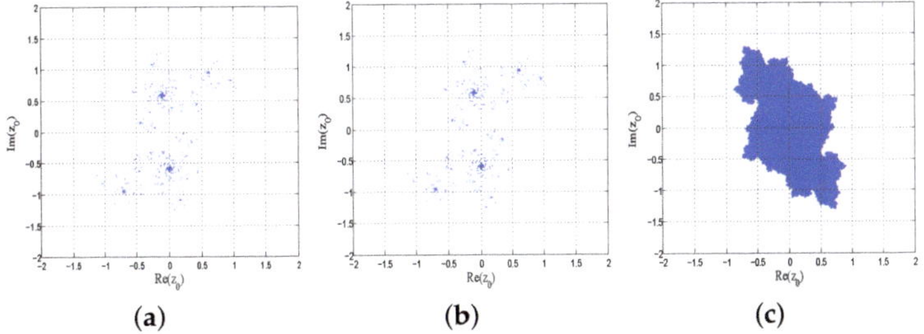

Figure 8. (**a**) $K(P_{c=0.35+0.59i})$; (**b**) $K(P_{c=0.35-0.59i})$; (**c**) $K(P_{c_1=0.35+0.59i,\, c_2=0.35-0.59i})$.

Based on the above research, in addition to "disconnected+disconnected=connected", the establishment condition of "connected+connected=disconnected" is that c_2 located between the blue and red boundary. According to the enlarged part in Figure 9, one can choose a proper point $c_2 = 0.39 + 0.32i$ (point θ_2 in Figure 9), connected filled superior Julia set $K(P_{c=0.40+0.35i})$, connected filled superior Julia set $K(P_{c=0.39+0.32i})$ and totally disconnected filled alternate superior Julia set $K(P_{c_1=0.40+0.35i,\,c_2=0.39+0.32i})$ are shown in Figure 10.

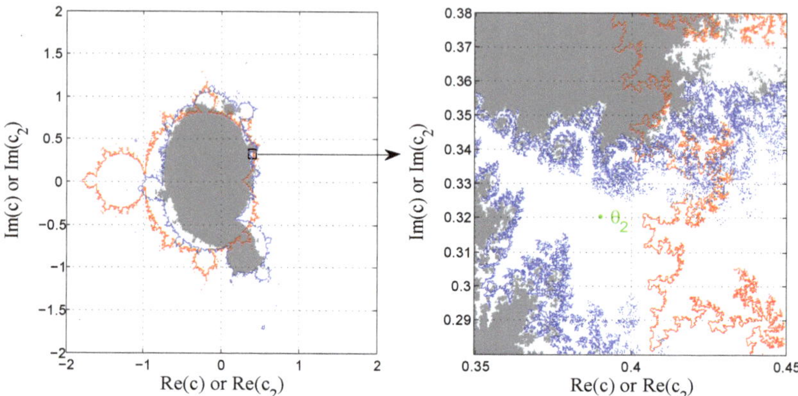

Figure 9. The detail of $\mathcal{M}(\mathcal{P})$ with $c_1 = 0.4 + 0.35i$.

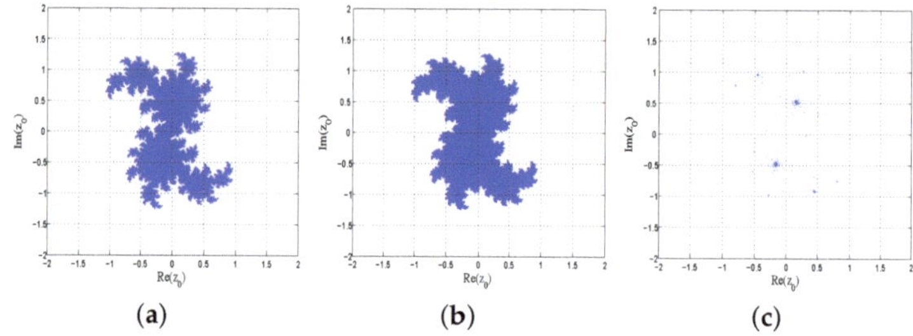

(a) (b) (c)

Figure 10. (a) $K(P_{c=0.40+0.35i})$; (b) $K(P_{c=0.39+0.32i})$; (c) $K(P_{c_1=0.40+0.35i,\,c_2=0.39+0.32i})$.

4. Conclusions

This paper demonstrates that "disconnected+disconnected=connected" and "connected+connected=disconnected" Parrondo's Paradox phenomena exist in an alternate superior system. As mentioned in the introduction section, superior Julia sets show higher stability in certain situations, and alternate systems have been widely applied to physics, biology, etc. This phenomenon, occurring in alternate superior systems may have potential applications in many fields. We hope that the result of this paper can provide a reference for future research. On the other hand, according to the introduction, there is a close relationship between fractal and fractional; therefore, future research may further expand the Parrondo's Paradox phenomenon to the fields, combining fractal and fractional.

Author Contributions: Conceptualization, D.W.; methodology, Y.Z. and D.W.; validation, Y.Z. and D.W.; formal analysis, Y.Z. and D.W.; writing—original draft preparation, Y.Z. and D.W.; writing—review and editing, Y.Z. and D.W.; visualization, Y.Z.; funding acquisition, D.W. All authors have read and agreed to the published version of the manuscript.

Funding: This research was funded by the China Postdoctoral Science Foundation Funded Project (No. 2017M612337), and the Scientific and Technological Planning Projects of Universities in Shandong Province (No. J18KB097).

Conflicts of Interest: The authors declare no conflict of interest.

References

1. Danca, M.F.; Romera, M.; Pastor, G. Alternated Julia sets and connectivity properties. *Int. J. Bifurc. Chaos* **2009**, *19*, 2123–2129. [CrossRef]
2. Danca, M.F.; Bourke, P.; Romera, M. Graphical exploration of the connectivity sets of alternated Julia sets. *Nonlinear Dyn.* **2013**, *73*, 1155–1163. [CrossRef]
3. Harmer, G.P.; Abbott, D. Parrondo's paradox. *Stat. Sci.* **1999**, *14*, 206–213.
4. Harmer, G.P.; Abbott, D. Game theory Losing strategies can win by Parrondo's paradox. *Nature* **1999**, *402*, 864. [CrossRef]
5. Minor, D.P. Parrondo's paradox–hope for losers! *Coll. Math. J.* **2003**, *34*, 15–20.
6. Chen, L.; Li, C.F.; Gong, M.; Guo, G.C. Quantum Parrondo game based on a quantum ratchet effect. *Phys. A Stat. Mech. Its Appl.* **2010**, *389*, 4071–4074. [CrossRef]
7. Mandelbrot, B.B. *Fractals: Form, Chance, and Dimension*; W. H. Freeman: San Francisco, CA, USA, 1977.
8. Julia, G. Mémoire sur l'itération des fonctions rationnelles. *J. De Mathématiques Pureset Appliquées* **1918**, *8*, 47–245.
9. Wang, X.Y.; Gu, L. Research fractal structures of generalized M-J sets using three algorithms. *Fractals* **2008**, *16*, 79–88. [CrossRef]
10. Nazeer, W.; Kang, S.M.; Tanveer, M.; Shahid, A.A. Fixed point results in the generation of Julia and Mandelbrot sets. *J. Inequalities Appl.* **2015**, *2015*, 1–16. [CrossRef]
11. Dhurandhar, S.V.; Bhavsar, V.C.; Gujar, U.G. Analysis of z-plane fractal images from $z \leftarrow z^\alpha + c$ for $\alpha < 0$. *Comput. Graph.* **1993**, *17*, 89–94.
12. Wang, X.Y.; Liu, X.D.; Zhu, W.Y.; Gu, S.S. Analysis of c-plane fractal images from $z \leftarrow z^\alpha + c$ for $(\alpha < 0)$. *Fractals* **2020**, *8*, 307–314.
13. Rani, M.; Kumar, V. Superior julia set. *J. Korea Soc. Math. Educ. Ser. D Res. Math. Educ.* **2004**, *8*, 261–277.
14. Andreadis, I.; Karakasidis, T.E. On a closeness of the Julia sets of noise-perturbed complex quadratic maps. *Int. J. Bifurc. Chaos* **2012**, *22*, 1250221. [CrossRef]
15. Wang, X.Y.; Ge, F.D. Quasi-sine Fibonacci M set with perturbation. *Nonlinear Dyn.* **2012**, *69*, 1765–1779. [CrossRef]
16. Wang, X.Y.; Wang, Z.; Lang, Y.H.; Zhang, Z.F. Noise perturbed generalized Mandelbrot sets. *J. Math. Anal. Appl.* **2008**, *347*, 179–187. [CrossRef]
17. Wang, X.Y.; Chang, P.J.; Gu, N.N. Additive perturbed generalized Mandelbrot-Julia sets. *Appl. Math. Comput.* **2007**, *189*, 754–765. [CrossRef]
18. Wang, X.Y.; Jia, R.H.; Zhang, Z.F. The generalized Mandelbrot set perturbed by composing noise of additive and multiplicative. *Appl. Math. Comput.* **2009**, *210*, 107–118.
19. Beck, C. Physical meaning for Mandelbrot and Julia sets. *Phys. D Nonlinear Phenom.* **1999**, *125*, 171–182. [CrossRef]
20. Wang, X.Y.; Meng, Q.Y. Study on the physical meaning for generalized Mandelbrot-Julia sets based on the Langevin problem. *Acta Phys. Sin.* **2004**, *53*, 388–395.
21. Wang, X.Y.; Liu, W.; Yu, X.J. Research on Brownian movement based on generalized Mandelbrot-Julia sets from a class complex mapping system. *Mod. Phys. Lett. B* **2007**, *21*, 1321–1341. [CrossRef]
22. Sun, W.H.; Zhang, Y.P.; Zhang, X. Fractal analysis and control in the predator-prey model. *Int. J. Comput. Math.* **2017**, *94*, 737–746. [CrossRef]
23. José M.; António M.L. Fractional Jensen–Shannon Analysis of the Scientific Output of Researchers in Fractional Calculus. *Entropy* **2017**, *19*, 127.
24. Ionescu, C.M. *The Human Respiratory System: An Analysis of the Interplay between Anatomy, Structure, Breathing and Fractal Dynamics*; Springer: New York, NY, USA, 2013.
25. Wang, Y.P.; Liu, S.T.; Li, H.; Wang, D. On the spatial Julia set generated by fractional Lotka-Volterra system with noise. *Chaos Solitons Fractals* **2019**, *128*, 129–138. [CrossRef]
26. Wang, Y.P.; Liu, S.T. Fractal analysis and control of the fractional Lotka–Volterra model. *Nonlinear Dyn.* **2019**, *95*, 1457–1470. [CrossRef]
27. Wang, Y.P.; Liu, S.T.; Wang, D. Fractal dimension analysis and control of Julia set generated by fractional Lotka–Volterra models. *Commun. Nonlinear Sci. Numer. Simul.* **2019**, *72*, 417–431. [CrossRef]
28. Yadav, M.P.; Agarwal, R. Numerical investigation of fractional-fractal Boussinesq equation. *Chaos Interdiscip. J. Nonlinear Sci.* **2019**, *29*, 013109. [CrossRef] [PubMed]
29. Mojica, N.S.; Navarro, J.; Marijuán, P.C.; Lahoz-Beltra, R. Cellular "bauplans": Evolving unicellular forms by means of Julia sets and Pickover biomorphs. *Biosystems* **2009**, *98*, 19–30. [CrossRef] [PubMed]
30. Wang, D.; Liu, S.T.; Zhao, Y. A preliminary study on the fractal phenomenon:" disconnected+disconnected=connected". *Fractals* **2017**, *25*, 1750004. [CrossRef]
31. Zou, C.; Shahid, A.A.; Tassaddiq, A.; Khan, A.; Ahmad, M. Mandelbrot sets and julia sets in picard-mann orbit. *IEEE Access* **2020**, *8*, 64411–64421. [CrossRef]
32. Kwun, Y.C.; Tanveer, M.; Nazeer, W.; Gdawiec, K.; Kang, S.M. Mandelbrot and Julia Sets via Jungck–CR Iteration with s–Convexity. *IEEE Access* **2020**, *7*, 12167–12176. [CrossRef]
33. Jolaoso, L.O.; Khan, S.H. Some Escape Time Results for General Complex Polynomials and Biomorphs Generation by a New Iteration Process. *Mathematics* **2020**, *8*, 2172. [CrossRef]

34. Zhou, H.; Tanveer, M.; Li, J. Comparative study of some fixed-point methods in the generation of Julia and mandelbrot sets. *J. Math.* **2020**, *2020*, 7020921. [CrossRef]
35. Rani, M.; Kumar, V. Superior mandelbrot set. *J. Korea Soc. Math. Educ. Ser. D Res. Math. Educ.* **2004**, *8*, 279–291.
36. Yadav, A.; Rani, M. Alternate superior Julia sets. *Chaos Solitons Fractals* **2015**, *73*, 1–9. [CrossRef]
37. Negi, A.; Rani, M. New approach to dynamic noise on superior Mandelbrot set. *Chaos Solitons Fractals* **2008**, *36*, 1089–1096. [CrossRef]
38. Rani, M.; Agarwal, R. Effect of stochastic noise on superior Julia sets. *J. Math. Imaging Vis.* **2010**, *36*, 63–68. [CrossRef]
39. Agarwal, R.; Agarwal, V. Dynamic noise perturbed generalized superior Mandelbrot sets. *Nonlinear Dyn.* **2012**, *67*, 1883–1891. [CrossRef]
40. Gdawiec, K.; Kotarski, W.; Lisowska, A. Biomorphs via modified iterations. *J. Nonlinear Sci. Appl.* **2016**, *9*, 2305–2315. [CrossRef]

fractal and fractional

Article

Fractional Integral Inequalities for Exponentially Nonconvex Functions and Their Applications

Hari Mohan Srivastava [1,2,3,4] , Artion Kashuri [5] , Pshtiwan Othman Mohammed [6,*] , Dumitru Baleanu [7,8,*] and Y. S. Hamed [9]

1. Department of Mathematics and Statistics, University of Victoria, Victoria, BC V8W 3R4, Canada; harimsri@math.uvic.ca
2. Department of Medical Research, China Medical University Hospital, China Medical University, Taichung 40402, Taiwan
3. Department of Mathematics and Informatics, Azerbaijan University, 71 Jeyhun Hajibeyli Street, Baku AZ1007, Azerbaijan
4. Section of Mathematics, International Telematic University Uninettuno, I-00186 Rome, Italy
5. Department of Mathematics, Faculty of Technical Science, University "Ismail Qemali", 9400 Vlora, Albania; artionkashuri@gmail.com
6. Department of Mathematics, College of Education, University of Sulaimani, Sulaimani 46001, Kurdistan Region, Iraq
7. Department of Mathematics, Faculty of Arts and Sciences, Cankaya University, Ankara TR-06530, Turkey
8. Institute of Space Sciences, R-76900 Magurele-Bucharest, Romania
9. Department of Mathematics and Statistics, College of Science, Taif University, P.O. Box 11099, Taif 21944, Saudi Arabia; yasersalah@tu.edu.sa
* Correspondence: pshtiwansangawi@gmail.com (P.O.M.); dumitru@cankaya.edu.tr (D.B.)

check for **updates**

Citation: Srivastava, H.M.; Kashuri, A.; Mohammed, P.O.; Baleanu, D.; Hamed, Y.S. Fractional Integral Inequalities for Exponentially Nonconvex Functions and Their Applications. *Fractal Fract.* **2021**, 5, 80. https://doi.org/10.3390/fractalfract 5030080

Academic Editors: António M. Lopes, Liping Chen and Bruce Henry

Received: 08 June 2021
Accepted: 27 July 2021
Published: 29 July 2021

Publisher's Note: MDPI stays neutral with regard to jurisdictional claims in published maps and institutional affiliations.

Abstract: In this paper, the authors define a new generic class of functions involving a certain modified Fox–Wright function. A useful identity using fractional integrals and this modified Fox–Wright function with two parameters is also found. Applying this as an auxiliary result, we establish some Hermite–Hadamard-type integral inequalities by using the above-mentioned class of functions. Some special cases are derived with relevant details. Moreover, in order to show the efficiency of our main results, an application for error estimation is obtained as well.

Keywords: Hermite–Hadamard inequality; exponentially nonconvex function; modified Fox–Wright function; fractional integrals; error estimation

1. Introduction and Preliminaries

In many problems in mathematics and its applications, fractional calculus has a crucial role (see [1–6]). The analysis of the uniqueness of fractional ordinary differential equations can be accomplished by using fractional integral inequalities (see [7–9]).

Integral inequalities play a major role in the fields of differential equations and applied mathematics (see [10,11]). Moreover, they are linked with such other areas as differential equations, difference equations, mathematical analysis, mathematical physics, convexity theory, and discrete fractional calculus (see [12–18]).

Convexity is a fascinating and natural concept; it is beneficial in optimization theory, the theory of inequalities, numerical analysis, economics, and in other fields of pure and applied mathematics.

The notion of the h–convex function is introduced below.

Definition 1 (see [19]). *Let* $h : [0,1] \to [0,\infty)$ *be a function. A function* $\psi : \mathrm{I} \to \mathbb{R}$ *is said to be* h–*convex if*

$$\psi(\imath\xi_1 + (1-\imath)\xi_2) \leqq h(\imath)\psi(\xi_1) + h(1-\imath)\psi(\xi_2)$$

holds true for every $\xi_1, \xi_2 \in \mathrm{I}$ *and* $\imath \in [0,1]$.

The following class of functions was introduced by Awan et al. (see [20]) and was demonstrated to play an important role in optimization theory and mathematical economics.

Definition 2. *A function* $\psi : I \subseteq \mathbb{R} \to \mathbb{R}$ *is called exponentially convex if*

$$\psi(\imath \xi_1 + (1 - \imath)\xi_2) \leqq \imath \frac{\psi(\xi_1)}{e^{\omega \xi_1}} + (1 - \imath)\frac{\psi(\xi_2)}{e^{\omega \xi_2}}$$

holds true for all $\xi_1, \xi_2 \in I$, $\omega \in \mathbb{R}$ *and* $\imath \in [0, 1]$.

The most significant inequality about a convex function ψ on the closed interval $[\xi_1, \xi_2]$ is the Hermite–Hadamard integral inequality (that is, the trapezium inequality). This two-sided inequality is expressed as follows:

$$\psi\left(\frac{\xi_1 + \xi_2}{2}\right) \leqq \frac{1}{\xi_2 - \xi_1} \int_{\xi_1}^{\xi_2} \psi(\imath)d\imath \leqq \frac{\psi(\xi_1) + \psi(\xi_2)}{2}. \tag{1}$$

The two-sided inequality (1) has become a very important foundation within the field of mathematical analysis and optimization. Several applications of inequalities of this type have been derived in a number of different settings (see [21–29]).

In the context of fractional calculus, the standard left and right-sided Riemann–Liouville (*RL*) fractional integrals of order $\alpha > 0$ are given, respectively, by

$$\mathcal{I}_{\xi_1+}^{\alpha} \psi(x) = \frac{1}{\Gamma(\alpha)} \int_{\xi_1}^{x} (x - \imath)^{\alpha - 1} \psi(\imath)d\imath \qquad (x > \xi_1)$$

and

$$\mathcal{I}_{\xi_2-}^{\alpha} \psi(x) = \frac{1}{\Gamma(\alpha)} \int_{x}^{\xi_2} (\imath - x)^{\alpha - 1} \psi(\imath)d\imath \qquad (x < \xi_2),$$

$$\tag{2}$$

where ψ is a function defined on the closed interval $[\xi_1, \xi_2]$ and $\Gamma(\cdot)$ is the classical (Euler's) gamma function.

Regarding information for some of the fractional integral operators, including those that are known as Erdélyi–Kober, Riemann–Liouville (*RL*), Weyl and Liouville–Caputo (*LC*) operators, see [30–34].

There are many directions in which one can introduce a new definition of fractional derivatives and fractional integrals, which are related to or inspired by (for example) the *RL* definitions (see [35,36]), with reference to some general classes into which such fractional calculus operators can be classified. In applied mathematics, it is important to consider particular types of fractional calculus operators which are suited to the fractional-order modeling of a given real-world problem.

We now recall the familiar Fox–Wright hypergeometric function $_p\Psi_q(z)$ (with p numerator and q denominator parameters), which is given by the following series (see [5] (p. 67, Equation (1.12(68)))) and [37] (p. 21, Equation (1.2(38))))):

$$_p\Psi_q\left[\begin{array}{c} (\alpha_1, \mathcal{U}_1), \cdots, (\alpha_p, \mathcal{U}_p); \\ \\ (\beta_1, \mathcal{V}_1), \cdots, (\beta_q, \mathcal{V}_q); \end{array} z\right] := \sum_{n=0}^{\infty} \frac{\prod\limits_{\ell=1}^{p} \Gamma(\alpha_\ell + n\mathcal{U}_\ell)}{\prod\limits_{j=1}^{q} \Gamma(\beta_j + n\mathcal{V}_j)} \frac{z^n}{n!}, \tag{3}$$

where the parameters

$$\alpha_\ell, \beta_j \in \mathbb{C} \quad (\ell = 1, \cdots, p; \; j = 1, \cdots, q)$$

and the coefficients

$$\mathcal{U}_1, \cdots, \mathcal{U}_p \in \mathbb{R}^+ \qquad \text{and} \qquad \mathcal{V}_1, \cdots, \mathcal{V}_q \in \mathbb{R}^+$$

satisfy the following condition:

$$1 + \sum_{j=1}^{q} \mathcal{V}_j - \sum_{\ell=1}^{p} \mathcal{U}_\ell \geq 0. \tag{4}$$

Here and in what follows, we have made use of the general Pochhammer symbol $(\eta)_\nu$ $(\eta, \nu \in \mathbb{C})$ defined by

$$(\eta)_\nu := \frac{\Gamma(\eta + \nu)}{\Gamma(\eta)} = \begin{cases} 1 & (\nu = 0; \ \eta \in \mathbb{C} \setminus \{0\}) \\ \eta(\eta+1)\cdots(\eta+n-1) & (\nu = n \in \mathbb{N}; \ \eta \in \mathbb{C}), \end{cases} \tag{5}$$

it being assumed *conventionally* that $(0)_0 := 1$ and understand *tacitly* that the Γ-quotient in (5) exists.

The following modified version of the Fox–Wright function $_p\Psi_q(z)$ in (3) was introduced, as long ago as 1940, by Wright [38] (p. 424), who partially and formally replaced the Γ-quotient in (3) by a sequence $\{\sigma(n)\}_{n=0}^{\infty}$ based upon a suitably-restricted function $\sigma(\tau)$ as follows (see also [39], where the same definition is reproduced without giving credit to Wright [38]):

$$\mathcal{F}_{\rho,\varsigma}^{\sigma}(z) = \mathcal{F}_{\rho,\varsigma}^{\sigma(0),\sigma(1),\cdots}(z) = \sum_{\ell=0}^{\infty} \frac{\sigma(\ell)}{\Gamma(\rho\ell + \varsigma)} z^\ell \qquad (\rho > 0; \ \varsigma > 0). \tag{6}$$

If, in Wright's definition (6) from 1940 (see [38] (p. 424)), we take $\rho = \varsigma = 1$ and

$$\sigma(\ell) = \frac{\prod\limits_{j=1}^{p} \Gamma(\alpha_j + \mathcal{U}_j\ell)}{\prod\limits_{k=1}^{q} \Gamma(\beta_k + \mathcal{V}_k\ell)} \qquad (\ell = 0, 1, 2, \cdots),$$

then Wright's definition (6) would immediately yield the familiar Fox–Wright hypergeometric function $_p\Psi_q(z)$ defined by (3). The one- and two-parameter Mittag–Leffler functions $E_\alpha(z)$ and $E_{\alpha,\beta}(z)$, and indeed also almost all of the parametric generalizations of the Mittag–Leffler type functions, can be deduced as obvious special cases of the Fox–Wright hypergeometric function $_p\Psi_q(z)$ defined by (3) (see [40] for details).

We are now in the position to introduce a new generic class of functions involving the modified Fox–Wright function $\mathcal{F}_{\rho,\varsigma}^{\sigma}(\cdot)$.

Definition 3. *Let $h_1, h_2 : [0, 1] \to [0, \infty)$ be two functions and $\psi : I \subseteq \mathbb{R} \to \mathbb{R}$. If ψ satisfies the following inequality,*

$$\psi\left(\xi_1 + \iota\mathcal{F}_{\rho,\varsigma}^{\sigma}(\xi_2 - \xi_1)\right) \leq h_1(\iota)\frac{\psi(\xi_1)}{e^{\omega_1\xi_1}} + h_2(\iota)\frac{\psi(\xi_2)}{e^{\omega_2\xi_2}}$$

for all $\iota \in [0,1]$, $\omega_1, \omega_2 \in \mathbb{R}$, and $\xi_1, \xi_2 \in I$, where $\mathcal{F}_{\rho,\varsigma}^{\sigma}(\xi_2 - \xi_1) > 0$, then ψ is called an exponentially $(\omega_1, \omega_2, h_1, h_2)$–nonconvex function.

Remark 1. *Upon setting*

$$\omega_1 = \omega_2 = \omega, \quad h_1(\iota) = 1 - \iota, \quad h_2(\iota) = \iota$$

and

$$\mathcal{F}_{\rho,\varsigma}^{\sigma}(\xi_2 - \xi_1) = \xi_2 - \xi_1$$

in Definition 3, we then obtain Definition 2.

Remark 2. *Some special cases of our Definition 3 are listed below:*

(I) *Taking $h_1(\iota) = h_2(\iota) = 1$, we have an exponentially (ω_1, ω_2, P)–nonconvex function.*

(II) *Choosing $h_1(\iota) = h(1 - \iota)$ and $h_2(\iota) = h(\iota)$, we obtain an exponentially (ω_1, ω_2, h)–nonconvex function.*

(III) *Setting $h_1(\iota) = (1 - \iota)^s$ and $h_2(\iota) = \iota^s$ for $s \in (0, 1]$, we obtain an exponentially (s, ω_1, ω_2)–Breckner-nonconvex function.*

(IV) *Putting $h_1(\iota) = (1 - \iota)^{-s}$ and $h_2(\iota) = \iota^{-s}$ for $s \in (0, 1)$, we obtain an exponentially (s, ω_1, ω_2)–Godunova–Levin–Dragomir-nonconvex function.*

(V) *Taking $h_1(\iota) = h_2(\iota) = \iota(1 - \iota)$, we obtain an exponentially $(\omega_1, \omega_2, tgs)$–nonconvex function.*

Our paper has the following structure: in Section 2, we first find a useful identity using fractional integrals with two parameters λ and μ involving the modified Fox–Wright function $\mathcal{F}_{\rho,\varsigma}^{\sigma}(\cdot)$. Applying this as an auxiliary result, we give some Hermite–Hadamard-type integral inequalities pertaining to exponentially $(\omega_1, \omega_2, h_1, h_2)$–nonconvex functions, and some special cases are derived in details. In Section 3, the efficiency of our main results is demonstrated with an application for error estimation. Section 4 presents the conclusion of this paper.

2. Main Results and Their Consequences

The following notations are used below:

$$\Delta := \left[\xi_1, \xi_1 + \mathcal{F}_{\rho,\varsigma}^{\sigma}(\xi_2 - \xi_1) \right],$$

where

$$\mathcal{F}_{\rho,\varsigma}^{\sigma}(\xi_2 - \xi_1) > 0$$

and Δ° is the interior of the closed interval Δ with $\omega_1, \omega_2 \in \mathbb{R}$. We denote by $\mathcal{L}_1(\Delta)$ the space of integrable functions over Δ. We need to prove the following basic lemma.

Lemma 1. *Let the function $\psi : \Delta \to \mathbb{R}$ be differentiable on Δ° and $\lambda, \mu \in (0, 1]$. If $\psi' \in \mathcal{L}_1(\Delta)$, then, for $\alpha > 0$,*

$$\frac{\mu^{\alpha}\psi(\xi_1) + \lambda^{\alpha}\psi\left(\xi_1 + \mathcal{F}_{\rho,\varsigma}^{\sigma}(\xi_2 - \xi_1)\right)}{(\lambda + \mu)^{\alpha}} - \frac{\Gamma(\alpha + 1)}{\left[\mathcal{F}_{\rho,\varsigma}^{\sigma}(\xi_2 - \xi_1)\right]^{\alpha}}$$

$$\cdot \left[\mathcal{I}_{\xi_1^+}^{\alpha}\psi\left(\xi_1 + \frac{\mu}{\lambda + \mu}\mathcal{F}_{\rho,\varsigma}^{\sigma}(\xi_2 - \xi_1)\right) + \mathcal{I}_{(\xi_1 + \mathcal{F}_{\rho,\varsigma}^{\sigma}(\xi_2 - \xi_1))^-}^{\alpha}\psi\left(\xi_1 + \frac{\mu}{\lambda + \mu}\mathcal{F}_{\rho,\varsigma}^{\sigma}(\xi_2 - \xi_1)\right) \right]$$

$$= \frac{\mathcal{F}_{\rho,\varsigma}^{\sigma}(\xi_2 - \xi_1)}{(\lambda + \mu)^{\alpha+1}}$$

$$\cdot \left[\int_0^{\lambda} \iota^{\alpha}\psi'\left(\xi_1 + \frac{\mu + \iota}{\lambda + \mu}\mathcal{F}_{\rho,\varsigma}^{\sigma}(\xi_2 - \xi_1)\right)d\iota - \int_0^{\mu} \iota^{\alpha}\psi'\left(\xi_1 + \frac{\mu - \iota}{\lambda + \mu}\mathcal{F}_{\rho,\varsigma}^{\sigma}(\xi_2 - \xi_1)\right)d\iota \right]. \quad (7)$$

Proof. We define

$$T_{\lambda,\mu}^{\alpha}(\xi_1, \xi_2) := \frac{\mathcal{F}_{\rho,\varsigma}^{\sigma}(\xi_2 - \xi_1)}{(\lambda + \mu)^{\alpha+1}}[\mathcal{I}_2 - \mathcal{I}_1], \quad (8)$$

where

$$\mathcal{I}_1 := \int_0^\mu \imath^\alpha \psi'\left(\xi_1 + \frac{\mu - \imath}{\lambda + \mu}\mathcal{F}_{\rho,\varsigma}^\sigma(\xi_2 - \xi_1)\right)d\imath,$$

and

$$\mathcal{I}_2 := \int_0^\lambda \imath^\alpha \psi'\left(\xi_1 + \frac{\mu + \imath}{\lambda + \mu}\mathcal{F}_{\rho,\varsigma}^\sigma(\xi_2 - \xi_1)\right)d\imath,$$

which, upon integrating by parts, would yield

$$\mathcal{I}_1 = -\frac{(\lambda + \mu)}{\mathcal{F}_{\rho,\varsigma}^\sigma(\xi_2 - \xi_1)}\imath^\alpha \psi\left(\xi_1 + \frac{\mu - \imath}{\lambda + \mu}\mathcal{F}_{\rho,\varsigma}^\sigma(\xi_2 - \xi_1)\right)\Bigg|_0^\mu$$

$$+ \frac{\alpha(\lambda + \mu)}{\mathcal{F}_{\rho,\varsigma}^\sigma(\xi_2 - \xi_1)}\int_0^\mu \imath^{\alpha-1} \psi\left(\xi_1 + \frac{\mu - \imath}{\lambda + \mu}\mathcal{F}_{\rho,\varsigma}^\sigma(\xi_2 - \xi_1)\right)d\imath$$

$$= \Gamma(\alpha + 1)\left(\frac{\lambda + \mu}{\mathcal{F}_{\rho,\varsigma}^\sigma(\xi_2 - \xi_1)}\right)^{\alpha+1}$$

$$\cdot \mathcal{I}_{\xi_1^+}^\alpha \psi\left(\xi_1 + \frac{\mu}{\lambda + \mu}\mathcal{F}_{\rho,\varsigma}^\sigma(\xi_2 - \xi_1)\right) - \frac{\mu^\alpha(\lambda + \mu)}{\mathcal{F}_{\rho,\varsigma}^\sigma(\xi_2 - \xi_1)}\psi(\xi_1). \quad (9)$$

Similarly, we find that

$$\mathcal{I}_2 = \frac{\lambda^\alpha(\lambda + \mu)}{\mathcal{F}_{\rho,\varsigma}^\sigma(\xi_2 - \xi_1)}\psi\left(\xi_1 + \mathcal{F}_{\rho,\varsigma}^\sigma(\xi_2 - \xi_1)\right)$$

$$- \Gamma(\alpha + 1)\left(\frac{\lambda + \mu}{\mathcal{F}_{\rho,\varsigma}^\sigma(\xi_2 - \xi_1)}\right)^{\alpha+1} \cdot \mathcal{I}_{(\xi_1 + \mathcal{F}_{\rho,\varsigma}^\sigma(\xi_2 - \xi_1))^-}^\alpha \psi\left(\xi_1 + \frac{\mu}{\lambda + \mu}\mathcal{F}_{\rho,\varsigma}^\sigma(\xi_2 - \xi_1)\right). \quad (10)$$

Substituting from (9) and (10) into (8), we obtain the desired result (7). $\quad\square$

From Lemma 1, we can derive the following case:

Remark 3. *Taking $\alpha = 1$ in Lemma 1, we have*

$$\frac{\mu\psi(\xi_1) + \lambda\psi\left(\xi_1 + \mathcal{F}_{\rho,\varsigma}^\sigma(\xi_2 - \xi_1)\right)}{\lambda + \mu} - \frac{1}{\mathcal{F}_{\rho,\varsigma}^\sigma(\xi_2 - \xi_1)}\int_{\xi_1}^{\xi_1 + \mathcal{F}_{\rho,\varsigma}^\sigma(\xi_2-\xi_1)} \psi(\imath)d\imath$$

$$= \frac{\mathcal{F}_{\rho,\varsigma}^\sigma(\xi_2 - \xi_1)}{(\lambda + \mu)^2}\left[\int_0^\lambda \imath\psi'\left(\xi_1 + \frac{\mu + \imath}{\lambda + \mu}\mathcal{F}_{\rho,\varsigma}^\sigma(\xi_2 - \xi_1)\right)d\imath\right.$$

$$\left. - \int_0^\mu \imath\psi'\left(\xi_1 + \frac{\mu - \imath}{\lambda + \mu}\mathcal{F}_{\rho,\varsigma}^\sigma(\xi_2 - \xi_1)\right)d\imath\right]. \quad (11)$$

Our first main result is stated as Theorem 1 below.

Theorem 1. *Assume that $h_1, h_2 : [0,1] \to [0,\infty)$ are two continuous functions and let $\psi : \Delta \to \mathbb{R}$ be a differentiable function on Δ° with $\lambda, \mu \in (0,1]$. Furthermore, let $\psi' \in \mathcal{L}_1(\Delta)$. If $|\psi'|^q$ is an exponentially $(\varpi_1, \varpi_2, h_1, h_2)$–nonconvex function, then, for $q > 1$, $\frac{1}{p} + \frac{1}{q} = 1$ and $\alpha > 0$, it is asserted that*

$$\left|\mathcal{T}^{\alpha}_{\lambda,\mu}(\xi_1,\xi_2)\right| \leq \frac{\mathcal{F}^{\sigma}_{\rho,\varsigma}(\xi_2-\xi_1)}{(\lambda+\mu)^{\alpha+1}}\left[\mathcal{A}_1^{\frac{1}{p}}\left(\frac{|\psi'(\xi_1)|^q}{e^{\omega_1\xi_1}}\mathcal{H}_{1,1}+\frac{|\psi'(\xi_2)|^q}{e^{\omega_2\xi_2}}\mathcal{H}_{1,2}\right)^{\frac{1}{q}}\right.$$

$$\left.+\mathcal{A}_2^{\frac{1}{p}}\left(\frac{|\psi'(\xi_1)|^q}{e^{\omega_1\xi_1}}\mathcal{H}_{2,1}+\frac{|\psi'(\xi_2)|^q}{e^{\omega_2\xi_2}}\mathcal{H}_{2,2}\right)^{\frac{1}{q}}\right], \quad (12)$$

where

$$\mathcal{A}_1 := \frac{\lambda^{p\alpha+1}}{p\alpha+1}, \quad \mathcal{A}_2 := \frac{\mu^{p\alpha+1}}{p\alpha+1},$$

$$\mathcal{H}_{1,1} := \int_0^{\lambda} h_1\left(\frac{\mu+\imath}{\lambda+\mu}\right)d\imath, \quad \mathcal{H}_{1,2} := \int_0^{\lambda} h_2\left(\frac{\mu+\imath}{\lambda+\mu}\right)d\imath,$$

$$\mathcal{H}_{2,1} := \int_0^{\mu} h_1\left(\frac{\mu-\imath}{\lambda+\mu}\right)d\imath \quad and \quad \mathcal{H}_{2,2} := \int_0^{\mu} h_2\left(\frac{\mu-\imath}{\lambda+\mu}\right)d\imath.$$

Proof. Applying Lemma 1, the property of the modulus, Hölder's inequality, and the exponential $(\omega_1,\omega_2,h_1,h_2)$–nonconvexity of $|\psi'|^q$, we have

$$\left|\mathcal{T}^{\alpha}_{\lambda,\mu}(\xi_1,\xi_2)\right| \leq \frac{\mathcal{F}^{\sigma}_{\rho,\varsigma}(\xi_2-\xi_1)}{(\lambda+\mu)^{\alpha+1}}\left[\int_0^{\lambda} \imath^{\alpha}\left|\psi'\left(\xi_1+\frac{\mu+\imath}{\lambda+\mu}\mathcal{F}^{\sigma}_{\rho,\varsigma}(\xi_2-\xi_1)\right)\right|d\imath\right.$$

$$\left.+\int_0^{\mu} \imath^{\alpha}\left|\psi'\left(\xi_1+\frac{\mu-\imath}{\lambda+\mu}\mathcal{F}^{\sigma}_{\rho,\varsigma}(\xi_2-\xi_1)\right)\right|d\imath\right]$$

$$\leq \frac{\mathcal{F}^{\sigma}_{\rho,\varsigma}(\xi_2-\xi_1)}{(\lambda+\mu)^{\alpha+1}}\left[\left(\int_0^{\lambda} \imath^{p\alpha}d\imath\right)^{\frac{1}{p}}\left(\int_0^{\lambda}\left|\psi'\left(\xi_1+\frac{\mu+\imath}{\lambda+\mu}\mathcal{F}^{\sigma}_{\rho,\varsigma}(\xi_2-\xi_1)\right)\right|^q d\imath\right)^{\frac{1}{q}}\right.$$

$$\left.+\left(\int_0^{\mu} \imath^{p\alpha}d\imath\right)^{\frac{1}{p}}\left(\int_0^{\mu}\left|\psi'\left(\xi_1+\frac{\mu-\imath}{\lambda+\mu}\mathcal{F}^{\sigma}_{\rho,\varsigma}(\xi_2-\xi_1)\right)\right|^q d\imath\right)^{\frac{1}{q}}\right]$$

$$\leq \frac{\mathcal{F}^{\sigma}_{\rho,\varsigma}(\xi_2-\xi_1)}{(\lambda+\mu)^{\alpha+1}}\left[\left(\int_0^{\lambda} \imath^{p\alpha}d\imath\right)^{\frac{1}{p}}\left(\int_0^{\lambda}\left\{h_1\left(\frac{\mu+\imath}{\lambda+\mu}\right)\frac{|\psi'(\xi_1)|^q}{e^{\omega_1\xi_1}}+h_2\left(\frac{\mu+\imath}{\lambda+\mu}\right)\frac{|\psi'(\xi_2)|^q}{e^{\omega_2\xi_2}}\right\}d\imath\right)^{\frac{1}{q}}\right.$$

$$\left.+\left(\int_0^{\mu} \imath^{p\alpha}d\imath\right)^{\frac{1}{p}}\left(\int_0^{\mu}\left\{h_1\left(\frac{\mu-\imath}{\lambda-\mu}\right)\frac{|\psi'(\xi_1)|^q}{e^{\omega_1\xi_1}}+h_2\left(\frac{\mu-\imath}{\lambda+\mu}\right)\frac{|\psi'(\xi_2)|^q}{e^{\omega_2\xi_2}}\right\}d\imath\right)^{\frac{1}{q}}\right]$$

$$= \frac{\mathcal{F}^{\sigma}_{\rho,\varsigma}(\xi_2-\xi_1)}{(\lambda+\mu)^{\alpha+1}}\left[\mathcal{A}_1^{\frac{1}{p}}\left(\frac{|\psi'(\xi_1)|^q}{e^{\omega_1\xi_1}}\mathcal{H}_{1,1}+\frac{|\psi'(\xi_2)|^q}{e^{\omega_2\xi_2}}\mathcal{H}_{1,2}\right)^{\frac{1}{q}}\right.$$

$$\left.+\mathcal{A}_2^{\frac{1}{p}}\left(\frac{|\psi'(\xi_1)|^q}{e^{\omega_1\xi_1}}\mathcal{H}_{2,1}+\frac{|\psi'(\xi_2)|^q}{e^{\omega_2\xi_2}}\mathcal{H}_{2,2}\right)^{\frac{1}{q}}\right],$$

which completes the proof of Theorem 1. $\square$

Some corollaries and consequences of Theorem 1 are listed below:

Corollary 1. *Upon setting $\alpha = 1$, Theorem 1 yields*

$$\left| \frac{\mu\psi(\xi_1) + \lambda\psi\left(\xi_1 + \mathcal{F}^\sigma_{\rho,\varsigma}(\xi_2 - \xi_1)\right)}{\lambda + \mu} - \frac{1}{\mathcal{F}^\sigma_{\rho,\varsigma}(\xi_2 - \xi_1)} \int_{\xi_1}^{\xi_1 + \mathcal{F}^\sigma_{\rho,\varsigma}(\xi_2 - \xi_1)} \psi(\iota)d\iota \right|$$

$$\leq \frac{\mathcal{F}^\sigma_{\rho,\varsigma}(\xi_2 - \xi_1)}{(\lambda + \mu)^2} \left[\left(\frac{\lambda^{p+1}}{p+1}\right)^{\frac{1}{p}} \left(\frac{|\psi'(\xi_1)|^q}{e^{\omega_1 \xi_1}} \mathcal{H}_{1,1} + \frac{|\psi'(\xi_2)|^q}{e^{\omega_2 \xi_2}} \mathcal{H}_{1,2} \right)^{\frac{1}{q}} \right.$$

$$\left. + \left(\frac{\mu^{p+1}}{p+1}\right)^{\frac{1}{p}} \left(\frac{|\psi'(\xi_1)|^q}{e^{\omega_1 \xi_1}} \mathcal{H}_{2,1} + \frac{|\psi'(\xi_2)|^q}{e^{\omega_2 \xi_2}} \mathcal{H}_{2,2} \right)^{\frac{1}{q}} \right]. \tag{13}$$

Corollary 2. *Choosing* $h_1(\iota) = h_2(\iota) = 1$ *in Theorem 1, it is asserted that*

$$\left| T^\alpha_{\lambda,\mu}(\xi_1, \xi_2) \right| \leq \frac{\mathcal{F}^\sigma_{\rho,\varsigma}(\xi_2 - \xi_1)}{(\lambda + \mu)^{\alpha+1}} \left(\mathcal{A}_1^{\frac{1}{p}} \lambda^{\frac{1}{q}} + \mathcal{A}_2^{\frac{1}{p}} \mu^{\frac{1}{q}} \right) \left[\frac{|\psi'(\xi_1)|^q}{e^{\omega_1 \xi_1}} + \frac{|\psi'(\xi_2)|^q}{e^{\omega_2 \xi_2}} \right]^{\frac{1}{q}}. \tag{14}$$

Corollary 3. *Choosing*

$$h_1(\iota) = (1 - \iota)^s \qquad and \qquad h_2(\iota) = \iota^s \qquad (s \in (0,1])$$

in Theorem 1, it is asserted that

$$\left| T^\alpha_{\lambda,\mu}(\xi_1, \xi_2) \right| \leq \frac{\mathcal{F}^\sigma_{\rho,\varsigma}(\xi_2 - \xi_1)}{(\lambda + \mu)^{\alpha+1}} \left[\mathcal{A}_1^{\frac{1}{p}} \left(\frac{|\psi'(\xi_1)|^q}{e^{\omega_1 \xi_1}} \mathcal{D}_{1,1} + \frac{|\psi'(\xi_2)|^q}{e^{\omega_2 \xi_2}} \mathcal{D}_{1,2} \right)^{\frac{1}{q}} \right.$$

$$\left. + \mathcal{A}_2^{\frac{1}{p}} \left(\frac{|\psi'(\xi_1)|^q}{e^{\omega_1 \xi_1}} \mathcal{D}_{2,1} + \frac{|\psi'(\xi_2)|^q}{e^{\omega_2 \xi_2}} \mathcal{D}_{2,2} \right)^{\frac{1}{q}} \right], \tag{15}$$

where

$$\mathcal{D}_{1,1} := \frac{\lambda^{s+1}}{(s+1)(\lambda + \mu)^s}, \qquad \mathcal{D}_{1,2} := \frac{(\lambda + \mu)^{s+1} - \mu^{s+1}}{(s+1)(\lambda + \mu)^s},$$

$$\mathcal{D}_{2,1} := \frac{(\lambda + \mu)^{s+1} - \lambda^{s+1}}{(s+1)(\lambda + \mu)^s} \qquad and \qquad \mathcal{D}_{2,2} := \frac{\mu^{s+1}}{(s+1)(\lambda + \mu)^s}.$$

Corollary 4. *Taking*

$$h_1(\iota) = (1 - \iota)^{-s} \qquad and \qquad h_2(\iota) = \iota^{-s} \qquad (s \in (0,1))$$

in Theorem 1, the following inequality is deduced:

$$\left| T^\alpha_{\lambda,\mu}(\xi_1, \xi_2) \right| \leq \frac{\mathcal{F}^\sigma_{\rho,\varsigma}(\xi_2 - \xi_1)}{(\lambda + \mu)^{\alpha+1}} \left[\mathcal{A}_1^{\frac{1}{p}} \left(\frac{|\psi'(\xi_1)|^q}{e^{\omega_1 \xi_1}} \mathcal{F}_{1,1} + \frac{|\psi'(\xi_2)|^q}{e^{\omega_2 \xi_2}} \mathcal{F}_{1,2} \right)^{\frac{1}{q}} \right.$$

$$\left. + \mathcal{A}_2^{\frac{1}{p}} \left(\frac{|\psi'(\xi_1)|^q}{e^{\omega_1 \xi_1}} \mathcal{F}_{2,1} + \frac{|\psi'(\xi_2)|^q}{e^{\omega_2 \xi_2}} \mathcal{F}_{2,2} \right)^{\frac{1}{q}} \right], \tag{16}$$

where

$$\mathcal{F}_{1,1} := \frac{\lambda^{1-s}(\lambda + \mu)^s}{1 - s}, \qquad \mathcal{F}_{1,2} := \frac{\left((\lambda + \mu)^{1-s} - \mu^{1-s}\right)(\lambda + \mu)^s}{1 - s},$$

$$\mathcal{F}_{2,1} := \frac{\left((\lambda + \mu)^{1-s} - \lambda^{1-s}\right)(\lambda + \mu)^s}{1 - s} \qquad and \qquad \mathcal{F}_{2,2} := \frac{\mu^{1-s}(\lambda + \mu)^s}{1 - s}.$$

Corollary 5. *For*

$$h_1(\iota) = h_2(\iota) = \iota(1 - \iota),$$

Theorem 1 yields

$$\left|\mathcal{T}_{\lambda,\mu}^{\alpha}(\xi_1,\xi_2)\right| \leqq \frac{\mathcal{F}_{\rho,\varsigma}^{\sigma}(\xi_2-\xi_1)}{(\lambda+\mu)^{\alpha+1}}\left(\mathcal{A}_1^{\frac{1}{p}}\mathcal{G}_1^{\frac{1}{q}}+\mathcal{A}_2^{\frac{1}{p}}\mathcal{G}_2^{\frac{1}{q}}\right)\left[\frac{|\psi'(\xi_1)|^q}{e^{\varpi_1\xi_1}}+\frac{|\psi'(\xi_2)|^q}{e^{\varpi_2\xi_2}}\right]^{\frac{1}{q}}, \tag{17}$$

where

$$\mathcal{G}_1 := \frac{\lambda^2(3\mu+\lambda)}{6(\lambda+\mu)^2} \quad and \quad \mathcal{G}_2 := \frac{\mu^2(3\lambda+\mu)}{6(\lambda+\mu)^2}.$$

Our second main result is stated as Theorem 2 below.

Theorem 2. *Assume that $h_1, h_2 : [0,1] \to [0,\infty)$ are two continuous functions and $\psi : \Delta \to \mathbb{R}$ is a differentiable function on Δ° with $\lambda, \mu \in (0,1]$. Furthermore, let $\psi' \in \mathcal{L}_1(\Delta)$. If $|\psi'|^q$ be an exponentially $(\varpi_1, \varpi_2, h_1, h_2)$–nonconvex function; then, for $q \geqq 1$ and $\alpha > 0$, it is asserted that*

$$\left|\mathcal{T}_{\lambda,\mu}^{\alpha}(\xi_1,\xi_2)\right| \leqq \frac{\mathcal{F}_{\rho,\varsigma}^{\sigma}(\xi_2-\xi_1)}{(\lambda+\mu)^{\alpha+1}}\left[\mathcal{B}_1^{1-\frac{1}{q}}\left(\frac{|\psi'(\xi_1)|^q}{e^{\varpi_1\xi_1}}\mathcal{S}_{1,1}+\frac{|\psi'(\xi_2)|^q}{e^{\varpi_2\xi_2}}\mathcal{S}_{1,2}\right)^{\frac{1}{q}}\right.$$

$$\left.+\mathcal{B}_2^{1-\frac{1}{q}}\left(\frac{|\psi'(\xi_1)|^q}{e^{\varpi_1\xi_1}}\mathcal{S}_{2,1}+\frac{|\psi'(\xi_2)|^q}{e^{\varpi_2\xi_2}}\mathcal{S}_{2,2}\right)^{\frac{1}{q}}\right], \tag{18}$$

where

$$\mathcal{B}_1 := \frac{\lambda^{\alpha+1}}{\alpha+1}, \qquad \mathcal{B}_2 := \frac{\mu^{\alpha+1}}{\alpha+1},$$

$$\mathcal{S}_{1,1} := \int_0^{\lambda} \iota^{\alpha} h_1\left(\frac{\mu+\iota}{\lambda+\mu}\right) d\iota, \qquad \mathcal{S}_{1,2} := \int_0^{\lambda} \iota^{\alpha} h_2\left(\frac{\mu+\iota}{\lambda+\mu}\right) d\iota,$$

$$\mathcal{S}_{2,1} := \int_0^{\mu} \iota^{\alpha} h_1\left(\frac{\mu-\iota}{\lambda+\mu}\right) d\iota \quad and \quad \mathcal{S}_{2,2} := \int_0^{\mu} \iota^{\alpha} h_2\left(\frac{\mu-\iota}{\lambda+\mu}\right) d\iota.$$

Proof. Applying Lemma 1, the property of the modulus, power-mean inequality and the exponential $(\varpi_1, \varpi_2, h_1, h_2)$-nonconvexity of $|\psi'|^q$, we obtain

$$\left|\mathcal{T}_{\lambda,\mu}^{\alpha}(\xi_1,\xi_2)\right| \leqq \frac{\mathcal{F}_{\rho,\varsigma}^{\sigma}(\xi_2-\xi_1)}{(\lambda+\mu)^{\alpha+1}}$$

$$\cdot\left[\int_0^{\lambda} \iota^{\alpha}\left|\psi'\left(\xi_1+\frac{\mu+\iota}{\lambda+\mu}\mathcal{F}_{\rho,\varsigma}^{\sigma}(\xi_2-\xi_1)\right)\right|d\iota + \int_0^{\mu} \iota^{\alpha}\left|\psi'\left(\xi_1+\frac{\mu-\iota}{\lambda+\mu}\mathcal{F}_{\rho,\varsigma}^{\sigma}(\xi_2-\xi_1)\right)\right|d\iota\right]$$

$$\leqq \frac{\mathcal{F}_{\rho,\varsigma}^{\sigma}(\xi_2-\xi_1)}{(\lambda+\mu)^{\alpha+1}}$$

$$\cdot\left[\left(\int_0^{\lambda} \iota^{\alpha}d\iota\right)^{1-\frac{1}{q}}\left(\int_0^{\lambda} \iota^{\alpha}\left|\psi'\left(\xi_1+\frac{\mu+\iota}{\lambda+\mu}\mathcal{F}_{\rho,\varsigma}^{\sigma}(\xi_2-\xi_1)\right)\right|^q d\iota\right)^{\frac{1}{q}}\right.$$

$$\left.+\left(\int_0^{\mu} \iota^{\alpha}d\iota\right)^{1-\frac{1}{q}}\left(\int_0^{\mu} \iota^{\alpha}\left|\psi'\left(\xi_1+\frac{\mu-\iota}{\lambda+\mu}\mathcal{F}_{\rho,\varsigma}^{\sigma}(\xi_2-\xi_1)\right)\right|^q d\iota\right)^{\frac{1}{q}}\right]$$

$$\leqq \frac{\mathcal{F}_{\rho,\varsigma}^{\sigma}(\xi_2-\xi_1)}{(\lambda+\mu)^{\alpha+1}}$$

$$\cdot\left[\left(\int_0^{\lambda} \iota^{\alpha}d\iota\right)^{1-\frac{1}{q}}\left(\int_0^{\lambda} \iota^{\alpha}\left\{h_1\left(\frac{\mu+\iota}{\lambda+\mu}\right)\frac{|\psi'(\xi_1)|^q}{e^{\varpi_1\xi_1}}+h_2\left(\frac{\mu+\iota}{\lambda+\mu}\right)\frac{|\psi'(\xi_2)|^q}{e^{\varpi_2\xi_2}}\right\}d\iota\right)^{\frac{1}{q}}\right.$$

$$+ \left(\int_0^\mu \imath^\alpha d\imath \right)^{1-\frac{1}{q}} \left(\int_0^\mu \imath^\alpha \left\{ h_1\left(\frac{\mu - \imath}{\lambda - \mu} \right) \frac{|\psi'(\xi_1)|^q}{e^{\omega_1 \tilde\xi_1}} + h_2\left(\frac{\mu - \imath}{\lambda + \mu} \right) \frac{|\psi'(\xi_2)|^q}{e^{\omega_2 \tilde\xi_2}} \right\} d\imath \right)^{\frac{1}{q}} \right]$$

$$= \frac{\mathcal{F}^\sigma_{\rho,\varsigma}(\xi_2 - \xi_1)}{(\lambda + \mu)^{\alpha + 1}}$$

$$\cdot \left[\mathcal{B}_1^{1-\frac{1}{q}} \left(\frac{|\psi'(\xi_1)|^q}{e^{\omega_1 \tilde\xi_1}} \mathcal{S}_{1,1} + \frac{|\psi'(\xi_2)|^q}{e^{\omega_2 \tilde\xi_2}} \mathcal{S}_{1,2} \right)^{\frac{1}{q}} + \mathcal{B}_2^{1-\frac{1}{q}} \left(\frac{|\psi'(\xi_1)|^q}{e^{\omega_1 \tilde\xi_1}} \mathcal{S}_{2,1} + \frac{|\psi'(\xi_2)|^q}{e^{\omega_2 \tilde\xi_2}} \mathcal{S}_{2,2} \right)^{\frac{1}{q}} \right].$$

The proof of Theorem 2 is completed. $\square$

We now state several corollaries and consequences of Theorem 2.

Corollary 6. *Upon setting $\alpha = 1$, Theorem 2 yields*

$$\left| \frac{\mu\psi(\xi_1) + \lambda\psi\left(\xi_1 + \mathcal{F}^\sigma_{\rho,\varsigma}(\xi_2 - \xi_1) \right)}{\lambda + \mu} - \frac{1}{\mathcal{F}^\sigma_{\rho,\varsigma}(\xi_2 - \xi_1)} \int_{\xi_1}^{\xi_1 + \mathcal{F}^\sigma_{\rho,\varsigma}(\xi_2 - \xi_1)} \psi(\imath) d\imath \right|$$

$$\leqq \frac{\mathcal{F}^\sigma_{\rho,\varsigma}(\xi_2 - \xi_1)}{(\lambda + \mu)^2} \left[\left(\frac{\lambda^2}{2} \right)^{1-\frac{1}{q}} \left(\frac{|\psi'(\xi_1)|^q}{e^{\omega_1 \tilde\xi_1}} \mathcal{M}_{1,1} + \frac{|\psi'(\xi_2)|^q}{e^{\omega_2 \tilde\xi_2}} \mathcal{M}_{1,2} \right)^{\frac{1}{q}} \right.$$

$$\left. + \left(\frac{\mu^2}{2} \right)^{1-\frac{1}{q}} \left(\frac{|\psi'(\xi_1)|^q}{e^{\omega_1 \tilde\xi_1}} \mathcal{M}_{2,1} + \frac{|\psi'(\xi_2)|^q}{e^{\omega_2 \tilde\xi_2}} \mathcal{M}_{2,2} \right)^{\frac{1}{q}} \right], \quad (19)$$

where

$$\mathcal{M}_{1,1} := \int_0^\lambda \imath h_1\left(\frac{\mu + \imath}{\lambda + \mu} \right) d\imath, \qquad \mathcal{M}_{1,2} := \int_0^\lambda \imath h_2\left(\frac{\mu + \imath}{\lambda + \mu} \right) d\imath,$$

$$\mathcal{M}_{2,1} := \int_0^\mu \imath h_1\left(\frac{\mu - \imath}{\lambda + \mu} \right) d\imath \quad and \quad \mathcal{M}_{2,2} := \int_0^\mu \imath h_2\left(\frac{\mu - \imath}{\lambda + \mu} \right) d\imath.$$

Corollary 7. *Choosing $h_1(\imath) = h_2(\imath) = 1$ in Theorem 2, the following inequality holds true:*

$$\left| \mathcal{T}^\alpha_{\lambda,\mu}(\xi_1, \xi_2) \right| \leqq \frac{\mathcal{F}^\sigma_{\rho,\varsigma}(\xi_2 - \xi_1)}{(\lambda + \mu)^{\alpha+1}} (\mathcal{B}_1 + \mathcal{B}_2) \left[\frac{|\psi'(\xi_1)|^q}{e^{\omega_1 \tilde\xi_1}} + \frac{|\psi'(\xi_2)|^q}{e^{\omega_2 \tilde\xi_2}} \right]^{\frac{1}{q}}. \quad (20)$$

Corollary 8. *Choosing*

$$h_1(\imath) = (1 - \imath)^s \quad and \quad h_2(\imath) = \imath^s \quad (s \in (0, 1]),$$

Theorem 2 is reduced to the following inequality:

$$\left| \mathcal{T}^\alpha_{\lambda,\mu}(\xi_1, \xi_2) \right| \leqq \frac{\mathcal{F}^\sigma_{\rho,\varsigma}(\xi_2 - \xi_1)}{(\lambda + \mu)^{\alpha + \frac{s}{q} + 1}} \left[\mathcal{B}_1^{1-\frac{1}{q}} \left(\frac{|\psi'(\xi_1)|^q}{e^{\omega_1 \tilde\xi_1}} \mathcal{P}_{1,1} + \frac{|\psi'(\xi_2)|^q}{e^{\omega_2 \tilde\xi_2}} \mathcal{P}_{1,2} \right)^{\frac{1}{q}} \right.$$

$$\left. + \mathcal{B}_2^{1-\frac{1}{q}} \left(\frac{|\psi'(\xi_1)|^q}{e^{\omega_1 \tilde\xi_1}} \mathcal{P}_{2,1} + \frac{|\psi'(\xi_2)|^q}{e^{\omega_2 \tilde\xi_2}} \mathcal{P}_{2,2} \right)^{\frac{1}{q}} \right], \quad (21)$$

where

$$\mathcal{P}_{1,1} := \int_0^\lambda \imath^\alpha (\lambda - \imath)^s d\imath, \qquad \mathcal{P}_{1,2} := \int_0^\lambda \imath^\alpha (\mu + \imath)^s d\imath,$$

$$\mathcal{P}_{2,1} := \int_0^\mu \imath^\alpha (\lambda + \imath)^s d\imath \quad and \quad \mathcal{P}_{2,2} := \int_0^\mu \imath^\alpha (\mu - \imath)^s d\imath.$$

Corollary 9. *By putting*

$$h_1(\imath) = (1 - \imath)^{-s} \qquad and \qquad h_2(\imath) = \imath^{-s} \qquad (s \in (0,1)),$$

Theorem 2 yields the following inequality:

$$\left| \mathcal{T}^{\alpha}_{\lambda,\mu}(\xi_1, \xi_2) \right| \leq \frac{\mathcal{F}^{\sigma}_{\rho,\varsigma}(\xi_2 - \xi_1)}{(\lambda + \mu)^{\alpha - \frac{s}{q} + 1}} \left[\mathcal{B}_1^{1 - \frac{1}{q}} \left(\frac{|\psi'(\xi_1)|^q}{e^{\omega_1 \tilde{\xi}_1}} \mathcal{R}_{1,1} + \frac{|\psi'(\xi_2)|^q}{e^{\omega_2 \tilde{\xi}_2}} \mathcal{R}_{1,2} \right)^{\frac{1}{q}} \right.$$
$$\left. + \mathcal{B}_2^{1 - \frac{1}{q}} \left(\frac{|\psi'(\xi_1)|^q}{e^{\omega_1 \tilde{\xi}_1}} \mathcal{R}_{2,1} + \frac{|\psi'(\xi_2)|^q}{e^{\omega_2 \tilde{\xi}_2}} \mathcal{R}_{2,2} \right)^{\frac{1}{q}} \right], \qquad (22)$$

where

$$\mathcal{R}_{1,1} := \int_0^{\lambda} \frac{\imath^{\alpha}}{(\lambda - \imath)^s} d\imath, \qquad \mathcal{R}_{1,2} := \int_0^{\lambda} \frac{\imath^{\alpha}}{(\mu + \imath)^s} d\imath,$$

$$\mathcal{R}_{2,1} := \int_0^{\mu} \frac{\imath^{\alpha}}{(\lambda + \imath)^s} d\imath \quad and \quad \mathcal{R}_{2,2} := \int_0^{\mu} \frac{\imath^{\alpha}}{(\mu - \imath)^s} d\imath.$$

Corollary 10. *Upon letting*

$$h_1(\imath) = h_2(\imath) = \imath(1 - \imath),$$

Theorem 2 yields the following inequality:

$$\left| \mathcal{T}^{\alpha}_{\lambda,\mu}(\xi_1, \xi_2) \right| \leq \frac{\mathcal{F}^{\sigma}_{\rho,\varsigma}(\xi_2 - \xi_1)}{(\lambda + \mu)^{\alpha + 1}} \left(\mathcal{B}_1^{1 - \frac{1}{q}} \mathcal{K}_1^{\frac{1}{q}} + \mathcal{B}_2^{1 - \frac{1}{q}} \mathcal{K}_2^{\frac{1}{q}} \right) \left[\frac{|\psi'(\xi_1)|^q}{e^{\omega_1 \tilde{\xi}_1}} + \frac{|\psi'(\xi_2)|^q}{e^{\omega_2 \tilde{\xi}_2}} \right]^{\frac{1}{q}}, \qquad (23)$$

where

$$\mathcal{K}_1 := \frac{1}{(\lambda + \mu)^2} \left[\frac{\mu \lambda^{\alpha + 2}}{\alpha + 1} - \frac{\mu \lambda^{\alpha + 2}}{\alpha + 2} + \frac{\lambda^{\alpha + 3}}{\alpha + 2} - \frac{\lambda^{\alpha + 3}}{\alpha + 3} \right],$$

and

$$\mathcal{K}_2 := \frac{1}{(\lambda + \mu)^2} \left[\frac{\lambda \mu^{\alpha + 2}}{\alpha + 1} - \frac{\lambda \mu^{\alpha + 2}}{\alpha + 2} + \frac{\mu^{\alpha + 3}}{\alpha + 2} - \frac{\mu^{\alpha + 3}}{\alpha + 3} \right].$$

Remark 4. *If we take* $\lambda = \mu = 1$ *or* $\mathcal{F}^{\sigma}_{\rho,\varsigma}(\xi_2 - \xi_1) = \xi_2 - \xi_1$ *or* $h_1(\imath) = h(1 - \imath)$ *and* $h_2(\imath) = h(\imath)$ *in Theorem 1 and Theorem 2, then we can obtain some interesting results immediately. We omit their proofs here, and the details are left to the interested reader.*

Remark 5. *If we choose* $\omega_1 = \omega_2 = 0$ *in our results in this paper, then all of the consequent results will hold true for the* (h_1, h_2)*–nonconvex functions.*

3. Application

In this section, we present an application involving a new error estimation for the trapezoidal formula by using the inequalities obtained in Section 2. We fix the parameters ρ and ς. We also suppose that the bounded sequence $\{\sigma(\ell)\}_{\ell=0}^{\infty}$ of positive real numbers is given.

Let

$$U : \xi_1 = \chi_0 < \chi_1 < \cdots < \chi_{n-1} < \chi_n = \xi_1 + \mathcal{F}^{\sigma}_{\rho,\varsigma}(\xi_2 - \xi_1)$$

be a partition of the closed interval Δ.

For $\lambda, \mu \in (0, 1]$, let us define

$$T(U, \psi) := \sum_{i=0}^{n-1} \left(\frac{\mu \psi(\chi_i) + \lambda \psi \left(\chi_i + \mathcal{F}^{\sigma}_{\rho,\varsigma}(\hbar_i) \right)}{\lambda + \mu} \right) \mathcal{F}^{\sigma}_{\rho,\varsigma}(\hbar_i),$$

and

$$\int_{\xi_1}^{\xi_1 + \mathcal{F}_{\rho,\varsigma}^{\sigma}(\xi_2 - \xi_1)} \psi(\iota)d\iota = \mathrm{T}(\mathrm{U}, \psi) + \mathrm{R}(\mathrm{U}, \psi),$$

where $\mathrm{R}(\mathrm{U}, \psi)$ is the remainder term and

$$\hbar_i = \chi_{i+1} - \chi_i \qquad (\forall\, i = 0, 1, 2, \cdots, n-1).$$

From the above notations, we can obtain some new bounds regarding error estimation.

Proposition 1. *Assume that $h_1, h_2 : [0,1] \to [0,\infty)$ are two continuous functions. Furthermore, let $\psi : \Delta \to \mathbb{R}$ be a differentiable function on Δ° with $\lambda, \mu \in (0,1]$. Suppose that $\psi' \in \mathcal{L}_1(\Delta)$ and that $|\psi'|^q$ is an exponentially $(\omega_1, \omega_2, h_1, h_2)$–nonconvex function. Then, for $q > 1$ and $\frac{1}{p} + \frac{1}{q} = 1$, it is asserted that*

$$|\mathrm{R}(\mathrm{U}, \psi)| \leqq \frac{1}{(\lambda + \mu)^2} \sum_{i=0}^{n-1} \left[\mathcal{F}_{\rho,\varsigma}^{\sigma}(\hbar_i)\right]^2$$

$$\cdot \left[\left(\frac{\lambda^{p+1}}{p+1}\right)^{\frac{1}{p}} \left(\frac{|\psi'(\chi_i)|^q}{e^{\omega_1 \chi_i}}\mathcal{H}_{1,1} + \frac{|\psi'(\chi_{i+1})|^q}{e^{\omega_2 \chi_{i+1}}}\mathcal{H}_{1,2}\right)^{\frac{1}{q}} \right.$$

$$\left. + \left(\frac{\mu^{p+1}}{p+1}\right)^{\frac{1}{p}} \left(\frac{|\psi'(\chi_i)|^q}{e^{\omega_1 \chi_i}}\mathcal{H}_{2,1} + \frac{|\psi'(\chi_{i+1})|^q}{e^{\omega_2 \chi_{i+1}}}\mathcal{H}_{2,2}\right)^{\frac{1}{q}}\right]. \tag{24}$$

Proof. Applying Theorem 1 on the subinterval $[\chi_i, \chi_{i+1}]$ of the closed interval Δ ($\forall\, i = 0, 1, 2, \cdots, n-1$), and taking $\alpha = 1$, we obtain

$$\left|\left(\frac{\mu\psi(\chi_i) + \lambda\psi\left(\chi_i + \mathcal{F}_{\rho,\varsigma}^{\sigma}(\hbar_i)\right)}{\lambda + \mu}\right)\mathcal{F}_{\rho,\varsigma}^{\sigma}(\hbar_i) - \int_{\chi_i}^{\chi_i + \mathcal{F}_{\rho,\varsigma}^{\sigma}(\hbar_i)}\psi(\iota)d\iota\right| \leqq \frac{\left[\mathcal{F}_{\rho,\varsigma}^{\sigma}(\hbar_i)\right]^2}{(\lambda + \mu)^2}$$

$$\cdot \left[\left(\frac{\lambda^{p+1}}{p+1}\right)^{\frac{1}{p}} \left(\frac{|\psi'(\chi_i)|^q}{e^{\omega_1 \chi_i}}\mathcal{H}_{1,1} + \frac{|\psi'(\chi_{i+1})|^q}{e^{\omega_2 \chi_{i+1}}}\mathcal{H}_{1,2}\right)^{\frac{1}{q}} \right.$$

$$\left. + \left(\frac{\mu^{p+1}}{p+1}\right)^{\frac{1}{p}} \left(\frac{|\psi'(\chi_i)|^q}{e^{\omega_1 \chi_i}}\mathcal{H}_{2,1} + \frac{|\psi'(\chi_{i+1})|^q}{e^{\omega_2 \chi_{i+1}}}\mathcal{H}_{2,2}\right)^{\frac{1}{q}}\right]. \tag{25}$$

Upon summing the inequality (25) over i from 0 to $n-1$ and using the property of the modulus, we obtain inequality (24). $\quad\square$

Proposition 2. *Assume that $h_1, h_2 : [0,1] \to [0,\infty)$ are two continuous functions. Furthermore, let $\psi : \Delta \to \mathbb{R}$ be a differentiable function on Δ° with $\lambda, \mu \in (0,1]$. Suppose that $\psi' \in \mathcal{L}_1(\Delta)$ and that $|\psi'|^q$ is an exponentially $(\omega_1, \omega_2, h_1, h_2)$–nonconvex function. Then, for $q \geqq 1$, the following inequality holds true:*

$$|\mathrm{R}(\mathrm{U}, \psi)| \leqq \frac{1}{(\lambda + \mu)^2} \sum_{i=0}^{n-1} \left[\mathcal{F}_{\rho,\varsigma}^{\sigma}(\hbar_i)\right]^2$$

$$\cdot \left[\left(\frac{\lambda^2}{2}\right)^{1-\frac{1}{q}} \left(\frac{|\psi'(\chi_i)|^q}{e^{\omega_1 \chi_i}}\mathcal{M}_{1,1} + \frac{|\psi'(\chi_{i+1})|^q}{e^{\omega_2 \chi_{i+1}}}\mathcal{M}_{1,2}\right)^{\frac{1}{q}} \right.$$

$$\left. + \left(\frac{\mu^2}{2}\right)^{1-\frac{1}{q}} \left(\frac{|\psi'(\chi_i)|^q}{e^{\omega_1 \chi_i}}\mathcal{M}_{2,1} + \frac{|\psi'(\chi_{i+1})|^q}{e^{\omega_2 \chi_{i+1}}}\mathcal{M}_{2,2}\right)^{\frac{1}{q}}\right]. \tag{26}$$

Proof. Choosing $\alpha = 1$ in Theorem 2 and using the same technique as in our demonstration of Proposition 1, we obtain the desired inequality (26). $\square$

Remark 6. *In view of Remark 2, we can establish new error estimations by using Proposition 1 and Proposition 2.*

4. Conclusions

In this paper, the authors have defined a new generic class of functions involving the modified Fox–Wright function $\mathcal{F}^{\sigma}_{\rho,\varsigma}(\cdot)$ as well as the so-called exponentially $(\omega_1, \omega_2, h_1, h_2)$-nonconvex function. A useful identity has also been found by using fractional integrals and the function $\mathcal{F}^{\sigma}_{\rho,\varsigma}(\cdot)$ with two parameters λ and μ. We have established some Hermite–Hadamard-type integral inequalities by using the above class of functions and the aforementioned identity as an auxiliary result. Several special cases have been deduced as corollaries including relevant details. We have also outlined the derivations of several other corollaries and consequences for the interested reader. The efficiency of our main results has been shown by proving an application for error estimation.

Author Contributions: Conceptualization, H.M.S., A.K., P.O.M., D.B.; methodology, H.M.S., P.O.M., Y.S.H.; software, P.O.M., Y.S.H., A.K.; validation, P.O.M., Y.S.H.; formal analysis, P.O.M., Y.S.H., D.B.; investigation, H.M.S., P.O.M.; resources, P.O.M., A.K.; data curation, D.B., Y.S.H.; writing—original draft preparation, H.M.S., A.K., P.O.M.; writing—review and editing, D.B., Y.S.H.; visualization, Y.S.H.; supervision, H.M.S., Y.S.H., D.B. All authors have read and agreed to the final version of the manuscript.

Funding: This research received no external funding.

Institutional Review Board Statement: Not applicable.

Informed Consent Statement: Not applicable.

Data Availability Statement: Not applicable.

Acknowledgments: This research was supported by Taif University Researchers Supporting Project (No. TURSP-2020/155), Taif University, Taif, Saudi Arabia.

Conflicts of Interest: The authors declare no conflict of interest.

References

1. Adjabi, Y.; Jarad, F.; Baleanu, D.; Abdeljawad, T. On Cauchy problems with Caputo Hadamard fractional derivatives. *Math. Methods Appl. Sci.* **2016**, *40*, 661–681.
2. Tan, W.; Jiang, F.L.; Huang, C.X.; Zhou, L. Synchronization for a class of fractional-order hyperchaotic system and its application. *J. Appl. Math.* **2012**, *2012*, 974639. [CrossRef]
3. Zhou, X.S.; Huang, C.X.; Hu, H.J.; Liu, L. Inequality estimates for the boundedness of multilinear singular and fractional integral operators. *J. Inequal. Appl.* **2013**, *2013*, 303. [CrossRef]
4. Liu, F.W.; Feng, L.B.; Anh, V.; Li, J. Unstructured-mesh Galerkin finite element method for the two-dimensional multi-term time-space fractional Bloch-Torrey equations on irregular convex domains. *Comput. Math. Appl.* **2019**, *78*, 1637–1650. [CrossRef]
5. Kilbas, A.A.; Srivastava, H.M.; Trujillo, J.J. *Theory and Applications of Fractional Differential Equations*; North-Holland Mathematical Studies; Elsevier: Amsterdam, The Netherlands; Science Publishers: Amsterdam, The Netherlands; London, UK; New York, NY, USA, 2006; Volume 204
6. Srivastava, H.M. Fractional-order derivatives and integrals: Introductory overview and recent developments. *Kyungpook Math. J.* **2020**, *60*, 73–116.
7. Cai, Z.W.; Huang, J.H.; Huang, L.H. Periodic orbit analysis for the delayed Filippov system. *Proc. Am. Math. Soc.* **2018**, *146*, 4667–4682. [CrossRef]
8. Chen, T.; Huang, L.H.; Yu, P.; Huang, W.T. Bifurcation of limit cycles at infinity in piecewise polynomial systems. *Nonlinear Anal. Real World Appl.* **2018**, *41*, 82–106. [CrossRef]
9. Wang, J.F.; Chen, X.Y.; Huang, L.H. The number and stability of limit cycles for planar piecewise linear systems of node-saddle type. *J. Math. Anal. Appl.* **2019**, *469*, 405–427. [CrossRef]
10. Houas, M. Certain weighted integral inequalities involving the fractional hypergeometric operators. *Sci. Ser. A Math. Sci.* **2016**, *27*, 87–97.
11. Houas, M. On some generalized integral inequalities for Hadamard fractional integrals. *Mediterr. J. Model. Simul.* **2018**, *9*, 43–52.

12. Baleanu, D.; Mohammed, P.O.; Vivas-Cortez, M.; Rangel-Oliveros, Y. Some modifications in conformable fractional integral inequalities. *Adv. Differ. Equ.* **2020**, *2020*, 374. [CrossRef]
13. Abdeljawad, T.; Mohammed, P.O.; Kashuri, A. New modified conformable fractional integral inequalities of Hermite-Hadamard type with applications. *J. Funct. Spaces* **2020**, *2020*, 4352357. [CrossRef]
14. Mohammed, P.O.; Abdeljawad, T. Integral inequalities for a fractional operator of a function with respect to another function with nonsingular kernel. *Adv. Differ. Equ.* **2020**, *2020*, 363. [CrossRef]
15. Mohammed, P.O.; Brevik, I. A new version of the Hermite-Hadamard inequality for Riemann-Liouville fractional integrals. *Symmetry* **2020**, *12*, 610. [CrossRef]
16. Cloud, M.J.; Drachman, B.C.; Lebedev, L. *Inequalities*, 2nd ed.; Springer: Cham, Switzerland, 2014.
17. Atici, F.M.; Yaldiz, H. Convex functions on discrete time domains. *Can. Math. Bull.* **2016**, *59*, 225–233. [CrossRef]
18. Tomar, M.; Agarwal, P.; Jleli, M.; Samet, B. Certain Ostrowski type inequalities for generalized s–convex functions. *J. Nonlinear Sci. Appl.* **2017**, *10*, 5947–5957. [CrossRef]
19. Varošanec, S. On h-convexity. *J. Math. Anal. Appl.* **2007**, *326*, 303–311. [CrossRef]
20. Awan, M.U.; Noor, M.A.; Noor, K.I. Hermite-Hadamard inequalities for exponentially convex functions. *Appl. Math. Inf. Sci.* **2018**, *12*, 405–409. [CrossRef]
21. Baleanu, D.; Kashuri, A.; Mohammed, P.O.; Meftah, B. General Raina fractional integral inequalities on coordinates of convex functions. *Adv. Differ. Equ.* **2021**, *2021*, 82. [CrossRef]
22. Mohammed, P.O.; Abdeljawad, T.; Zeng, S.; Kashuri, A. Fractional Hermite-Hadamard integral inequalities for a new class of convex functions. *Symmetry* **2020**, *12*, 1485. [CrossRef]
23. Kashuri, A.; Liko, R. Some new Hermite-Hadamard type inequalities and their applications. *Stud. Sci. Math. Hung.* **2019**, *56*, 103–142. [CrossRef]
24. Alqudah, M.A.; Kashuri, A.; Mohammed, P.O.; Abdeljawad, T.; Raees, M.; Anwar, M.; Hamed, Y.S. Hermite-Hadamard integral inequalities on coordinated convex functions in quantum calculus. *Adv. Differ. Equ.* **2021**, *2021*, 264. [CrossRef]
25. Delavar, M.R.; De La Sen, D. Some generalizations of Hermite–Hadamard type inequalities. *SpringerPlus* **2016**, *5*, 1661. [CrossRef] [PubMed]
26. Khan, M.B.; Mohammed, P.O.; Noor, B.; Hamed, Y.S. New Hermite-Hadamard inequalities in fuzzy-interval fractional calculus and related inequalities. *Symmetry* **2021**, *13*, 673. [CrossRef]
27. Mohammed, P.O.; Abdeljawad, T.; Alqudah, M.A.; Jarad, F. New discrete inequalities of Hermite-Hadamard type for convex functions. *Adv. Differ. Equ.* **2021**, *2021*, 122. [CrossRef]
28. Srivastava, H.M.; Zhang, Z.-H.; Wu, Y.-D. Some further refinements and extensions of the Hermite-Hadamard and Jensen inequalities in several variables. *Math. Comput. Model.* **2011**, *54*, 2709–2717. [CrossRef]
29. Mohammed, P.O. New generalized Riemann-Liouville fractional integral inequalities for convex functions. *J. Math. Inequal.* **2021**, *15*, 511–519. [CrossRef]
30. Miller, K.S.; Ross, B. *An Introduction to the Fractional Calculus and Fractional Differential Equations*; Wiley: New York, NY, USA, 1993.
31. Samko, S.G.; Kilbas, A.A.; Marichev, O.I. *Fractional Integrals and Derivatives: Theory and Applications*; Gordon & Breach Science Publishers: Yverdon, Switzerland, 1993.
32. Herrmann, R. *Fractional Calculus: An Introduction for Physicists*; World Scientific Publishing Company: Singapore; Hackensack, NJ, USA; London, UK; Hong Kong, China, 2011
33. Katugampola, U.N. A new approach to generalized fractional derivatives. *Bull. Math. Anal. Appl.* **2014**, *6*, 1–15.
34. Katugampola, U.N. New fractional integral unifying six existing fractional integrals. *arXiv* **2016**, arXiv:1612.08596.
35. Baleanu, D.; Fernandez, A. On fractional operators and their classifications. *Mathematics* **2019**, *7*, 830. [CrossRef]
36. Hilfer, R.; Luchko, Y. Desiderata for fractional derivatives and integrals. *Mathematics* **2019**, *7*, 149. [CrossRef]
37. Srivastava, H.M.; Karlsson, P.W. *Multiple Gaussian Hypergeometric Series*; Halsted Press, Ellis Horwood Limited: Chichester, UK; John Wiley and Sons: New York, NY, USA; Chichester, UK; Brisbane, Australia; Toronto, ON, Canada, 1985
38. Wright, E.M. The asymptotic expansion of integral functions defined by Taylor series. *Philos. Trans. R. Soc. Lond. Ser. A Math. Phys. Sci.* **1940**, *238*, 423–451.
39. Raina, R.K. On generalized Wright's hypergeometric functions and fractional calculus operators. *East Asian Math. J.* **2005**, *21*, 191–203.
40. Srivastava, H.M.; Bansal, M.K.; Harjule, P. A study of fractional integral operators involving a certain generalized multi-index Mittag-Leffler function. *Math. Meth. Appl. Sci.* **2018**, *41*, 6108–6121. [CrossRef]

fractal and fractional

Article

Guaranteed Cost Leaderless Consensus Protocol Design for Fractional-Order Uncertain Multi-Agent Systems with State and Input Delays

Yingming Tian [1,2], Qin Xia [1,3], Yi Chai [1,*], Liping Chen [4], António M. Lopes [5] and Yangquan Chen [6]

1   College of Automation, Chongqing University, Chongqing 400044, China; tianyingming@cqcy.com (Y.T.); xiaqincqu@126.com (Q.X.)
2   Chongqing Chuanyi Automation Co. Ltd., Chongqing 401123, China
3   China Automotive Engineering Research Institute Co. Ltd., Chongqing 401122, China
4   School of Electrical Engineering and Automation, Hefei University of Technology, Hefei 230009, China; lip_chen@hfut.edu.cn
5   LAETA/INEGI, Faculty of Engineering, University of Porto, Rua Dr. Roberto Frias, 4200-465 Porto, Portugal; aml@fe.up.pt
6   Mechatronics, Embedded Systems and Automation Lab, University of California, Merced, CA 95343, USA; ychen53@ucmerced.edu
*   Correspondence: chaiyicqu@126.com

Abstract: This paper addresses the guaranteed cost leaderless consensus of delayed fractional-order (FO) multi-agent systems (FOMASs) with nonlinearities and uncertainties. A guaranteed cost function for FOMAS is proposed to simultaneously consider consensus performance and energy consumption. By employing the linear matrix inequality approach and the FO Razumikhin theorem, a delay-dependent and order-dependent consensus protocol is formulated for FOMASs with input delay. The proposed protocol not only guarantees the robust stability of the closed-loop system error but also ensures that the performance degradation caused by the system uncertainty is lesser than that obtained with other approaches. Two numerical examples are provided in order to verify the effectiveness and accuracy of the proposed protocol.

Keywords: fractional-order system; multi-agent; guaranteed cost consensus; delay-dependent

Citation: Tian, Y.; Xia, Q.; Chai, Y.; Chen, L.; Lopes, A.M.; Chen, Y. Guaranteed Cost Leaderless Consensus Protocol Design for Fractional-Order Uncertain Multi-Agent Systems with State and Input Delays. *Fractal Fract.* **2021**, *5*, 141. https://doi.org/10.3390/fractalfract5040141

Academic Editor: Amar Debbouche

Received: 25 July 2021
Accepted: 26 September 2021
Published: 28 September 2021

Publisher's Note: MDPI stays neutral with regard to jurisdictional claims in published maps and institutional affiliations.

1. Introduction

In recent years, there has been increasing interest in the coordination of multi-agent systems (MASs) that have a variety of applications. For example, we can mention the distributed consensus behavior in sensor networks [1], satellite formation flying [2] and cooperative control of unmanned aerial vehicles rendezvous [3]. Consensus, as a critical dynamic behaviour in MASs, has been focused on integer-order (IO) MASs, where every agent is described by classical IO dynamics [4–8]. It has been shown that many phenomena can be explained naturally by coordinated behavior of agents with FO dynamics [9–11]. This includes flocking movement and food searching by means of individual secretions and microbial secretions, submarine underwater robots exploring seawater with a large number of microorganisms and viscous substances and operating unmanned aerial vehicles in complex space environments. Therefore, the question of how to achieve consensus for FOMASs has received much attention, and important developments involving leader-following group, cluster, finite-time, bipartite, group multiple lag and others have been presented. For example, nonlinear FOMASs with distributed input delays were considered in [12], a delay-dependent consensus condition for a class of linear FOMASs with distributed control containing input time-delay was proposed in [13], the event-triggered consensus for general linear FOMASs was investigated in [14,15] and the consensus of FOMASs without delay terms was studied in [16,17].

In practical applications, control systems are subject to time delays caused by the limited speed at which signals propagate [18]. Time delays may degrade system performance and robustness and even cause instability. Generally, consensus conditions of delayed FOMASs are divided into the categories of delay dependent and delay independent based on whether the consensus criteria depend on the delay or not. Usually, delay independent criteria are excessively conservative in comparison with delay dependent ones, particularly when the time delay is small. On the other hand, despite actual physical systems being nonlinear, there are few stability results for nonlinear delayed FOMASs. Therefore, addressing such systems is fundamental [19–22].

It should be mentioned that all works mentioned above focus only on the consensus regulation performance for FOMASs with the existence of time delays or/and nonlinearities [12–17,19–22]. However, energy consumption is an issue, and the so-called guaranteed cost control approach to tackle this problem, which considers the consensus regulation performance and the energy consumption at the same time, was proposed. The guaranteed cost consensus of MASs has increasingly attracted the attention of researchers. In [23], the event-triggered guaranteed cost consensus for nonlinear MASs with time delay and uncertain parameters was addressed. In [24], the guaranteed performance consensus for MASs with Lipschitz nonlinear dynamics was investigated. In [25,26], the guaranteed cost consensus for MASs was also investigated. However, it should be pointed out that most research has been focused on IO MASs instead of FOMASs and, in particular, the guaranteed cost consensus of FOMASs with state and input time-delay receiveed limited attention.

Motivated by the above discussion, a guaranteed cost leaderless consensus protocol for uncertain nonlinear delayed FOMASs with input time delay is proposed in this paper. The main contributions are the following: (1) to address the guaranteed cost consensus for nonlinear FOMASs with state and input time delay; (2) to establish in terms of linear matrix inequality (LMI) a delay-dependent and order-dependent sufficient condition for guaranteed cost leaderless consensus protocol; and (3) to obtain a guaranteed cost leaderless consensus protocol less conservative than the ones already proposed in the literature.

The rest of this paper is organized as follows. Section 2 introduces some fundamental concepts and lemmas necessary for theoretical development. Section 3 presents the main results and discusses the most relevant details. Section 4 demonstrates the effectiveness of the novel procedure with two numerical examples. Finally, Section 5 outlines the main conclusions.

Standard notation is used in the sequel. The symbols R^{n*m}, $\| * \|$ and $\otimes$ represent the set of real matrices, the Euclidean norm of a vector or the derived two-norm of a matrix and the Kronecker product, respectively. The symbol I_N is an identity matrix, and diag$\{ * \}$ denotes the diagonal matrix. The expression $A > 0(\geq 0)$ represents a symmetric positive definite (semi-definite) matrix. The matrices A^T and A^{-1} denote the transpose and inverse of A, respectively.

2. Preliminaries and Problem Formulation

In this section, we introduce basic concepts of graph theory, definitions related to fractional calculus, guaranteed cost function related to FOMASs and some useful lemmas.

2.1. Graph Theory

An undirected graph G is a tuple (V, E) in which $V = \{v_1, v_2, \cdots, v_N\}$ denotes the set of nodes, and $E \subseteq V \times V$ is the set of edges of G. Any edge connecting nodes v_i and v_j is represented by $e_{ij} = (v_i, v_j)$ or $e_{ji} = (v_j, v_i)$ since we have $e_{ij} \in E \Leftrightarrow e_{ji} \in E$. For example, the tuple (V, E) with $V = \{v_1, v_2, v_3, v_4\}$ and $E = \{(v_1, v_2), (v_2, v_2), (v_2, v_3), (v_1, v_3), (v_3, v_4)\}$, represents an undirected graph with four nodes and five edges. The number of edges associated with a node v_i is called degree of the node, e.g., deg$(v_1) = 2$ means that there are 2 edges associated with v_1. The adjacency matrix of the graph is $A = (a_{ij})_{N \times N}$, where a_{ij} denotes the weight of edge (i, j), the degree matrix corresponds to $D = \text{diag}\{d_1, d_2, \cdots, d_N\}$

where the elements are defined by $d_i = \sum_j a_{ij}$ and the Laplacian of the weighted digraph G is defined as $L = D - A$, with each element in L expressed as follows.

$$
l_{ij} = \begin{cases} -a_{ij}, i \neq j, \\ \sum_{j=1, j \neq i}^{N} a_{ij}, i = j. \end{cases}
$$

2.2. Useful Lemmas

Some useful lemmas are presented in the follow up.

Lemma 1. *All eigenvalues of $\tilde{L}$ are greater than or equal to 0 if and only if the graph G is connected, where the following is the case.*

$$
\tilde{L} = \begin{bmatrix} d_2 + a_{12} & a_{13} - a_{23} & \cdots & a_{1N} - a_{2N} \\ a_{12} - a_{32} & d_3 + a_{13} & \cdots & a_{1N} - a_{3N} \\ \vdots & \vdots & \vdots & \vdots \\ a_{12} - a_{N2} & a_{13} - a_{N3} & \cdots & d_N + a_{1N} \end{bmatrix}.
$$

Proof. Let $1_{n-1}^T = \underbrace{[1, \cdots, 1]}_{n-1}$, $0_{N-1} = \underbrace{[0, \cdots, 0]}_{n-1}$ and the following be the case:

$$
Q = \begin{bmatrix} 1 & 0_{N-1} \\ -1_{n-1} & I_{n-1} \end{bmatrix},
$$

where I_{n-1} is an identity matrix with dimension $n-1$. Then, we have the following:

$$
QLQ^{-1} = \begin{bmatrix} 0 & a \\ 0_{n-1}^T & \tilde{L} \end{bmatrix},
$$

where $a = [-a_{12}, \cdots, -a_{1n}]$. Since all eigenvalues of matrix L are greater than or equal to 0 if and only if G is connected, then Lemma 1 holds. $\square$

Lemma 2. *The Laplacian L of the undirected graph obeys the following [27]:*

$$
x^T(t)Lx(t) = \frac{1}{2} \sum_{i=1}^{N} \sum_{j=1}^{N} a_{ij}(x_i(t) - x_j(t))^T (x_i(t) - x_j(t)),
$$

and $L = L^T \geq 0$.

Lemma 3. *For given matrices $Q = Q^T$, H, M and $R = R^T$ with appropriate dimensions [28], the following condition:*

$$
Q + HNM + M^T N^T H^T < 0
$$

is verified for $N(t)N^T(t) \leq R$ if and only if there exists some $\lambda > 0$ such that the following is the case.

$$
Q + \lambda HH^T + \lambda^{-1} M^T RM < 0.
$$

Lemma 4. *When $x(t) \in R^n$ is a differentiable vector-value function, $P = P^{\mathrm{T}} > 0$ and $\forall \alpha \in (0,1)$ [29]. We can obtain the following.*

$$_{t_0}^{C}D_t^{\alpha}(x^{\mathrm{T}}(t)Px(t)) \le (x^{\mathrm{T}}(t)P)_{t_0}^{C}D_t^{\alpha}x(t) + (_{t_0}^{C}D_t^{\alpha}x(t))^{\mathrm{T}}Px(t).$$

Lemma 5. *For any real vectors with the same dimension x and y, the following inequality is verified [30]:*

$$2x^T y \le \varepsilon x^T x + \varepsilon^{-1} y^T y,$$

where ε is a positive number.

2.3. Problem Statement

The i-th agent can be modeled as follows:

$$_{t_0}^{C}D_t^{\alpha}x_i(t) = (A + \Delta A)x_i(t) + (A_\tau + \Delta A_\tau)x_i(t - \tau) + f(x_i(t)) + (B + \Delta B)u_i(t), \quad i = 1, 2 \cdots N, \quad (1)$$

where $x_i(t) = [x_{i1}(t), x_{i2}(t) \cdots x_{in}(t)]^{\mathrm{T}}$, $_{t_0}^{C}D_t^{\alpha}x(t) = \frac{1}{\Gamma(n-\alpha)} \int_{t_0}^{t} \frac{x^n(s)}{(t-s)^{\alpha-n+1}}ds$, and $\Gamma(s) = \int_0^\infty t^{s-1}e^{-t}dt$ is the Gamma function. The variable $u_i(t)$ represents the control input and A, and $A_\tau \in R^{n \times n}$ and $B \subset R^{n \times m}$ are known constant matrices. The symbols ΔA, ΔA_τ and ΔB represent uncertain matrices given by the following:

$$\begin{bmatrix} \Delta A & \Delta A_\tau & \Delta B \end{bmatrix} = EH(t)\begin{bmatrix} F_1 & F_2 & F_3 \end{bmatrix},$$

where E, F_1, F_2 and F_3 are known constant real matrices with appropriate dimensions, and $H(t)$ is the unknown time-varying matrix satisfying $H(t)H^{\mathrm{T}}(t) \le I$. Moreover, $f : R^n \to R^n$ is a continuous function that satisfies the Lipschitz condition. There is a positive constant l such that the following is the case.

$$\|f(x) - f(y)\| \le l\|x - y\|, \ \forall x, y \in R^n.$$

Remark 1. *If $f(x_i(t)) = 0$, then the model of the i-th agent can be expressed as follows.*

$$_{t_0}^{C}D_t^{\alpha}x_i(t) = (A + \Delta A)x_i(t) + (A_\tau + \Delta A_\tau)x_i(t - \tau) + (B + \Delta B)u_i(t), \quad i = 1, 2 \cdots N. \quad (2)$$

The control protocol will be designed as follows:

$$u_i(t) = -K\left(\sum_{j \in N_i} (a_{ij}(x_i(t - \tau) - x_j(t - \tau))\right). \quad (3)$$

where K is feedback control gain.

Definition 1. *The consensus of MASs without a leader can be achieved if and only if the following is the case [30].*

$$\lim_{t \to \infty} \|x_i(t) - x_1(t)\| = 0.$$

Let $e_i(t) = x_i(t) - x_1(t)$. By Definition 1, if $\lim_{t \to \infty} \|e_i(t)\| = 0$, then consensus for system (1) can be achieved. It follows from system (1) and (2) that the following error systems can be obtained.

$$
\begin{aligned}
{}^{C}_{t_0}D^{\alpha}_t e_i(t) &= (A+\Delta A)e_i(t) + (A_\tau+\Delta A_\tau)e_i(t-\tau) + (B+\Delta B)u_i(t) - (B+\Delta B)u_1(t) \\
&+ f(x_i(t)) - f(x_1(t)),
\end{aligned}
\tag{4}
$$

$$
{}^{C}_{t_0}D^{\alpha}_t e_i(t) = (A+\Delta A)e_i(t) + (A_\tau+\Delta A_\tau)e_i(t-\tau) + (B+\Delta B)u_i(t) - (B+\Delta B)u_1(t).
\tag{5}
$$

Definition 2. *The guaranteed cost function associated with FOMASs ($0 \le \alpha \le 1$) is defined as follows:*

$$
J = \frac{1}{\Gamma(\alpha)} \int_0^t (t-s)^{\alpha-1}(J_x(s) + J_u(s))ds,
\tag{6}
$$

where

$$
\begin{aligned}
J_x(t) &= \sum_{i=1}^{N}\sum_{j=1}^{N} a_{ij}(x_i(t)-x_j(t))^{\mathrm{T}}Q_1(x_i(t)-x_j(t)), \\
J_u(t) &= \sum_{i=1}^{N} u_i^{\mathrm{T}}(t)Q_2 u_i(t),
\end{aligned}
$$

with Q_1 and Q_2 representing two given symmetric positive matrices.

Remark 2. *In (6), $J_x(t)$ and $J_u(t)$ represent the consensus error performance and the control energy consumption for the FOMASs. Reference [31] addressed the guaranteed cost of control for a single system. The works [23–26,32,33] proposed a guaranteed cost function related to MASs, but that cannot be applied to the guaranteed cost consensus of FOMASs. Therefore, the definition of the guaranteed cost function (6) related to FOMASs is given.*

3. Main Results

In this section, a delay-dependent sufficient condition for the guaranteed cost consensus protocol is established in terms of LMI, and its guaranteed cost is derived.

Theorem 1. *For given positive definite symmetric matrices Q_1 and Q_2, if there exist a symmetric positive definite matrix $\bar{P}$, a matrix Y and the constant positive scalars λ such that the following is the case:*

$$
\Delta =
\begin{bmatrix}
\Delta_{11} & \Delta_{12} & \Delta_{13} & \Delta_{14} & \Delta_{15} & \Delta_{16} & \Delta_{17} & \Delta_{18} & 0 \\
* & \Delta_{22} & \Delta_{23} & \Delta_{24} & 0 & 0 & 0 & 0 & \Delta_{29} \\
* & * & \Delta_{33} & 0 & \Delta_{35} & 0 & 0 & 0 & 0 \\
* & * & * & \Delta_{44} & 0 & 0 & 0 & 0 & 0 \\
* & * & * & * & \Delta_{55} & 0 & 0 & 0 & 0 \\
* & * & * & * & * & \Delta_{66} & 0 & 0 & 0 \\
* & * & * & * & * & * & \Delta_{77} & 0 & 0 \\
* & * & * & * & * & * & * & \Delta_{88} & 0 \\
* & * & * & * & * & * & * & * & \Delta_{99}
\end{bmatrix}
< 0,
\tag{7}
$$

where

$$
\begin{aligned}
\Delta_{11} &= I_{N-1} \otimes (\bar{P}A^{\mathrm{T}} + A\bar{P} + \bar{P} + I), \\
\Delta_{12} &= I_{N-1} \otimes A_\tau \bar{P} - \tilde{L} \otimes BY, \\
\Delta_{13} &= \tau^\alpha \alpha^{-1} I_{N-1} \otimes \bar{P}A^{\mathrm{T}}, \Delta_{14} = I_{N-1} \otimes \bar{P}F_1^{\mathrm{T}}, \\
\Delta_{15} &= I_{N-1} \otimes \lambda E, \Delta_{16} = I_{N-1} \otimes l^2 \bar{P}, \\
\Delta_{17} &= I_{N-1} \otimes \tau^\alpha \alpha^{-1} \bar{P}, \Delta_{18} = 2\lambda_{max}(L) I_{N-1} \otimes \bar{P}Q_1, \\
\Delta_{22} &= -I_{N-1} \otimes \bar{P}, \\
\Delta_{23} &= \tau^\alpha \alpha^{-1} (I_{N-1} \otimes \bar{P}A_\tau^{\mathrm{T}} - \tilde{L}^{\mathrm{T}} \otimes Y^{\mathrm{T}} B^{\mathrm{T}}), \\
\Delta_{24} &= I_{N-1} \otimes \bar{P}F_2^{\mathrm{T}} - \tilde{L}^{\mathrm{T}} \otimes Y^{\mathrm{T}} F_3^{\mathrm{T}}, \\
\Delta_{29} &= \lambda_{max}^2(L) I_{N-1} \otimes Y^{\mathrm{T}} Q_2, \\
\Delta_{33} &= \Omega_{77} = -I_{N-1} \otimes \tau^\alpha \alpha^{-1} I, \Omega_{35} = \tau^\alpha \alpha^{-1} I_{N-1} \otimes \lambda E, \\
\Delta_{44} &= \Omega_{55} = -I_{N-1} \otimes \lambda I, \Delta_{66} = -I_{N-1} \otimes l^2 I, \\
\Delta_{88} &= -2\lambda_{max}(L) I_{N-1} \otimes Q_1, \\
\Delta_{99} &= -\lambda_{max}^2(L) I_{N-1} \otimes Q_2,
\end{aligned}
$$

then the consensus of the FOMASs (1) with control protocol (3) is achieved. Moreover, the feedback gain K is given by the following:

$$
K = Y\bar{P}^{-1},
$$

and the guaranteed cost defined as follows.

$$
J^* = \lambda_{max}(I_{N-1} \otimes P)\|e(0)\|^2.
$$

Proof. According to (3) and (5), one obtains the following:

$$
\begin{aligned}
{}_{t_0}^{C} D_t^\alpha e_i(t) &= (A + \Delta A)e_i(t) + (A_\tau + \Delta A_\tau)e_i(t - \tau) - (B + \Delta B)K\Big(\sum_{j \in N_i} (a_{ij}(x_i(t - \tau) - x_j(t - \tau)) \\
&\quad + (B + \Delta B)K\Big(\sum_{j \in N_i} (a_{1j}(x_1(t - \tau) - x_j(t - \tau))) + f(x_i(t)) - f(x_1(t)),
\end{aligned}
$$

which can be written as follows:

$$
\begin{aligned}
{}_{t_0}^{C} D_t^\alpha e(t) &= I_{N-1} \otimes (A + \Delta A)e(t) + I_{N-1} \otimes (A_\tau + \Delta A_\tau)e(t - \tau) \\
&\quad - \tilde{L} \otimes (BK + \Delta BK)e(t - \tau) + F(x(t)),
\end{aligned}
$$

where $e(t) = [e_2^T(t), \cdots, e_n^T(t)]^T$, $e(t - \tau) = [e_2^T(t - \tau), \cdots, e_n^T((t - \tau)]^T$ and $F(x(t)) = [(f(x_2(t)) - f(x_1(t)))^T, (f(x_3(t)) - f(x_1(t)))^T, \cdots, (f(x_N(t)) - f(x_1(t)))^T]^T$.

From the definition of guaranteed cost function, the following is the case:

$$
\begin{aligned}
&\bar{e}^T(t)(2L \otimes Q_1)\bar{e}(t) + \bar{e}^T(t - \tau)(L^{\mathrm{T}}L \otimes K^{\mathrm{T}} Q_2 K)\bar{e}(t - \tau) \\
&\leq e^T(t)(2\lambda_{max}(L) I_{N-1} \otimes Q_1)e(t) + e^T(t - \tau)(\lambda_{max}^2(L) I_{N-1} \otimes K^{\mathrm{T}} Q_2 K)e(t - \tau),
\end{aligned}
$$

where $\bar{e}(t) = [e_1^T(t), \cdots, e_n^T(t)]^T$ and $\bar{e}(t - \tau) = [e_1^T(t - \tau), \cdots, e_n^T((t - \tau)]^T$.
Let us select a Lyapunov function.

$$
V(t) = e^T(t)(I_{N-1} \otimes P)e(t).
$$

Taking the α-order derivative and using Lemma 4 results in the following.

$$
\begin{aligned}
{}^{C}_{t_0}D^{\alpha}_t V(t) \;+\; & e^{\mathrm{T}}(t)(2L \otimes Q_1)e(t) + e^{\mathrm{T}}(t)(L^{\mathrm{T}}L \otimes K^{\mathrm{T}}Q_2K)e(t) \\
\leq\; & e^{\mathrm{T}}(t)(I_{N-1} \otimes P){}^{C}_{t_0}D^{\alpha}_t e(t) + ({}^{C}_{t_0}D^{\alpha}_t e(t))^{\mathrm{T}}(I_{N-1} \otimes P)e(t) \\
+\; & e^{\mathrm{T}}(t)(2L \otimes Q_1)e(t) + e^{\mathrm{T}}(t-\tau)(\tilde{L}^{\mathrm{T}}\tilde{L} \otimes K^{\mathrm{T}}Q_2K)e(t-\tau) \\
\leq\; & e^{\mathrm{T}}(t)(I_{N-1} \otimes (A^{\mathrm{T}}P + PA + \Delta A^{\mathrm{T}}P + P\Delta A))e(t) \\
+\; & e^{\mathrm{T}}(t-\tau)(I_{N-1} \otimes (A^{\mathrm{T}}_\tau P + \Delta A^{\mathrm{T}}_\tau P) \\
-\; & \tilde{L}^{\mathrm{T}} \otimes (K^{\mathrm{T}}B^{\mathrm{T}}P + K^{\mathrm{T}}\Delta B^{\mathrm{T}}P))e(t) + e^{\mathrm{T}}(t)(I_{N-1} \otimes (PA_\tau + P\Delta A_\tau) \\
-\; & \tilde{L} \otimes (PBK + P\Delta BK))e(t-\tau) + e^{\mathrm{T}}(t)(2\lambda_{max}(L)I_{N-1} \otimes Q_1)e(t) \\
+\; & e^{\mathrm{T}}(t-\tau)(\lambda^2_{max}(L)I_{N-1} \otimes K^{\mathrm{T}}Q_2K)e(t-\tau) + e^{\mathrm{T}}(t)(I_{N-1} \otimes P)F(x(t)) \\
+\; & F^{\mathrm{T}}(x(t))(I_{N-1} \otimes P)e(t).
\end{aligned}
\tag{8}
$$

It follows from Lemma 5 that there exist two positive constants ϵ and ε such that the following is the case.

$$
\begin{aligned}
e^{\mathrm{T}}(t)(I_{N-1} \otimes P)F(x(t)) \;+\; & F^{\mathrm{T}}(x(t))(I_N \otimes P)e(t) \\
\leq\; & e^{\mathrm{T}}(t)\left(\frac{1}{2}(\epsilon^{-1} + \varepsilon)l^2 I_{N-1} \otimes I \right. \\
+\; & \left. \frac{1}{2}(\epsilon + \varepsilon^{-1})I_{N-1} \otimes P^2 \right)e(t).
\end{aligned}
$$

For the analysis, let us consider $\epsilon = \varepsilon = 1$. Then, one obtains the following:

$$
\begin{aligned}
e^{\mathrm{T}}(t)(I_{N-1} \otimes P)F(x(t)) \;+\; & F^{\mathrm{T}}(x(t))(I_N \otimes P)e(t) \\
\leq\; & e^{\mathrm{T}}(t)(I_{N-1} \otimes l^2 I + I_{N-1} \otimes P^2)e(t).
\end{aligned}
\tag{9}
$$

while $-\tau \leq \theta \leq 0$, $e(t)$ satisfies the following.

$$
V(t+\theta, e(t+\theta)) < \mu V(t, e(t))
$$

When $\mu > 1$, one can obtain the following.

$$
\mu e^{\mathrm{T}}(t)(I_{N-1} \otimes P)e(t) - e^{\mathrm{T}}(t-\tau)(I_{N-1} \otimes P)e(t-\tau) \geq 0.
\tag{10}
$$

Combining (8) with (10) yields the following.

$$
\begin{aligned}
{}^{C}_{t_0}D^{\alpha}_t V(t) \;+\; & e^{\mathrm{T}}(t)(2\lambda_{max}(L)I_{N-1} \otimes Q_1)e(t) + e^{\mathrm{T}}(t-\tau)(\lambda^2_{max}(L)I_{N-1} \otimes K^{\mathrm{T}}Q_2K)e(t-\tau) \\
\leq\; & e^{\mathrm{T}}(t)(I_{N-1} \otimes (A^{\mathrm{T}}P + PA + \Delta A^{\mathrm{T}}P + P\Delta A + l^2 I \\
+\; & P^2 + \mu P + 2\lambda_{max}(L)Q_1))e(t) + e^{\mathrm{T}}(t-\tau)(I_{N-1} \otimes (A^{\mathrm{T}}_\tau P + \Delta A^{\mathrm{T}}_\tau P) \\
-\; & \tilde{L}^{\mathrm{T}} \otimes (K^{\mathrm{T}}B^{\mathrm{T}}P + K^{\mathrm{T}}\Delta B^{\mathrm{T}}P))e(t) + e^{\mathrm{T}}(t)(I_{N-1} \otimes (PA_\tau + P\Delta A_\tau) \\
-\; & \tilde{L} \otimes (PBK + P\Delta BK))e(t-\tau) \\
-\; & e^{\mathrm{T}}(t-\tau)(I_{N-1} \otimes P + \lambda^2_{max}(L)K^{\mathrm{T}}Q_2K)e(t-\tau).
\end{aligned}
\tag{11}
$$

Moreover, for symmetric real matrices $X = X^{\mathrm{T}}$, $Z = Z^{\mathrm{T}}$ and matrix W, we have the following.

$$
\Omega = \begin{bmatrix} X & W \\ W^{\mathrm{T}} & Z \end{bmatrix} \geq 0.
$$

Then, the following inequality holds:

$$
\tau^{\alpha}\alpha^{-1}\zeta^{\mathrm{T}}(t)\Omega\zeta(t) - \int_{t-\tau(t)}^{t}(t-s)^{\alpha-1}\zeta^{\mathrm{T}}(t)\Omega\zeta(t)ds \geq 0,
\tag{12}
$$

where $\zeta(t) = [e^{\mathrm{T}}(t), ({}^{C}_{t_0}D^{\alpha}_t V(t))^{\mathrm{T}}]^{\mathrm{T}}$. Let $X = Z = I_{N-1} \otimes I_n, W = 0_{(N-1)n}$ for simplicity. According to (11) and (12), one obtains the following:

$$
\begin{aligned}
{}^{C}_{t_0}D^{\alpha}_t V(t) \; + \; & e^{\mathrm{T}}(t)(2\lambda_{max}(L)I_{N-1} \otimes Q_1)e(t) + e^{\mathrm{T}}(t-\tau)(\lambda^2_{max}(L)I_{N-1} \otimes K^{\mathrm{T}}Q_2 K)e(t-\tau) \\
\leq \; & e^{\mathrm{T}}(t)(I_{N-1} \otimes (A^{\mathrm{T}}P + PA + \Delta A^{\mathrm{T}}P + P\Delta A + l^2 I \\
+ \; & P^2 + \mu P + 2\lambda_{max}(L)Q_1))e(t) + e^{\mathrm{T}}(t-\tau)(I_{N-1} \otimes (A^{\mathrm{T}}_{\tau}P + \Delta A^{\mathrm{T}}_{\tau}P) \\
- \; & \tilde{L}^{\mathrm{T}} \otimes (K^{\mathrm{T}}B^{\mathrm{T}}P + K^{\mathrm{T}}\Delta B^{\mathrm{T}}P))e(t) + e^{\mathrm{T}}(t)(I_{N-1} \otimes (PA_{\tau} + P\Delta A_{\tau}) \\
- \; & \tilde{L} \otimes (PBK + P\Delta BK))e(t-\tau) \\
- \; & e^{\mathrm{T}}(t-\tau)(I_{N-1} \otimes P + \lambda^2_{max}(L)K^{\mathrm{T}}Q_2 K)e(t-\tau) \\
+ \; & \tau^{\alpha}\alpha^{-1}\zeta^{\mathrm{T}}(t)\Omega\zeta(t) - \int_{t-\tau(t)}^{t}(t-s)^{\alpha-1}\zeta^{\mathrm{T}}(t)\Omega\zeta(t)ds \\
= \; & \eta^{\mathrm{T}}(t)\Theta\eta(t) - \int_{t-\tau(t)}^{t}(t-s)^{\alpha-1}\zeta^{\mathrm{T}}(t)\Omega\zeta(t)ds,
\end{aligned}
$$

where $\eta(t) = [e^{\mathrm{T}}(t), e^{\mathrm{T}}(t-\tau)]^{\mathrm{T}}$ and

$$
\Theta = \begin{bmatrix} \Theta_{11} & \Theta_{12} \\ * & \Theta_{22} \end{bmatrix},
$$

with the following.

$$
\begin{aligned}
\Theta_{11} \; = \; & I_{N-1} \otimes (A^{\mathrm{T}}P + PA + \Delta A^{\mathrm{T}}P + P\Delta A + \mu P + l^2 I \\
+ \; & P^2 + \tau^{\alpha}\alpha^{-1}I + 2\lambda_{max}(L)Q_1) + \tau^{\alpha}\alpha^{-1}(I_{N-1} \otimes (A^{\mathrm{T}} \\
+ \; & \Delta A^{\mathrm{T}}))(I_{N-1} \otimes (A + \Delta A)), \\
\Theta_{12} \; = \; & I_{N-1} \otimes (PA_{\tau} + P\Delta A_{\tau}) - \tilde{L} \otimes (PBK + P\Delta BK)) \\
+ \; & \tau^{\alpha}\alpha^{-1}(I_{N-1} \otimes (A^{\mathrm{T}} + \Delta A^{\mathrm{T}}))(I_{N-1} \otimes (A_{\tau} + \Delta A_{\tau}) - \tilde{L} \otimes (BK + \Delta BK)), \\
\Theta_{22} \; = \; & -I_{N-1} \otimes P + \lambda^2_{max}(L)I_{N-1} \otimes K^{\mathrm{T}}Q_2 K + \tau^{\alpha}\alpha^{-1}(I_{N-1} \otimes (A^{\mathrm{T}}_{\tau} \\
+ \; & \Delta A^{\mathrm{T}}_{\tau}) - \tilde{L}^{\mathrm{T}} \otimes (K^{\mathrm{T}}B^{\mathrm{T}} + K^{\mathrm{T}}\Delta B^{\mathrm{T}}))(I_{N-1} \otimes (A_{\tau} + \Delta A_{\tau}) - \tilde{L} \otimes (BK + \Delta BK)).
\end{aligned}
$$

It is straightforward to verify that $\Theta \leq 0$ can be written as follows.

$$
\Theta = \begin{bmatrix} Y_{11} & Y_{12} \\ * & Y_{22} \end{bmatrix} - \begin{bmatrix} Y_{13} \\ Y_{23} \end{bmatrix}Y_{33}^{-1}\begin{bmatrix} Y_{31} & Y_{32} \end{bmatrix} < 0.
$$

By employing the Schur Complement, we can obtain the following:

$$
\begin{bmatrix} Y_{11} & Y_{12} & Y_{13} \\ * & Y_{22} & Y_{23} \\ * & * & Y_{33} \end{bmatrix} < 0, \tag{13}
$$

where

$$
\begin{aligned}
Y_{11} \; = \; & I_{N-1} \otimes (A^{\mathrm{T}}P + PA + \Delta A^{\mathrm{T}}P + P\Delta A + \mu P + l^2 I \\
+ \; & P^2 + \tau^{\alpha}\alpha^{-1}I + 2\lambda_{max}(L)Q_1) \\
Y_{12} \; = \; & I_{N-1} \otimes (PA_{\tau} + P\Delta A_{\tau}) - \tilde{L} \otimes (PBK + P\Delta BK), \\
Y_{22} \; = \; & -I_{N-1} \otimes P + \lambda^2_{max}(L)K^{\mathrm{T}}Q_2 K, Y_{13} = \tau^{\alpha}\alpha^{-1}I_{N-1} \otimes (A^{\mathrm{T}} + \Delta A^{\mathrm{T}}) \\
Y_{23} \; = \; & \tau^{\alpha}\alpha^{-1}(I_{N-1} \otimes (A^{\mathrm{T}}_{\tau} + \Delta A^{\mathrm{T}}_{\tau}) - \tilde{L} \otimes (K^{\mathrm{T}}B^{\mathrm{T}} + K^{\mathrm{T}}\Delta B^{\mathrm{T}})), \\
Y_{33} \; = \; & -I_{N-1} \otimes \tau^{\alpha}\alpha^{-1}I.
\end{aligned}
$$

Then, Expression (13) can be rewritten as follows:

$$
\begin{bmatrix}
\Sigma_{11} & \Sigma_{12} & \tau^\alpha \alpha^{-1} I_{N-1} \otimes A^{\mathrm{T}} \\
* & -I_{N-1} \otimes P + \lambda_{max}^2(L) K^{\mathrm{T}} Q_2 K & \Sigma_{23} \\
* & * & -I_{N-1} \otimes \tau^\alpha \alpha^{-1} I
\end{bmatrix}
$$

$$
+ \begin{bmatrix}
I_{N-1} \otimes PE \\
0 \\
\tau^\alpha \alpha^{-1} I_{N-1} \otimes E
\end{bmatrix} H(t)
\begin{bmatrix} I_{N-1} \otimes F_1 & I_{N-1} \otimes F_2 - \tilde{L} \otimes F_3 K & 0 \end{bmatrix}
$$

$$
+ \begin{bmatrix}
I_{N-1} \otimes F_1^{\mathrm{T}} \\
I_{N-1} \otimes F_2^{\mathrm{T}} - \tilde{L}^{\mathrm{T}} \otimes K^{\mathrm{T}} F_3^{\mathrm{T}} \\
0
\end{bmatrix} H^{\mathrm{T}}(t)
\begin{bmatrix} I_{N-1} \otimes E^{\mathrm{T}} P & 0 & \tau^\alpha \alpha^{-1} I_{N-1} \otimes E^{\mathrm{T}} \end{bmatrix} < 0,
$$

where

$$
\begin{aligned}
\Sigma_{11} &= I_{N-1} \otimes (A^{\mathrm{T}} P + PA + \mu P + \tau^\alpha \alpha^{-1} I + 2\lambda_{max}(L) Q_1 \\
&\quad + l^2 I + P^2), \\
\Sigma_{12} &= I_{N-1} \otimes PA_\tau - \tilde{L} \otimes PBK, \\
\Sigma_{23} &= \tau^\alpha \alpha^{-1} (I_{N-1} \otimes A_\tau^{\mathrm{T}} - \tilde{L} \otimes K^{\mathrm{T}} B^{\mathrm{T}}),
\end{aligned}
$$

which is equivalent to the following inequality

$$
\Pi = \begin{bmatrix}
\Pi_{11} & \Pi_{12} & \Pi_{13} & I_N \otimes F_1^{\mathrm{T}} & I_N \otimes \lambda PE \\
* & \Pi_{22} & \Pi_{23} & I_N \otimes F_2^{\mathrm{T}} - L^{\mathrm{T}} \otimes K^{\mathrm{T}} F_3^{\mathrm{T}} & 0 \\
* & * & \Pi_{33} & 0 & \Pi_{35} \\
* & * & * & -I_{N-1} \otimes \lambda I & 0 \\
* & * & * & * & -I_{N-1} \otimes \lambda I
\end{bmatrix} < 0, \tag{14}
$$

where

$$
\begin{aligned}
\Pi_{11} &= I_{N-1} \otimes (A^{\mathrm{T}} P + PA + \mu P + \tau^\alpha \alpha^{-1} I + 2\lambda_{max}(L) Q_1 + l^2 I + P^2) \\
\Pi_{12} &= I_{N-1} \otimes PA_\tau - \tilde{L} \otimes PBK, \\
\Pi_{13} &= \tau^\alpha \alpha^{-1} I_{N-1} \otimes A^{\mathrm{T}}, \Pi_{22} = -I_{N-1} \otimes P + \lambda_{max}^2(L) K^{\mathrm{T}} Q_2 K, \\
\Pi_{23} &= \tau^\alpha \alpha^{-1} (I_{N-1} \otimes A_\tau^{\mathrm{T}} - \tilde{L}^{\mathrm{T}} \otimes K^{\mathrm{T}} B^{\mathrm{T}}), \\
\Pi_{33} &= -\tau^\alpha \alpha^{-1} I_{N-1} \otimes I, \Pi_{35} = \tau^\alpha \alpha^{-1} I_{N-1} \otimes \lambda E.
\end{aligned}
$$

Usingt the Schur complement theorem once again yields the following:

$$
\begin{bmatrix}
\Omega_{11} & \Omega_{12} & \Omega_{13} & \Omega_{14} & \Omega_{15} & \Omega_{16} & \Omega_{17} & \Omega_{18} & 0 \\
* & \Omega_{22} & \Omega_{23} & \Omega_{24} & 0 & 0 & 0 & 0 & \Omega_{29} \\
* & * & \Omega_{33} & 0 & \Omega_{35} & 0 & 0 & 0 & 0 \\
* & * & * & \Omega_{44} & 0 & 0 & 0 & 0 & 0 \\
* & * & * & * & \Omega_{55} & 0 & 0 & 0 & 0 \\
* & * & * & * & * & \Omega_{66} & 0 & 0 & 0 \\
* & * & * & * & * & * & \Omega_{77} & 0 & 0 \\
* & * & * & * & * & * & * & \Omega_{88} & 0 \\
* & * & * & * & * & * & * & * & \Omega_{99}
\end{bmatrix} < 0,
$$

where the following is the case.

$$
\begin{aligned}
\Omega_{11} &= I_{N-1} \otimes (A^\mathrm{T} P + PA + \mu P + \tau^\alpha \alpha^{-1} I + 2\lambda_{max}(L)Q_1 + l^2 I + P^2), \\
\Omega_{12} &= I_{N-1} \otimes PA_\tau - \tilde{L} \otimes PBK, \\
\Omega_{13} &= \tau^\alpha \alpha^{-1} I_{N-1} \otimes A^\mathrm{T}, \Omega_{14} = I_{N-1} \otimes F_1^\mathrm{T}, \\
\Omega_{15} &= I_{N-1} \otimes \lambda PE, \Omega_{16} = I_{N-1} \otimes l^2 I, \\
\Omega_{17} &= I_{N-1} \otimes \tau^\alpha \alpha^{-1} I, \Omega_{18} = 2\lambda_{max}(L) I_{N-1} \otimes Q_1, \\
\Omega_{22} &= -I_{N-1} \otimes P, \Omega_{23} = \tau^\alpha \alpha^{-1}(I_{N-1} \otimes A_\tau^\mathrm{T} - \tilde{L}^\mathrm{T} \otimes K^\mathrm{T} B^\mathrm{T}), \\
\Omega_{24} &= I_{N-1} \otimes F_2^\mathrm{T} - \tilde{L}^\mathrm{T} \otimes K^\mathrm{T} F_3^\mathrm{T}, \\
\Omega_{29} &= \lambda_{max}^2(H) I_{N-1} \otimes K^\mathrm{T} Q_2, \\
\Omega_{33} &= \Omega_{77} = -I_{N-1} \otimes \tau^\alpha \alpha^{-1} I, \Omega_{35} = \tau^\alpha \alpha^{-1} I_{N-1} \otimes \lambda E, \\
\Omega_{44} &= \Omega_{55} = -I_{N-1} \otimes \lambda I, \Omega_{66} = -I_{N-1} \otimes l^2 I, \\
\Omega_{88} &= -2\lambda_{max}(L) I_{N-1} \otimes Q_1, \\
\Omega_{99} &= -\lambda_{max}^2(L) I_{N-1} \otimes Q_2.
\end{aligned}
$$

By multiplying both sides of the previous equation by the diagonal matrix $\{I_{N-1} \otimes P^{-1}, I_{N-1} \otimes P^{-1}, I_{N-1} \otimes I, I_{N-1} \otimes I, I_{N-1} \otimes I, I_{N-1} \otimes I, I_{N-1} \otimes I, I_{N-1} \otimes I, I_{N-1} \otimes I\}$, we yield the following:

$$
\Xi =
\begin{bmatrix}
\Xi_{11} & \Xi_{12} & \Xi_{13} & \Xi_{14} & \Xi_{15} & \Xi_{16} & \Xi_{17} & \Xi_{18} & 0 \\
* & \Xi_{22} & \Xi_{23} & \Xi_{24} & 0 & 0 & 0 & 0 & \Xi_{29} \\
* & * & \Xi_{33} & 0 & \Xi_{35} & 0 & 0 & 0 & 0 \\
* & * & * & \Xi_{44} & 0 & 0 & 0 & 0 & 0 \\
* & * & * & * & \Xi_{55} & 0 & 0 & 0 & 0 \\
* & * & * & * & * & \Xi_{66} & 0 & 0 & 0 \\
* & * & * & * & * & * & \Xi_{77} & 0 & 0 \\
* & * & * & * & * & * & * & \Xi_{88} & 0 \\
* & * & * & * & * & * & * & * & \Xi_{99}
\end{bmatrix} < 0,
\tag{15}
$$

where the following is the case.

$$
\begin{aligned}
\Xi_{11} &= I_{N-1} \otimes (P^{-1} A^\mathrm{T} + AP^{-1} + \mu P^{-1} + I), \\
\Xi_{12} &= I_{N-1} \otimes A_\tau P^{-1} - \tilde{L} \otimes BKP^{-1}, \\
\Xi_{13} &= \tau^\alpha \alpha^{-1} I_{N-1} \otimes P^{-1} A^\mathrm{T}, \Xi_{14} = I_{N-1} \otimes P^{-1} F_1^\mathrm{T}, \\
\Xi_{15} &= I_{N-1} \otimes \lambda E, \Xi_{16} = I_{N-1} \otimes l^2 P^{-1}, \\
\Xi_{17} &= I_{N-1} \otimes \tau^\alpha \alpha^{-1} P^{-1}, \Xi_{18} = 2\lambda_{max}(L) I_{N-1} \otimes P^{-1} Q_1, \\
\Xi_{22} &= -I_{N-1} \otimes P^{-1}, \\
\Xi_{23} &= \tau^\alpha \alpha^{-1}(I_{N-1} \otimes P^{-1} A_\tau^\mathrm{T} - \tilde{L}^\mathrm{T} \otimes P^{-1} K^\mathrm{T} B^\mathrm{T}), \\
\Xi_{24} &= I_{N-1} \otimes P^{-1} F_2^\mathrm{T} - \tilde{L}^\mathrm{T} \otimes P^{-1} K^\mathrm{T} F_3^\mathrm{T}, \\
\Xi_{29} &= \lambda_{max}^2(L) I_{N-1} \otimes P^{-1} K^\mathrm{T} Q_2, \\
\Xi_{33} &= \Omega_{77} = -I_{N-1} \otimes \tau^\alpha \alpha^{-1} I, \Omega_{35} = \tau^\alpha \alpha^{-1} I_{N-1} \otimes \lambda E, \\
\Xi_{44} &= \Omega_{55} = -I_{N-1} \otimes \lambda I, \Xi_{66} = -I_{N-1} \otimes l^2 I, \\
\Xi_{88} &= -2\lambda_{max}(L) I_N \otimes Q_1, \\
\Xi_{99} &= -\lambda_{max}^2(L) I_N \otimes Q_2.
\end{aligned}
$$

Let $\bar{P} = P^{-1}, Y = KP^{-1}$ and $\mu \to 1$. Then, Expression (15) can be described as (7), and one can obtain the following.

$$
\begin{aligned}
{}_{t_0}^{C}D_t^\alpha V(t) \leq{}& \eta^{\mathrm{T}}(t)\Delta\eta(t) - e^{\mathrm{T}}(t)(2\lambda_{max}(L)I_{N-1} \otimes Q_1)e(t) \\
& - e^{\mathrm{T}}(t)(\lambda_{max}^2(L)I_{N-1} \otimes K^{\mathrm{T}}Q_2K)e(t) \\
& - \int_{t-\tau(t)}^{t}(t-s)^{\alpha-1}\zeta^{\mathrm{T}}(t)\Omega\zeta(t)ds < 0.
\end{aligned}
\tag{16}
$$

It follows from the Razumikhin theorem [34] that the error system (3) is asymptotically stable. According to Definition 1, the consensus of the original system (1) can be achieved. Furthermore, from Definition 2, the upper bound of the guaranteed cost function can be found as follows.

$$
\begin{aligned}
J(t) ={}& \frac{1}{\Gamma(1-\alpha)}\int_0^t (t-s)^\alpha (e^{\mathrm{T}}(s)(2L \otimes Q_1)e(s) \\
& + e^{\mathrm{T}}(s-\tau)(L^{\mathrm{T}}L \otimes K^{\mathrm{T}}Q_2K)e(s-\tau))ds.
\end{aligned}
$$

From (16), we obtain the following.

$$
\begin{aligned}
{}_{t_0}^{C}D_t^\alpha V(t) \leq{}& -2\lambda_{max}(L)e^{\mathrm{T}}(t)(I_N \otimes Q_1)e(t) - \lambda_{max}(L)^2 e^{\mathrm{T}}(t-\tau)(I_N \otimes K^{\mathrm{T}}Q_2K)e(t-\tau) \\
\leq{}& -\bar{e}^{\mathrm{T}}(t)(2L \otimes Q_1)e(t) - \bar{e}^{\mathrm{T}}(t-\tau)(L^{\mathrm{T}}L \otimes K^{\mathrm{T}}Q_2K)\bar{e}(t-\tau).
\end{aligned}
\tag{17}
$$

By applying integration of order α on both sides of (17) and considering $V(t) > 0$, one obtains the following.

$$
\begin{aligned}
J \leq{}& V(0) - V(t) \leq V(0) = e(0)^{\mathrm{T}}(I_{N-1} \otimes P)e(0) \\
\leq{}& \lambda_{max}(I_{N-1} \otimes P)\|e(0)\|^2 = J^*.
\end{aligned}
$$

This ends the proof. $\quad\square$

Remark 3. *It can be noted that the consensus condition obtained in this paper is delay-and order-dependent for FOMAS. It is obvious that the consensus conditions proposed in [12,14–19,22,35] do not apply herein.*

Remark 4. *The stability of the MASs including FOMASs is the primary requirement for designing a control protocol. Moreover, it is also desirable that the control system can not only preserve stability but also guarantee an adequate level of performance. Since each agent may only have limited energy supplies to carry out some tasks, such as perception, communication, and movement, energy consumption is a real problem that should be considered critically. The guaranteed cost control method has been proved capable of meeting both requirements.*

Remark 5. *Both MASs and FOMASs are usually implemented by large-scale integrated circuits. Thus, signal propagation inevitably introduces time delays, which can result in oscillation, chaos and even instability phenomena. References [12,13,18,36,37] investigated the consensus of FOMASs considering merely input delays, that is, without addressing state delays. Herein, we consider both state-delays and input-delays.*

Remark 6. *When $f(x_i(t)) = 0$, the following corollary of Theorem 1 can be obtained.*

Corollary 1. *For known positive definite symmetric matrices Q_1 and Q_2, if there exist a symmetric positive definite matrix $\bar{P}$, a matrix Y and constant positive scalars λ such that the following is the case:*

$$\Delta = \begin{bmatrix} \Delta_{11} & \Delta_{12} & \Delta_{13} & \Delta_{14} & \Delta_{15} & \Delta_{16} & \Delta_{17} & 0 \\ * & \Delta_{22} & \Delta_{23} & \Delta_{24} & 0 & 0 & 0 & \Delta_{28} \\ * & * & \Delta_{33} & 0 & \Delta_{35} & 0 & 0 & 0 \\ * & * & * & \Delta_{44} & 0 & 0 & 0 & 0 \\ * & * & * & * & \Delta_{55} & 0 & 0 & 0 \\ * & * & * & * & * & \Delta_{66} & 0 & 0 \\ * & * & * & * & * & * & \Delta_{77} & 0 \\ * & * & * & * & * & * & * & \Delta_{88} \end{bmatrix} < 0, \tag{18}$$

where

$$\begin{aligned} \Delta_{11} &= I_{N-1} \otimes (\bar{P}A^{\mathrm{T}} + A\bar{P} + \mu\bar{P}), \\ \Delta_{12} &= I_{N-1} \otimes A_\tau \bar{P} - \tilde{L} \otimes BY, \\ \Delta_{13} &= \tau^\alpha \alpha^{-1} I_{N-1} \otimes \bar{P}A^{\mathrm{T}}, \Delta_{14} = I_{N-1} \otimes \bar{P}F_1^{\mathrm{T}}, \\ \Delta_{15} &= I_{N-1} \otimes \lambda E, \Delta_{16} = I_{N-1} \otimes \tau^\alpha \alpha^{-1}\bar{P}, \\ \Delta_{17} &= 2\lambda_{max}(L)I_{N-1} \otimes \bar{P}Q_1, \Delta_{22} = I_{N-1} \otimes \bar{P}, \\ \Delta_{23} &= \tau^\alpha \alpha^{-1}(I_{N-1} \otimes \bar{P}A_\tau^{\mathrm{T}} - \tilde{L}^{\mathrm{T}} \otimes Y^{\mathrm{T}}B^{\mathrm{T}}), \\ \Delta_{24} &= I_{N-1} \otimes \bar{P}F_2^{\mathrm{T}} - \tilde{L}^{\mathrm{T}} \otimes Y^{\mathrm{T}}F_3^{\mathrm{T}}, \Delta_{28} = \lambda_{max}^2(L)I_{N-1} \otimes Y^{\mathrm{T}}Q_2, \\ \Delta_{33} &= \Omega_{66} = -I_{N-1} \otimes \tau^\alpha \alpha^{-1}I, \Omega_{35} = \tau^\alpha \alpha^{-1}I_{N-1} \otimes \lambda E, \\ \Delta_{44} &= \Omega_{55} = -I_{N-1} \otimes \lambda I, \Delta_{77} = -2\lambda_{max}(L)I_{N-1} \otimes Q_1, \\ \Delta_{88} &= -\lambda_{max}^2(L)I_{N-1} \otimes Q_2, \end{aligned}$$

then the consensus of the FOMASs (2) with control protocol (3) is achieved. Moreover, the feedback gain K is given by the following:

$$K = Y\bar{P}^{-1},$$

and the guaranteed cost is stated as follows.

$$J^* = \lambda_{max}(I_{N-1} \otimes P)\|e(0)\|^2.$$

Proof. The proof is similar to that of Theorem 1, so we omit it herein. □

4. Numerical Simulations

In this section two examples are presented to verify the applicability and effectiveness of the scheme proposed.

Example 1. *Consider the undirected graph topology depicted in Figure 1. The matrices L and $\tilde{L}$ are given by the following:*

$$L = \begin{bmatrix} 1 & -1 & 0 & 0 \\ -1 & 3 & -1 & -1 \\ 0 & -1 & 2 & -1 \\ 0 & -1 & -1 & 2 \end{bmatrix}, \tilde{L} = \begin{bmatrix} 4 & -1 & -1 \\ 0 & 2 & -1 \\ 0 & -1 & 2 \end{bmatrix}$$

and the parameters of each multi-agent are stated as follows.

$$A \;=\; \begin{bmatrix} -3 & -2 \\ 1 & -4 \end{bmatrix}, A_\tau = \begin{bmatrix} 0.2 & 0.1 \\ -0.1 & 0 \end{bmatrix}, B = \begin{bmatrix} 0.2 \\ -0.2 \end{bmatrix},$$

$$E_1 \;=\; \begin{bmatrix} 0.1 \\ 0.1 \end{bmatrix}, F_1 = \begin{bmatrix} 0.2 & -0.3 \end{bmatrix}, F_2 = \begin{bmatrix} -0.1 & 0.2 \end{bmatrix},$$

$$\alpha \;=\; 0.9, \tau = 0.1, F_3 = -0.1, f(x(t)) = \sin(x(t)).$$

It follows from Theorem 1 that the matrix $\bar{P}$, constant λ and gain matrix K can be obtained as the following.

$$\bar{P} \;=\; \begin{bmatrix} 0.4441 & -0.0221 \\ -0.0221 & 0.3075 \end{bmatrix}, \lambda = 1.1251,$$

$$K = Y\bar{P}^{-1} = \begin{bmatrix} 0.0556 & 0.0003 \end{bmatrix}.$$

Let us choose $h(t) = \cos(t)$ and the initial states $x_1(0) = [1,2]^T$, $x_2(0) = [3,0]^T$, $x_3(0) = [5,1]^T$ and $x_4(0) = [4,4]^T$. For different orders α, we also carried out the simulations and gave the corresponding error trajectories of this system. When order $\alpha = 0.9, 0.8, 0.7$, we show the consensus errors versus time of the agents in Figures 2–7, respectively. From the numerical results, we verify that $e_{1i}(t)$ and $e_{2i}(t)$ tend fast to 0, which means that the guaranteed cost consensus of the system (1) can be achieved. The upper bound of the guaranteed cost function $J^* = 17.0088(\alpha = 0.9)$ can be obtained. Moreover, by denoting $\|e(t)\| = \sqrt{\sum_i^n e_{ij}^2(t)}\,(j = 1,2,\cdots,n)$, we also carried out the curve of $\|e(t)\|$. From Figure 8, we could note that the order influence on consensus property with varying orders. With the higher orders, the system consensus's error will be achieved more rapidly.

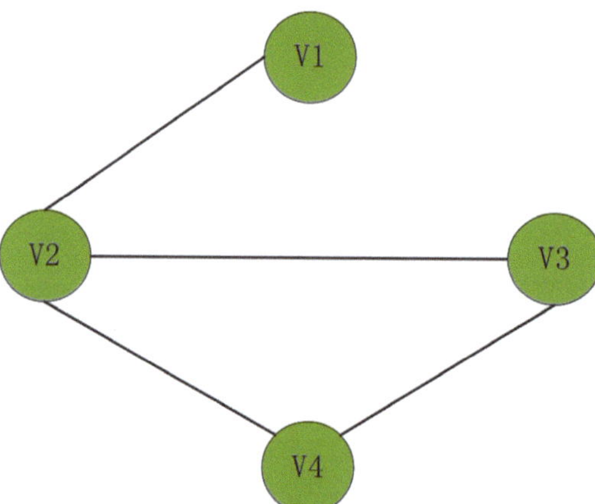

Figure 1. Topology of the system in Example 1.

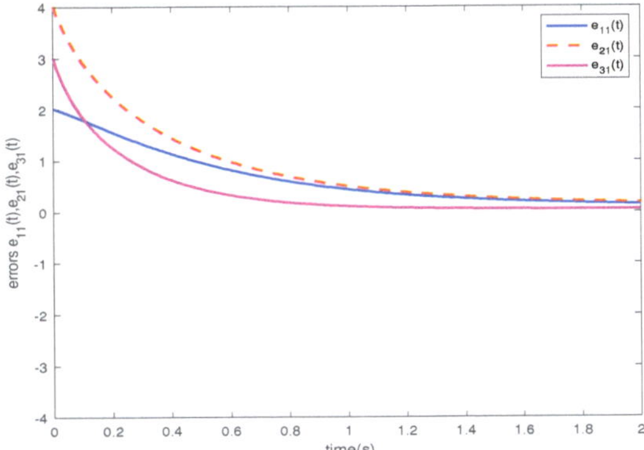

Figure 2. Error response $e_{i1}(t)$ of the system in Example 1 with $\alpha = 0.9$.

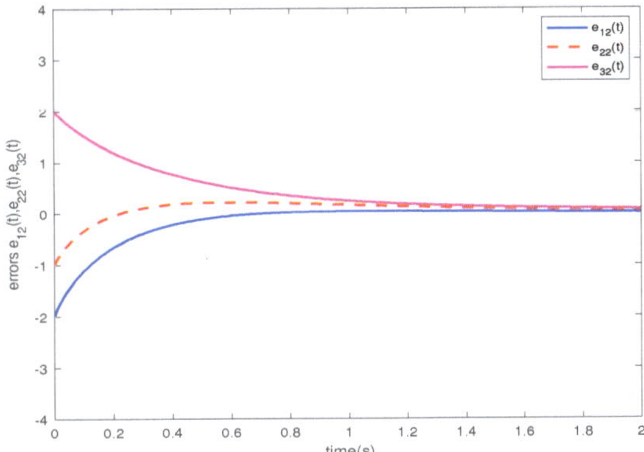

Figure 3. Error response $e_{i2}(t)$ of the system in Example 1 with $\alpha = 0.9$.

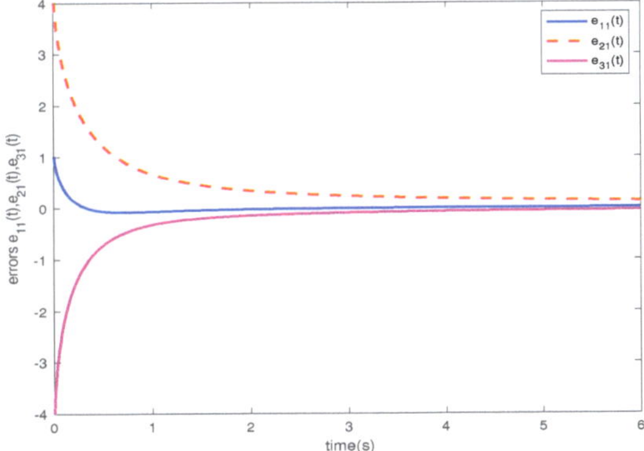

Figure 4. Error response $e_{i1}(t)$ of the system in Example 1 with $\alpha = 0.8$.

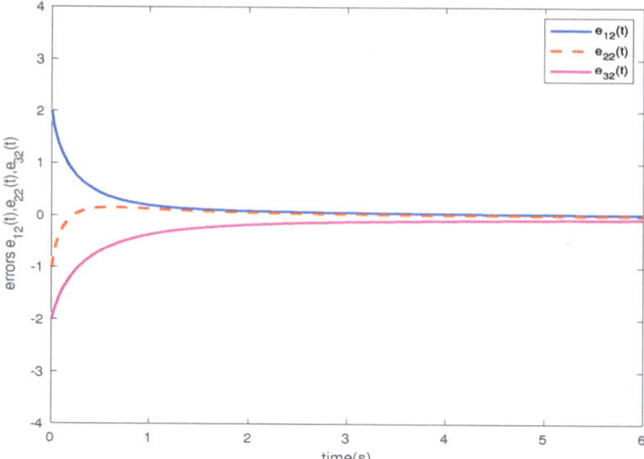

Figure 5. Error response $e_{i2}(t)$ of the system in Example 1 with $\alpha = 0.8$.

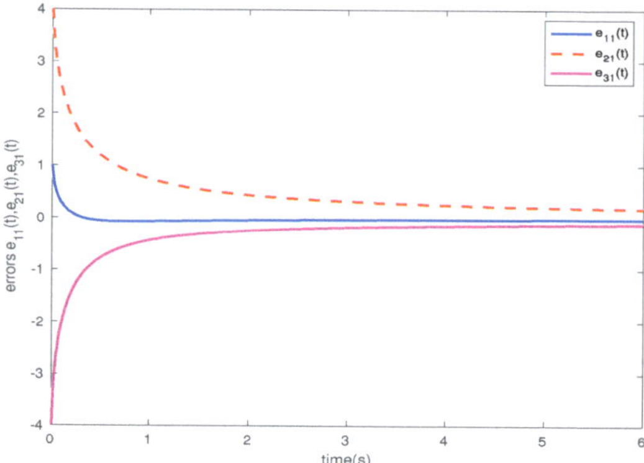

Figure 6. Error response $e_{i1}(t)$ of the system in Example 1 with $\alpha = 0.7$.

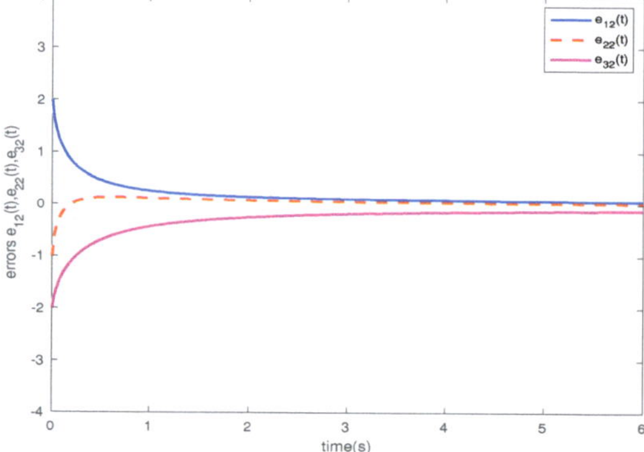

Figure 7. Error response $e_{i2}(t)$ of the system in Example 1 with $\alpha = 0.7$.

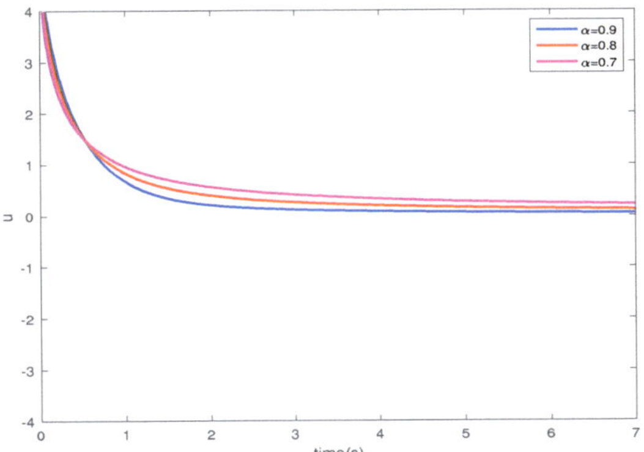

Figure 8. Error response $\|e(t)\|$ of the system in Example 1.

Example 2. *Consider the undirected graph represented in Figure 9. The matrices L and $\tilde{L}$ are the following:*

$$
L = \begin{bmatrix}
3 & -1 & -1 & -1 & 0 & 0 & 0 & 0 & 0 & 0 \\
-1 & 3 & -1 & 0 & 0 & 0 & -1 & 0 & 0 & 0 \\
-1 & -1 & 5 & -1 & 0 & -1 & 0 & -1 & 0 & 0 \\
-1 & 0 & -1 & 3 & -1 & 0 & 0 & 0 & 0 & 0 \\
0 & 0 & 0 & -1 & 3 & -1 & 0 & 0 & 0 & -1 \\
0 & 0 & -1 & 0 & -1 & 4 & 0 & 0 & -1 & -1 \\
0 & -1 & 0 & 0 & 0 & 0 & 2 & -1 & 0 & 0 \\
0 & 0 & -1 & 0 & 0 & 0 & -1 & 3 & -1 & 0 \\
0 & 0 & 0 & 0 & 0 & -1 & 0 & -1 & 3 & -1 \\
0 & 0 & 0 & 0 & -1 & -1 & 0 & 0 & -1 & 3
\end{bmatrix},
$$

$$
\tilde{L} = \begin{bmatrix}
2 & -1 & 0 & 0 & 0 & 1 & 0 & 0 & 0 \\
0 & 4 & 1 & 0 & 1 & 0 & 1 & 0 & 0 \\
0 & 0 & 2 & 1 & 0 & 0 & 0 & 0 & 0 \\
-1 & -1 & 0 & 3 & 1 & 0 & 0 & 0 & 1 \\
-1 & 0 & -1 & 1 & 4 & 0 & 0 & 1 & 1 \\
0 & -1 & -1 & 0 & 0 & 2 & 1 & 0 & 0 \\
-1 & 0 & -1 & 0 & 0 & 1 & 3 & 1 & 0 \\
-1 & -1 & -1 & 0 & 1 & 0 & 1 & 3 & 1 \\
-1 & -1 & -1 & 1 & 1 & 0 & 0 & 1 & 3
\end{bmatrix}
$$

and the parameters of every agent are given by the following.

$$
A = \begin{bmatrix} -7 & -5 \\ -1 & -6 \end{bmatrix}, A_\tau = \begin{bmatrix} 0.1 & 0.3 \\ -0.2 & 0.1 \end{bmatrix}, B = \begin{bmatrix} 0.1 \\ -0.4 \end{bmatrix},
$$

$$
E_1 = \begin{bmatrix} 0.4 \\ 0.1 \end{bmatrix}, F_1 = \begin{bmatrix} 0.1 & -0.2 \end{bmatrix}, F_2 = \begin{bmatrix} 0.3 & -0.1 \end{bmatrix},
$$

$$
\alpha = 0.8, \tau = 0.1, F_3 = -0.4.
$$

From Corollary 1, the matrix $\bar{P}$, constant λ and gain matrix K can be obtained as follows.

$$
\bar{P} = \begin{bmatrix} 0.1318 & 0.0187 \\ 0.0187 & 0.0284 \end{bmatrix}, \lambda = 0.8642
$$

$$K = Y \bar{P}^{-1} = \begin{bmatrix} 0.0918 & 0.0062 \end{bmatrix}.$$

Let us select $h(t) = \sin(t)$ and the initial states $x_1(0) = [1,0]^T$, $x_2(0) = [2,2]^T$, $x_3(0) = [5,-1]^T$ and $x_4(0) = [-3,-2]^T$. Similar with Example 1, we also carried out consensus error curve of system with different order. The simulation results are shown in the Figures 10–15, respectively. From these numerical results, we verify that $e_{1i}(t)$ and $e_{2i}(t)$ approach 0 very fast, meaning that the guaranteed cost consensus of the system (2) is obtained. Additionally, the upper bound of the guaranteed cost function is $J^* = 0.5502 (\alpha = 0.8)$. By plotting the curve of $\|e(t)\|$ shown in Figure 16, one can note that the system will achieve consensus more rapidly when the order increases.

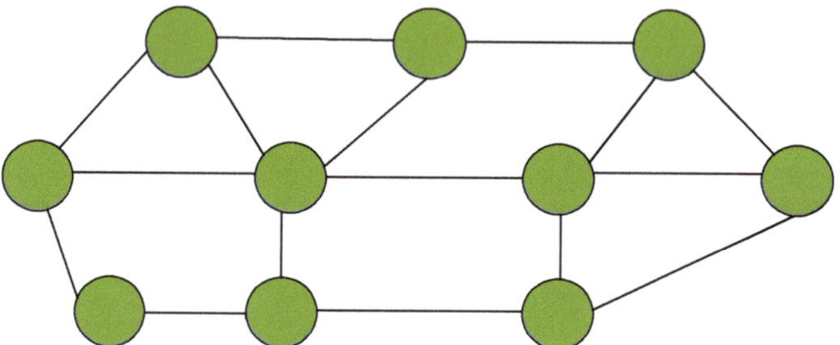

Figure 9. Topology of the system in Example 2.

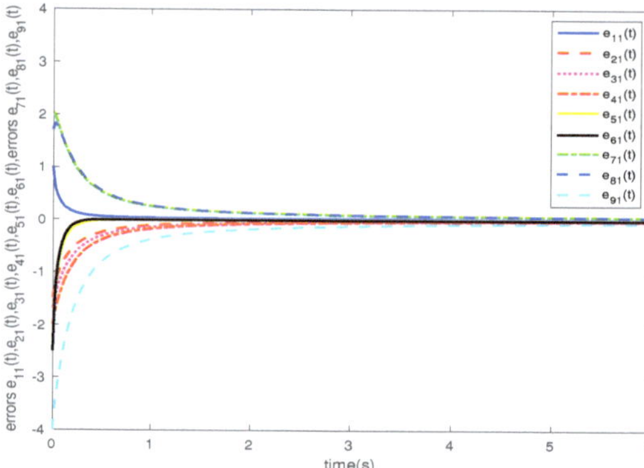

Figure 10. Error response $e_{i1}(t)$ of the system in Example 2 with $\alpha = 0.8$.

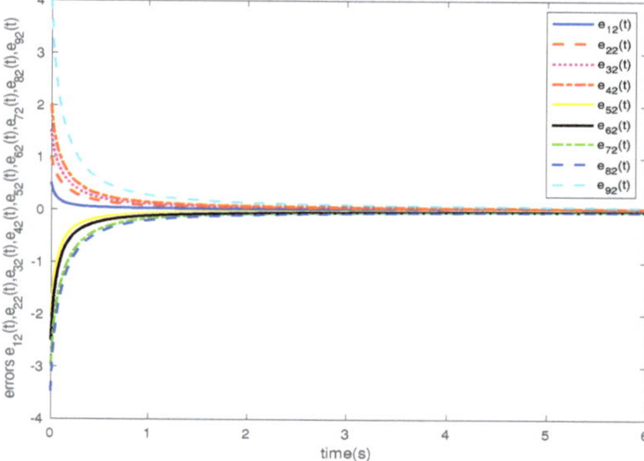

Figure 11. Error response $e_{i2}(t)$ of the system in Example 2 with $\alpha = 0.8$.

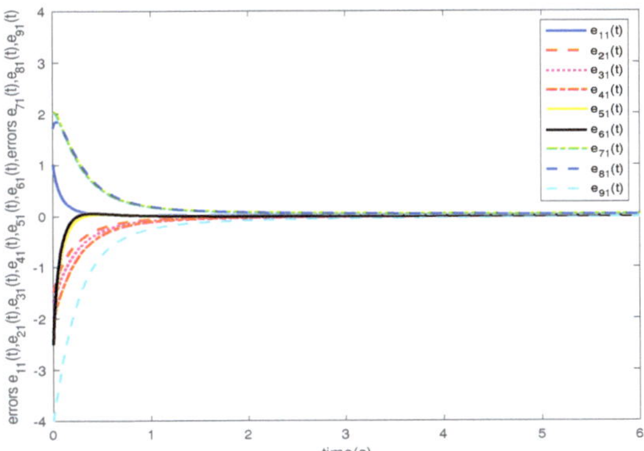

Figure 12. Error response $e_{i1}(t)$ of the system in Example 2 with $\alpha = 0.9$.

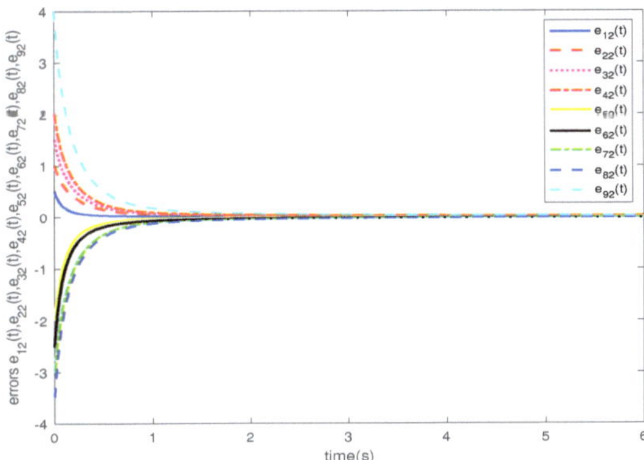

Figure 13. Error response $e_{i2}(t)$ of the system in Example 2 with $\alpha = 0.9$.

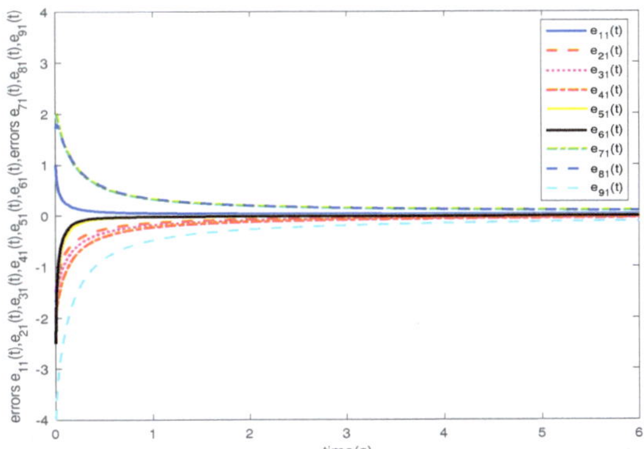

Figure 14. Error response $e_{i1}(t)$ of the system in Example 2 with $\alpha = 0.7$.

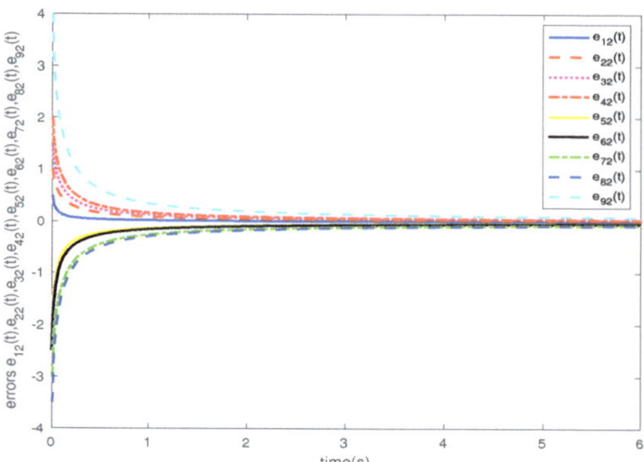

Figure 15. Error response $e_{i2}(t)$ of the system in Example 2 with $\alpha = 0.7$.

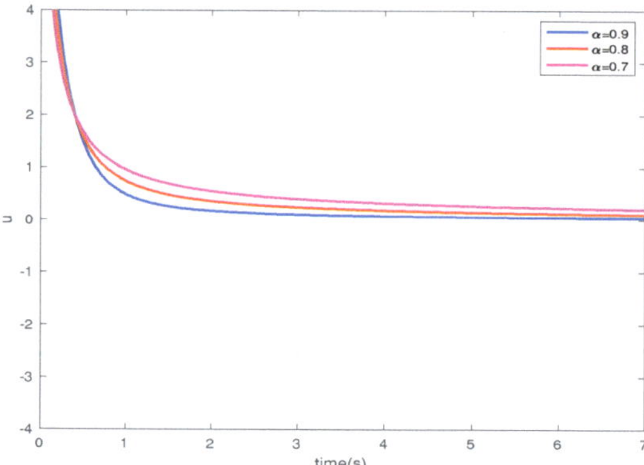

Figure 16. Error response $\|e(t)\|$ of the system in Example 2.

5. Conclusions

The guaranteed cost consensus of uncertain delayed FOMASs with input delay was addressed in this paper. A guaranteed cost function related to FOMASs was proposed in order to consider the consensus regulation performance and the energy consumption simultaneously. By employing the FO Razumikhin theorem and the LMI approach, sufficient conditions on guaranteed cost and upper bounds for the guaranteed cost function were obtained. The proposed approach is order-dependent and delay-dependent, which results in less conservative conditions than those presented in alternative methods. It should be mentioned that taking the state and input delay as identical is unreasonable in real-world applications. However, since stability results for fractional-order systems with multiple time delays are unavailable, we considered this simplified case. Therefore, further work is needed to solve this problem. In addition, we will consider the guaranteed cost consensus of uncertain delayed FOMASs with order lying in (1,2) in our next work.

Author Contributions: Methodology, Y.T.; software, L.C.; validation, Q.X.; writing—original draft preparation, Y.T.; writing—review and editing, A.M.L., L.C. and Y.C.; supervision and funding, Y.C. All authors have read and agreed to the published version of the manuscript.

Funding: This research was funded by the National Natural Science Funds of China (Nos, 61633005; 62073114; 11971032) and National Key RD Program of China (No. 2020YFB2009400).

Institutional Review Board Statement: Not applicable.

Informed Consent Statement: Not applicable.

Data Availability Statement: Not applicable.

Conflicts of Interest: The authors declare no conflicts of interest.

References

1. Yu, W.; Chen, G.; Wang, Z.; Yang, W. Distributed consensus filtering in sensor networks. *IEEE Trans. Syst. Man Cybern. Part B (Cybern.)* **2009**, *39*, 1568–1577.
2. Carpenter, J.R. A preliminary investigation of decentralized control far satellite formations. In Proceedings of the 2000 IEEE Aerospace Conference. Proceedings (Cat. No. 00TH8484), Big Sky, MT, USA, 25 March 2000; IEEE: Piscataway, NJ, USA, 2000; Volume 7, pp. 63–74.
3. McLain, T.W.; Chandler, P.R.; Rasmussen, S.; Pachter, M. Cooperative control of UAV rendezvous. In Proceedings of the 2001 American Control Conference, (Cat. No. 01CH37148), Arlington, VA, USA, 25–27 June 2001; IEEE: Piscataway, NJ, USA, 2001; Volume 3, pp. 2309–2314.
4. Hong, Y.; Hu, J.; Gao, L. Tracking control for multi-agent consensus with an active leader and variable topology. *Automatica* **2006**, *42*, 1177–1182. [CrossRef]
5. Tang, Y.; Gao, H.; Zhang, W.; Kurths, J. Leader-following consensus of a class of stochastic delayed multi-agent systems with partial mixed impulses. *Automatica* **2015**, *53*, 346–354. [CrossRef]
6. Kim, H.; Shim, H.; Seo, J.H. Output consensus of heterogeneous uncertain linear multi-agent systems. *IEEE Trans. Autom. Control* **2010**, *56*, 200–206. [CrossRef]
7. Zhu, W.; Jiang, Z.P.; Feng, G. Event-based consensus of multi-agent systems with general linear models. *Automatica* **2014**, *50*, 552–558. [CrossRef]
8. Zheng, Y.; Ma, J.; Wang, L. Consensus of hybrid multi-agent systems. *IEEE Trans. Neural Netw. Learn. Syst.* **2017**, *29*, 1359–1365. [CrossRef]
9. Yang, H.y.; Zhu, X.l.; Cao, K.c. Distributed coordination of fractional order multi-agent systems with communication delays. *Fract. Calc. Appl. Anal.* **2014**, *17*, 23–37. [CrossRef]
10. Peng, Z.; Wang, D.; Wang, H.; Wang, W. Distributed coordinated tracking of multiple autonomous underwater vehicles. *Nonlinear Dyn.* **2014**, *78*, 1261–1276. [CrossRef]
11. Chen, L.; Yin, H.; Huang, T.; Yuan, L.; Zheng, S.; Yin, L. Chaos in fractional-order discrete neural networks with application to image encryption. *Neural Netw.* **2020**, *125*, 174–184. [CrossRef]
12. Yang, R.; Liu, S.; Tan, Y.Y.; Zhang, Y.J.; Jiang, W. Consensus analysis of fractional-order nonlinear multi-agent systems with distributed and input delays. *Neurocomputing* **2019**, *329*, 46–52. [CrossRef]
13. Shi, M.; Yu, Y.; Xu, Q. Delay-dependent consensus condition for a class of fractional-order linear multi-agent systems with input time-delay. *Int. J. Syst. Sci.* **2019**, *50*, 669–678. [CrossRef]
14. Ren, G.; Yu, Y.; Xu, C.; Hai, X. Consensus of fractional multi-agent systems by distributed event-triggered strategy. *Nonlinear Dyn.* **2019**, *95*, 541–555. [CrossRef]
15. Hu, T.; He, Z.; Zhang, X.; Zhong, S. Leader-following consensus of fractional-order multi-agent systems based on event-triggered control. *Nonlinear Dyn.* **2020**, *99*, 2219–2232. [CrossRef]
16. Gong, P.; Wang, K. Output feedback consensus control for fractional-order nonlinear multi-agent systems with directed topologies. *J. Frankl. Inst.* **2020**, *357*, 1473–1493. [CrossRef]
17. Gong, P.; Wang, K.; Lan, W. Fully distributed robust consensus control of multi-agent systems with heterogeneous unknown fractional-order dynamics. *Int. J. Syst. Sci.* **2019**, *50*, 1902–1919. [CrossRef]
18. Hu, W.; Wen, G.; Rahmani, A.; Bai, J.; Yu, Y. Leader-following consensus of heterogenous fractional-order multi-agent systems under input delays. *Asian J. Control* **2020**, *22*, 2217–2228. [CrossRef]
19. Ren, G.; Yu, Y. Consensus of fractional multi-agent systems using distributed adaptive protocols. *Asian J. Control* **2017**, *19*, 2076–2084. [CrossRef]
20. Wyrwas, M.; Mozyrska, D.; Girejko, E. Fractional discrete-time consensus models for single-and double-summator dynamics. *Int. J. Syst. Sci.* **2018**, *49*, 1212–1225. [CrossRef]
21. Yaghoubi, Z.; Talebi, H.A. Consensus tracking for nonlinear fractional-order multi-agent systems using adaptive sliding mode controller. *Mechatron. Syst. Control (Former Control Intell. Syst.)* **2019**, *47*, 194–200. [CrossRef]
22. Wei, Z.; Bo, C. Leader-following consensus of fractional-order multi-agent systems with nonlinear models. *Appl. Math. Mech.* **2015**, *36*, 555–562.
23. Luo, Y.; Xiao, X.; Cao, J.; Li, A. Event-triggered guaranteed cost consensus for uncertain nonlinear multi-agent systems with time delay. *Neurocomputing* **2020**, *394*, 13–26. [CrossRef]
24. Wang, Z.; Fan, Z.; Liu, G. Guaranteed performance consensus problems for nonlinear multi-agent systems with directed topologies. *Int. J. Control* **2019**, *92*, 2952–2962. [CrossRef]
25. Wang, Z.; Xi, J.; Yao, Z.; Liu, G. Guaranteed cost consensus for multi-agent systems with fixed topologies. *Asian J. Control* **2015**, *17*, 729–735. [CrossRef]
26. Wang, Z.; Xi, J.; Yu, Z.; Liu, G. Guaranteed cost consensus for multi-agent systems with time delays. *J. Frankl. Inst.* **2015**, *352*, 3612–3627. [CrossRef]

27. Liu, J.; Zhang, Y.; Sun, C.; Yu, Y. Fixed-time consensus of multi-agent systems with input delay and uncertain disturbances via event-triggered control. *Inf. Sci.* **2019**, *480*, 261–272. [CrossRef]
28. Xie, L. Output feedback H_∞ control of systems with parameter uncertainty. *Int. J. Control* **1996**, *63*, 741–750. [CrossRef]
29. Liang, S.; Wu, R.; Chen, L. Comparison principles and stability of nonlinear fractional-order cellular neural networks with multiple time delays. *Neurocomputing* **2015**, *168*, 618–625. [CrossRef]
30. Ni, W.; Cheng, D. Leader-following consensus of multi-agent systems under fixed and switching topologies. *Syst. Control Lett.* **2010**, *59*, 209–217. [CrossRef]
31. Chen, L.; Li, T.; Wu, R.; Lopes, A.M.; Machado, J.T.; Wu, K. Output-feedback guaranteed-cost control of fractional-order uncertain linear delayed systems. *Comput. Appl. Math.* **2020**, *39*, 210. [CrossRef]
32. Shen, B.; Wang, Z.; Tan, H. Guaranteed cost control for uncertain nonlinear systems with mixed time-delays: The discrete-time case. *Eur. J. Control* **2018**, *40*, 62–67. [CrossRef]
33. Xi, J.; Fan, Z.; Liu, H.; Zheng, T. Guaranteed-cost consensus for multiagent networks with Lipschitz nonlinear dynamics and switching topologies. *Int. J. Robust Nonlinear Control* **2018**, *28*, 2841–2852. [CrossRef]
34. Chen, B.; Chen, J. Razumikhin-type stability theorems for functional fractional-order differential systems and applications. *Appl. Math. Comput.* **2015**, *254*, 63–69. [CrossRef]
35. Chen, Y.; Wen, G.; Peng, Z.; Huang, T.; Yu, Y. Necessary and sufficient conditions for group consensus of fractional multiagent systems under fixed and switching topologies via pinning control. *IEEE Trans. Cybern.* **2019**, *51*, 28–39. [CrossRef] [PubMed]
36. Zhu, W.; Chen, B.; Yang, J. Consensus of fractional-order multi-agent systems with input time delay. *Fract. Calc. Appl. Anal.* **2017**, *20*, 52. [CrossRef]
37. Wang, X.; Li, X.; Huang, N.; O'Regan, D. Asymptotical consensus of fractional-order multi-agent systems with current and delay states. *Appl. Math. Mech.* **2019**, *40*, 1677–1694. [CrossRef]

fractal and fractional

Article

Jacobi Spectral Collocation Technique for Time-Fractional Inverse Heat Equations

Mohamed A. Abdelkawy [1,2,*], **Ahmed Z. M. Amin** [3], **Mohammed M. Babatin** [1], **Abeer S. Alnahdi** [1], **Mahmoud A. Zaky** [4] and **Ramy M. Hafez** [5]

1   Department of Mathematics and Statistics, College of Science, Imam Mohammad Ibn Saud Islamic University, Riyadh 13318, Saudi Arabia; mmbabatin@imamu.edu.sa (M.M.B.); ASAlnahdi@imamu.edu.sa (A.S.A.)
2   Department of Mathematics, Faculty of Science, Beni-Suef University, Beni-Suef 65211, Egypt
3   Department of Mathematical Sciences, Faculty of Science, Technology, Universiti Kebangsaan Malaysia (UKM), Bangi 43600, Malaysia; P108214@siswa.ukm.edu.my
4   Department of Applied Mathematics, National Research Centre, Cairo 12622, Egypt; ma.zaky@yahoo.com
5   Faculty of Education, Matrouh University, Matrouh 51512, Egypt; r_mhafez@yahoo.com
*   Correspondence: maohamed@imamu.edu.sa

Abstract: In this paper, we introduce a numerical solution for the time-fractional inverse heat equations. We focus on obtaining the unknown source term along with the unknown temperature function based on an additional condition given in an integral form. The proposed scheme is based on a spectral collocation approach to obtain the two independent variables. Our approach is accurate, efficient, and feasible for the model problem under consideration. The proposed Jacobi spectral collocation method yields an exponential rate of convergence with a relatively small number of degrees of freedom. Finally, a series of numerical examples are provided to demonstrate the efficiency and flexibility of the numerical scheme.

Keywords: inverse problem; spectral collocation method; fractional diffusion; fractional calculus

MSC: 35R30; 65M70; 35R11

check for updates

Citation: Abdelkawy, M.A.; Amin, A.Z.M.; Babatin, M.M.; Alnahd, A.S.; Zaky, M.A.; Hafez, R.M. Jacobi Spectral Collocation Technique for Time-Fractional Inverse Heat Equations. *Fractal Fract.* **2021**, *5*, 115. https://doi.org/10.3390/fractalfract5030115

Academic Editors: Ivanka Stamova, António M. Lopes and Liping Chen

Received: 8 July 2021
Accepted: 7 September 2021
Published: 9 September 2021

Publisher's Note: MDPI stays neutral with regard to jurisdictional claims in published maps and institutional affiliations.

1. Introduction

The concept of fractional derivatives has become one of the key aspects of applied mathematics because it is more suitable for modelling many real-world problems than the local derivative. As a result, the fractional derivative has received considerable attention and development in a wide range of fields [1–5]. Fractional derivatives are defined in a variety of ways in the mathematical literature, including Riemann–Liouville and Caputo fractional derivatives. Hence, fractional differential equations have attracted the attention of researchers in recent years. The main reason for this is that they are commonly used in chemistry [6], mathematical biology systems [7], electrical engineering [8], systems identification, control theory [9], signal processing, mechanical engineering [10–12], finance and fractional dynamics and so on.

Direct fractional-order diffusion equations have been extensively discussed in the literature; see [13–15]. Often, for many practical studies, there is an unknown parameter that is found in the initial or boundary data or the source term that requires an additional measurement. The inverse fractional-order case introduces an appropriate instrument for describing anomalous diffusion phenomena appeared in chemical [16], biological [17,18], hydrological [19], physical [20,21] and engineering [22,23] fields. In contrast to those classical problems, the studies of inverse problems have not satisfactorily been studied. The mathematical problem of studying inverse problems with non-Fourier heat-conduction constitutive models is extremely novel. The goal of inverse problems for heat-conduction process is to set unknown ingredients of the conduction system from some measurement data, which is of major importance in the applied area. Hung and Lin [24] solved the hyperbolic inverse heat-conduction problem. Yang [25] solved the two-dimensional inverse hyperbolic heat problem by modified Newton–Raphson method. Tang and Araki [26] estimated thermal diffusivity and the relaxation parameters for solving the inverse heat equation. Wang and Liu [27] used the total variation regularization method for solving backward time-fractional diffusion problem. Zhang and Xu [28] solved the inverse source problem of the fractional diffusion equation.

Spectral methods are powerful tools for solving different types of differential and integral equations that arise in various fields of science and engineering. In recent decades, they have been adopted in a variety of notable areas [29–38]. In the numerical solutions of fractional differential equations, a variety of spectral methods have recently been used [39,40]. Their major advantages are exponential convergence rates, high accuracy level, and low computational costs. The spectral methods are distinguished over finite difference, finite element, and finite volume in their global, high-accuracy numerical results and have applicability to most problems with integer or fractional orders; see [36,41–43]. Because explicit analytical solutions of space and/or time-fractional differential equations are difficult to obtain in most cases, developing efficient numerical schemes is a very important demand. In various applications, many efficient numerical techniques have been introduced to treat various problems. Presently there is a wide and constantly increasing range of spectral methods and there has been significant growth in fractional differential and integral equations [44] due to their high-order accuracy. Compared to the effort put into analyzing finite difference schemes in the literature for solving fractional-order differential equations, only a little research has been made in developing and analyzing global spectral schemes.

Our main aim in this paper is to provide shifted Jacobi Gauss–Lobatto and shifted Jacobi Gauss–Radau collection schemes for solving fractional inverse heat equations (IHEs). The unknown solution is approximated using the shifted Jacobi polynomials as a truncated series. The collocation technique is provided along with appropriate treatment for addressing the extra condition. This procedure allows us to exclude the unknown function $\mathcal{Q}(\tau)$ from the equation under consideration. As a result, this problem is reduced to a system of algebraic equations by employing the spectral collocation approach. Finally, in terms of shifted Jacobi polynomials, we can extend the unknown functions $\mathcal{U}(\xi, \tau)$ and $\mathcal{Q}(\tau)$. To the best of our knowledge, there are no numerical results on the spectral collocation method for treating the IHEs.

This paper is organized as follows. We introduce some mathematical preliminaries in Section 2. In Section 3.2, we construct the numerical scheme to solve the fractional IHEs with initial-boundary conditions and nonlocal conditions. In Section 4, we solve and analyze some examples to illustrate the efficiency and accuracy of the method. In Section 5, we provide the main conclusions.

2. Preliminaries and Notations

This section introduces some properties of the shifted Jacobi polynomials. The Jacobi polynomials are defined as follows:

$$\mathcal{G}_{k+1}^{(\sigma_1,\varrho_1)}(y) = (a_k^{(\sigma_1,\varrho_1)}y - b_k^{(\sigma_1,\varrho_1)})\mathcal{G}_k^{(\sigma_1,\varrho_1)}(y) - c_k^{(\sigma_1,\varrho_1)}\mathcal{G}_{k-1}^{(\sigma_1,\varrho_1)}(y), \quad k \geq 1,$$

$$\mathcal{G}_0^{(\sigma_1,\varrho_1)}(y) = 1, \quad \mathcal{G}_1^{(\sigma_1,\varrho_1)}(y) = \frac{1}{2}(\sigma_1 + \varrho_1 + 2)y + \frac{1}{2}(\sigma_1 - \varrho_1),$$

$$\mathcal{G}_k^{(\sigma_1,\varrho_1)}(-y) = (-1)^k \mathcal{G}_k^{(\sigma_1,\varrho_1)}(y), \quad \mathcal{G}_k^{(\sigma_1,\varrho_1)}(-1) = \frac{(-1)^k \Gamma(k + \varrho_1 + 1)}{k!\,\Gamma(\varrho_1 + 1)}, \tag{1}$$

where $\sigma_1, \varrho_1 > -1$, $y \in (-1, 1)$ and

$$a_k^{(\sigma_1,\varrho_1)} = \frac{(2k + \sigma_1 + \varrho_1 + 1)(2k + \sigma_1 + \varrho_1 + 2)}{2(k + 1)(k + \sigma_1 + \varrho_1 + 1)},$$

$$b_k^{(\sigma_1,\varrho_1)} = \frac{(\varrho_1^2 - \sigma_1^2)(2k + \sigma_1 + \varrho_1 + 1)}{2(k + 1)(k + \sigma_1 + \varrho_1 + 1)(2k + \sigma_1 + \varrho_1)},$$

$$c_k^{(\sigma_1,\varrho_1)} = \frac{(k + \sigma_1)(k + \varrho_1)(2k + \sigma_1 + \varrho_1 + 2)}{(k + 1)(k + \sigma_1 + \varrho_1 + 1)(2k + \sigma_1 + \varrho_1)}.$$

The nth-order derivative (n is an integer) of $\mathcal{G}_j^{(\sigma_1,\varrho_1)}(y)$ can also be obtained from

$$D^n \mathcal{G}_j^{(\sigma_1,\varrho_1)}(y) = \frac{\Gamma(j + \sigma_1 + \varrho_1 + q + 1)}{2^n \Gamma(j + \sigma_1 + \varrho_1 + 1)} \mathcal{G}_{j-n}^{(\sigma_1+n,\varrho_1+n)}(y). \tag{2}$$

The analytic form of the shifted Jacobi polynomial $\mathcal{G}_{L,k}^{(\sigma_1,\varrho_1)}(y) = \mathcal{G}_k^{(\sigma_1,\varrho_1)}(\frac{2y}{L} - 1)$, $L > 0$, is written as

$$\begin{aligned}
\mathcal{G}_{L,k}^{(\sigma_1,\varrho_1)}(y) &= \sum_{j=0}^{k} (-1)^{k-j} \frac{\Gamma(k + \varrho_1 + 1)\Gamma(j + k + \sigma_1 + \varrho_1 + 1)}{\Gamma(j + \varrho_1 + 1)\Gamma(k + \sigma_1 + \varrho_1 + 1)(k - j)! j! L^j} y^j \\
&= \sum_{j=0}^{k} \frac{\Gamma(k + \sigma_1 + 1)\Gamma(k + j + \sigma_1 + \varrho_1 + 1)}{j!(k - j)! \Gamma(j + \sigma_1 + 1)\Gamma(k + \sigma_1 + \varrho_1 + 1) L^j} (y - L)^j.
\end{aligned} \tag{3}$$

As a result, for any integer n, we can derive the following properties

$$\mathcal{G}_{L,k}^{(\sigma_1,\varrho_1)}(0) = (-1)^k \frac{\Gamma(k + \varrho_1 + 1)}{\Gamma(\varrho_1 + 1) \, k!},$$

$$\mathcal{G}_{L,k}^{(\sigma_1,\varrho_1)}(L) = \frac{\Gamma(k + \sigma_1 + 1)}{\Gamma(\sigma_1 + 1) \, k!}, \tag{4}$$

$$D^n \mathcal{G}_{L,k}^{(\sigma_1,\varrho_1)}(0) = \frac{(-1)^{k-n} \Gamma(k + \varrho_1 + 1)(k + \sigma_1 + \varrho_1 + 1)_n}{L^n \Gamma(k - n + 1)\Gamma(n + \varrho_1 + 1)}, \tag{5}$$

$$D^n \mathcal{G}_{L,k}^{(\sigma_1,\varrho_1)}(L) = \frac{\Gamma(k + \sigma_1 + 1)(k + \sigma_1 + \varrho_1 + 1)_n}{L^n \Gamma(k - n + 1)\Gamma(n + \sigma_1 + 1)}, \tag{6}$$

$$D^n \mathcal{G}_{L,k}^{(\sigma_1,\varrho_1)}(y) = \frac{\Gamma(n + k + \sigma_1 + \varrho_1 + 1)}{L^n \Gamma(k + \sigma_1 + \varrho_1 + 1)} \mathcal{G}_{L,k-n}^{(\sigma_1+n,\varrho_1+n)}(y). \tag{7}$$

Let $w_L^{(\sigma_1,\varrho_1)}(y) = (L - y)^{\sigma_1} y^{\varrho_1}$. Then, we define

$$(u, v)_{w_L^{(\sigma_1,\varrho_1)}} = \int_0^L u(y)\, v(y)\, w_L^{(\sigma_1,\varrho_1)}(y)\, dy, \quad \|v\|_{w_L^{(\sigma_1,\varrho_1)}} = (v, v)_{w_L^{(\sigma_1,\varrho_1)}}^{\frac{1}{2}}. \tag{8}$$

The set of the shifted Jacobi polynomials forms a complete $L_{w_L^{(\sigma_1,\varrho_1)}}^2 [0, L]$-orthogonal system. Furthermore, and as a result of (8), we have

$$\|\mathcal{G}_{L,k}^{(\sigma_1,\varrho_1)}\|_{w_L^{(\sigma_1,\varrho_1)}}^2 = \left(\frac{L}{2}\right)^{\sigma_1+\varrho_1+1} h_k^{(\sigma_1,\varrho_1)} = h_{L,k}^{(\sigma_1,\varrho_1)}, \tag{9}$$

where

$$h_n^{(\sigma_1,\varrho_1)} = \frac{2^{\sigma_1+\varrho_1+1} \Gamma(n + \varrho_1 + 1)\Gamma(n + \sigma_1 + 1)}{n! \Gamma(n + \sigma_1 + \varrho_1 + 1)(2n + \sigma_1 + \varrho_1 + 1)}.$$

We denote $y_{N,j}^{(\sigma_1,\varrho_1)}$ and $\varpi_{N,j}^{(\sigma_1,\varrho_1)}$, $0 \le j \le N$, the nodes and Christoffel numbers on the interval $[-1,1]$. For the shifted Jacobi on the interval $[0,L]$, we obtain

$$y_{L,N,j}^{(\sigma_1,\varrho_1)} = \frac{L}{2}(y_{N,j}^{(\sigma_1,\varrho_1)} + 1),$$

$$\varpi_{L,N,j}^{(\sigma_1,\varrho_1)} = \left(\frac{L}{2}\right)^{\sigma_1+\varrho_1+1} \varpi_{N,j}^{(\sigma_1,\varrho_1)}, \ 0 \le j \le N.$$

Applying the quadrature rule, for $\phi \in S_{2N+1}[0,L]$, we obtain

$$
\begin{aligned}
\int_0^L (L-y)^{\sigma_1} y^{\varrho_1} \phi(y) dy &= \left(\frac{L}{2}\right)^{\sigma_1+\varrho_1+1} \int_{-1}^1 (1-y)^{\sigma_1}(1+y)^{\varrho_1} \phi\left(\frac{L}{2}(y+1)\right) dy \\
&= \left(\frac{L}{2}\right)^{\sigma_1+\varrho_1+1} \sum_{j=0}^N \varpi_{N,j}^{(\sigma_1,\varrho_1)} \phi\left(\frac{L}{2}(y_{N,j}^{(\sigma_1,\varrho_1)}+1)\right) \\
&= \sum_{j=0}^N \varpi_{L,N,j}^{(\sigma_1,\varrho_1)} \phi\left(y_{L,N,j}^{(\sigma_1,\varrho_1)}\right),
\end{aligned}
\tag{10}
$$

where $S_N[0,L]$ is the set of polynomials of degree at most N.

3. Fully Spectral Collocation Treatment

3.1. Initial-Boundary Conditions

First, we developed a numerical technique for dealing with the time-fractional IHEs of the form:

$$\frac{\partial^\nu}{\partial\tau^\nu}\left(\mathcal{U}(\xi,\tau) - \mathcal{U}(\xi,0)\right) = \frac{\partial^2}{\partial\xi^2}(\mathcal{U}(\xi,\tau)) + \mathcal{Q}(\tau)\Delta(\xi,\tau), \quad (\xi,\tau) \in \Lambda^\bullet \times \Lambda^\circ, \tag{11}$$

$$
\begin{aligned}
\mathcal{U}(\xi,0) &= \lambda_1(\xi), \quad \xi \in \Lambda^\bullet, \\
\mathcal{U}(0,\tau) &= \lambda_2(\tau), \quad \mathcal{U}(\xi_{end},\tau) = \lambda_3(\tau), \quad \tau \in \Lambda^\circ,
\end{aligned}
\tag{12}
$$

where ξ and τ are used for spatial and temporal variables, respectively. The fractional derivative term $\frac{\partial^\nu}{\partial\tau^\nu}\left(\mathcal{U}(\xi,\tau) - \mathcal{U}(\xi,0)\right)$ instead of $\frac{\partial^\nu}{\partial\tau^\nu}\left(\mathcal{U}(\xi,\tau)\right)$ is not only to eschew the singularity at zero, but also provide a significative initial condition, where fractional integral is not included [45].

Where $\frac{\partial^\nu}{\partial\tau^\nu}$ is the fractional temporal derivative in Riemann–Liouville sense,

$$\frac{\partial^\nu \mathcal{U}(\xi,\tau)}{\partial\tau^\nu} = \frac{1}{\Gamma(1-\nu)}\frac{\partial}{\partial\tau}\int_0^\tau \frac{\mathcal{U}(\xi,s)}{(s-\tau)^\mu}ds,$$

and $\Lambda^\bullet \equiv [0,\xi_{end}]$, $\Lambda^\circ \equiv [0,\tau_{end}]$, $\mathcal{U}(\xi,\tau)$ and $\mathcal{Q}(\tau)$ are unknown functions, while $\Delta(\xi,\tau)$ is a given function. The complexity of the suggested problem is that the function $\mathcal{Q}(\tau)$ is unknown, which necessitates the determination of an additional condition. To resolve this problem, we use the following energy condition

$$\int_0^1 \mathcal{U}(\xi,\tau)d\xi = \mathcal{E}(\tau). \tag{13}$$

Here, the shifted Jacobi Gauss–Lobatto collection method and the shifted Jacobi Gauss–Radau collection scheme are applied to convert the IHEs into a linear system of algebraic equations. We approximate the solution as,

$$\mathcal{U}_{\mathcal{N},\mathcal{M}}(\xi,\tau) = \sum_{\substack{r_1=0,\ldots,\mathcal{N}\\ r_2=0,\ldots,\mathcal{M}}} \varsigma_{r_1,r_2} \mathcal{G}^{\sigma_1,\varrho_1}_{\xi_{end},r_1}(\xi)\mathcal{G}^{\sigma_2,\varrho_2}_{\tau_{end},r_2}(\tau), \tag{14}$$

where $\mathcal{G}^{\sigma,\varsigma}_{\xi_{end},s}(\zeta)$ is used for shifted Jacobi polynomials on $[0,\xi_{end}]$.

The first derivative $\frac{\partial}{\partial x}(\mathcal{U}_{\mathcal{N},\mathcal{M}}(\xi,\tau))$ is given as

$$\frac{\partial}{\partial\xi}(\mathcal{U}_{\mathcal{N},\mathcal{M}}(\xi,\tau)) = \sum_{\substack{r_1=0,\ldots,\mathcal{N}\\ r_2=0,\ldots,\mathcal{M}}} \varsigma_{r_1,r_2} \widetilde{\mathcal{G}}^{\sigma_1,\varrho_1}_{\xi_{end},r_1}(\xi)\mathcal{G}^{\sigma_2,\varrho_2}_{\tau_{end},r_2}(\tau), \tag{15}$$

where $\widetilde{\mathcal{G}}^{\sigma_1,\varrho_1}_{\xi_{end},r_1}(\xi) = \frac{\partial}{\partial x}(\mathcal{G}^{\sigma_1,\varrho_1}_{\xi_{end},r_1}(\xi))$. Similarly, we find

$$\frac{\partial^2}{\partial x^2}(\mathcal{U}_{\mathcal{N},\mathcal{M}}(\xi,\tau)) = \sum_{\substack{r_1=0,\ldots,\mathcal{N}\\ r_2=0,\ldots,\mathcal{M}}} \varsigma_{r_1,r_2} \widehat{\mathcal{G}}^{\sigma_1,\varrho_1}_{\xi_{end},r_1}(\xi)\mathcal{G}^{\sigma_2,\varrho_2}_{\tau_{end},r_2}(\tau), \tag{16}$$

where $\widehat{\mathcal{G}}^{\sigma_1,\varrho_1}_{\xi_{end},r_1}(\xi) = \frac{\partial^2}{\partial\xi^2}(\mathcal{G}^{\sigma_1,\varrho_1}_{\xi_{end},r_1}(\xi))$. Please note that $\frac{\partial}{\partial\xi}(\mathcal{G}^{\sigma_1,\varrho_1}_{\xi_{end},r_1}(\xi))$ and $\frac{\partial^2}{\partial\xi^2}(\mathcal{G}^{\sigma_1,\varrho_1}_{\xi_{end},r_1}(\xi))$ can be directly computed using (7). However, the fractional temporal derivative in Riemann–Liouville sense is computed as

$$\frac{\partial^\nu}{\partial t^\nu}(\mathcal{U}(\xi,\tau)) = \sum_{\substack{r_1=0,\ldots,\mathcal{N}\\ r_2=0,\ldots,\mathcal{M}}} \varsigma_{r_1,r_2} \mathcal{G}^{\sigma_1,\varrho_1}_{\xi_{end},r_1}(\xi)\check{\mathcal{G}}^{\sigma_2,\varrho_2}_{\tau_{end},r_2}(\tau), \tag{17}$$

where $\check{\mathcal{G}}^{\sigma_2,\varrho_2}_{\tau_{end},r_2}(\tau) = \frac{\partial^\nu}{\partial\tau^\nu}(\mathcal{G}^{\sigma_2,\varrho_2}_{\tau_{end},r_2}(\tau))$. Using (3), we obtain

$$\begin{aligned}\frac{\partial^\nu}{\partial\tau^\nu}(\mathcal{G}^{\sigma_2,\varrho_2}_{\tau_{end},r_2}(\tau)) &= \sum_{k=0}^{r_2} (-1)^{r_2-k} \frac{\Gamma(r_2+\varrho_2+1)\Gamma(k+r_2+\sigma_2+\varrho_2+1)}{\Gamma(k+\varrho_2+1)\Gamma(r_2+\sigma_2+\varrho_2+1)(r_2-k)!k!\tau^k_{end}} \frac{\partial^\nu}{\partial\tau^\nu}(\tau^k)\\ &= \sum_{k=0}^{r_2} (-1)^{r_2-k} \frac{\Gamma(r_2+\varrho_2+1)\Gamma(k+r_2+\sigma_2+\varrho_2+1)}{\Gamma(k+\varrho_2+1)\Gamma(r_2+\sigma_2+\varrho_2+1)(r_2-k)!k!\tau^k_{end}}\delta(\tau),\end{aligned} \tag{18}$$

where $\delta(\tau) = \frac{\Gamma(k+1)\tau^{k-\nu}}{\Gamma(k-\nu+1)}$.

When we differentiate, of order ν, Equation (13) with respect to τ, we obtain

$$\int_0^1 \frac{\partial^\nu}{\partial t^\nu}\Big(\mathcal{U}(\xi,\tau) - \mathcal{U}(\xi,0)\Big)d\xi = \frac{\partial^\nu}{\partial t^\nu}\Big(\mathcal{E}(\tau) - \mathcal{E}(0)\Big), \tag{19}$$

merging the previous equation with (11), we obtain

$$\frac{\partial}{\partial\xi}(\mathcal{U}(\xi,\tau))_{\xi=\xi_{end}} - \frac{\partial}{\partial\xi}(\mathcal{U}(\xi,\tau))_{\xi=0} = \frac{\partial^\nu}{\partial t^\nu}(\mathcal{E}(\tau) - \mathcal{E}(0)) - \mathcal{Q}(\tau)\int_0^1 \Delta(\xi,\tau)d\xi, \tag{20}$$

yields,

$$\mathcal{Q}(\tau) = \frac{\Theta(\tau)}{\int_0^1 \Delta(\xi,\tau)d\xi}, \tag{21}$$

where

$$\Theta(\tau) = \mathcal{E}^{(\nu)}(\tau) - \mathcal{E}^{(\nu)}(0) - \sum_{\substack{r_1=0,\ldots,\mathcal{N} \\ r_2=0,\ldots,\mathcal{M}}} \varsigma_{r_1,r_2} \widetilde{\mathcal{G}}_{\xi_{end},r_1}^{\sigma_1,\varrho_1}(1)\mathcal{G}_{\tau_{end},r_2}^{\sigma_2,\varrho_2}(\tau) + \sum_{\substack{r_1=0,\ldots,\mathcal{N} \\ r_2=0,\ldots,\mathcal{M}}} \varsigma_{r_1,r_2} \widetilde{\mathcal{G}}_{\xi_{end},r_1}^{\sigma_1,\varrho_1}(0)\mathcal{G}_{\tau_{end},r_2}^{\sigma_2,\varrho_2}(\tau),$$

and $\mathcal{E}^{(\nu)}(\tau) = \frac{\partial^\nu \mathcal{E}(\tau)}{\partial t^\nu}$. The previous derivatives of spatial and temporal variables are computed at specific nodes as

$$\left(\frac{\partial^\nu}{\partial \tau^\nu}(\mathcal{U}(\xi,\tau))\right)_{\tau=\tau_{\tau_{end},\mathcal{M}}^{\sigma_2,\varrho_2,m}}^{\xi=\xi_{\xi_{end},\mathcal{N}}^{\sigma_1,\varrho_1,n}} = \sum_{\substack{r_1=0,\ldots,\mathcal{N} \\ r_2=0,\ldots,\mathcal{M}}} \varsigma_{r_1,r_2} \mathcal{G}_{\xi_{end},r_1}^{\sigma_1,\varrho_1}(\xi_{\xi_{end},\mathcal{N}}^{\sigma_1,\varrho_1,n}) \widetilde{\mathcal{G}}_{\tau_{end},r_2}^{\sigma_2,\varrho_2}(\tau_{\tau_{end},\mathcal{M}}^{\sigma_2,\varrho_2,m}),$$

$$\left(\frac{\partial^\nu}{\partial \tau^\nu}(\mathcal{U}(\xi,0))\right)_{\tau=\tau_{\tau_{end},\mathcal{M}}^{\sigma_2,\varrho_2,m}}^{\xi=\xi_{\xi_{end},\mathcal{N}}^{\sigma_1,\varrho_1,n}} = \sum_{\substack{r_1=0,\ldots,\mathcal{N} \\ r_2=0,\ldots,\mathcal{M}}} \varsigma_{r_1,r_2} \mathcal{G}_{\xi_{end},r_1}^{\sigma_1,\varrho_1}(\xi_{\xi_{end},\mathcal{N}}^{\sigma_1,\varrho_1,n}) \left(\frac{d^\nu}{d\tau^\nu}(\mathcal{G}_{\tau_{end},r_2}^{\sigma_2,\varrho_2}(0))\right)_{\tau=\tau_{\tau_{end},\mathcal{M}}^{\sigma_2,\varrho_2,m}},$$

$$\left(\frac{\partial}{\partial \xi}(\mathcal{U}_{\mathcal{N},\mathcal{M}}(\xi,\tau))\right)_{\tau=\tau_{\tau_{end},\mathcal{M}}^{\sigma_2,\varrho_2,m}}^{\xi=\xi_{end}} = \sum_{\substack{r_1=0,\ldots,\mathcal{N} \\ r_2=0,\ldots,\mathcal{M}}} \varsigma_{r_1,r_2} \widetilde{\mathcal{G}}_{\xi_{end},r_1}^{\sigma_1,\varrho_1}(\xi_{end})\mathcal{G}_{\tau_{end},r_2}^{\sigma_2,\varrho_2}(\tau_{\tau_{end},\mathcal{M}}^{\sigma_2,\varrho_2,m}), \tag{22}$$

$$\left(\frac{\partial}{\partial \xi}(\mathcal{U}_{\mathcal{N},\mathcal{M}}(\xi,\tau))\right)_{\tau=\tau_{\tau_{end},\mathcal{M}}^{\sigma_2,\varrho_2,m}}^{\xi=0} = \sum_{\substack{r_1=0,\ldots,\mathcal{N} \\ r_2=0,\ldots,\mathcal{M}}} \varsigma_{r_1,r_2} \widetilde{\mathcal{G}}_{\xi_{end},r_1}^{\sigma_1,\varrho_1}(0)\mathcal{G}_{\tau_{end},r_2}^{\sigma_2,\varrho_2}(\tau_{\tau_{end},\mathcal{M}}^{\sigma_2,\varrho_2,m}),$$

$$\left(\frac{\partial^2}{\partial \xi^2}(\mathcal{U}(\xi,\tau))\right)_{\tau=\tau_{\tau_{end},\mathcal{M}}^{\sigma_2,\varrho_2,m}}^{\xi=\xi_{\xi_{end},\mathcal{N}}^{\sigma_1,\varrho_1,n}} = \sum_{\substack{r_1=0,\ldots,\mathcal{N} \\ r_2=0,\ldots,\mathcal{M}}} \varsigma_{r_1,r_2} \widehat{\mathcal{G}}_{\xi_{end},r_1}^{\sigma_1,\varrho_1}(\xi_{\xi_{end},\mathcal{N}}^{\sigma_1,\varrho_1,n})\mathcal{G}_{\tau_{end},r_2}^{\sigma_2,\varrho_2}(\tau_{\tau_{end},\mathcal{M}}^{\sigma_2,\varrho_2,m}).$$

Additionally, we obtain

$$\mathcal{Q}(\tau_{\tau_{end},\mathcal{M}}^{\sigma_2,\varrho_2,m}) = \frac{\Theta(\tau_{\tau_{end},\mathcal{M}}^{\sigma_2,\varrho_2,m})}{\int_0^1 \Delta(\xi,\tau_{\tau_{end},\mathcal{M}}^{\sigma_2,\varrho_2,m})d\xi}, \tag{23}$$

where $n = 0,1,\cdots,\mathcal{N}, \quad m = 0,1,\cdots,\mathcal{M}$.

For the proposed spectral collocation technique, Equation (11) is enforced to be zero at $(\mathcal{N}-1) \times (\mathcal{M})$ points. Therefore, adapting (11)–(23), obtain linear system of algebraic equations

$$\begin{pmatrix} \aleph_{1,1} & \aleph_{1,2} & \cdots & \cdots & \aleph_{1,\mathcal{M}} \\ \aleph_{2,1} & \aleph_{2,2} & \cdots & \cdots & \aleph_{2,\mathcal{M}} \\ \cdots & \vdots & \vdots & \vdots & \cdots \\ \cdots & \vdots & \vdots & \vdots & \cdots \\ \aleph_{\mathcal{N},1} & \aleph_{\mathcal{N},2} & \cdots & \cdots & \aleph_{\mathcal{N},\mathcal{M}} \end{pmatrix} = \begin{pmatrix} \wp_{1,1} & \wp_{1,2} & \cdots & \cdots & \wp_{1,\mathcal{M}} \\ \wp_{2,1} & \wp_{2,2} & \cdots & \cdots & \wp_{2,\mathcal{M}} \\ \cdots & \vdots & \vdots & \vdots & \cdots \\ \cdots & \vdots & \vdots & \vdots & \cdots \\ \wp_{\mathcal{N},1} & \wp_{\mathcal{N},2} & \cdots & \cdots & \wp_{\mathcal{N},\mathcal{M}} \end{pmatrix} \tag{24}$$

where

$$
\aleph_{n,m} = \begin{cases}
\Upsilon(\xi_{\xi_{end},\mathcal{N}}^{\sigma_1,\varrho_1,n}, \tau_{\tau_{end},\mathcal{M}}^{\sigma_2,\varrho_2,m}), & n = 1, \cdots, \mathcal{N}-1, \quad m = 1, \cdots, \mathcal{M}; \\[2mm]
\sum_{\substack{r_1=0,\ldots,\mathcal{N} \\ r_2=0,\ldots,\mathcal{M}}} \varsigma_{r_1,r_2} \mathcal{G}_{\xi_{end},r_1}^{\sigma_1,\varrho_1}(\xi_{\xi_{end},\mathcal{N}}^{\sigma_1,\varrho_1,n}) \mathcal{G}_{\tau_{end},r_2}^{\sigma_2,\varrho_2}(0), & m = 0, n = 1, \cdots, \mathcal{N}-1; \\[2mm]
\sum_{\substack{r_1=0,\ldots,\mathcal{N} \\ r_2=0,\ldots,\mathcal{M}}} \varsigma_{r_1,r_2} \mathcal{G}_{\xi_{end},r_1}^{\sigma_1,\varrho_1}(0) \mathcal{G}_{\tau_{end},r_2}^{\sigma_2,\varrho_2}(\tau_{\tau_{end},\mathcal{M}}^{\sigma_2,\varrho_2,m}), & n = 0, m = 0, \cdots, \mathcal{M}; \\[2mm]
\sum_{\substack{r_1=0,\ldots,\mathcal{N} \\ r_2=0,\ldots,\mathcal{M}}} \varsigma_{r_1,r_2} \mathcal{G}_{\xi_{end},r_1}^{\sigma_1,\varrho_1}(\xi_{end}) \mathcal{G}_{\tau_{end},r_2}^{\sigma_2,\varrho_2}(\tau_{\tau_{end},\mathcal{M}}^{\sigma_2,\varrho_2,m}), & n = \mathcal{N}, = 0, \cdots, \mathcal{M}.
\end{cases}
$$

$$
\wp_{n,m} = \begin{cases}
\dfrac{\Theta(\tau_{\tau_{end},\mathcal{M}}^{\sigma_2,\varrho_2,m})}{\int_0^1 \Delta(\xi, \tau_{\tau_{end},\mathcal{M}}^{\sigma_2,\varrho_2,m}) d\xi} \Delta(\xi_{\xi_{end},\mathcal{N}}^{\sigma_1,\varrho_1,n}, \tau_{\tau_{end},\mathcal{M}}^{\sigma_2,\varrho_2,m}), & n = 1, \cdots, \mathcal{N}-1, \quad m = 1, \cdots, \mathcal{M}; \\[3mm]
\lambda_1(\xi_{\xi_{end},\mathcal{N}}^{\sigma_1,\varrho_1,n}), & m = 0, n = 1, \cdots, \mathcal{N}-1; \\[2mm]
\lambda_2(\tau_{\tau_{end},\mathcal{M}}^{\sigma_2,\varrho_2,m}), & n = 0, m = 0, \cdots, \mathcal{M}; \\[2mm]
\lambda_3(\tau_{\tau_{end},\mathcal{M}}^{\sigma_2,\varrho_2,m}), & n = \mathcal{N}, = 0, \cdots, \mathcal{M}.
\end{cases}
$$

where

$$
\begin{aligned}
\Upsilon(\xi_{\xi_{end},\mathcal{N}}^{\sigma_1,\varrho_1,n}, \tau_{\tau_{end},\mathcal{M}}^{\sigma_2,\varrho_2,m}) = {} & \sum_{\substack{r_1=0,\ldots,\mathcal{N} \\ r_2=0,\ldots,\mathcal{M}}} \varsigma_{r_1,r_2} \mathcal{G}_{\xi_{end},r_1}^{\sigma_1,\varrho_1}(\xi_{\xi_{end},\mathcal{N}}^{\sigma_1,\varrho_1,n}) \tilde{\mathcal{G}}_{\tau_{end},r_2}^{\sigma_2,\varrho_2}(\tau_{\tau_{end},\mathcal{M}}^{\sigma_2,\varrho_2,m}) - \\[2mm]
& \sum_{\substack{r_1=0,\ldots,\mathcal{N} \\ r_2=0,\ldots,\mathcal{M}}} \varsigma_{r_1,r_2} \mathcal{G}_{\xi_{end},r_1}^{\sigma_1,\varrho_1}(\xi_{\xi_{end},\mathcal{N}}^{\sigma_1,\varrho_1,n}) \left(\frac{d^\nu}{d\tau^\nu}(\mathcal{G}_{\tau_{end},r_2}^{\sigma_2,\varrho_2}(0)) \right)_{\tau = \tau_{\tau_{end},\mathcal{M}}^{\sigma_2,\varrho_2,m}} - \\[2mm]
& \sum_{\substack{r_1=0,\ldots,\mathcal{N} \\ r_2=0,\ldots,\mathcal{M}}} \varsigma_{r_1,r_2} \widehat{\mathcal{G}}_{\xi_{end},r_1}^{\sigma_1,\varrho_1}(\xi_{\xi_{end},\mathcal{N}}^{\sigma_1,\varrho_1,n}) \mathcal{G}_{\tau_{end},r_2}^{\sigma_2,\varrho_2}(\tau_{\tau_{end},\mathcal{M}}^{\sigma_2,\varrho_2,m}).
\end{aligned}
$$

3.2. Nonlocal Conditions

Here, we develop a numerical scheme to deal with the time-fractional IHEs of the form

$$
\frac{\partial^\nu}{\partial t^\nu}\Big(\mathcal{U}(\xi,\tau) - \mathcal{U}(\xi,0)\Big) = \frac{\partial^2}{\partial x^2}(\mathcal{U}(\xi,\tau)) + \mathcal{Q}(\tau)\Delta(\xi,\tau), \quad (\xi,\tau) \in \Lambda^\bullet \times \Lambda^\circ, \tag{25}
$$

$$
\mathcal{U}(\xi,0) = \lambda_1(\xi), \quad \xi \in \Lambda^\bullet, \tag{26}
$$
$$
\mathcal{U}(0,\tau) + \alpha_1 \mathcal{U}(\xi_{end},\tau) = \lambda_2(\tau), \quad \mathcal{U}_\xi(0,\tau) + \alpha_2 \mathcal{U}(\xi_{end},\tau) = \lambda_3(\tau), \quad \tau \in \Lambda^\circ,
$$

where $\Lambda^\bullet \equiv [0,\xi_{end}]$, $\Lambda^\circ \equiv [0,\tau_{end}]$, $\mathcal{U}(\xi,\tau)$ and $\mathcal{Q}(\tau)$ are unknown functions, while $\Delta(\xi,\tau)$ is a given one. The energy condition is given by

$$
\int_0^1 \mathcal{U}(\xi,\tau) d\xi = \mathcal{E}(\tau). \tag{27}
$$

Using the previous results, we obtain the following linear system of algebraic equations

$$
\begin{pmatrix}
\aleph_{1,1} & \aleph_{1,2} & \cdots & \cdots & \aleph_{1,\mathcal{M}} \\
\aleph_{2,1} & \aleph_{2,2} & \cdots & \cdots & \aleph_{2,\mathcal{M}} \\
\cdots & \vdots & \vdots & \vdots & \cdots \\
\cdots & \vdots & \vdots & \vdots & \cdots \\
\aleph_{\mathcal{N},1} & \aleph_{\mathcal{N},2} & \cdots & \cdots & \aleph_{\mathcal{N},\mathcal{M}}
\end{pmatrix}
=
\begin{pmatrix}
\wp_{1,1} & \wp_{1,2} & \cdots & \cdots & \wp_{1,\mathcal{M}} \\
\wp_{2,1} & \wp_{2,2} & \cdots & \cdots & \wp_{2,\mathcal{M}} \\
\cdots & \vdots & \vdots & \vdots & \cdots \\
\cdots & \vdots & \vdots & \vdots & \cdots \\
\wp_{\mathcal{N},1} & \wp_{\mathcal{N},2} & \cdots & \cdots & \wp_{\mathcal{N},\mathcal{M}}
\end{pmatrix}, \tag{28}
$$

where

$$
\aleph_{n,m} =
\begin{cases}
\Upsilon(\xi_{\xi_{end},\mathcal{N}}^{\sigma_1,\varrho_1,n},\tau_{\tau_{end},\mathcal{M}}^{\sigma_2,\varrho_2,m}), & n=1,\cdots,\mathcal{N}-1,\quad m=1,\cdots,\mathcal{M}; \\[4pt]
\sum_{\substack{r_1=0,\dots,\mathcal{N}\\ r_2=0,\dots,\mathcal{M}}} \varsigma_{r_1,r_2}\,\mathcal{G}_{\xi_{end},r_1}^{\sigma_1,\varrho_1}(\xi_{\xi_{end},\mathcal{N}}^{\sigma_1,\varrho_1,n})\,\mathcal{G}_{\tau_{end},r_2}^{\sigma_2,\varrho_2}(0), & m=0,\ n=1,\cdots,\mathcal{N}-1; \\[4pt]
\sum_{\substack{r_1=0,\dots,\mathcal{N}\\ r_2=0,\dots,\mathcal{M}}} \varsigma_{r_1,r_2}\,\mathcal{G}_{\xi_{end},r_1}^{\sigma_1,\varrho_1}(0)\,\mathcal{G}_{\tau_{end},r_2}^{\sigma_2,\varrho_2}(\tau_{\tau_{end},\mathcal{M}}^{\sigma_2,\varrho_2,m})+ \\
\alpha_1 \sum_{\substack{r_1=0,\dots,\mathcal{N}\\ r_2=0,\dots,\mathcal{M}}} \varsigma_{r_1,r_2}\,\mathcal{G}_{\xi_{end},r_1}^{\sigma_1,\varrho_1}(\xi_{end})\,\mathcal{G}_{\tau_{end},r_2}^{\sigma_2,\varrho_2}(\tau_{\tau_{end},\mathcal{M}}^{\sigma_2,\varrho_2,m}), & n=0,\ m=0,\cdots,\mathcal{M}; \\[4pt]
\sum_{\substack{r_1=0,\dots,\mathcal{N}\\ r_2=0,\dots,\mathcal{M}}} \varsigma_{r_1,r_2}\,\widetilde{\mathcal{G}}_{\xi_{end},r_1}^{\sigma_1,\varrho_1}(0)\,\mathcal{G}_{\tau_{end},r_2}^{\sigma_2,\varrho_2}(\tau_{\tau_{end},\mathcal{M}}^{\sigma_2,\varrho_2,m})+ \\
\alpha_2 \sum_{\substack{r_1=0,\dots,\mathcal{N}\\ r_2=0,\dots,\mathcal{M}}} \varsigma_{r_1,r_2}\,\mathcal{G}_{\xi_{end},r_1}^{\sigma_1,\varrho_1}(0)\,\mathcal{G}_{\tau_{end},r_2}^{\sigma_2,\varrho_2}(\tau_{\tau_{end},\mathcal{M}}^{\sigma_2,\varrho_2,m}), & n=\mathcal{N},=0,\cdots,\mathcal{M}.
\end{cases}
$$

$$
\wp_{n,m} =
\begin{cases}
\dfrac{\Theta(\tau_{\tau_{end},\mathcal{M}}^{\sigma_2,\varrho_2,m})}{\int_0^1 \Delta(\xi,\tau_{\tau_{end},\mathcal{M}}^{\sigma_2,\varrho_2,m})d\xi}\,\Delta(\xi_{\xi_{end},\mathcal{N}}^{\sigma_1,\varrho_1,n},\tau_{\tau_{end},\mathcal{M}}^{\sigma_2,\varrho_2,m}), & n=1,\cdots,\mathcal{N}-1,\quad m=1,\cdots,\mathcal{M}; \\[12pt]
\lambda_1(\xi_{\xi_{end},\mathcal{N}}^{\sigma_1,\varrho_1,n}), & m=0,\ n=1,\cdots,\mathcal{N}-1; \\[4pt]
\lambda_2(\tau_{\tau_{end},\mathcal{M}}^{\sigma_2,\varrho_2,m}), & n=0,\ m=0,\cdots,\mathcal{M}; \\[4pt]
\lambda_3(\tau_{\tau_{end},\mathcal{M}}^{\sigma_2,\varrho_2,m}), & n=\mathcal{N},=0,\cdots,\mathcal{M}.
\end{cases}
$$

where

$$
\Upsilon(\xi_{\xi_{end},\mathcal{N}}^{\sigma_1,\varrho_1,n},\tau_{\tau_{end},\mathcal{M}}^{\sigma_2,\varrho_2,m}) = \sum_{\substack{r_1=0,\dots,\mathcal{N}\\ r_2=0,\dots,\mathcal{M}}} \varsigma_{r_1,r_2}\,\mathcal{G}_{\xi_{end},r_1}^{\sigma_1,\varrho_1}(\xi_{\xi_{end},\mathcal{N}}^{\sigma_1,\varrho_1,n})\,\widetilde{\mathcal{G}}_{\tau_{end},r_2}^{\sigma_2,\varrho_2}(\tau_{\tau_{end},\mathcal{M}}^{\sigma_2,\varrho_2,m})-
$$

$$
\sum_{\substack{r_1=0,\dots,\mathcal{N}\\ r_2=0,\dots,\mathcal{M}}} \varsigma_{r_1,r_2}\,\mathcal{G}_{\xi_{end},r_1}^{\sigma_1,\varrho_1}(\xi_{\xi_{end},\mathcal{N}}^{\sigma_1,\varrho_1,n})\left(\frac{d^\nu}{d\tau^\nu}\big(\mathcal{G}_{\tau_{end},r_2}^{\sigma_2,\varrho_2}(0)\big)\right)_{\tau=\tau_{\tau_{end},\mathcal{M}}^{\sigma_2,\varrho_2,m}}-
$$

$$
\sum_{\substack{r_1=0,\dots,\mathcal{N}\\ r_2=0,\dots,\mathcal{M}}} \varsigma_{r_1,r_2}\,\widehat{\mathcal{G}}_{\xi_{end},r_1}^{\sigma_1,\varrho_1}(\xi_{\xi_{end},\mathcal{N}}^{\sigma_1,\varrho_1,n})\,\mathcal{G}_{\tau_{end},r_2}^{\sigma_2,\varrho_2}(\tau_{\tau_{end},\mathcal{M}}^{\sigma_2,\varrho_2,m}).
$$

$$
\Upsilon(\xi_{\xi_{end},\mathcal{N}}^{\sigma_1,\varrho_1,n},\tau_{\tau_{end},\mathcal{M}}^{\sigma_2,\varrho_2,m}) = \frac{\Theta(\tau_{\tau_{end},\mathcal{M}}^{\sigma_2,\varrho_2,m})}{\int_0^1 \Delta(\xi,\tau_{\tau_{end},\mathcal{M}}^{\sigma_2,\varrho_2,m})d\xi}\,\Delta(\xi_{\xi_{end},\mathcal{N}}^{\sigma_1,\varrho_1,n},\tau_{\tau_{end},\mathcal{M}}^{\sigma_2,\varrho_2,m}), \tag{29}
$$

4. Numerical Results

This section is devoted to providing some numerical results to show the robustness and the accuracy of the spectral collocation schemes presented in this work.

Example 1. *We consider the following IHEs*

$$
\frac{\partial^\nu}{\partial t^\nu}\big(\mathcal{U}(\xi,\tau)-\mathcal{U}(\xi,0)\big) = \frac{\partial^2}{\partial x^2}(\mathcal{U}(\xi,\tau)) + \mathcal{Q}(\tau)e^{-\tau^2}\sin(\pi\xi)\left(\frac{2(-\nu+\tau+2)\tau^{1-\nu}}{\Gamma(3-\nu)}+\pi^2(\tau+1)^2\right), \tag{30}
$$

$$
(\xi,\tau)\in[0,1]\times[0,1],
$$

with the local conditions

$$
\mathcal{U}(\xi,0)=\sin(\pi\xi),\quad \xi\in[0,1],
$$
$$
\mathcal{U}(0,\tau)=0,\quad \mathcal{U}(1,\tau)=0,\quad \tau\in[0,1], \tag{31}
$$

and the extra energy condition

$$
\int_0^1 \mathcal{U}(\xi,\tau)d\xi = \frac{2(\tau+1)^2}{\pi}, \tag{32}
$$

the exact solution and unknown source function are given by $\mathcal{U}(\xi,\tau) = (\tau+1)^2 \sin(\pi\xi)$, $\mathcal{Q}(\tau) = e^{\tau^2}$.

The absolute errors $E_\mathcal{U}$ and $E_\mathcal{Q}$ are defined as

$$E_\mathcal{U}(\xi,\tau) = |\mathcal{U}(\xi,\tau) - \mathcal{U}_{Approx}(\xi,\tau)|,$$
$$E_\mathcal{Q}(\tau) = |\mathcal{Q}(\tau) - \mathcal{Q}_{Approx}(\tau)|.$$

Moreover, the maximum absolute errors $M_{E_\mathcal{U}}$ and $M_{E_\mathcal{Q}}$ are defined as

$$M_{E_\mathcal{U}} = MAX E_\mathcal{U}(\xi,\tau)(\xi,\tau) \in \Lambda^\bullet \times \Lambda^\diamond,$$
$$M_{E_\mathcal{Q}} = MAX E_\mathcal{Q}(\tau)\tau \in \Lambda^\diamond.$$

Tables 1 and 2 provide the maximum absolute errors $M_{E_\mathcal{U}}$ and $M_{E_\mathcal{Q}}$ of the approximate solution at various values of parameters. From these results, the proposed scheme provides better numerical results. It is also observed that excellent approximations with a few collocation points are achieved. In Figures 1 and 2, with values of parameters listed in their captions, the numerical solution and its absolute errors functions are displayed, respectively. Additionally, the exact and approximate solutions are readily displayed in Figures 3 and 4 for $\mathcal{Q}(\tau)$ and temperature function $\mathcal{U}(\xi,\tau)$, respectively. However, absolute errors functions of the temperature and $\mathcal{Q}(\tau)$ are displayed in Figures 5–7. Moreover, rate of convergence is displayed in Figures 8 and 9. The exponential convergence of our method is observed in these graphs.

Table 1. $M_{E_\mathcal{U}}$ for problem (30).

ν	$(\mathcal{N},\mathcal{M})$	CPU Time	$(0, 0, 0, 0)$	$(0, -0.5, 0, 0.5)$	$(-0.5, -0.5, 0, 0)$	$(-0.5, -0.5, 0.5, 0.5)$
0.5	(4,4)	3.874	5.54207×10^{-1}	4.50114×10^{-1}	3.30903×10^{-1}	3.30301×10^{-1}
	(8,8)	10.937	1.069×10^{-4}	7.62826×10^{-5}	4.23773×10^{-5}	4.23533×10^{-5}
	(12,12)	55.062	2.84348×10^{-9}	1.93179×10^{-9}	9.20255×10^{-10}	9.20021×10^{-10}
	(16,16)	232.329	6.33922×10^{-14}	9.6867×10^{-14}	3.18634×10^{-14}	9.12603×10^{-14}
0.9	(4,4)	5.751	4.51528×10^{-1}	1.78626×10^{-1}	2.64455×10^{-1}	2.64435×10^{-1}
	(8,8)	12.657	8.30291×10^{-5}	5.92663×10^{-5}	3.28873×10^{-5}	328844×10^{-5}
	(12,12)	61.278	2.20829×10^{-9}	1.50057×10^{-9}	7.14412×10^{-10}	7.14381×10^{-10}
	(16,16)	239.312	2.17604×10^{-14}	1.3467×10^{-13}	5.29576×10^{-14}	6.1945×10^{-14}
1.0	(4,4)	3.39	4.24550×10^{-1}	3.41265×10^{-1}	2.47565×10^{-1}	2.476208×10^{-1}
	(8,8)	8.812	7.71774×10^{-5}	5.510238×10^{-5}	3.05607×10^{-5}	3.05625×10^{-5}
	(12,12)	58.25	2.05212×10^{-9}	1.39463×10^{-9}	6.638062×10^{-10}	6.63822×10^{-10}
	(16,16)	235.514	1.5614×10^{-14}	1.02934×10^{-14}	4.37859×10^{-15}	4.378644×10^{-15}

Table 2. $M_{E_\mathcal{Q}}$ for problem (30).

ν	$(\mathcal{N},\mathcal{M})$	$(0, 0, 0, 0)$	$(0, -0.5, 0, 0.5)$	$(-0.5, -0.5, 0, 0)$	$(-0.5, -0.5, 0.5, 0.5)$
0.5	(4,4)	2.37763×10^{-1}	1.93756×10^{-1}	1.47532×10^{-1}	1.47488×10^{-1}
	(8,8)	5.79538×10^{-5}	4.13846×10^{-5}	2.31711×10^{-5}	2.3159×10^{-5}
	(12,12)	1.68474×10^{-9}	1.1447×10^{-9}	5.47435×10^{-10}	5.47225×10^{-10}
	(16,16)	1.83853×10^{-13}	5.48228×10^{-13}	1.16351×10^{-13}	1.28564×10^{-13}
0.9	(4,4)	1.84722×10^{-1}	1.49677×10^{-1}	1.13969×10^{-1}	1.13951×10^{-1}
	(8,8)	4.46693×10^{-5}	3.19146×10^{-5}	1.78888×10^{-5}	1.78816×10^{-5}
	(12,12)	1.31451×10^{-9}	8.93408×10^{-10}	4.27471×10^{-10}	4.27336×10^{-10}
	(16,16)	1.7919×10^{-13}	3.80598×10^{-14}	2.26041×10^{-13}	2.14051×10^{-13}
1.0	(4,4)	1.725×10^{-1}	1.39696×10^{-1}	1.06467×10^{-1}	1.06484×10^{-1}
	(8,8)	4.16394×10^{-5}	2.97589×10^{-5}	1.66843×10^{-5}	1.66798×10^{-5}
	(12,12)	1.22865×10^{-9}	8.35128×10^{-10}	3.99611×10^{-10}	3.99526×10^{-10}
	(16,16)	9.76996×10^{-15}	6.21725×10^{-15}	2.66454×10^{-15}	2.66454×10^{-15}

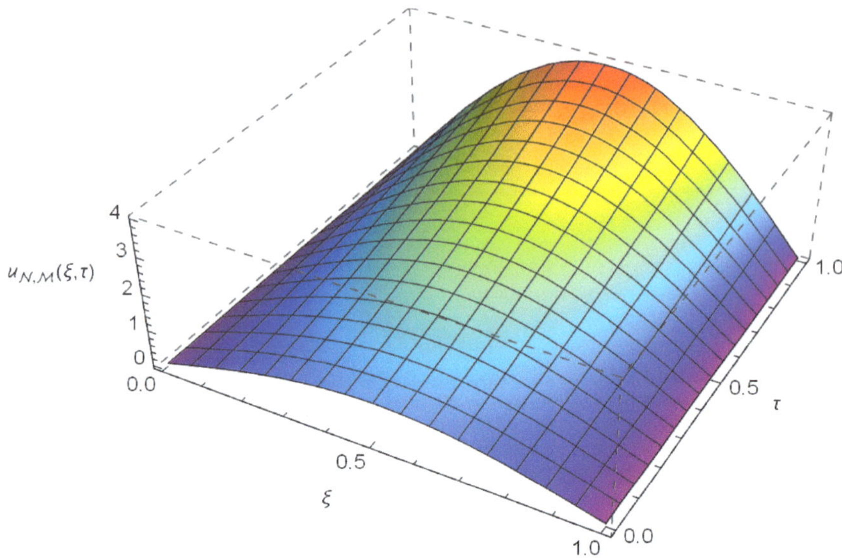

Figure 1. Numerical solution of the problem (30), where $\sigma_1 = \varrho_1 = \sigma_2 = \varrho_2 = 0$, $\nu = 0.5$ and $\mathcal{N} = \mathcal{M} = 16$.

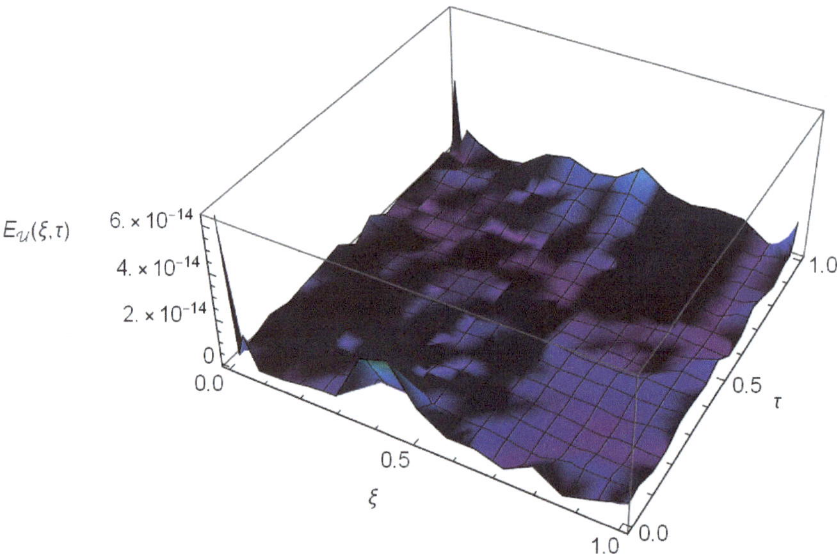

Figure 2. Absolute errors graph of the problem (30), where $\sigma_1 = \varrho_1 = \sigma_2 = \varrho_2 = 0$, $\nu = 0.5$ and $\mathcal{N} = \mathcal{M} = 16$.

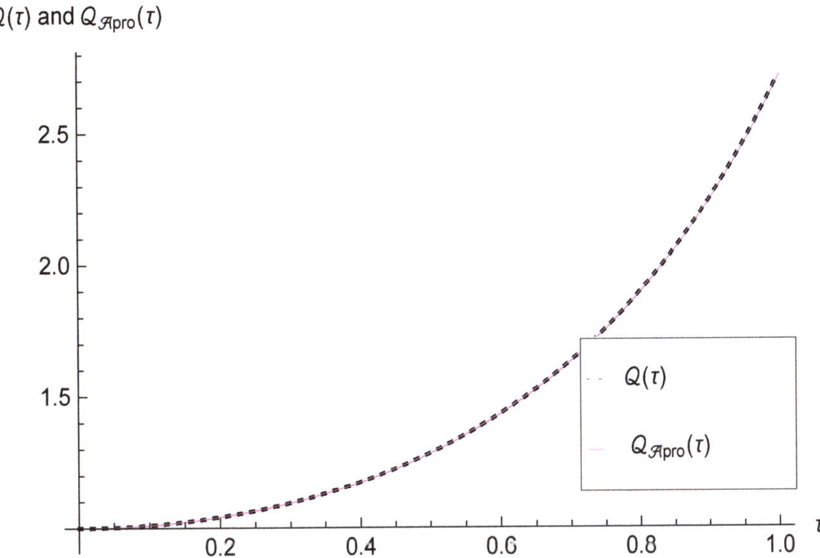

Figure 3. Curves of the exact and numerical solutions of $\mathcal{Q}(\tau)$ of the problem (30), where $\sigma_1 = \varrho_1 = \sigma_2 = \varrho_2 = 0$, $\nu = 0.5$ and $\mathcal{N} = \mathcal{M} = 16$.

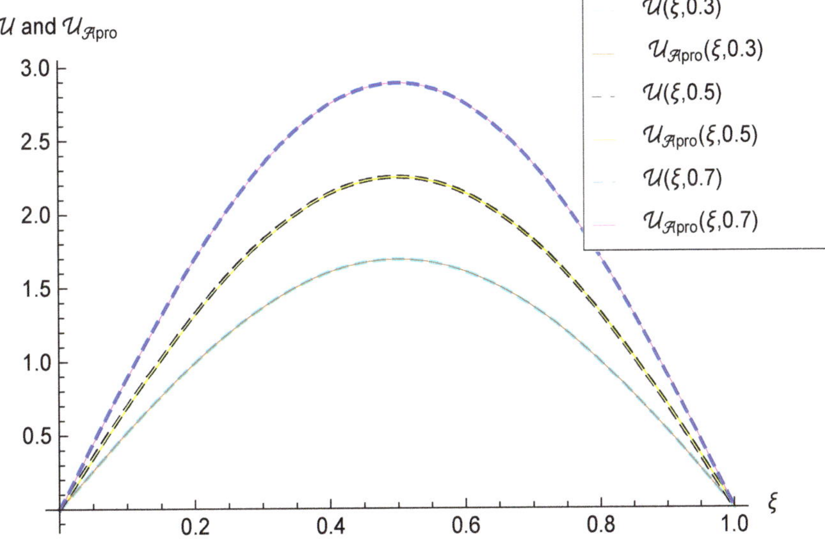

Figure 4. x-Curves of the exact and numerical solutions of $\mathcal{U}(\xi, \tau)$ of the problem (30), where $\sigma_1 = \varrho_1 = \sigma_2 = \varrho_2 = 0$, $\nu = 0.5$ and $\mathcal{N} = \mathcal{M} = 16$.

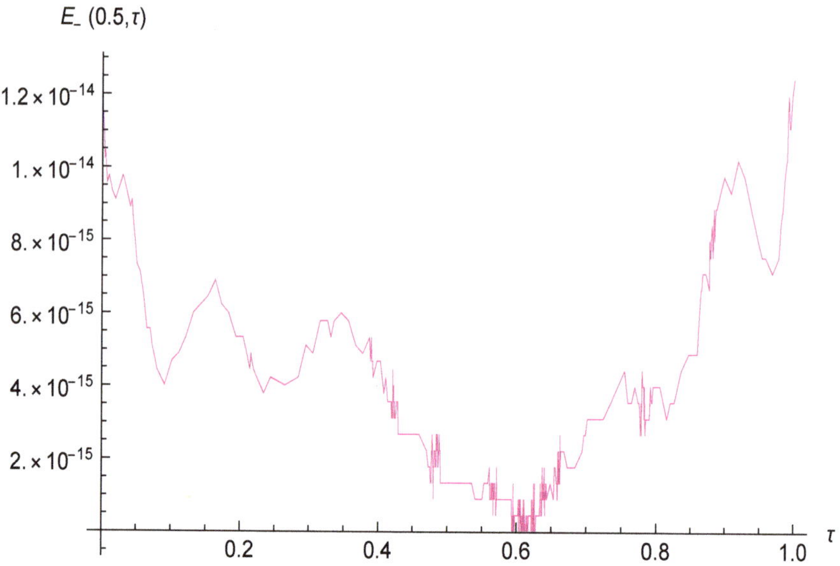

Figure 5. τ-Absolute errors $E_{\mathcal{U}}(0.5, \tau)$ graph of the problem (30), where $\sigma_1 = \varrho_1 = \sigma_2 = \varrho_2 = 0$, $\nu = 0.5$ and $\mathcal{N} = \mathcal{M} = 16$.

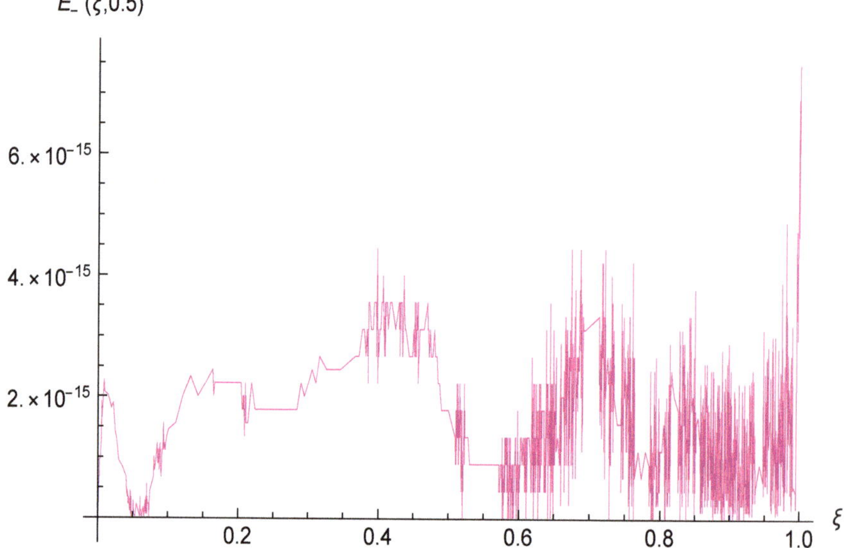

Figure 6. ξ-Absolute errors graph $E_{\mathcal{U}}(\xi, 0.5)$ of the problem (30), where $\sigma_1 = \varrho_1 = \sigma_2 = \varrho_2 = 0$, $\nu = 0.5$ and $\mathcal{N} = \mathcal{M} = 16$.

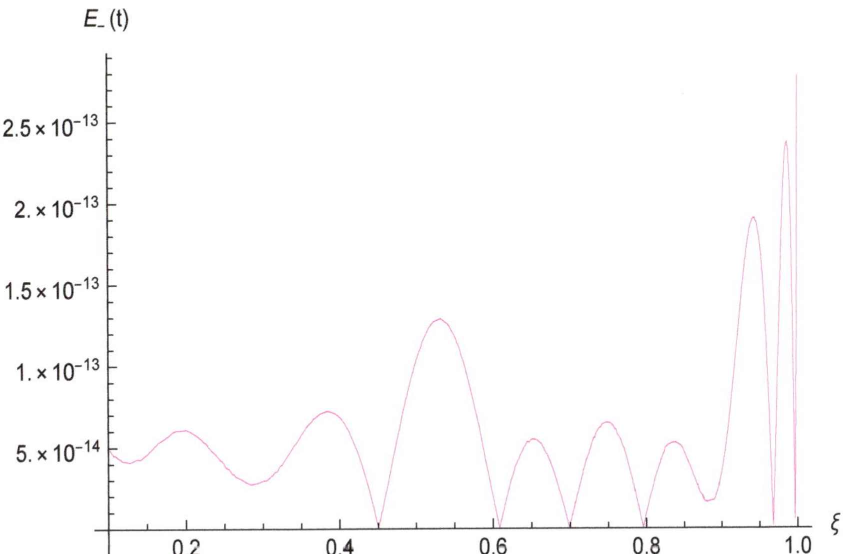

Figure 7. Absolute errors $E_Q(\tau)$ graph of the problem (30), where $\sigma_1 = \varrho_1 = \sigma_2 = \varrho_2 = 0$, $\nu = 0.5$ and $\mathcal{N} = \mathcal{M} = 16$.

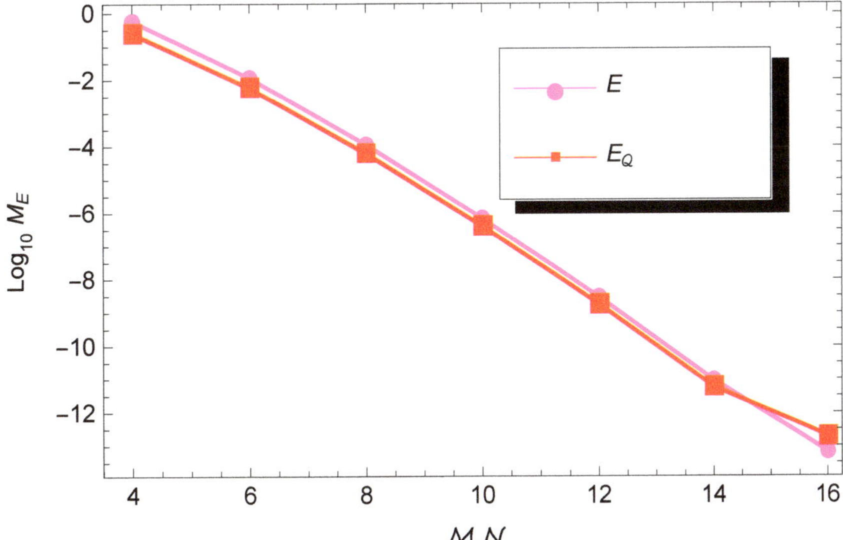

Figure 8. M_E convergence of problem (30), where $\sigma_1 = \varrho_1 = \sigma_2 = \varrho_2 = 0$, $\nu = 0.5$.

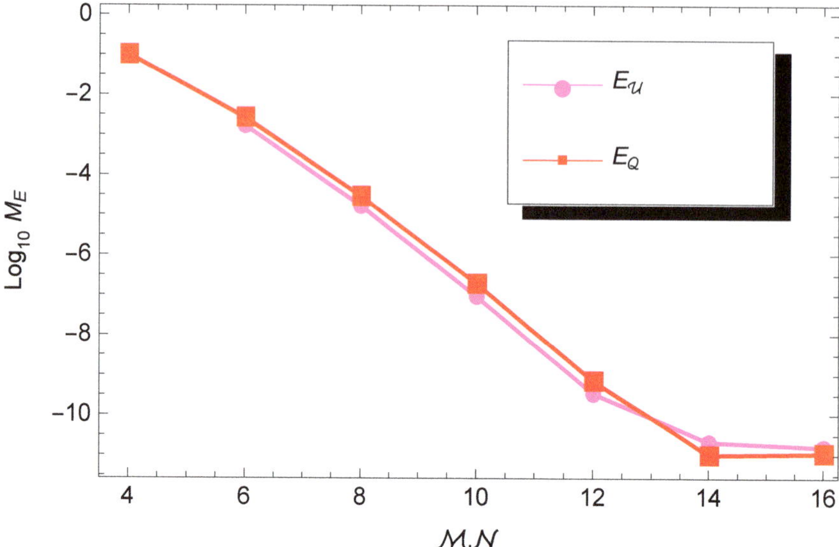

Figure 9. M_E convergence of problem (30), where $\sigma_1 = \varrho_1 = -0.5$, $\sigma_2 = \varrho_2 = 0.5$, $\nu = 0.9$.

Example 2. *We consider the IHEs*

$$\frac{\partial^\nu}{\partial t^\nu}\Big(\mathcal{U}(\xi,\tau) - \mathcal{U}(\xi,0)\Big) = \frac{\partial^2}{\partial x^2}(\mathcal{U}(\xi,\tau)) + \mathcal{Q}(\tau)\cos\left(\pi\left(\xi + \frac{1}{4}\right)\right)\left(\frac{2\tau^{-\nu}}{\Gamma(3-\nu)} + \pi^2\right), \tag{33}$$
$$(\xi,\tau) \in [0,1] \times [0,1],$$

with the nonlocal conditions

$$\mathcal{U}(\xi,0) = \sin(\pi\xi), \quad \xi \in [0,1],$$
$$\mathcal{U}(0,\tau) = \mathcal{U}(1,\tau), \quad \frac{\partial}{\partial x}\Big(\mathcal{U}(\xi,\tau)\Big)_{\xi=0} + \pi\mathcal{U}(0,\tau) = 0, \quad \tau \in [0,1], \tag{34}$$

and the extra energy condition

$$\int_0^1 \mathcal{U}(\xi,\tau)d\xi = -\frac{\sqrt{2}\tau^2}{\pi}, \tag{35}$$

the exact solution and unknown source function are given by $\mathcal{U}(\xi,\tau) = \tau^2 \cos\left(\pi\xi + \frac{\pi}{4}\right)$, $\mathcal{Q}(\tau) = \tau^2$.

Tables 3 and 4 display the maximum absolute errors $M_{E_\mathcal{U}}$ and $M_{E_\mathcal{Q}}$ of the approximate solution at different values of parameters, respectively. In Figures 10 and 11, with values of parameters listed in their captions, numerical solution and absolute errors graphs are displayed, respectively. Additionally, the exact and approximate solutions are displayed in Figures 12 and 13 for $\mathcal{Q}(\tau)$ and temperature function $\mathcal{U}(\xi,\tau)$, respectively. However, absolute errors curves of the temperature and $\mathcal{Q}(\tau)$ functions are displayed in Figures 14–16.

Table 3. M_{E_u} for problem (33).

ν	$(\mathcal{N}, \mathcal{M})$	$(0, 0, 0, 0)$	$(0, -0.5, 0, 0.5)$	$(-0.5, -0.5, 0, 0)$	$(-0.5, -0.5, 0.5, 0.5)$
0.5	(4,4)	1.27607×10^{-1}	3.15438×10^{-2}	8.46812×10^{-2}	8.46738×10^{-2}
	(8,8)	4.7083×10^{-5}	3.90995×10^{-5}	1.91081×10^{-5}	1.91082×10^{-5}
	(12,12)	1.66851×10^{-9}	1.97664×10^{-9}	5.41362×10^{-10}	5.41373×10^{-10}
	(16,16)	2.0849×10^{-14}	3.87468×10^{-14}	1.69864×10^{-14}	3.38618×10^{-14}
0.9	(4,4)	1.17572×10^{-1}	3.60552×10^{-2}	7.75196×10^{-2}	7.74936×10^{-2}
	(8,8)	4.15201×10^{-5}	2.09117×10^{-5}	1.65554×10^{-5}	1.67636×10^{-5}
	(12,12)	1.63111×10^{-9}	1.97664×10^{-9}	5.35284×10^{-10}	3.46596×10^{-10}
	(16,16)	9.74665×10^{-11}	7.43456×10^{-11}	1.87779×10^{-10}	1.60306×10^{-11}

Table 4. M_{E_ϱ} for problem (33).

ν	$(\mathcal{N}, \mathcal{M})$	$(0, 0, 0, 0)$	$(0, -0.5, 0, 0.5)$	$(-0.5, -0.5, 0, 0)$	$(-0.5, -0.5, 0.5, 0.5)$
0.5	(4,4)	1.81999×10^{-1}	6.99633×10^{-3}	1.22372×10^{-1}	0.122373×10^{-1}
	(8,8)	8.67215×10^{-5}	4.68354×10^{-5}	3.46128×10^{-5}	3.46127×10^{-5}
	(12,12)	3.25847×10^{-9}	2.82092×10^{-9}	1.05757×10^{-9}	1.05754×10^{-9}
	(16,16)	7.70495×10^{-14}	6.28386×10^{-14}	4.31877×10^{-14}	4.75175×10^{-14}
0.9	(4,4)	1.52889×10^{-1}	2.92712×10^{-2}	1.01293×10^{-1}	1.0128×10^{-1}
	(8,8)	7.33228×10^{-5}	1.97339×10^{-5}	2.9295×10^{-5}	2.93316×10^{-5}
	(12,12)	2.13166×10^{-9}	2.82092×10^{-9}	6.91775×10^{-10}	7.22927×10^{-10}
	(16,16)	6.70036×10^{-11}	5.29218×10^{-11}	1.33594×10^{-10}	1.13941×10^{-11}

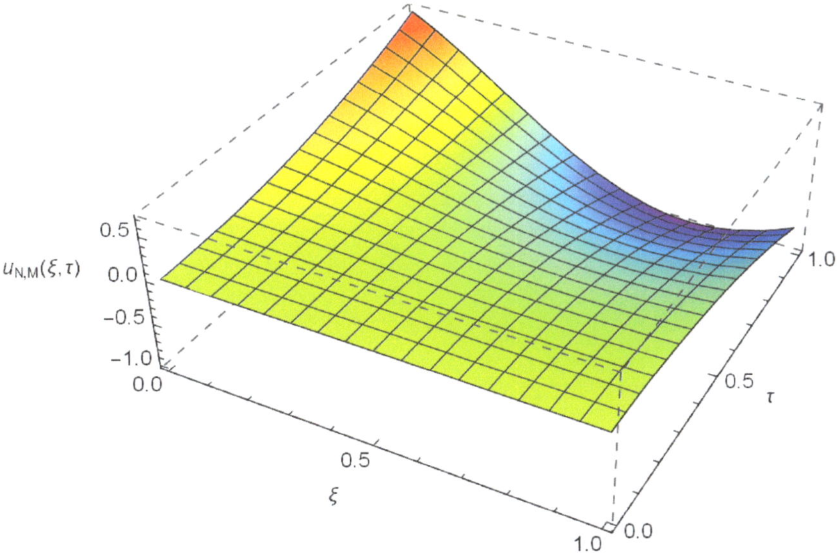

Figure 10. Numerical solution of the problem (33), where $\sigma_1 = \varrho_1 = \sigma_2 = \varrho_2 = 0$, $\nu = 0.9$ and $\mathcal{N} = \mathcal{M} = 16$.

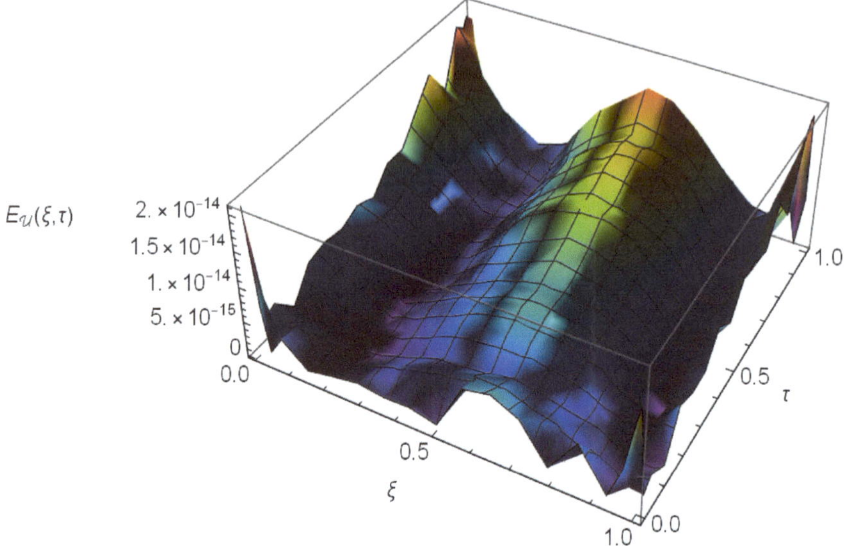

Figure 11. Absolute errors graph of the problem (33), where $\sigma_1 = \varrho_1 = \sigma_2 = \varrho_2 = 0$, $\nu = 0.9$ and $\mathcal{N} = \mathcal{M} = 16$.

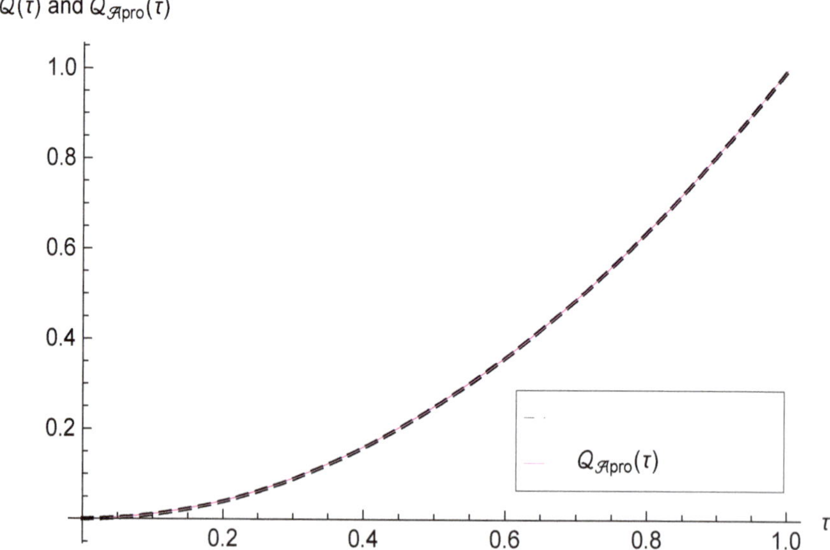

Figure 12. Curves of the exact and numerical solutions of $\mathcal{Q}(\tau)$ of the problem (33), where $\sigma_1 = \varrho_1 = \sigma_2 = \varrho_2 = 0$, $\nu = 0.9$ and $\mathcal{N} = \mathcal{M} = 16$.

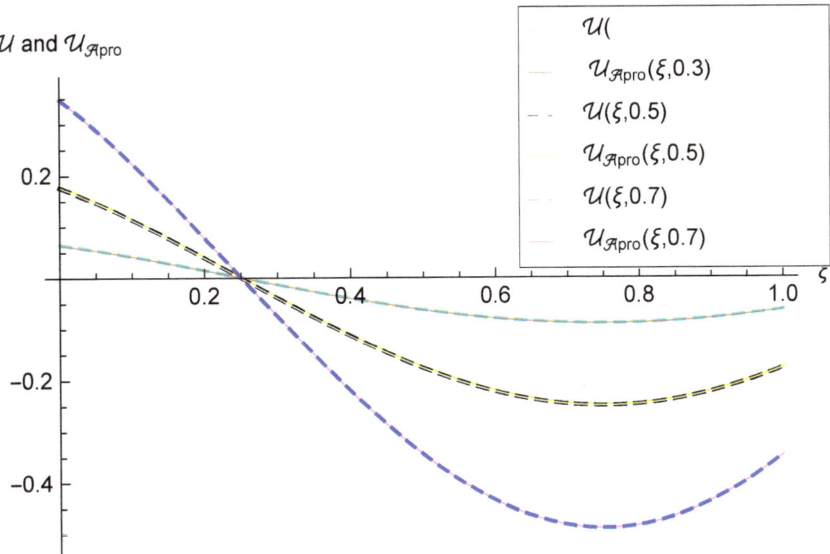

Figure 13. *x*-Curves of the exact and numerical solutions of $\mathcal{U}(\xi, \tau)$ of the problem (33), where $\sigma_1 = \varrho_1 = \sigma_2 = \varrho_2 = 0$, $\nu = 0.9$ and $\mathcal{N} = \mathcal{M} = 16$.

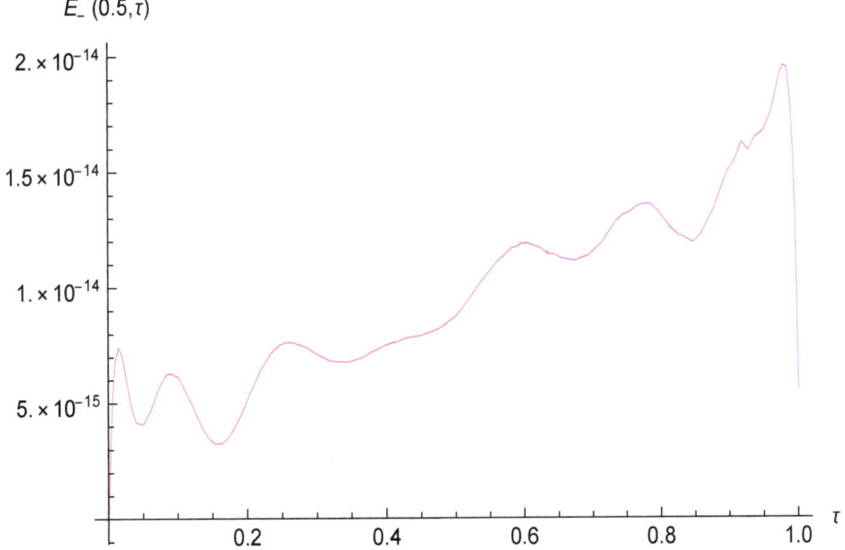

Figure 14. τ-Absolute errors $E_{\mathcal{U}}(0.5, \tau)$ graph of the problem (33), where $\sigma_1 = \varrho_1 = \sigma_2 = \varrho_2 = 0$, $\nu = 0.9$ and $\mathcal{N} = \mathcal{M} = 16$.

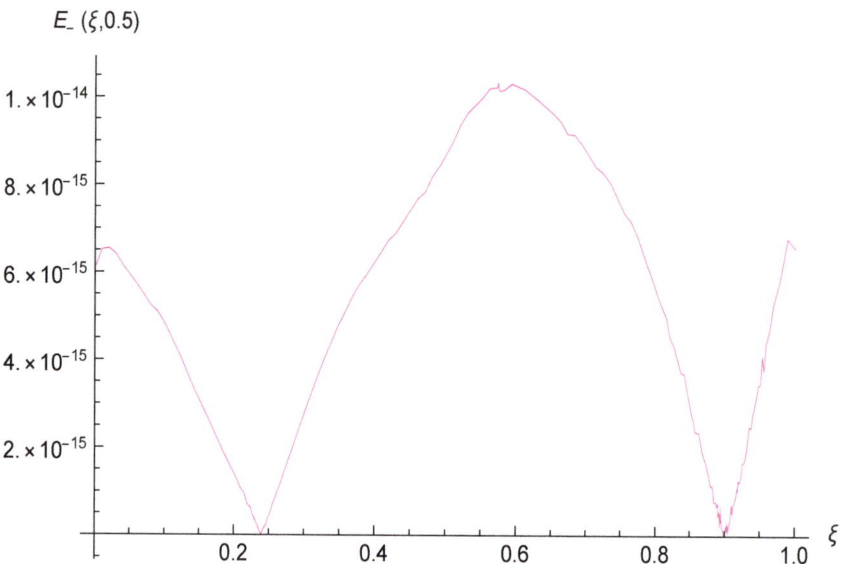

Figure 15. ξ-Absolute errors graph $E_{\mathcal{U}}(\xi, 0.5)$ of the problem (33), where $\sigma_1 = \varrho_1 = \sigma_2 = \varrho_2 = 0$, $\nu = 0.9$ and $\mathcal{N} = \mathcal{M} = 16$.

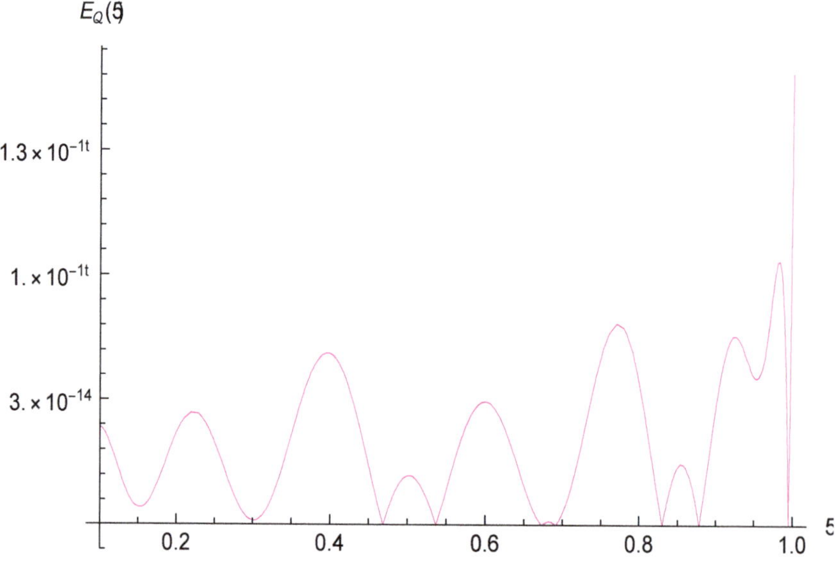

Figure 16. Absolute errors $E_{\mathcal{Q}}(\tau)$ graph of the problem (33), where $\sigma_1 = \varrho_1 = \sigma_2 = \varrho_2 = 0$, $\nu = 0.9$ and $\mathcal{N} = \mathcal{M} = 16$.

5. Conclusions

We have constructed fully shifted Jacobi collocation schemes to study the time-fractional IHEs. Various orthogonal polynomials can be acquired as a particular case of the shifted Jacobi polynomials, such as the shifted Chebyshev of the first or second or third or fourth kind, shifted Legendre, and shifted Gegenbauer. Recently, shifted Jacobi polynomials have been used for solving fractional problems via collocation techniques and have acquired growing popularity due to the ability to obtained the approximate solution depends on the shifted Jacobi parameters σ and ϱ. The powerful proposed approach yielded impressive numerical results that demonstrate the algorithm's great efficiency. The study was treated with both local and nonlocal conditions. The algorithm's results open the way for more studies in this field to be conducted in the future to display additional results in the future.

Author Contributions: Data curation, M.A.A., A.Z.M.A., M.M.B., A.S.A., M.A.Z. and R.M.H.; Formal analysis, M.A.A., A.Z.M.A., M.M.B., A.S.A., M.A.Z. and R.M.H.; Funding acquisition, M.A.A.; Methodology, M.A.A., A.Z.M.A., M.A.Z. and R.M.H.; Software, M.A.A., A.Z.M.A., A.S.A. and M.A.Z.; Writing—original draft, M.A.A., A.Z.M.A., M.M.B., A.S.A., M.A.Z. and R.M.H.; Writing—review and editing, M.A.A., A.Z.M.A., M.A.Z. and R.M.H. All authors contributed equally. All authors have read and agreed to the published version of the manuscript.

Funding: This research was supported by the Deanship of Scientific Research at Imam Mohammad Ibn Saud Islamic University through Research Group no. RG-21-09-05.

Institutional Review Board Statement: Not applicable.

Informed Consent Statement: Not applicable.

Data Availability Statement: Not applicable.

Acknowledgments: The authors extend their appreciation to the Deanship of Scientific Research at Imam Mohammad Ibn Saud Islamic University for funding this work

Conflicts of Interest: The authors declare no conflict of interest.

References

1. Herrmann, R. *Fractional Calculus: An Introduction for Physicists*; World Scientific: Singapore, 2011.
2. Tarasov, V.E. *Fractional Dynamics: Applications of Fractional Calculus to Dynamics of Particles, Fields and Media*; Springer Science & Business Media: Berlin/Heidelberg, Germany, 2011.
3. West, B.J. *Fractional Calculus View of Complexity: Tomorrow's Science*; CRC Press: Boca Raton, FL, USA, 2016.
4. Kilbas, A.A.; Srivastava, H.M.; Trujillo, J.J. *Theory and Applications of Fractional Differential Equations*; Elsevier: Amsterdam, The Netherlands, 2006; Volume 204.
5. West, B.J. *Nature's Patterns and the Fractional Calculus*; De Gruyter: Berlin, Germany, 2017.
6. Seki, K.; Wojcik, M.; Tachiya, M. Fractional reaction-diffusion equation. *J. Chem. Phys.* **2003**, *119*, 2165–2170. [CrossRef]
7. Baleanu, D.; Magin, R.L.; Bhalekar, S.; Daftardar-Gejji, V. Chaos in the fractional order nonlinear Bloch equation with delay. *Commun. Nonlinear Sci. Numer. Simul.* **2015**, *25*, 41–49. [CrossRef]
8. Shen, S.; Liu, F.; Liu, Q.; Anh, V. Numerical simulation of anomalous infiltration in porous media. *Numer. Algorithms* **2015**, *68*, 443–454. [CrossRef]
9. Podlubny, I. *Fractional Differential Equations: An Introduction to Fractional Derivatives, Fractional Differential Equations, to Methods of Their Solution and Some of Their Applications*; Elsevier: Amsterdam, The Netherlands, 1998.
10. Sumelka, W.; Łuczak, B.; Gajewski, T.; Voyiadjis, G. Modelling of AAA in the framework of time-fractional damage hyperelasticity. *Int. J. Solids Struct.* **2020**, *206*, 30–42. [CrossRef]
11. Podlubny, I. *Fractional Differential Equations*; Academic Press: San Diego, CA, USA; Boston, MA, USA, 1999.
12. Zhou, Y.; Zhang, Y. Noether symmetries for fractional generalized Birkhoffian systems in terms of classical and combined Caputo derivatives. *Acta Mech.* **2020**, *231*, 3017–3029. [CrossRef]
13. Eidelman, S.D.; Kochubei, A.N. Cauchy problem for fractional diffusion equations. *J. Differ. Equ.* **2004**, *199*, 211–255. [CrossRef]
14. Luchko, Y. Some uniqueness and existence results for the initial-boundary-value problems for the generalized time-fractional diffusion equation. *Comput. Math. Appl.* **2010**, *59*, 1766–1772. [CrossRef]
15. Metzler, R.; Klafter, J. Boundary value problems for fractional diffusion equations. *Phys. A Stat. Mech. Appl.* **2000**, *278*, 107–125. [CrossRef]
16. Yuste, S.; Acedo, L.; Lindenberg, K. Reaction front in an A + B $\rightarrow$ C reaction-subdiffusion process. *Phys. Rev. E* **2004**, *69*, 036126. [CrossRef] [PubMed]
17. Liu, F.; Burrage, K. Novel techniques in parameter estimation for fractional dynamical models arising from biological systems. *Comput. Math. Appl.* **2011**, *62*, 822–833. [CrossRef]
18. Yu, B.; Jiang, X.; Wang, C. Numerical algorithms to estimate relaxation parameters and Caputo fractional derivative for a fractional thermal wave model in spherical composite medium. *Appl. Math. Comput.* **2016**, *274*, 106–118. [CrossRef]
19. Liu, F.; Anh, V.; Turner, I. Numerical solution of the space fractional Fokker–Planck equation. *J. Comput. Appl. Math.* **2004**, *166*, 209–219. [CrossRef]

20. Metzler, R.; Klafter, J. The random walk's guide to anomalous diffusion: a fractional dynamics approach. *Phys. Rep.* **2000**, *339*, 1–77. [CrossRef]
21. Fan, W.; Jiang, X.; Qi, H. Parameter estimation for the generalized fractional element network Zener model based on the Bayesian method. *Phys. A Stat. Mech. Appl.* **2015**, *427*, 40–49. [CrossRef]
22. Cannon, J.R.; van der Hoek, J. Diffusion subject to the specification of mass. *J. Math. Anal. Appl.* **1986**, *115*, 517–529. [CrossRef]
23. Cannon, J.R.; Esteva, S.P.; Van Der Hoek, J. A Galerkin procedure for the diffusion equation subject to the specification of mass. *SIAM J. Numer. Anal.* **1987**, *24*, 499–515. [CrossRef]
24. Huang, C.H.; Lin, C.Y. Inverse hyperbolic conduction problem in estimating two unknown surface heat fluxes simultaneously. *J. Thermophys. Heat Transf.* **2008**, *22*, 766–774. [CrossRef]
25. Yang, C.y. Direct and inverse solutions of the two-dimensional hyperbolic heat conduction problems. *Appl. Math. Model.* **2009**, *33*, 2907–2918. [CrossRef]
26. Tang, D.; Araki, N. An Inverse Analysis to Estimate Relaxation Parameters and Thermal Diffusivity with a Universal Heat Conduction Equation 1. *Int. J. Thermophys.* **2000**, *21*, 553–561. [CrossRef]
27. Wang, L.; Liu, J. Total variation regularization for a backward time-fractional diffusion problem. *Inverse Probl.* **2013**, *29*, 115013. [CrossRef]
28. Zhang, Y.; Xu, X. Inverse source problem for a fractional diffusion equation. *Inverse Probl.* **2011**, *27*, 035010. [CrossRef]
29. Bhrawy, A.H. An efficient Jacobi pseudospectral approximation for nonlinear complex generalized Zakharov system. *Appl. Math. Comput.* **2014**, *247*, 30–46. [CrossRef]
30. Bhrawy, A.H. A Jacobi spectral collocation method for solving multi-dimensional nonlinear fractional sub-diffusion equations. *Numer. Algorithms* **2016**, *73*, 91–113. [CrossRef]
31. Bhrawy, A.H.; Abdelkawy, M.A.; Baleanu, D.; Amin, A.Z. A spectral technique for solving two-dimensional fractional integral equations with weakly singular kernel. *Hacet. J. Math. Stat.* **2018**, *47*, 553–566. [CrossRef]
32. Abdelkawy, M.A.; Doha, E.H.; Bhrawy, A.H.; Amin, A.Z. Efficient pseudospectral scheme for 3D integral equations. *Proc. Rom. Acad. Ser. A Math. Phys. Tech. Sci. Inf. Sci* **2017**, *18*, 199–206.
33. Ameen, I.G.; Zaky, M.A.; Doha, E.H. Singularity preserving spectral collocation method for nonlinear systems of fractional differential equations with the right-sided Caputo fractional derivative. *J. Comput. Appl. Math.* **2021**, *392*, 113468. [CrossRef]
34. Abdelkawy, M.A.; Amin, A.Z.; Bhrawy, A.H.; Machado, J.A.T.; Lopes, A.M. Jacobi collocation approximation for solving multi-dimensional Volterra integral equations. *Int. J. Nonlinear Sci. Numer. Simul.* **2017**, *18*, 411–425. [CrossRef]
35. Doha, E.; Abdelkawy, M.; Amin, A.; Baleanu, D. Spectral technique for solving variable-order fractional Volterra integro-differential equations. *Numer. Methods Partial. Differ. Equ.* **2018**, *34*, 1659–1677. [CrossRef]
36. Doha, E.H.; Bhrawy, A.H.; Abdelkawy, M.A.; Van Gorder, R.A. Jacobi–Gauss–Lobatto collocation method for the numerical solution of 1+ 1 nonlinear Schrödinger equations. *J. Comput. Phys.* **2014**, *261*, 244–255. [CrossRef]
37. Doha, E.H.; Abdelkawy, M.A.; Amin, A.Z.; Baleanu, D. Approximate solutions for solving nonlinear variable-order fractional Riccati differential equations. *Nonlinear Anal. Model. Control.* **2019**, *24*, 176–188. [CrossRef]
38. Bhrawy, A.H.; Zaky, M.A.; Baleanu, D. New numerical approximations for space-time fractional Burgers' equations via a Legendre spectral-collocation method. *Rom. Rep. Phys.* **2015**, *67*, 340–349.
39. Bhrawy, A.H.; Zaky, M.A. Highly accurate numerical schemes for multi-dimensional space variable-order fractional Schrödinger equations. *Comput. Math. Appl.* **2017**, *73*, 1100–1117. [CrossRef]
40. Bhrawy, A.; Zaky, M. An improved collocation method for multi-dimensional space–time variable-order fractional Schrödinger equations. *Appl. Numer. Math.* **2017**, *111*, 197–218. [CrossRef]
41. Bhrawy, A.; Doha, E.; Baleanu, D.; Ezz-Eldien, S.; Abdelkawy, M. An accurate numerical technique for solving fractional optimal control problems. *Differ. Equ.* **2015**, *15*, 23.
42. Bhrawy, A.H.; Abdelkawy, M.A. A fully spectral collocation approximation for multi-dimensional fractional Schrödinger equations. *J. Comput. Phys.* **2015**, *294*, 462–483. [CrossRef]
43. Abdelkawy, M.; Zaky, M.A.; Bhrawy, A.H.; Baleanu, D. Numerical simulation of time variable fractional order mobile-immobile advection-dispersion model. *Rom. Rep. Phys.* **2015**, *67*, 773–791.
44. Hendy, A.S.; Zaky, M.A.; Hafez, R.M.; De Staelen, R.H. The impact of memory effect on space fractional strong quantum couplers with tunable decay behavior and its numerical simulation. *Sci. Rep.* **2021**, *11*, 1–15. [CrossRef]
45. Diethelm, K.; Ford, N.J. Analysis of fractional differential equations. *J. Math. Anal. Appl.* **2002**, *265*, 229–248. [CrossRef]

fractal and fractional

Article

State of Charge Estimation of Lithium-Ion Batteries Based on Fuzzy Fractional-Order Unscented Kalman Filter

Liping Chen [1,*], Yu Chen [1], António M. Lopes [2], Huifang Kong [1] and Ranchao Wu [3]

1   School of Electrical Engineering and Automation, Hefei University of Technology, Hefei 230009, China; 2019170408mail@hfut.edu.cn (Y.C.); 1989800024@hfut.edu.cn (H.K.)
2   LAETA/INEGI, Faculty of Engineering, University of Porto, Rua Dr. Roberto Frias, 4200-465 Porto, Portugal; aml@fe.up.pt
3   School of Mathematics, Anhui University, Hefei 230039, China; rcwu@ahu.edu.cn
*   Correspondence: lip_chen@hfut.edu.cn

Abstract: The covariance matrix of measurement noise is fixed in the Kalman filter algorithm. However, in the process of battery operation, the measurement noise is affected by different charging and discharging conditions and the external environment. Consequently, obtaining the noise statistical characteristics is difficult, which affects the accuracy of the Kalman filter algorithm. In order to improve the estimation accuracy of the state of charge (SOC) of lithium-ion batteries under actual working conditions, a fuzzy fractional-order unscented Kalman filter (FFUKF) is proposed. The algorithm combines fuzzy inference with fractional-order unscented Kalman filter (FUKF) to infer the measurement noise in real time and take advantage of fractional calculus in describing the dynamic behavior of the lithium batteries. The accuracy of the SOC estimation under different working conditions at three different temperatures is verified. The results show that the accuracy of the proposed algorithm is superior to those of the FUKF and extended Kalman filter (EKF) algorithms.

Keywords: Kalman filter; state of charge; fuzzy inference; lithium-ion batteries

Citation: Chen, L.; Chen, Y.; Lopes, A.M.; Kong, H.; Wu, R. State of Charge Estimation of Lithium-Ion Batteries Based on Fuzzy Fractional-Order Unscented Kalman Filter. *Fractal Fract.* **2021**, *5*, 91. https://doi.org/10.3390/fractalfract5030091

Academic Editor: Ivanka Stamova

Received: 25 June 2021
Accepted: 6 August 2021
Published: 8 August 2021

Publisher's Note: MDPI stays neutral with regard to jurisdictional claims in published maps and institutional affiliations.

1. Introduction

The automobile industry has payed extensive attention to new energies to reduce the emissions of greenhouse gases [1]. An important component of the new energy vehicles is the power battery system. Lithium-ion batteries have the advantage of high energy density and excellent performance cycles [2]. However, their safe and effective management are crucial. The battery management system (BMS) is critical and the state of charge (SOC) estimation plays a vital role in the BMS [3,4]. However, the SOC of the power battery cannot be measured directly, and some efficient and accurate estimation methods must be employed. Compared with electrochemical and data-driven models, the equivalent circuit model (ECM) was widely adopted in recent years, which uses ideal resistors, capacitors, constant voltage sources and other circuit devices to form a circuit network that describes the characteristics of power batteries [5]. In addition, to obtain a reliable battery model, the SOC estimation also requires a high precision algorithm. Recently, effective estimation methods have been presented, such as the open circuit voltage (OCV) method [6], ampere-hour integration method [7,8], Kalman filter algorithm, neural network method [9,10], sliding mode observer [11,12], H_∞ filter [13,14], adaptive particle filter [15] and others. Each of these methods presents advantages and disadvantages. For example, the OCV is the most direct method, but it requires the battery to stand for a long enough time. The ampere-hour integration method is classic, easy and widely used, but its initial SOC is difficult to obtain. The neural network method is popular with high estimation accuracy, but it requires a large amount of experimental data as prior knowledge, and these data can fully reflect the characteristics of the battery.

The Kalman filter is currently the most used estimation algorithm. It includes the extended Kalman filter (EKF) [16–18], unscented Kalman filter (UKF) [19,20], adaptive Kalman filter [21–23], fuzzy unscented Kalman filter [24] and other variants. The EKF uses the Taylor expansion to linearize high-order terms, resulting in error accumulation during the iterative process. The UKF uses the unscented transformation to linearize the nonlinear function of random variables by linear regression. Generally speaking, the unscented transformation is more accurate than the Taylor series approach. However, the statistical characteristics of the measurement noise are vital for the UKF, being difficult to obtain accurately. Indeed, the statistical characteristics of the noise are affected by uncertain factors, such as the system noise, which causes the UKF to converge slowly and even to diverge. The adaptive Kalman filter method can estimate the process and the observation noise online, improving the accuracy of the estimation [25]. However, when the nonlinearity is strong, the estimation accuracy is limited.

Recently, it was found that fractional-order ECM (FECM), where constant phase elements (CPE) are used instead of ideal capacitors [26], have advantages for describing the dynamic behavior of lithium batteries. The FECM can accurately simulate the double-layer effect of the battery electrode. Therefore, SOC algorithms based on FECM, such as fractional-order unscented Kalman filter (FUKF) [27,28] and fractional-order extended Kalman filter (FEKF) [29–31] have been proposed. Experimental results have also shown that these methods improve the accuracy of the estimation when compared with the ECM. However, the statistical characteristics of the measurement noise are still hard to obtain accurately [32] and affect greatly the accuracy of the SOC estimation. In order to mitigate this shortcoming, a new fuzzy fractional-order unscented Kalman filter (FFUKF) that combines fuzzy inference and FUKF is proposed. This method can infer the measurement noise in real time and has higher accuracy compared with traditional algorithms, according to the difference between the actual and the theoretical value of the noise measurements. The covariance matrix of the measurement noise is adjusted continuously to make the FUKF more adaptive and accurate.

The main objective of this paper is (1) to propose a fractional-order second-order RC equivalent circuit model of lithium batteries based on particle swarm optimization (PSO), (2) to derive a FFUKF to solve the influence of measurement noise on SOC estimation accuracy, (3) to test the FFUKF under different working conditions and compare its performance with the FUKF and EKF.

The paper is organized as follows. Section 2 introduces the fractional-order model and its parameter identification. Section 3 presents the fuzzy controller. Section 4 lists the steps of the FFUKF algorithm. Section 5 compares the results with those obtained with existing algorithms. Section 6 outlines the main conclusions.

2. Theory and Method Research

2.1. Fractional-Order Calculus

In contrast with the integer-order derivative, the fractional-order derivative have many definitions, such as the Grünwald-Letnikov (GL), Riemann-Liouville (RL), and Caputo formulations. Here, the GL definition is used [33]:

$$
{}_{t_0}D_t^\alpha x(t) = \lim_{\Delta T \to 0} \left(\frac{1}{\Delta T^\alpha} \right) \sum_{j=0}^{[t/\Delta T]} (-1)^j \binom{\alpha}{j} x(t - j\Delta T), \tag{1}
$$

$$
\binom{\alpha}{j} = \frac{\Gamma(\alpha + 1)}{\Gamma(j + 1) \cdot \Gamma(\alpha - j + 1)}. \tag{2}
$$

with $\Gamma(\alpha)$ given by:

$$
\Gamma(\alpha) = \int_0^{+\infty} \xi^{\alpha - 1} e^{-\xi} d\xi, \tag{3}
$$

where t denotes the variable, with lower bound t_0, ΔT stands for the sampling time, and $[t/\Delta T]$ is the memory length. The continuous-time GL derivative ${}_{t_0}D_t^\alpha$ can be discretized with a fixed memory length L, yielding:

$$D^\alpha x_{k+1} = \frac{1}{\Delta T^\alpha} x_{k+1} + \left(\frac{1}{\Delta T^\alpha}\right) \sum_{j=1}^{L+1} (-1)^j \binom{\alpha}{j} x_{k+1-j}. \tag{4}$$

2.2. Fractional-Order Model

In order to describe accurately the internal electrochemical reaction that occurs in a battery, and to design an accurate and reliable lithium-ion battery SOC estimation algorithm, an accurate model is necessary. It has been shown that the ECM describes well the battery characteristics and that a second-order RC model yields good results in terms of accuracy and computational complexity [34]. Usually, the ECM includes two ideal resistors. However, for the complex electrochemical reactions inside the battery, the ideal capacitance cannot be simulated reasonably. Therefore, the CPE has been used instead of ideal capacitors, which resulted in the FECM. The impedance of a CPE is given by:

$$Z(s) = \frac{1}{Cs^\alpha}. \tag{5}$$

The fractional-order RC circuit used here is shown in Figure 1, where U_{oc} stands for the open circuit voltage, R_0 is the ohmic internal resistance, and R_1 and R_2 are the electrochemical polarization and concentration polarization resistances, respectively. The CPE_1 and CPE_2 stand for the fractional capacitors, V_0 represents the terminal voltage of the battery, and I corresponds to the load current. If we denote by V_1 and V_2 the voltages on the two parallel associations, respectively, then the dynamic equations can be expressed as:

$$\begin{cases} D^\alpha V_1(t) = -\dfrac{V_1(t)}{R_1 C_1} + \dfrac{I(t)}{C_1}, \\[2mm] D^\beta V_2(t) = -\dfrac{V_2(t)}{R_2 C_2} + \dfrac{I(t)}{C_2}. \end{cases} \tag{6}$$

where $\alpha, \beta \in (0,1)$ are the fractional orders of CPE_1 and CPE_2, respectively. The variable Q_n is the nominal capacity of the lithium-ion battery and η is the Coulomb efficiency. The SOC of the lithium battery can be written as:

$$\frac{dSOC(t)}{dt} = -\frac{\eta}{Q_n} I(t). \tag{7}$$

It follows from the Kirchhoff's voltage law that the output equation is given by:

$$V_0(t) = U_{oc} - I(t)R_0 - V_1(t) - V_2(t), \tag{8}$$

where OCV is a nonlinear function of the SOC. Equation (8) has been used to describe the OCV-SOC relationship [35,36], which is usually expressed as:

$$f[\theta(t)] = U_{oc} = \sum_{i=0}^{4} a_i SOC(t), \tag{9}$$

where a_i $(i = 0, \cdots, 4)$ are polynomial coefficients.

Further, the state space equation of the lithium-ion battery can be established as:

$$\begin{cases} D^\eta x(t) = Ax(t) + Bu(t), \\ y(t) = Cx(t) + Du(t), \end{cases} \tag{10}$$

where $\eta = [\alpha, \beta, 1]^T$ represents the incommensurate order vector, $x(t) = [V_1, V_2, SOC]^T$ is the state vector, $u(t)$ denotes the system input (battery current $I(t)$) and $y(t)$ represents the system output (battery terminal voltage V_0). The matrices A, B, C and D are given as:

$$A = \begin{bmatrix} -\frac{1}{R_1 C_1} & 0 & 0 \\ 0 & -\frac{1}{R_2 C_2} & 0 \\ 0 & 0 & 0 \end{bmatrix}, B = \begin{bmatrix} \frac{1}{C_1} \\ \frac{1}{C_2} \\ -\frac{\eta}{Q_n} \end{bmatrix}, \tag{11}$$

$$C = \begin{bmatrix} -1 & -1 & \frac{f[\theta(t)]}{SOC(t)} \end{bmatrix}, D = -R_0. \tag{12}$$

With Equation (4) in mind, model (10) can be written in discrete time:

$$\begin{cases} x_{k+1} = A_1 x_k + B_1 u_k - \displaystyle\sum_{j=2}^{L+1} (-1)^j \gamma_j^\eta x_{k+1-j}, \\ y_k = f(\theta_k) - V_{1k} - V_{2k} - R_0 I_k. \end{cases} \tag{13}$$

The matrices of A_1, B_1 and γ_j^η are as follows:

$$A_1 = \mathrm{diag}((\Delta T)^\alpha, (\Delta T)^\beta, (\Delta T))A + \mathrm{diag}(\alpha, \beta, 1), \tag{14}$$

$$B_1 = \mathrm{diag}((\Delta T)^\alpha, (\Delta T)^\beta, (\Delta T))B, \tag{15}$$

$$\gamma_j^\eta = \mathrm{diag}\left(\begin{pmatrix} \alpha \\ j \end{pmatrix}, \begin{pmatrix} \beta \\ j \end{pmatrix}, \begin{pmatrix} 1 \\ j \end{pmatrix}\right). \tag{16}$$

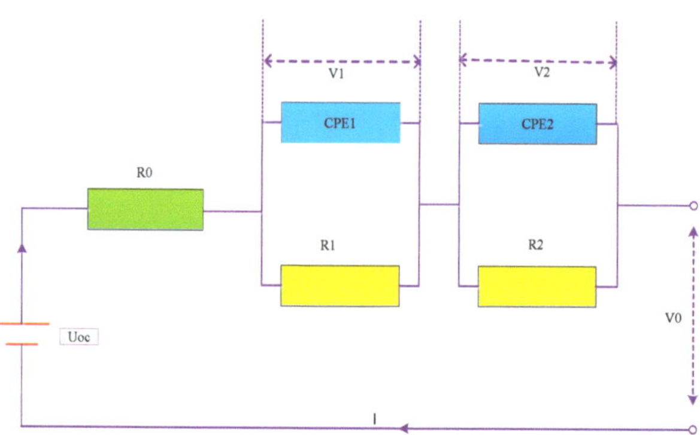

Figure 1. Equivalent circuit model of a lithium battery.

2.3. Model Parameter Identification and Validation

The main current and voltage data of the battery INR 18650-20R with a capacity 2000 mAH are provided by the CALCE Battery Research Group. The experimental platform is composed by the test samples, a thermal chamber (Weiss-Voetsch, Germany), an Arbin BT2000 battery test system (ARBIN, TX, USA), and a PC with Arbin software (V4.27, Caltest Instruments Ltd, Petersfield, UK) to give orders to the test system and monitor data information. All tests were performed for 0.8 and 0.5 battery level at 0 °C, 25 °C and 45 °C. We use three typical current and voltage test data sets of the vehicle operating conditions: Dynamic Stress Test (DST), Federal Urban Driving Schedule (FUDS) and Beijing Dynamic Stress Test (BJDST). Through the analysis of the established fractional-order model, we need to identify twelve parameters, which are $a_0, a_1, a_2, a_3, a_4, R_0, R_1, C_1, R_2, C_2, \alpha$ and β. Here, a PSO is used as the identification algorithm, but other methods are possible to estimate the parameters, such as, for example, the observer method [37,38]. The PSO originated in the study of the behavior of birds. The basic idea of the algorithm is to find optimal solutions through collaboration and information sharing between individuals in a group. The advantage of PSO is that it is simple and easy to implement with a limited number of parameter adjustments. Here, we set the goal of minimizing the root mean square error (RMSE) between the measured and the estimated voltages. Therefore, we define the objective function E as:

$$minE = \sum_{k=1}^{n} [V_o(k) - \widehat{V}_o(k)]^2, \tag{17}$$

where $V_o(k)$ and $\widehat{V}_o(k)$ are the measured and estimated voltages, respectively, and n is the number of the sampling points.

Table 1 shows the results of the parameter identification of the fractional-order model. For model validation, the DST is used. The current and voltage profiles of the DST at a temperature of 25 °C are shown in Figure 2. We verify that the DST condition is composed of many small cycles, each with a duration of 350 s [39]. Here, to reduce the complexity, a cycle is selected for parameter identification. Figure 3 presents the current and voltage profiles of a cycle. Figures 4 and 5 show the accuracy of the fractional-order model, which is also compared with an integer-order model. From Figure 5, we observe that the error of the fractional-order model can be kept within 40 mV. However, the maximum error of the integer-order model is 80 mV. The RMSE of the two models is 0.0125 and 0.0573, respectively. Therefore, from Figures 3–5, one can see that the fractional-order model can perform better than the integer-order one in modeling the change of terminal voltage, and that the fractional-order model is more accurate.

Table 1. The results of the fractional-model parameter identification.

a_0	a_1	a_2	a_3	a_4	R_0
2.4877	1.8243	0.6608	1.1131	-3.2348	0.0687

R_1	C_1	R_2	C_2	α	β
0.5975	264.25	1.2679	448.54	0.4325	0.4380

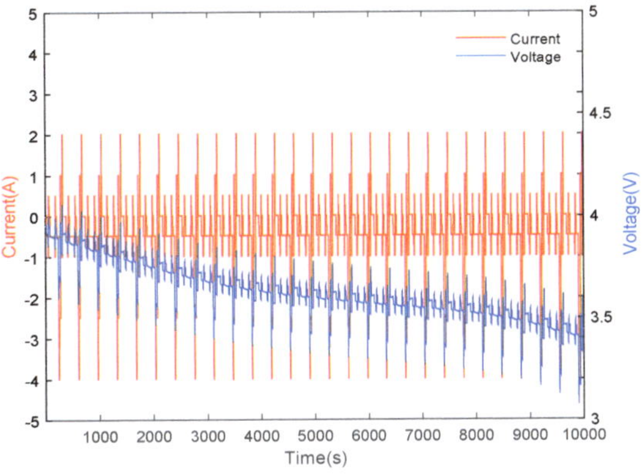

Figure 2. Current and voltage profiles of the operation conditions: DST.

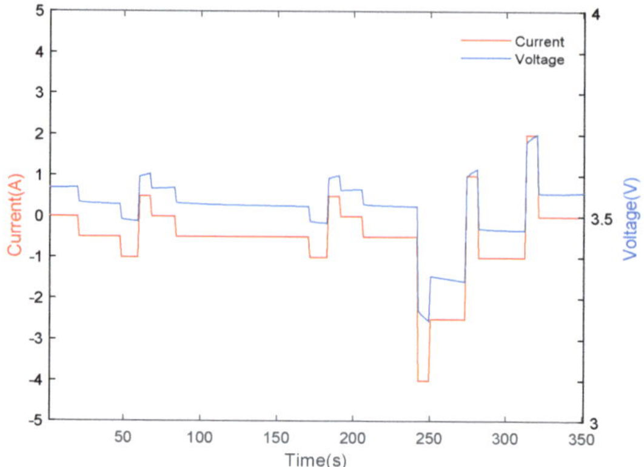

Figure 3. Current and voltage data of a DST cycle.

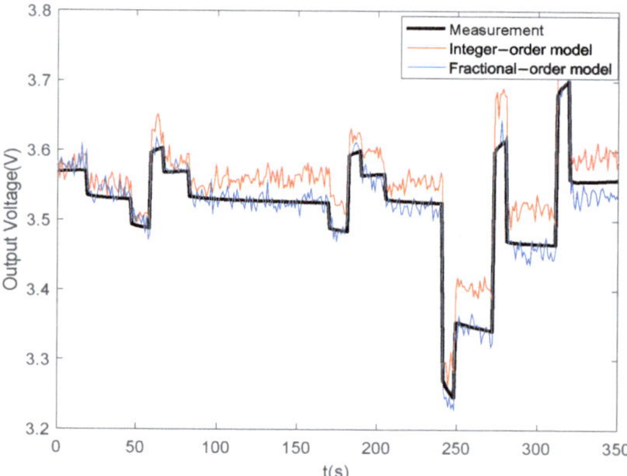

Figure 4. Accuracy verification of the model.

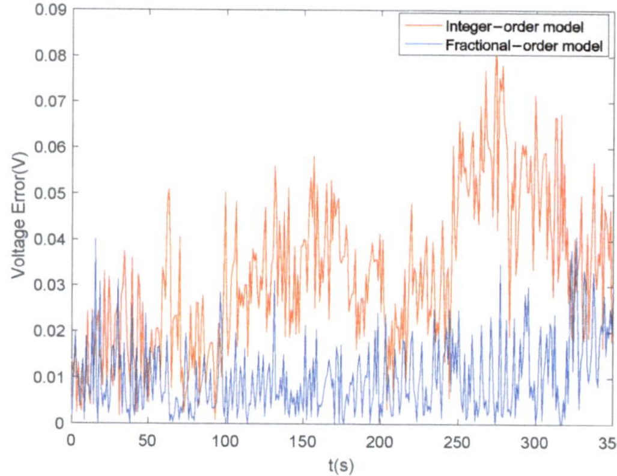

Figure 5. Output terminal voltage error curve.

3. Fuzzy Controller

The selection of the membership functions is very significant for the performance of a fuzzy controller. There is no ready-made rule for the establishment of the membership functions, and most of the methods are still based on experience and experimentation. The membership functions used here are shown in the Figure 6.

Fuzzy control is an effective method to solve the influence of measurement noise on the accuracy of SOC estimation in a complex environment. The fuzzy controller includes three main parts, as illustrated in Figure 7. First, we start the fuzzy processing on the input value G_k, based on the input membership function, to obtain the corresponding fuzzy index, where G_k is the difference between the theoretical and actual covariances M_k and N_k. Second, we establish the fuzzy rules as shown in Table 2. Large observation noise leads to changes in the actual covariance N_k, while the theoretical covariance M_k is affected by changes in the observation noise variance V_k. To maintain the consistency regarding changes between M_k and N_k, when the observation noise is large (small), we adjust the output value μ_k to expand (reduce) V_k so that G_k is close to 0. Finally, we perform the inverse fuzzy processing, according to the output membership function, to obtain μ_k. Therefore, we can get a new noise variance $\widehat{V_k}$ to perform the update of the observation noise variance adaptively.

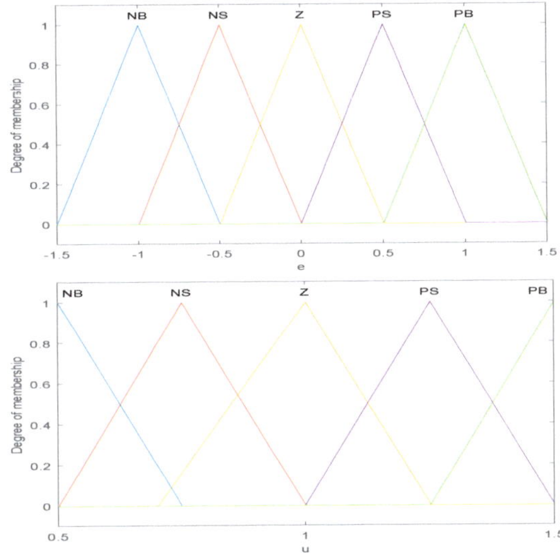

Figure 6. Input and output membership functions.

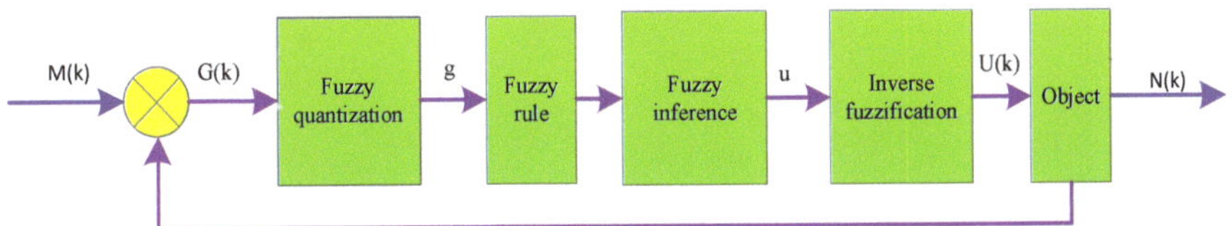

Figure 7. Fuzzy controller diagram.

Table 2. Fuzzy rules.

	NB	NS	Z	PS	PB
Input fuzziness	NB	NS	Z	PS	PB
Output fuzziness	NB	NS	Z	PS	PB

4. SOC Estimation

Firstly, the observability of the battery model is analyzed. A method to determine the observability of continuous time multi-order fractional-order systems was proposed in [40].

The fractional-order system is observable if its observability matrix is full rank. According to the Equations (10)–(12), we can get that the observability matrix O of the system is:

$$O = \begin{bmatrix} 1 & 1 & -\frac{U_{oc}}{SOC} \\ -\frac{1}{R_1 C_1} & 0 & 0 \\ 0 & -\frac{1}{R_2 C_2} & 0 \end{bmatrix}. \tag{18}$$

It is easy to see that this matrix is full rank. Therefore, the second-order FECM is observable. Compared to the FUKF, the FFUKF reveals higher accuracy and efficiency for SOC estimation. In this section, we discuss in detail the main steps of the FFUKF algorithm. The fractional-order system is given by:

$$\begin{cases} D^{\eta} x_{k+1} = f(x_k, u_k) + \omega_k, \\ x_{k+1} = D^{\eta} x_{k+1} - \sum_{j=1}^{L+1} (-1)^j \gamma_j x_{k+1-j}, \\ y_k = h(x_k) + V_k, \end{cases} \tag{19}$$

where x_k represents the system state variable, u_k and y_k denote the system input and output, respectively, $f(x_k, u_k)$ stands for the system process model, and $h(x_k)$ is the system measurement model. The symbol ω_k represents a Gaussian process noise and V_k corresponds to measurement noise. The variables Q and R represent the covariance matrices of ω_k and V_k, respectively.

The flow chart of the FFUKF is illustrated in Figure 8. The detailed FFUKF steps are presented as follows:

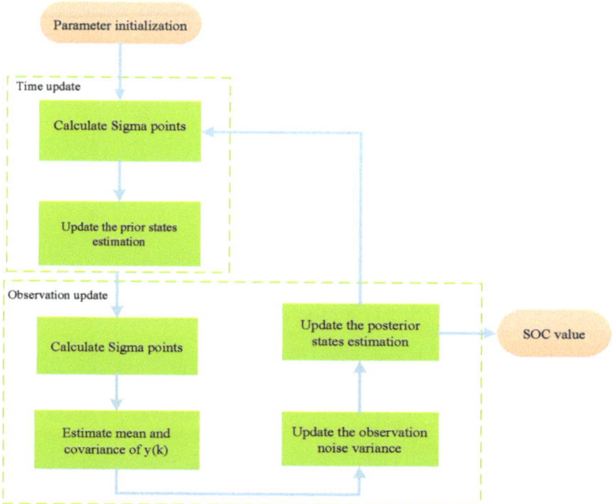

Figure 8. Flow chart of the FFUKF.

1   **Initialization**

     (1)   Give the initial state x_0, Q, R and state error covariance P.

2   **Time updating**

     (1)   Calculate sigma points using the singular value decomposition:

$$\begin{cases} P_{k-1|k-1} = U_{k-1}S_{k-1}V_{k-1}^T, \\ x_{0,k-1|k-1} = \widehat{x}_{k-1|k-1}, \\ x_{i,k-1|k-1} = \widehat{x}_{k-1|k-1} + \rho U_i \sqrt{s_i}, \\ \qquad i = 1,2,\cdots,n \\ x_{i,k-1|k-1} = \widehat{x}_{k-1|k-1} - \rho U_i \sqrt{s_i}, \\ \qquad i = n+1, n+2, \cdots, 2n, \end{cases} \tag{20}$$

where ρ is a scale coefficient that we can set equal to 1. The symbols s_i and U_i are the ith eigenvalue and eigenvector of S_{k-1} and $U_{(k-1)}$. The weight of sigma points can be calculated by the formula:

$$\begin{cases} \omega_m^0 = \dfrac{\lambda}{n+\lambda}, \\ \omega_c^0 = \dfrac{\lambda}{n+\lambda} + (1-\alpha^2+\beta), \\ \omega_m^i = \omega_c^i = \dfrac{1}{2(n+\lambda)}, i = 1,2,\cdots,2n, \end{cases} \tag{21}$$

where λ denotes $\alpha^2(n+k) - n$, α and k represent scaling and tuning parameters, respectively, n is the dimension of the state vector x, and β is a parameter related to the noise type.

(2) Transform the sigma sampling points using the nonlinear function $f(\cdot)$:

$$\begin{cases} \phi_{i,k-1|k-1} = f(x_{i,k-1|k-1}, u_{k-1}), i = 0,1,\ldots 2n, \\ D^\eta \widehat{x}_{k|k-1} = \displaystyle\sum_{i=0}^{2n} \omega_m^i \phi_{i,k-1|k-1}. \end{cases} \tag{22}$$

(3) Update the prior states estimation. The mean and covariance of $D^\eta x_k$ and x_k can be calculated by:

$$\begin{cases} P_{k|k-1}^{\Delta\Delta} = Cov[D^\eta x_{k|y_{k-1}}], \\ \qquad = \displaystyle\sum_{i=0}^{2n} \omega_c^j (\phi_{i,k-1|k-1} - D^\eta \widehat{x}_{k|k-1}), \\ \qquad \times (\phi_{i,k-1|k-1} - D^\eta \widehat{x}_{k|k-1})^T + Q, \\ P_{k|k-1}^{\Delta\Delta} = Cov[x_{k-1}, D^\eta x_{k|y_{k-1}}], \\ \qquad = \displaystyle\sum_{i=0}^{2n} \omega_c^j (\phi_{i,k-1|k-1} - D^\eta \widehat{x}_{k|k-1}), \\ \qquad \times (\phi_{i,k-1|k-1} - D^\eta \widehat{x}_{k|k-1})^T, \end{cases} \tag{23}$$

$$\begin{cases} \widehat{x}_{k|k-1} = D^\eta \widehat{x}_{k|k-1} - \displaystyle\sum_{j=1,}^{k} (-1)^j \gamma_j \widehat{x}_{k-j|k-j}, \\ P_{k|k-1} = P_{k|k-1}^{\Delta\Delta} + \gamma_1 P_{k|k-1}^{x\Delta} \\ \qquad + P_{k|k-1}^{\Delta x} \gamma_1 + \displaystyle\sum_{j=1}^{k} \gamma_j P_{k-j|k-j} \gamma_j. \end{cases} \tag{24}$$

3 Observation updating

(1) Calculate sigma points using the singular value decomposition. The weight of the sigma points is obtained using (20):

$$\begin{cases} P_{k|k-1} = U_{k-1}S_{k-1}V_{k-1}^{T}, \\ x_{0,k-1|k-1} = \widehat{x}_{k-1|k-1}, \\ x_{i,k-1|k-1} = \widehat{x}_{k-1|k-1} + \rho U_i \sqrt{s_i}, \\ \qquad i = 1, 2, \cdots, n \\ x_{i,k-1|k-1} = \widehat{x}_{k-1|k-1} + \rho U_i \sqrt{s_i}, \\ \qquad i = n+1, n+2, \cdots, 2n. \end{cases} \tag{25}$$

(2) Transform the sigma sampling points using the nonlinear function $h(\cdot)$:

$$\begin{cases} \theta_{i,k|k-1} = h(x_{i,k|k-1}), i = 0, 1, \cdots, 2n, \\ \widehat{y}_{k|k-1} = \displaystyle\sum_{i=0}^{2n} \omega_m^i \theta_{i,k|k-1}. \end{cases} \tag{26}$$

(3) Estimate the observation-error covariance matrix:

$$\begin{cases} P_{k|k-1}^{yy} = Cov[y_k|y_{k-1}], \\ \qquad = \displaystyle\sum_{i=0}^{2n} \omega_c^j (\theta_{i,k|k-1} - \widehat{y}_{k|k-1}), \\ \qquad \times (\theta_{i,k|k-1} - \widehat{y}_{k|k-1})^T + R, \\ P_{k|k-1}^{xy} = Cov[x_k, y_k|y_{k-1}], \\ \qquad = \displaystyle\sum_{i=0}^{2n} \omega_c^j (\theta_{i,k|k-1} - \widehat{y}_{k|k-1}), \\ \qquad \times (\theta_{i,k|k-1} - \widehat{y}_{k|k-1})^T. \end{cases} \tag{27}$$

(4) Calculate the theoretical and actual covariances:

$$\begin{cases} M_k = \displaystyle\sum_{i=0}^{2n} \omega_c^j (\theta_{i,k|k-1} - \widehat{y}_{k|k-1}), \\ \qquad \times (\theta_{i,k|k-1} - \widehat{y}_{k|k-1})^T + R, \\ N_k = \dfrac{1}{n} \displaystyle\sum_{i}^{k} [y_i - y_{i|i-1}][y_i - y_{i|i-1}]^T. \\ \qquad i = k - n + 1. \end{cases} \tag{28}$$

(5) Update the observation noise variance:

$$\begin{cases} G_k = M_k - N_k, \\ \widehat{V_k} = \mu_k V_k, \end{cases} \tag{29}$$

where G_k is the input value of the fuzzy controller and μ_k is the output value as an adjusted factor through the fuzzy inference system. We can then obtain the new $P_{k|k-1}^{yy}$.

(6)   Update the posterior states estimation:

$$
\begin{cases}
K_k = P^{xy}_{k|k-1}\left(P^{yy}_{k|k-1}\right)^{-1}, \\
\widehat{x}_{k|k} = \widehat{x}_{k|k-1} + K(y_k - \widehat{y}_{k|k-1}), \\
P_{k|k} = P_{k|k-1} - K_k P^{yy}_{k|k-1} K_k^{T},
\end{cases}
\tag{30}
$$

where K_k is the Kalman filter gain. With the update of V_k, we can get the updated Kalman filter gain K_k and the state error covariance matrix $P_{k|k}$.

5. Numerical Verification and Discussion

The current and voltage data under FUDS and BJDST conditions are used to verify the accuracy of the SOC estimation algorithm. The corresponding current and voltage profiles at the temperature of 25 °C are shown in Figure 9. Due to limited space, we omit the current and voltage data at 0 °C and 45 °C, available at the CALCE Battery Research Group. To verify the validity and feasibility of the proposed method, we also compare the FFUKF with the EKF and FUKF algorithms.

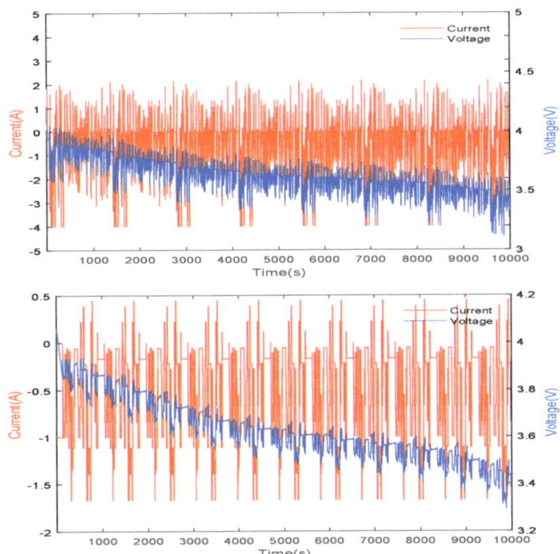

Figure 9. Current and voltage profiles of the operation conditions: FUDS and BJDST.

5.1. Experimental Results at 25 °C

Figures 10 and 11 show the SOC estimation results and estimation error, respectively. The blue line corresponds to the FFUKF. The red and magenta lines stand for FUKF and EKF, respectively. Also, the three algorithms are compared with the reference value represented by a black line. The closer to the reference, the higher the estimation accuracy of the algorithm. In order to see the differences between each algorithm more clearly, Figure 10 is partially magnified. We verify that the estimation results of the FFUKF are closer to the reference value. Under the two operating conditions, one can note that the FECM-based (FFUKF and FUKF) algorithm has higher accuracy than the ECM-based (EKF). Also, the FFUKF is more accurate than the FUKF. From Figure 11, the absolute estimation error of SOC of the FFUKF is no more than 0.005 under the two operating conditions, but the error of the other two algorithms is above 0.005 during the whole cycle. Even if disturbed by the noise environment, the FFUKF can still maintain high accuracy without large fluctuation, which shows that the proposed algorithm is stable to a certain extent. Additionally, it is clear that the error of the FFUKF is smoother than the one of the EKF, which confirms the superiority of the FFUKF in noisy environment. Table 3 summarizes the RMSE of the EKF, FUKF and FFUKF at 25 °C. Under the two operation conditions, the RMSE of the proposed algorithm is bellow 0.20%. However, the RMSE of the other two algorithms are 0.68% and 1.95%, respectively. This more clearly shows that our proposed algorithm has high accuracy over traditional methods.

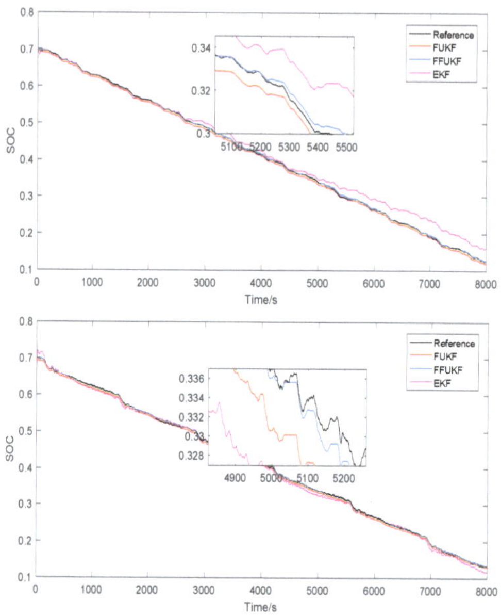

Figure 10. The SOC estimation curves under BJDST and FUDS at 25 °C.

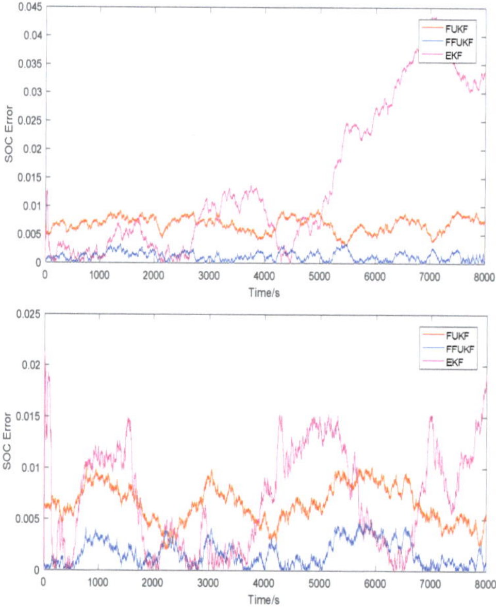

Figure 11. The SOC estimation error curves under BJDST and FUDS at 25 °C.

Table 3. The RMSE under BJDST and FUDS at 25 °C.

RMSE	EKF	FUKF	FFUKF
FUDS	0.87%	0.67%	0.20%
BJDST	1.95%	0.68%	0.13%

Because SOC estimation of lithium-ion batteries is affected by temperature, preserving the SOC estimation accuracy at different temperatures is crucial. As such, we carried out two sets of experiments at 0 °C and 45 °C to demonstrate the robustness of the proposed method at different temperatures.

5.2. Experimental Results at 0 °C

Figures 12 and 13 show the SOC estimation results and the estimation error under two cases at 0 °C, respectively. The ECM-based algorithm is obviously much worse than the FECM-based one in terms of accuracy. Figure 12 is partially magnified in order to highlight the differences. According to Figure 12, one can see that the blue line, representing the FFUKF, is closer to the reference value, which also shows that the estimation accuracy of the FFUKF is higher. From Figure 13, it is easy to see that the FFUKF can also maintain high accuracy at low temperatures. Most of the time, the estimation error of the FFUKF is kept within 0.005. The estimation errors of the other two algorithms are more than 0.01 in most of the time. In addition, the estimation error of the EKF fluctuates greatly, which shows that the EKF is very unstable in low temperature environment. The FFUKF has small fluctuation, which shows that it can maintain good estimation accuracy and has a certain stability even at low temperature. Table 4 gives a more intuitive explanation through the RMSE. Under FUDS, the RMSE of the FFUKF is 0.20%. However, the RMSE of the other two methods are more than 0.85%. Meanwhile, the RMSE of the FFUKF is lower than that presented by traditional algorithm. This proves once again that our method is superior to the traditional methods in low temperature environment. Under the two conditions, although the result is worse than that under the temperature of 25 °C, the maximum error of the FFUKF is still less than 0.32%.

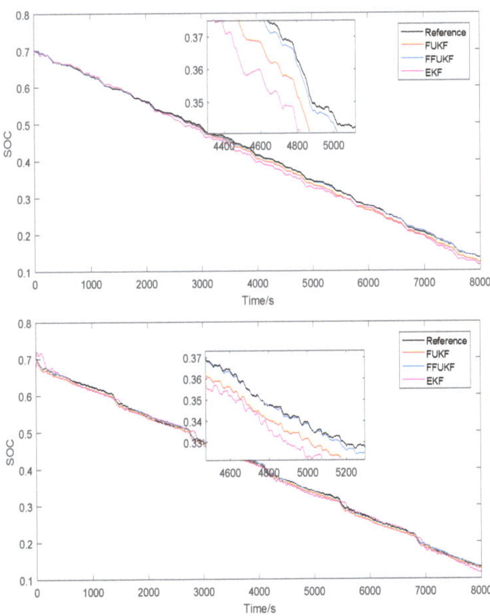

Figure 12. The SOC estimation curves under BJDST and FUDS at 0 °C.

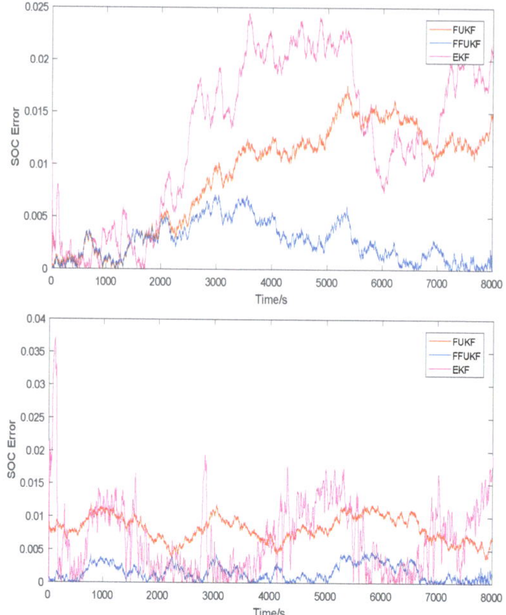

Figure 13. The SOC estimation error curves under BJDST and FUDS at 0 °C.

Table 4. The RMSE under BJDST and FUDS at 0 °C.

RMSE	EKF	FUKF	FFUKF
FUDS	0.88%	0.85%	0.20%
BJDST	1.49%	1.04%	0.32%

5.3. Experimental Results at 45 °C

Figures 14 and 15 show the SOC estimation results and the estimation error under FUDS and BJDST at 45 °C, respectively. Figure 14 is also partially magnified so that we can more clearly observe which line is closer to the reference value represented by the black line. Undoubtedly, compared with the other two traditional algorithms, the FFUKF represented by the blue line is closer to the reference and has very high accuracy. Also, It follows from Figure 14, that the ECM-based algorithm (EKF) is much worse than the FECM-based (FUKF and FFUKF) one in terms of accuracy. At a higher temperature (45 °C), the FECM-based algorithm still reveals smaller error. From Figure 15, under the two working conditions, the SOC estimation error of our algorithm does not exceed 0.01 in most of the time, but the estimation error of the other two algorithms are much more than 0.01. Especially, in the case of FUDS, the error of the EKF varies quickly. The fluctuation of the FFUKF is smaller than that of the EKF. This further verifies that our algorithm based on the FFUKF has a certain stability at high temperature. The FFUKF still yields higher accuracy at a higher temperature. Table 5 gives a more intuitive explanation through RMSE. In both cases, the RMSE of the EKF and FUKF exceeds 1%, but the RMSE of the FFUKF does not exceed 0.58%. This not only shows that the FFUKF is superior to the other two traditional algorithms, but also maintains a certain accuracy under the condition of high temperature and noise. However, compared with low temperature 0 °C and normal temperature 25 °C, the accuracy is not good enough, which may be because the battery model we established is vulnerable to high temperature.

From the above three sets of experiments, one can conclude that the accuracy of the FFUKF is always better than the one obtained with the other two methods in all operating conditions (FUDS and BJDST). Moreover, although the operating conditions are poor, the RMSE can almost be kept within 0.58%. It can maintain a certain stability even under the conditions of ambient temperature and noise. More important, the FFUKF solves the problem of low estimation accuracy caused by noise in practical operation. Obviously, the estimation accuracy of the algorithm is relatively higher at the same temperature.

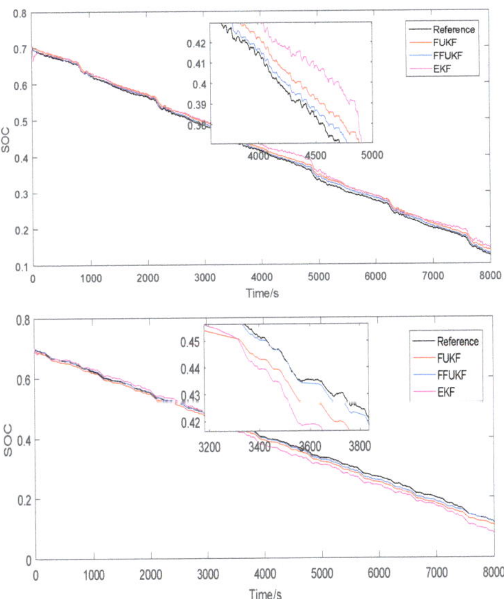

Figure 14. The SOC estimation curves under BJDST and FUDS at 45 °C.

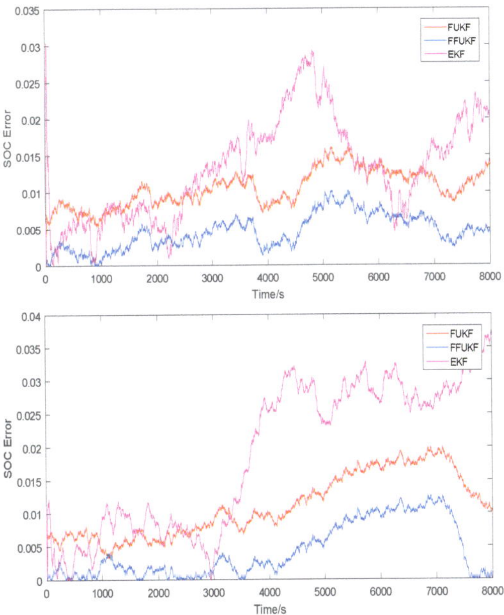

Figure 15. The SOC estimation error curves under BJDST and FUDS at 45 °C.

Table 5. The RMSE under BJDST and FUDS at 45 °C.

RMSE	EKF	FUKF	FFUKF
FUDS	1.49%	1.08%	0.51%
BJDST	2.18%	1.20%	0.58%

6. Conclusions

In this paper a new SOC estimation algorithm named fuzzy fractional-order unscented Kalman filter was proposed to estimate the SOC of lithium-ion batteries accurately. The method can infer the measurement noise in real time, so as to improve the influence of the measurement noise on the estimation results as the working conditions change. Compared with the EKF and FUKF algorithms, the experimental results indicated that the proposed method has better performance during the working conditions of BJDST and FUDS. It was also verified that the accuracy of the proposed algorithm is better than the EKF and FUKF at different temperatures.

Author Contributions: Methodology, L.C.; software, H.K.; validation and data curation, R.W.; writing—original draft preparation, Y.C.; writing—review and editing, A.M.L. and L.C. All authors have read and agreed to the published version of the manuscript.

Funding: This research was funded by the National Natural Science Funds of China (No. 62073114, No. 11971032) and Key Research and Development Project of Anhui Province (202104a05020035).

Institutional Review Board Statement: Not applicable.

Informed Consent Statement: Not applicable

Data Availability Statement: The data that support the findings of this study are available which has included references to the CALCE article that describes the experiments conducted for generating the data.

Conflicts of Interest: The authors declare no conflict of interest.

References

1. Fernández, R.Á. A more realistic approach to electric vehicle contribution to greenhouse gas emissions in the city. *J. Clean. Prod.* **2018**, *172*, 949–959. [CrossRef]
2. Larcher, D.; Tarascon, J.M. Towards greener and more sustainable batteries for electrical energy storage. *Nat. Chem.* **2015**, *7*, 19–29. [CrossRef] [PubMed]
3. Cheng, K.W.E.; Divakar, B.; Wu, H.; Ding, K.; Ho, H.F. Battery-management system (BMS) and SOC development for electrical vehicles. *IEEE Trans. Veh. Technol.* **2010**, *60*, 76–88. [CrossRef]
4. Zhang, Q.; Li, Y.; Shang, Y.; Duan, B.; Cui, N.; Zhang, C. A fractional-order kinetic battery model of lithium-ion batteries considering a nonlinear capacity. *Electronics* **2019**, *8*, 394. [CrossRef]
5. He, H.; Xiong, R.; Guo, H.; Li, S. Comparison study on the battery models used for the energy management of batteries in electric vehicles. *Energy Convers. Manag.* **2012**, *64*, 113–121. [CrossRef]
6. Chen, Y.; Yang, G.; Liu, X.; He, Z. A time-efficient and accurate open circuit voltage estimation method for lithium-ion batteries. *Energies* **2019**, *12*, 1803. [CrossRef]
7. Aylor, J.H.; Thieme, A.; Johnso, B. A battery state-of-charge indicator for electric wheelchairs. *IEEE Trans. Ind. Electron.* **1992**, *39*, 398–409. [CrossRef]
8. Feng, F.; Lu, R.; Zhu, C. A combined state of charge estimation method for lithium-ion batteries used in a wide ambient temperature range. *Energies* **2014**, *7*, 3004–3032. [CrossRef]
9. Waag, W.; Fleischer, C.; Sauer, D.U. Critical review of the methods for monitoring of lithium-ion batteries in electric and hybrid vehicles. *J. Power Sources* **2014**, *258*, 321–339. [CrossRef]
10. Tian, J.; Xiong, R.; Shen, W.; Lu, J. State-of-charge estimation of LiFePO4 batteries in electric vehicles: A deep-learning enabled approach. *Appl. Energy* **2021**, *291*, 116812. [CrossRef]
11. Belhani, A.; M'Sirdi, N.K.; Naamane, A. Adaptive sliding mode observer for estimation of state of charge. *Energy Procedia* **2013**, 377–386. [CrossRef]
12. Nath, A.; Gupta, R.; Mehta, R.; Bahga, S.S.; Gupta, A.; Bhasin, S. Attractive ellipsoid sliding mode observer design for state of charge estimation of lithium-ion cells. *IEEE Trans. Veh. Technol.* **2020**, *69*, 14701–14712. [CrossRef]

MDPI AG
Grosspeteranlage 5
4052 Basel
Switzerland
Tel.: +41 61 683 77 34

Fractal and Fractional Editorial Office
E-mail: fractalfract@mdpi.com
www.mdpi.com/journal/fractalfract

13. Chen, Z.; Zhou, J.; Zhou, F.; Xu, S. State-of-charge estimation of lithium-ion batteries based on improved H infinity filter algorithm and its novel equalization method. *J. Clean. Prod.* **2021**, *290*, 125180. [CrossRef]
14. Li, L.; Hu, M.; Xu, Y.; Fu, C.; Jin, G.; Li, Z. State of charge estimation for lithium-ion power battery based on H-infinity filter Algorithm. *Appl. Sci.* **2020**, *10*, 6371. [CrossRef]
15. Tao, J.; Zhu, D.; Sun, C.; Chu, D.; Ma, Y.; Li, H.; Li, Y.; Xu, T. A novel method of SOC estimation for electric vehicle based on adaptive particle filter. *Autom. Control. Comput. Sci.* **2020**, *54*, 412–422.
16. Simon, D. *Optimal State Estimation: Kalman, H Infinity, and Nonlinear Approaches*; John Wiley & Sons: Hoboken, NJ, USA, 2006.
17. Xiao, R.; Shen, J.; Li, X.; Yan, W.; Pan, E.; Chen, Z. Comparisons of modeling and state of charge estimation for lithium-ion battery based on fractional order and integral order methods. *Energies* **2016**, *9*, 184. [CrossRef]
18. Yang, S.; Zhou, S.; Hua, Y.; Zhou, X.; Liu, X.; Pan, Y.; Ling, H.; Wu, B. A parameter adaptive method for state of charge estimation of lithium-ion batteries with an improved extended Kalman filter. *Sci. Rep.* **2021**, *11*, 1–15.
19. Julier, S.J.; Uhlmann, J.K. Unscented filtering and nonlinear estimation. *Proc. IEEE* **2004**, *92*, 401–422. [CrossRef]
20. Zhang, S.; Guo, X.; Zhang, X. An improved adaptive unscented kalman filtering for state of charge online estimation of lithium-ion battery. *J. Energy Storage* **2020**, *32*, 101980. [CrossRef]
21. Han, J.; Kim, D.; Sunwoo, M. State-of-charge estimation of lead-acid batteries using an adaptive extended Kalman filter. *J. Power Sources* **2009**, *188*, 606–612. [CrossRef]
22. Sun, F.; Hu, X.; Zou, Y.; Li, S. Adaptive unscented Kalman filtering for state of charge estimation of a lithium-ion battery for electric vehicles. *Energy* **2011**, *36*, 3531–3540. [CrossRef]
23. Zhang, Z.; Jiang, L.; Zhang, L.; Huang, C. State-of-charge estimation of lithium-ion battery pack by using an adaptive extended Kalman filter for electric vehicles. *J. Energy Storage* **2021**, *37*, 102457. [CrossRef]
24. Zeng, M.; Zhang, P.; Yang, Y.; Xie, C.; Shi, Y. SOC and SOH joint estimation of the power batteries based on fuzzy unscented Kalman filtering algorithm. *Energies* **2019**, *12*, 3122. [CrossRef]
25. Lai, X.; Qiao, D.; Zheng, Y.; Zhou, L. A fuzzy state-of-charge estimation algorithm combining ampere-hour and an extended Kalman filter for Li-ion batteries based on multi-model global identification. *Appl. Sci.* **2018**, *8*, 2028. [CrossRef]
26. Victor, S.; Malti, R.; Garnier, H.; Oustaloup, A. Parameter and differentiation order estimation in fractional models. *Automatica* **2013**, *49*, 926–935. [CrossRef]
27. Chen, Y.; Huang, D.; Zhu, Q.; Liu, W.; Liu, C.; Xiong, N. A new state of charge estimation algorithm for lithium-ion batteries based on the fractional unscented Kalman filter. *Energies* **2017**, *10*, 1313. [CrossRef]
28. Xiong, R.; Tian, J.; Shen, W.; Sun, F. A novel fractional order model for state of charge estimation in lithium ion batteries. *IEEE Trans. Veh. Technol.* **2018**, *68*, 4130–4139. [CrossRef]
29. Sabatier, J.; Cugnet, M.; Laruelle, S.; Grugeon, S.; Sahut, B.; Oustaloup, A.; Tarascon, J. A fractional order model for lead-acid battery crankability estimation. *Commun. Nonlinear Sci. Numer. Simul.* **2010**, *15*, 1308–1317. [CrossRef]
30. Liu, C.; Liu, W.; Wang, L.; Hu, G.; Ma, L.; Ren, B. A new method of modeling and state of charge estimation of the battery. *J. Power Sources* **2016**, *320*, 1–12. [CrossRef]
31. Xu, J.; Mi, C.C.; Cao, B.; Cao, J. A new method to estimate the state of charge of lithium-ion batteries based on the battery impedance model. *J. Power Sources* **2013**, *233*, 277–284. [CrossRef]
32. Wei, Z.; Zou, C.; Leng, F.; Soong, B.H.; Tseng, K.J. Online model identification and state-of-charge estimate for lithium-ion battery with a recursive total least squares-based observer. *IEEE Trans. Ind. Electron.* **2017**, *65*, 1336–1346. [CrossRef]
33. Monje, C.A.; Chen, Y.; Vinagre, B.M.; Xue, D.; Feliu-Batlle, V. *Fractional-Order Systems and Controls: Fundamentals and Applications*; Springer Science & Business Media: Berlin/Heidelberg, Germany, 2010.
34. Hu, X.; Li, S.; Peng, H. A comparative study of equivalent circuit models for Li-ion batteries. *J. Power Sources* **2012**, *198*, 359–367. [CrossRef]
35. Wang, B.; Liu, Z.; Li, S.E.; Moura, S.J.; Peng, H. State-of-Charge estimation for lithium-ion batteries based on a nonlinear fractional model. *IEEE Trans. Control Syst. Technol.* **2017**, *25*, 3–11. [CrossRef]
36. Hu, X.; Yuan, H.; Zou, C.; Li, Z.; Zhang, L. Co-estimation of state of charge and state of health for lithium-ion batteries based on fractional-order calculus. *IEEE Trans. Veh. Technol.* **2018**, *67*, 10319–10329. [CrossRef]
37. Aggab, T.; Avila, M.; Vrignat, P.; Kratz, F. Unifying model-based prognosis with learning-based time-series prediction methods: application to Li-Ion battery. *IEEE Syst. J.* **2021**. [CrossRef]
38. Coronel-Escamilla, A.; Gómez-Aguilar, J.; Torres-Jiménez, J.; Mousa, A.; Elagan, S. Fractional synchronization involving fractional derivatives with nonsingular kernels: Application to chaotic systems. *Math. Methods Appl. Sci.* **2021**. [CrossRef]
39. Zheng, F.; Xing, Y.; Jiang, J.; Sun, B.; Kim, J.; Pecht, M. Influence of different open circuit voltage tests on state of charge online estimation for lithium-ion batteries. *Applied Energy* **2016**, *183*, 513–525. [CrossRef]
40. Tavakoli, M.; Tabatabaei, M. Controllability and observability analysis of continuous-time multi-order fractional systems. *Multidimens. Syst. Signal Process.* **2017**, *28*, 427–450. [CrossRef]